Further Algebra and Applications

P.M. Cohn

Further Algebra and Applications

With 27 Figures

Springer

P.M. Cohn, MA, PhD, FRS
Department of Mathematics, University College London,
Gower Street, London WC1E 6BT

Cohn, P.M.: Further Algebra and Applications

First Indian Reprint 2004
Third Indian Reprint 2013

ISBN 978-81-8128-223-1

This edition is manufactured in India for sale in India, Pakistan, Bangladesh, Nepal and Sri Lanka and any other country as authorized by the publisher.

This edition is published by Springer (India) Private Limited,
A part of Springer Science+Business Media, Registered Office: 7th Floor, Vijaya Building, 17 Barakhamba Road, New Delhi – 110 001, India.

Printed in India by Rashtriya Printers, Delhi.

Contents

Preface

This volume follows on the subject matter treated in *Basic Algebra* and together with that volume represents the contents of volumes 2 and 3 of my book on algebra, now out of print; the topics have been rearranged a little, with most of the applications in the present volume, while the basic theories (groups, rings, fields) are pursued further in the earlier book. In any case all parts of volumes 2 and 3 are represented. The whole text has been revised, some exercises have been added and of course errors have been corrected; I am grateful to a number of readers for bringing such errors to my attention.

Chapter 1 presents the basic notions of universal algebra: the isomorphism theorems, free algebras and varieties, with the natural numbers, viewed as algebra with a unary operator as an application, as well as the ultraproduct theorem and the diamond lemma. The introduction to homological algebra in Chapter 2 goes as far as derived functors and global dimension, with the case of polynomial rings and free algebras as an application. Chapter 3, on group theory, discusses some items of general interest and importance (group extensions, Hall subgroups, transfer), but also topics which find an echo elsewhere in the book, such as free groups and linear groups. Chapter 4, on algebras, deals with the Krull–Schmidt theorem, projective covers, Morita equivalence and related matters, but stops short of the representation theory of algebras, which would have required more space than was available. This is followed by an account of central simple algebras (Chapter 5), introducing the Brauer group and crossed products. The representation theory of finite groups in Chapter 6 presents the standard facts on representations and characters and illustrates this work by the symmetric group. The next two chapters return to rings; Chapter 7 presents topics on Noetherian rings such as Goldie's theory, as well as polynomial identities and central polynomials, while Chapter 8 deals with the general density theorem, the various radicals and non-unital algebras. Chapter 9, on skew fields, gives a simplified treatment of the Dieudonné determinant and establishes the existence of 'free fields'. Its proof is based on the specialization lemma, which is of independent interest.

The final two chapters are applications of a different kind. Chapter 10 is an introduction to block codes, in particular linear codes and cyclic codes, as well as some other kinds. Chapter 11 deals with algebraic language theory and the related topics of variable-length codes, automata and power series rings. In both chapters it is only possible to take the first steps in the subject, but we go far enough to show how techniques from coding theory are used in the study of free algebras.

The text assumes an acquaintance with much of *Basic Algebra*, to which reference is made in the form 'BA' followed by the section number. Definitions and key properties are usually recalled in some detail, but not necessarily on their first occurrence; the reader can easily trace explanations through the index. As before, there are occasional historical references and numerous exercises, often with hints, though no solutions.

A number of colleagues and friends have made comments on the earlier edition and I would like to express my thanks to them here. My thanks also go to the staff of Springer-Verlag London and to Mrs Lyn Imeson for the efficient way they have carried out their task.

University College London P.M. Cohn
October 2002

Conventions on Terminology and notes to the reader

References to *Basic Algebra* are in the form BA, followed by the section number.

A property is said to hold for *almost all* members of a set if it holds for all but a finite number. The complement of a subset Y in a set X is written $X\backslash Y$. As a rule mappings are written on the right; in particular this is done when mappings have to be composed, so that $\alpha\beta$ means: first α, then β. If α is a mapping from a set X and Y is a subset of X, then the restriction of α to Y is written $\alpha|Y$.

All rings and monoids have a unit element or one, which acts as neutral element for multiplication, usually denoted by 1; by contrast an algebra (over a coefficient ring) need not have a one. A ring is *trivial* or the *zero ring* if it consists of 0 alone; this happens just when $1 = 0$. An element a of a ring is called a *zero-divisor* if $a \neq 0$ and $ab = 0$ or $ba = 0$ for some $b \neq 0$; if a is neither 0 nor a zero-divisor, it is said to be *regular* (see Section 7.1). A non-trivial ring without zero-divisors is called an *integral domain*; this term is not taken to imply commutativity. A ring in which the non-zero elements form a group under multiplication is called a *skew field*; in the commutative case this reduces to a field, but sometimes (in Chapter 9) this term is also used in the general case. In any ring R, the set of all non-zero elements is denoted by $R^{\times}$; this notation is mainly used for integral domains, where $R^{\times}$ is a monoid. A skew field finite-dimensional over its centre is called a division algebra, but the term 'algebra' by itself is not taken to imply finite dimensionality. A ring is said to have *invariant basis number* (IBN) if any two bases of a free module have the same number of elements, or equivalently, if any matrix with a two-sided inverse is square (see BA, Section 4.6).

References to the bibliography are by name of author and date in round brackets for books and square brackets for papers. As in BA, all results in a section are numbered consecutively; further we abbreviate 'if and only if' by iff (except in enunciations) and use ■ to indicate the end (or absence) of a proof.

The chapters are to a large extent independent, so no interdependence chart has been given, but the reader may have to turn back for the occasional result; this is usually clearly indicated.

1 Universal algebra

Most algebraic systems such as groups, vector spaces, rings, lattices etc. can be regarded from a common point of view as sets with operations defined on them, subject to certain laws. This is done in Section 1.1 and it allows many basic results, such as the isomorphism theorems, to be stated and proved quite generally, as we shall see in Section 1.2. Of the general theory of universal algebra (by now quite extensive), we shall need very little, this forms the subject of Section 1.3; in addition to the basic concepts we define the notion of an algebraic variety, i.e. a class of algebraic systems defined by identical relations, or laws. But there are one or two other topics, not strictly part of the subject that are needed: the diamond lemma forms the subject of Section 1.4, while dependence relations have already been discussed in BA (Section 11.1). There is also the ultraproduct theorem in Section 1.5, a result from logic with many uses (see Chapter 7). The chapter ends in Section 1.6 with an axiomatic development of the natural numbers, regarded as an algebraic system, in an account following Leon Henkin [1960] (see also Cohn (1981)).

1.1 Algebras and homomorphisms

Algebraic structures show certain common features: they have operations defined on them, which satisfy laws such as the associative law. These operations are mostly binary, like addition or multiplication, but sometimes they are unary, e.g. taking the inverse of a number, or ternary, e.g. the basic operation in a ternary ring, occurring in the study of projective planes (see M. Hall (1959)), or even noughtary, like the neutral element in a group. For any integer $n \geq 0$ we define an *n-ary operation* on a set S to be a mapping of S^n into S. The number n is called the *arity* of the operation and we say *unary* for 1-ary, *binary* for 2-ary, *ternary* for 3-ary and *finitary* to mean n-ary for some natural number n. A 0-ary operation on S is just a particular element of S; this is also called a *constant operation.*

An *algebra* is to be thought of as a set with certain finitary operations defined on it, but in order to compare different algebras we need to establish a correspondence between their sets of operations. This is done by indexing the operations in each algebra by a given index set, which is kept fixed in any discussion. Its elements are called *operators*, each with a given arity.

Thus by an *operator domain* we understand a set Ω and a mapping $a : \Omega \to \mathbf{N}_0$. The elements of Ω are called operators; if $\omega \in \Omega$, then $a(\omega) \in \mathbf{N}_0$ is called the *arity* of ω. We shall write $\Omega(n) = \{\omega \in \Omega | a(\omega) = n\}$, and refer to the members of $\Omega(n)$ as *n-ary* operators.

An Ω-algebra is defined as a pair (A, Ω) consisting of a set A together with a family of operations indexed by Ω:

$$\omega : A^n \to A \quad \text{for each } \omega \in \Omega(n), n = 0, 1, 2, \ldots. \tag{1.1.1}$$

The set A is called the *carrier* of the algebra. Strictly speaking we should denote the algebra by (A, Ω, φ), where φ is the family of mappings $\varphi_n : \Omega(n) \to \mathrm{Map}(A^n, A)$ defined by (1.1.1), but usually we shall not distinguish notationally between an algebra and its carrier. The set Ω is called the *operator domain*, or also the *signature* of the algebra. We give some examples.

1. Groups. A group $(G, \cdot, ^{-1}, 1)$ is given by a set with a binary operation (multiplication), a unary operation (inversion) and a constant operation (the neutral element), satisfying certain laws which are familiar to the reader (see Section 1.3 below).
2. Rings. A ring $(R, +, -, \times, 0, 1)$ is given by set with two binary operations $+, \times$, two constant operations $0, 1$ and a unary operation $-$, again satisfying well-known laws.
3. Lattices. A lattice may be defined as a partially ordered set in which each pair of elements has a supremum and an infimum, or as an algebra $(L, \vee, \wedge)$ with two binary operations satisfying certain laws (see BA, Section 3.1). For Boolean algebras we require in addition a constant operation 0 and a unary operation $'$, which leads to another constant operation $1 = 0'$, an instance of a derived operation (see Section 1.3).
4. Vector spaces. Let k be a field. A vector space over k is an algebra $(V, +, 0, k)$ with a binary operation $+$, a constant operation 0 and a family of unary operations indexed by $k : \omega_\alpha : u \mapsto \alpha u (u \in V, \alpha \in k)$, satisfying the laws familiar from linear algebra. For an infinite field k this is an example of an algebra with an infinite signature.
5. A 1-element set has a unique Ω-algebra structure for any Ω. This is called the *trivial* Ω-algebra.
6. The empty set is an Ω-algebra precisely when Ω has no constant operators.

Given an Ω-algebra A and $\omega \in \Omega(n)$, we can apply ω to any n-tuple $a_1, \ldots, a_n \in A$ and obtain another element of A which is written $a_1 \ldots a_n\omega$. In the case $n = 0$ this just singles out an element of A, denoted by ω; the zero element in a ring is an example.

Many algebraic concepts can be formulated for general Ω-algebras. Thus given an Ω-algebra A, an Ω-subalgebra is an Ω-algebra B whose carrier is a subset of that of A and which is closed under the operations of Ω, as defined in A. It is clear from the definition that a given subset of A can be defined as a subalgebra of A in at most one way. To give an example, the ring $\mathbf{Z}$ of integers has no proper subrings, because the constant operation 1 already generates the whole of **Z**. The subset $\{0\}$ is again a ring, but it is not a subring because the operation 1 has different values on **Z** and on $\{0\}$.

It is not hard to see that the intersection of any family of subalgebras of a given algebra A is again a subalgebra of A. Hence for any subset X of A we can form the intersection of all subalgebras containing X. This is called the subalgebra of A *generated* by X; it may also be obtained by applying the operations of Ω to the elements of X and repeating this operation a finite number of times. If the subalgebra generated by X is the whole of A, then X is called a *generating set* of A. Clearly every algebra A has a generating set, e.g. A itself.

A mapping $f : A \to B$ between Ω-algebras A, B is said to be *compatible* with $\omega \in \Omega(n)$ if for all $a_1, \dots, a_n \in A$,

$$(a_1 f) \dots (a_n f)\omega = (a_1 \dots a_n \omega)f. \tag{1.1.2}$$

If f is compatible with each $\omega \in \Omega$, it is called a *homomorphism* from A to B. If a homomorphism from A to B has an inverse which is again a homomorphism, it is called an *isomorphism*, and A, B are then said to be *isomorphic*. For example, all 1-element Ω-algebras are isomorphic. As in more special cases, an isomorphism of an algebra with itself is called an *automorphism* and a homomorphism of an algebra into itself is an *endomorphism*.

We observe that a homomorphism is determined once it is known on a generating set. This is the content of

Proposition 1.1.1. *Let $f, g : A \to B$ be two homomorphisms between Ω-algebras A and B. If f and g agree on a generating set of A, then they are equal.*

Proof. The set $\{x \in A | xf = xg\}$ is easily seen to be a subalgebra of A. By hypothesis it contains a generating set of A, hence it is the whole of A and so $f = g$, as we had to show. ■

From any family $(A_i)_{i \in I}$ of Ω-algebras we can form the *direct product* $P = \prod A_i$; its carrier is the Cartesian product of the A_i, and the operations are carried out componentwise. Thus if $\pi_i : P \to A_i$ are the projections from the Cartesian product to the factors, then any $\omega \in \Omega(n)$ is defined on P by the equation

$$(a_1 \dots a_n \omega)\pi_i = (a_1 \pi_i) \dots (a_n \pi_i)\omega. \tag{1.1.3}$$

It is easily checked that this defines an Ω-algebra structure on P and the form of (1.1.3) shows that the projection π_i is a homomorphism from P to A_i.

Of course the A_i need not be all distinct. If for example $A_i = A$ for all $i \in I$, we obtain the *direct power* of A indexed by I, which is denoted by A^I. Its members may be regarded as functions $f : I \to A$ and the operations are defined componentwise; e.g. if an addition is defined on A, then in A^I we have

$$(f + g)(i) = f(i) + g(i), \quad i \in I.$$

Exercises

1. Show that the set of all subalgebras of an Ω-algebra is a complete lattice (i.e. a lattice in which every subset has a sup and an inf, see BA, Section 3.1).
2. Verify the equivalence of the two definitions of subalgebra generated by X, given in the text, i.e. show that the set obtained from X by repeatedly applying Ω is the least subalgebra containing X.
3. Show that if Ω is finite, then there are only finitely many Ω-algebras on a given finite set as carrier. Is there a bound in terms of the size of the carrier alone?
4. Show that every homomorphism which is bijective is an isomorphism.
5. Let A be an Ω-algebra with a carrier of n elements. Show that A has at most $n!$ automorphisms and at most n^n endomorphisms. Find bounds on the number of automorphisms and endomorphisms if Ω includes a constant operator. Find bounds if A has an r-element generating set.
6. Let A be an Ω-algebra. Show that the set $\mathrm{Map}(A)$ of all mappings of A into itself may be regarded as an Ω-algebra. Further show that $\mathrm{End}(A)$, the set of all endomorphisms, is a subalgebra of $\mathrm{Map}(A)$ provided the following condition is satisfied by A: Given $\theta \in \Omega(m)$, $\omega \in \Omega(n)$ and any $m \times n$ matrix over A, the element obtained by applying θ to each column and ω to the result is the same as the element obtained by applying ω to each row and θ to the result.

1.2 Congruences and the isomorphism theorems

Let A and B be any sets. By a *correspondence* from A to B we understand a subset of the Cartesian product $A \times B$. For example, a mapping $f : A \to B$ may be defined as a correspondence Γ_f from A to B which has the properties of being (i) everywhere defined and (ii) single-valued:

(i) *for each* $a \in A$ *there exists* $b \in B$ *such that* $(a, b) \in \Gamma_f$,
(ii) *if* $(a, b), (a, b') \in \Gamma_f$, *then* $b = b'$.

This correspondence is sometimes called the *graph* of the mapping f.

We shall define two operations on correspondences. For any $\Gamma \subseteq A \times B$ we have the *inverse*, defined as

$$\Gamma^{-1} = \{(b, a) \in B \times A | (a, b) \in \Gamma\};$$

next, if $\Gamma \subseteq A \times B$ and $\Delta \subseteq B \times C$, then their *composition* is given by

$$\Gamma \circ \Delta = \{(a, c) \in A \times C | (a, x) \in A \times B \text{ and } (x, c) \in B \times C \text{ for some } x \in B\}.$$

Further, if $\Gamma \subseteq A \times B$ and $A' \subseteq A$, we define $A'\Gamma = \{b \in B | (a, b) \in \Gamma \text{ for some } a \in A\}$.

On every set we have the identity correspondence, also called the *diagonal*, $1_A = \{(a, a) | a \in A\}$ and the *universal correspondence* $A^2 = \{(x, y) | x, y \in A\}$. For example, the above conditions (i), (ii) on Γ_f to be the graph of a mapping can be expressed as follows:

$$\Gamma_f \circ \Gamma_f^{-1} \supseteq 1_A, \quad \Gamma_f^{-1} \circ \Gamma_f \subseteq 1_B.$$

To give another example, an equivalence on A may be defined as a subset Γ of A^2 with the properties

(i) $\Gamma \circ \Gamma \subseteq \Gamma$ (transitivity)
(ii) $\Gamma^{-1} = \Gamma$ (symmetry)
(iii) $\Gamma \supseteq 1_A$ (reflexivity).

To use correspondences in the study of Ω-algebras, we shall need to know their behaviour as subalgebras.

Lemma 1.2.1. *Let A, B, C be Ω-algebras and Γ, Δ subalgebras of $A \times B$, $B \times C$ respectively. Then Γ^{-1} is a subalgebra of $B \times A$, $\Gamma \circ \Delta$ is a subalgebra of $A \times C$ and for any subalgebra A' of A, $A'\Gamma$ is a subalgebra of B.*

Proof. Take $\omega \in \Omega(n)$ and $(a_i, c_i) \in \Gamma \circ \Delta (i = 1, \dots, n)$, say $(a_i, b_i) \in \Gamma$, $(b_i, c_i) \in \Delta$, and put $a_1 \dots a_n\omega = a$, $b_1 \dots b_n\omega = b$, $c_1 \dots c_n\omega = c$. The since Γ, Δ are subalgebras, we have $(a, b) \in \Gamma$, $(b, c) \in \Delta$, hence $(a, c) \in \Gamma \circ \Delta$ and since this holds for all $\omega \in \Omega$, it shows $\Gamma \circ \Delta$ to be a subalgebra. The proof for Γ^{-1} and $A'\Gamma$ is quite similar and may be left to the reader. ■

Let S, T be any sets and $f : S \to T$ a mapping between them. Then the image of f is defined as $S\Gamma_f$, also written im f or Sf; the *kernel* of f is defined as the correspondence

$$\ker f = \{(x, y) \in S^2 | xf = yf\}. \tag{1.2.1}$$

In terms of the graph Γ_f of f we have

$$\ker f = \Gamma_f \circ \Gamma_f^{-1}.$$

Clearly it is an equivalence on S; the different equivalence classes are just the inverse images of elements in the image, sometimes called the *fibres* of f.

Let us consider how the above definition is related to the kernel of a homomorphism of groups. Given a group homomorphism $f : G \to H$, the kernel of f in the usual sense is the inverse image under f of the unit element of H; this is a normal subgroup N of G and the different cosets of N in G are just the fibres of f. So ker f as defined in (1.2.1) is the set of cosets of N in G. Since this collection is entirely determined by N, it makes sense to replace it by N, which is what is usually done in group theory. But for arbitrary sets we shall need the whole correspondence ker f as defined above and we cannot replace it by anything simpler. This is still true when we come to study kernels of homomorphisms of Ω-algebras.

Let S, T be any sets and Γ a correspondence from S to T. We shall use Γ to define systems of subsets of S, T between which there is an inclusion-reversing bijection, as follows.

For any subset X of S we define a subset X^* of T by

$$X^* = \{y \in T | (x, y) \in \Gamma \quad \text{for all} \quad x \in X\},$$

and similarly, for any subset Y of T we define a subset Y^* of S by

$$Y^* = \{x \in S | (x, y) \in \Gamma \quad \text{for all} \quad y \in Y\}.$$

We thus have mappings

$$X \mapsto X^*, \quad Y \mapsto Y^* \tag{1.2.2}$$

of $\mathscr{P}(X)$, $\mathscr{P}(Y)$ into each other with the properties

$$X_1 \subseteq X_2 \Rightarrow X_1^* \supseteq X_2^*, \quad Y_1 \subseteq Y_2 \Rightarrow Y_1^* \supseteq Y_2^*, \tag{1.2.3}$$

$$X \subseteq X^{**}, \quad Y \subseteq Y^{**} \tag{1.2.4}$$

$$X^{***} = X^*, \quad Y^{***} = Y^*. \tag{1.2.5}$$

Conditions (1.2.3) and (1.2.4) are immediate from the definitions. If (1.2.3) is applied to (1.2.4), we get $X^* \supseteq X^{***}$ and (1.2.4) applied with X^* in place of X gives $X^* \subseteq X^{***}$. Hence $X^{***} = X^*$ and similarly for Y^*. This proves (1.2.5) as a consequence of (1.2.3) and (1.2.4) alone.

A pair of mappings (1.2.2) between $\mathscr{P}(S)$ and $\mathscr{P}(T)$ satisfying (1.2.3), (1.2.4) and hence (1.2.5) is called a *Galois connexion*. An obvious example, which also accounts for the name, is the situation in field theory. If F is a field and G the group of all its automorphisms, then the pairs $(x, \alpha) \in F \times G$ such that $x^\alpha = x$ form a correspondence which establishes a Galois connexion between certain subfields of F and subgroups of G. If G is a finite group of automorphisms of F and k is the subfield of elements left fixed by G, then there is a correspondence between all subgroups of G and all fields between k and F (see BA, Section 7.6 and Section 11.8).

Let us define a *congruence* on an Ω-algebra A as an equivalence on A^2 which is also a subalgebra of A^2. For example, 1_A and A^2 are congruences on A, and every other congruence $\mathfrak{q}$ lies between these two: $1_A \subseteq \mathfrak{q} \subseteq A^2$. The congruence on $\mathbf{Z}$ (as ring) determined by a given positive integer m consists of the residue classes mod m, i.e. the sets of numbers leaving a given remainder after division by m. As in this example, we shall sometimes, for any congruence $\mathfrak{q}$ on A, write $a \equiv b \pmod{\mathfrak{q}}$ to mean $(a, b) \in \mathfrak{q}$.

The next two results explain the significance of congruences for algebras.

Theorem 1.2.2. *Let $f : A \to B$ be a homomorphism of Ω-algebras. Then* im f *is a subalgebra of B and* ker f *is a congruence on A.*

Proof. It is easily checked that the graph Γ_f of f is a subalgebra of $A \times B$. By Lemma 1.2.1, im $f = A\Gamma_f$ is a subalgebra of B and ker $f = \Gamma_f \circ \Gamma_f^{-1}$ is a subalgebra of A^2, therefore it is a congruence. ∎

Given a group G with a normal subgroup N, we can put a group structure on the set G/N such that the natural mapping $G \to G/N$ is a homomorphism. In the same way we can, for any congruence $\mathfrak{q}$ on an Ω-algebra A, define an algebra structure on the set of $\mathfrak{q}$-classes $A/\mathfrak{q}$ such that the natural mapping $A \to A/\mathfrak{q}$ is a homomorphism with kernel $\mathfrak{q}$. This is the content of

Theorem 1.2.3. *Let A be an Ω-algebra and $\mathfrak{q}$ a congruence on A. Then there exists a unique Ω-algebra structure on the set $A/\mathfrak{q}$ of all $\mathfrak{q}$-classes such that the natural mapping $\nu : A \to A/\mathfrak{q}$ is a homomorphism.*

Proof. The natural mapping $\nu : A \to A/\mathfrak{q}$ is well-defined because $\mathfrak{q}$ is an equivalence. It induces the mapping $\nu_n : A^n \to (A/\mathfrak{q})^n$ for $n = 0, 1, \ldots$ in an obvious fashion. Let us write $[x]$ for the residue class (mod $\mathfrak{q}$) of $x \in A$. To complete the proof we must show that for each $\omega \in \Omega(n)$ there is just one way to complete the diagram

$$\begin{array}{ccc} A^n & \xrightarrow{\nu_n} & (A/\mathfrak{q})^n \\ \downarrow \omega & & \downarrow \omega' \\ A & \xrightarrow{\nu_n} & A/\mathfrak{q} \end{array}$$

to a commutative square; thus we have to find a map $\omega' : (A/\mathfrak{q})^n \to A/\mathfrak{q}$ such that

$$[a_1] \ldots [a_n]\omega' = [a_1 \ldots a_n\omega]. \tag{1.2.6}$$

This equation defines ω' uniquely if we can show that the right-hand side is independent of the choice of a_i in its $\mathfrak{q}$-class. Let $(a_i, a_i') \in \mathfrak{q}$; since $\mathfrak{q}$ is a subalgebra, we have $(a_1 \ldots a_n\omega, a_1' \ldots a_n'\omega) \in \mathfrak{q}$, i.e.

$$[a_1 \ldots a_n\omega] = [a_1' \ldots a_n'\omega],$$

and this is what we had to show. ∎

The algebra so defined on $A/\mathfrak{q}$ is again denoted by $A/\mathfrak{q}$ and is called the *quotient algebra* of A by $\mathfrak{q}$, with the natural homomorphism $\nu : A \to A/\mathfrak{q}$. For example, as we have seen, A always has the congruences $1_A, A^2$; the corresponding quotients are A and the trivial Ω-algebra consisting of a single element. An Ω-algebra is said to be *simple* if it is non-trivial and has no quotients other than itself and the trivial algebra. It follows that an algebra A is simple iff it is non-trivial and has no congruences other than 1_A or A^2.

The isomorphism theorems for groups have precise analogues for Ω-algebras. We begin with the factor theorem, which is also familiar in the case of groups (BA, Theorem 2.3.1).

Theorem 1.2.4 (Factor theorem). *Let $f : A \to B$ be a homomorphism of Ω-algebras and $\mathfrak{q}$ a congruence on A such that $\mathfrak{q} \subseteq \ker f$. Then there is a unique homomorphism $f' : A/\mathfrak{q} \to B$ such that $f = \nu f'$, where ν is the natural homomorphism from A to $A/\mathfrak{q}$. Further, f' is injective if and only if $\mathfrak{q} = \ker f$.*

Proof. Let us again write $[x]$ for the $\mathfrak{q}$-class of $x \in A$. If a homomorphism f' with the stated properties exists, then it must satisfy

$$[a]f' = af, \quad a \in A. \tag{1.2.7}$$

Thus there can be at most one such mapping. To show that there is one we have to verify that the right-hand side of (1.2.7) depends only on the $\mathfrak{q}$-class containing a and not on a itself. Let $(a, a') \in \mathfrak{q}$; then $(a, a') \in \ker f$, hence $af = a'f$ as claimed. Thus there is a unique well-defined mapping f' to satisfy (1.2.7) and it only remains to show that f' is a homomorphism. Given $a_i \in A$, $\omega \in \Omega(n)$, we have $(a_1 \dots a_n\omega)f = (a_1 f)\dots(a_n f)\omega$, hence by (1.2.7),

$$[a_1 \dots a_n\omega]f' = [a_1]f' \dots [a_n]f'\omega,$$

and this shows f' to be a homomorphism. It is injective iff no two distinct $\mathfrak{q}$-classes are identified by f' and this is just the condition $\mathfrak{q} = \ker f$. ∎

Theorem 1.2.5 (First isomorphism theorem). *Let $f : A \to B$ be a homomorphism of Ω-algebras. Then*

$$A/\ker f \cong \operatorname{im} f. \tag{1.2.8}$$

Thus f may be factorized as $f = \nu f_1 \mu$, where $\nu : A \to A/\ker f$ is the natural homomorphism, f_1 is the isomorphism (1.2.8) and $\mu : \operatorname{im} f \to B$ is the inclusion mapping.

Proof. By applying the factor theorem with $\mathfrak{q} = \ker f$ we find $f' : A/\ker f \to B$ such that $f = \nu f'$, where f' is injective. Its image is im f, so there is an isomorphism $f_1 : A/\ker f \to \operatorname{im} f$ to satisfy $f = \nu f_1 \mu$, as claimed. ∎

Theorem 1.2.6 (Second isomorphism theorem). *Let A be an Ω-algebra, A_1 a subalgebra of A and $\mathfrak{q}$ a congruence on A. Then the union of all $\mathfrak{q}$-classes meeting A_1 is a subalgebra $A_1^{\mathfrak{q}}$ of A, $\mathfrak{q}_1 = \mathfrak{q} \cap A_1^2$ is a congruence on A_1 and we have an isomorphism*

$$A_1/\mathfrak{q}_1 \cong A_1^{\mathfrak{q}}/\mathfrak{q}. \tag{1.2.9}$$

Proof. Let $\nu : A \to A/\mathfrak{q}$ be the natural homomorphism and ν_1 its restriction to A_1. Then ν_1 is a homomorphism of A_1 into $A/\mathfrak{q}$; its image is the set of $\mathfrak{q}$-classes meeting A_1, namely $A_1^{\mathfrak{q}}/\mathfrak{q}$, and its kernel is $\mathfrak{q} \cap A_1^2 = \mathfrak{q}_1$. Applying Theorem 1.2.5 we obtain (1.2.9). ∎

Similarly, by applying the factor theorem with $B = A/\mathfrak{t}$ and the natural homomorphism $\nu_{\mathfrak{t}} : A \to A/\mathfrak{t}$ for f, we obtain

Theorem 1.2.7 (Third isomorphism theorem). *Let A be an Ω-algebra and $\mathfrak{q}, \mathfrak{t}$ congruences on A such that $\mathfrak{q} \subseteq \mathfrak{t}$. Then there is a unique homomorphism $\theta : A/\mathfrak{q} \to A/\mathfrak{t}$ such that $\nu_{\mathfrak{q}}\theta = \nu_{\mathfrak{t}}$. Further,* ker θ *is the set of pairs of $\mathfrak{q}$-classes that are identified in $A/\mathfrak{t}$. Denoting this set by $\mathfrak{t}/\mathfrak{q}$, we find that $\mathfrak{t}/\mathfrak{q}$ is a congruence on $A/\mathfrak{q}$ and θ induces an isomorphism*

$$\theta' : (A/\mathfrak{q})/(\mathfrak{t}/\mathfrak{q}) \to A/\mathfrak{t}, \tag{1.2.10}$$

such that $\theta = \nu_{\mathfrak{t}/\mathfrak{q}}\theta'$. ∎

If we fix $\mathfrak{q}$ and vary $\mathfrak{t}$, we obtain

Corollary 1.2.8. *Let A be an Ω-algebra and $\mathfrak{q}$ a congruence on A. There is a natural bijection between the set of congruences on $A/\mathfrak{q}$ and the set of congruences on A which contain $\mathfrak{q}$, and if $\mathfrak{t}/\mathfrak{q}$, $\mathfrak{t}$ correspond in this way, then*

$$A/\mathfrak{t} \cong (A/\mathfrak{q})/(\mathfrak{t}/\mathfrak{q}).$$

■

In particular, we see that $A/\mathfrak{q}$ is simple iff $\mathfrak{q}$ is a maximal proper congruence on A. We note the standard application of Zorn's lemma to obtain maximal subalgebras:

Theorem 1.2.9. *Let A be an Ω-algebra, A' a subalgebra and S a subset of A. Then there exists a subalgebra C of A which is maximal subject to the conditions $C \supseteq A'$, $C \cap S = A' \cap S$.*

Proof. The family of all subalgebras C such that $C \supseteq A'$, $C \cap S = A' \cap S$ is easily seen to be inductive; hence by Zorn's lemma there is a subalgebra which is maximal subject to these conditions. ■

Since congruences on A are certainly subalgebras of A^2 and the collection of all congruences is inductive, we obtain

Corollary 1.2.10. *Let A be an Ω-algebra, Γ a correspondence on A and $\mathfrak{q}$ a congruence on A. Then there exists a congruence $\mathfrak{q}^*$ on A which is maximal subject to the conditions $\mathfrak{q}^* \supseteq \mathfrak{q}$, $\mathfrak{q}^* \cap \Gamma = \mathfrak{q} \cap \Gamma$.* ■

We conclude this section with a construction which is often used, the subdirect product. Let us again take a direct product of Ω-algebras: $P = \prod A_i$, with projections $\pi_i : P \to A_i$. It is easily seen that P may be characterized by the properties:

(i) *for any $x, y \in P$, if $x\pi_i = y\pi_i$ for all i, then $x = y$,*
(ii) *given any family (a_i), where $a_i \in A_i$, there exists $x \in P$ such that $x\pi_i = a_i$.*

Often one encounters situations where only (i) holds. This means that we are dealing essentially with a certain subalgebra of the direct product $\prod A_i$, with projections $\pi_i : P \to A_i$. An algebra A is called a *subdirect product* of the A_i if there is an embedding of A in the direct product P such that the image is mapped by π_i onto A_i, for all i. We remark that any subalgebra A of $\prod A_i$ is a subdirect product of the family A'_i, where A'_i is the image of the restriction map $\pi_i|A$. Subdirect products usually arise as follows.

Proposition 1.2.11. *Let A be an Ω-algebra and $(\mathfrak{q}_i)$ a family of congruences on A. Put $\mathfrak{q} = \cap\mathfrak{q}_i$, $A_i = A/\mathfrak{q}_i$. Then $A/\mathfrak{q}$ is a subdirect product of the family (A_i).*

Proof. The map $\theta : A \to \prod A_i$ defined by

$$a\theta = (a_i), \quad \text{where } a_i \text{ is the } \mathfrak{q}_i\text{-class of } a, \tag{1.2.11}$$

is a homomorphism and its kernel is clearly $\cap\mathfrak{q}_i = \mathfrak{q}$. Dividing by $\mathfrak{q}$, we obtain an embedding $A/\mathfrak{q} \to \prod A_i$ and by (1.2.11), $\theta\pi_i$ is surjective, hence $A/\mathfrak{q}$ is a subdirect product of the A_i. ■

Of course if the congruences $\mathfrak{q}_i$ intersect in 1_A, then A itself is a subdirect product of the A_i.

This proposition may also be expressed as follows. A congruence $\mathfrak{q}$ on A is said to *separate* $a, b \in A$ if $(a, b) \notin \mathfrak{q}$. Given a class $\mathscr{P}$ of Ω-algebras, an algebra A is said to be *residually-$\mathscr{P}$* if for each pair of distinct elements a, b of A there is a congruence $\mathfrak{q}$ separating a and b such that $A/\mathfrak{q} \in \mathscr{P}$. Now Proposition 1.2.11 tells us that an algebra is a subdirect product of $\mathscr{P}$-algebras iff it is residually-$\mathscr{P}$.

A family $(\mathfrak{q}_i)$ of congruences on A is said to be *separating* if $\cap \mathfrak{q}_i = 1$. This just means that any pair of distinct elements is separated by some $\mathfrak{q}_i$. It is clear that any family which includes 1_A is separating. If the set of all congruences $\neq 1_A$ on A is separating, A is called *subdirectly reducible*; otherwise A is *subdirectly irreducible*. We note that the trivial algebra is subdirectly reducible according to this definition. Let us also remark that A is subdirectly irreducible iff in every subdirect product representation of A at least one of the homomorphisms to the factors is an isomorphism. With this definition we have the following theorem, due to Garrett Birkhoff, 1944.

Theorem 1.2.12. *Every Ω-algebra is a subdirect product of subdirectly irreducible Ω-algebras which are homomorphic images of A.*

Proof. Since the trivial algebra may be written as the empty product, we may take A to be non-trivial. If $a, b \in A$, $a \neq b$, then by Corollary 1.2.10, there is a maximal congruence $\mathfrak{q}_0$ not containing (a, b). Thus $(a, b) \notin \mathfrak{q}_0$ but $(a, b) \in \mathfrak{q}'$ for all $\mathfrak{q}' \supset \mathfrak{q}_0$; hence $A/\mathfrak{q}_0$ is subdirectly irreducible. Moreover, if $(\mathfrak{q}_i)$ is the family of congruences formed for all such pairs a, b, then $\cap \mathfrak{q}_i = 1$ because any pair $a \neq b$ is separated by some $\mathfrak{q}_i$. Thus by Proposition 1.2.11, A is a subdirect product of subdirectly irreducible algebras $A/\mathfrak{q}_i$, each a homomorphic image of A. ∎

Exercises

1. Write down the conditions for a correspondence from sets S to T to be a bijection.
2. Describe a partial ordering on a set S in terms of the correspondence $\{(a, b) \in S^2 | a \leq b\}$.
3. Fill in the details in the proof of Lemma 1.2.1.
4. Verify that the kernel of a ring homomorphism (in the sense defined in the text) is the equivalence whose classes are the cosets of an ideal. Consider the isomorphism theorems of the text in the case of rings.
5. Verify that sets (without structure) can be regarded as the special case of Ω-algebras with $\Omega = \emptyset$. Interpret the factor theorem and the isomorphism theorems for sets in this way.
6. Show that $\mathbf{Z}$ as a ring is a subdirect product of the fields $\mathbf{F}_p$, where p runs over all primes. Do the same for $\mathbf{F}_q$ where $q = p^n$ for a fixed prime and $n = 1, 2, \ldots$.

1.3 Free algebras and varieties

In order to study Ω-algebras we shall need to form expressions in indeterminates, just as polynomials in one or more indeterminates are used to study rings. Let $X = \{x_1, x_2, \dots\}$ be any set, our alphabet, usually taken to be countably infinite, and Ω any operator domain. We define an Ω-algebra $W(\Omega; X)$, the algebra of all Ω-rows in X, as follows: An *Ω-row* in X is a finite sequence of elements in the set $\Omega \cup X$ (where X is assumed disjoint from Ω). The action of Ω is by juxtaposition; thus if $\omega \in \Omega(n)$ and $u_1, \dots, u_n \in W(\Omega; X)$, then the effect of ω on the n-tuple $(u_1, \dots, u_n)$ is the row

$$u_1 u_2 \dots u_n \omega.$$

Clearly X is a subset of $W(\Omega; X)$; the subalgebra generated by X is called the *Ω-word algebra* on X and is denoted by $W_\Omega(X)$. Its elements are Ω-words in the alphabet X. For example, if there is one binary operation α, then $x_1x_2x_3\alpha x_4\alpha\alpha$ is an Ω-word, while $x_1\alpha\alpha x_2\alpha x_3$ is an Ω-row which is not an Ω-word.

We shall need a simple test for finding which Ω-rows are words. For this purpose we associate two integers with each Ω-row. The length of $w \in W(\Omega; X)$, written $|w|$, is the number of terms in w; thus if $w = c_1 \dots c_N$, where $c_i \in \Omega \cup X$, then $|w| = N$. Secondly we define the *valency* of w as $v(w) = \sum_i v(c_i)$, where

$$v(c_i) = \begin{cases} 1 & \text{if } c_i \in X, \\ 1 - n & \text{if } c_i \in \Omega(n). \end{cases}$$

Intuitively the valency represents the element-balance: thus if $\omega \in \Omega(n)$, then ω requires an input of n elements and has an output of one element, so that $v(\Omega) = \text{output} - \text{input}$. This idea is exploited in the following result (due to Karl Schröter and independently, Philip Hall), which provides a criterion for an Ω-row to be a word, using the notion of a *prefix*, i.e. a left-hand factor:

Proposition 1.3.1. *An Ω-row $w = c_1 \dots c_N$ in X is an Ω-word if and only if every prefix $w_i = c_1 \dots c_i$ of w satisfies*

$$v(w_i) > 0 \quad (i = 1, 2, \dots, N), \tag{1.3.1}$$

and

$$v(w) = 1. \tag{1.3.2}$$

Moreover, each word can be obtained in just one way from its constituents.

Proof. We shall show more generally, by induction on the length $|w|$, that w is a sequence of r words if (1.3.1) holds and

$$v(w) = r. \tag{1.3.3}$$

This includes the assertion of the theorem for $r = 1$. When $|w| = 1$, (1.3.1) implies that $v(w) = 1$, so $w \in X \cup \Omega(0)$, and conversely, so the result holds in this case; we may therefore take $|w| > 1$.

Suppose first that w is an Ω-word, say $w = u_1 \dots u_n\omega$, where $u_i \in W_\Omega(X)$ and $w \in \Omega(n)$. By the induction hypothesis, $v(u_i) = 1$ and $v(\omega) = 1 - n$, so $v(w) = n + 1 - n = 1$. Moreover, every prefix of each u_i has positive valency, hence the same is true of w. When w is a sequence of r words, (1.3.1) again holds and (1.3.3) follows by addition.

Conversely, let w be an Ω-row satisfying (1.3.1) and $|w| > 1$, $v(w) = r > 0$. We write $w = w'c$, where $c \in \Omega \cup X$ and $v(w') = r' > 0$ by (1.3.1). By induction on the length, w' is then a sequence of r' Ω-words. Now either $c \in X \cup \Omega(0)$, and then w is a sequence of $r' + 1$ words and $v(w) = r' + 1$; or $c \in \Omega(n)$, where $n > 0$, and then, since $v(w) = r > 0$, we have $r' + 1 - n = r > 0$, hence $n = r' + 1 - r$ and c is applied to the last n words of w' to produce a single word, so w is a sequence of $r' - (n - 1) = r$ Ω-words, as we had to show. This analysis of w also shows that it is built up from its constituents in just one way. ■

The uniqueness statement in Proposition 1.3.1 means that it is never necessary to insert brackets, because each expression is defined unambiguously. To give an example, let $+$ be a binary operation. Then the associative law may be written

$$x_1x_2 + x_3 + = x_1x_2x_3 + +.$$

If λ is a second binary operation, then the familiar distributive laws take the form

$$x_1x_2 + x_3\lambda = x_1x_3\lambda x_2x_3\lambda +, \quad x_1x_2x_3 + \lambda = x_1x_2\lambda x_1x_3\lambda + .$$

It is essential to write the operation symbols on one side of the variables, say on the right, as has been done here. Equivalently the operation symbols can all be written on the left (the Łukasiewicz prefix notation). But with the usual infix notation $x_1 + x_2$ an ambiguity arises as soon as we form $x_1 + x_2 + x_3$.

Let A be an Ω-algebra. If in an element w of $W = W_\Omega(X)$ we replace each element of X by an element of A we obtain a unique element of A. For $|w| = 1$ this is clear, so assume that $|w| > 1$ and use induction. We have $w = u_1 \dots u_n\omega (u_i \in W, w \in \Omega(n))$, where the u_i are uniquely determined once w is given, by Proposition 1.3.1. By induction each u_i becomes some $a_i \in A$ when we replace the elements of X by elements of A; hence w becomes $a_1 \dots a_n\omega$, another element of A. This remark can be used to establish the universal property of the Ω-word algebra.

Theorem 1.3.2. *Let A be an Ω-algebra and X a set. Then every mapping $\theta : X \to A$ extends in just one way to a homomorphism $\theta^* : W_\Omega(X) \to A$.*

Proof. Every Ω-word is of the form $w = c_1 \dots c_N$, where $c_i \in \Omega \cup X$. We write $w\theta^* = c'_1 \dots c'_N$, where

$$c' = \begin{cases} c & \text{if } c \in \Omega, \\ c\theta & \text{if } c \in X. \end{cases}$$

Thus $w\theta^*$ is just the unique element of A obtained by replacing each $x \in X$ by $x\theta$. The remark preceding the theorem shows that θ^* is well-defined, and it is easily seen to be a homomorphism extending θ, which is unique by Proposition 1.1.1. ■

The content of this theorem is also expressed by saying that $W(X)$ is the *free* Ω-algebra on X as free generating set. Soon we shall meet free algebras in varieties of algebras; the free groups encountered in BA Section 3.3, free modules of BA Section 4.6 and the free associative algebras of BA Section 5.1 are examples.

Given any Ω-algebra A, we can take a generating set X of A and apply the construction of Theorem 1.3.2. This yields

Corollary 1.3.3. *Any Ω-algebra A can be expressed as a homomorphic image of an Ω-word algebra $W_\Omega(X)$, for a suitable set X. Here X can be taken to be any set corresponding to a generating set of A.* ∎

The Ω-words may also be thought of as operations. Any word in $x_1, \ldots, x_m \in X$ (and elements of Ω) may be regarded as an m-ary operation, called a *derived operation*. For example, in groups the commutator $(x, y) = x^{-1}y^{-1}xy$ is a derived operation. The derived operations include the original operations $\omega \in \Omega$, in the form $x_1 \ldots x_n\omega$, as well as m operations $x_i (i = 1, \ldots, m)$. They are the *projection operators*

$$x_1 \ldots x_m\delta_i = x_i. \tag{1.3.4}$$

Moreover, we have a composition of operations: if $f_1, \ldots, f_n$ are any words in $x_1, \ldots, x_m$ and g is a word in n variables, then $f_1 \ldots f_n g$ is a word in $x_1, \ldots, x_m$ obtained by *composition* from $f_1, \ldots, f_n, g$.

On any set A we can consider families of operations which include all projections and are closed under composition. Such a family is called a *clone* of operations on A. For example, if A is an Ω-algebra we have the clone generated by Ω; this is the smallest clone including Ω and is obtained by repeatedly composing the elements of Ω and the projections.

In studying Ω-algebras we are often not interested in the precise operations Ω, but merely in the clone they generate, and they may be replaced by any other set of operations generating the same clone. For example, groups may be defined in terms of a constant operation e and the single binary operation xy^{-1}, or in terms of e and the single ternary operation $xy^{-1}z$, or even as a non-empty algebra with the single binary operation xy^{-1}, besides the usual ways. This raises the question of finding relatively simple sets of operations.

Consider first unary operations. An operation $\omega : A \to A$ will be called *essentially unary* if it depends on at most one argument. More precisely, in terms of projections this means that there is a unary operation $f : A \to A$ and i, $1 \le i \le n$, such that $\omega = \delta_i f$. For example, each projection operator is essentially unary. It is not hard to verify that the clone generated by any set of essentially unary operations consists entirely of essentially unary operations. For an Ω-algebra this case arises when $\Omega = \Omega(0) \cup \Omega(1)$. Such algebras can also be characterized by the fact that the union of any two subalgebras is again a subalgebra. This shows for example that groups cannot be defined by unary operations alone (since the union of two subgroups is not usually a subgroup).

The distinction between binary and higher operations is much less precise, for as the next result, due to Wacław Sierpiński [1945] shows, every finitary operation can be composed from binary ones.

Theorem 1.3.4. *Let A be a finite set. Then every finitary operation on A can be obtained by composition of binary operations on A.*

Proof. Suppose that $|A| = n$; we may without loss of generality regard A as the ring of integers mod n, $\mathbf{Z}/n$. The ring operations on $\mathbf{Z}/n$ are at most binary, so it will be enough to show that every operation can be expressed in terms of the ring operations and the δ-function

$$\delta_r(x) = \begin{cases} 1 & \text{if } x = r, \\ 0 & \text{if } x \neq r. \end{cases} \tag{1.3.5}$$

This can be accomplished by a multivariable analogue of the Lagrange interpolation formula: given $f(x_1, \ldots, x_k)$, we have

$$f(x_1, \ldots, x_k) = \sum f(a_1, \ldots, a_k)\delta_{a_1}(x_1) \ldots \delta_{a_k}(x_k), \tag{1.3.6}$$

where the summation is over all k-tuples $(a_1, \ldots, a_k)$. It is of course important to realize that the a_i on the right of (1.3.6) are parameters, not variables; thus ax, for any $a \in \mathbf{Z}/n$, can be built up from x by repeated addition and so is (for any given a) a unary operation. ■

The theorem still holds when A is infinite, but the proof in that case is quite different and is based on the fact that there is then a bijection from A^2 to A, which can be used to reduce n-ary operations to binary ones (see Cohn (1981) and Exercise 3).

When we come to define a concrete class of algebras such as groups, we do so by specifying its operations: μ binary, ν unary and ε 0-ary. The axioms for groups in terms of these operations take the form:

(associativity) $$x_1x_2\mu x_3\mu = x_1x_2x_3\mu\mu, \tag{1.3.7}$$

(neutral) $$x\varepsilon\mu = \varepsilon x\mu = x, \tag{1.3.8}$$

(inverse) $$xx\nu\mu = x\nu x\mu = \varepsilon. \tag{1.3.9}$$

Actually these laws as stated are redundant: parts of (1.3.8) and (1.3.9) follow from the rest. This point is well known and does not concern us here.

We see that the axioms take the form of equations holding identically for all values of the variables. Generally, by an *identity* or *law* over Ω in X we understand a pair $(u, v) \in W^2$, or sometimes the equation formed from the pair:

$$u = v. \tag{1.3.10}$$

We shall say that the law (1.3.10) *holds* in the Ω-algebra A or that A *satisfies* (1.3.10) if every homomorphism $W \to A$ maps u and v to the same element of A, in other words, if u and v define the same derived operation on A.

The relation between laws and algebras establishes a Galois connexion between the set of all sets of laws in the given alphabet X and the class of all sets of Ω-algebras. Given any set $\sum$ of laws, we can form $\mathscr{V}_\Omega(\sum)$, the class of all Ω-algebras satisfying all the laws in $\sum$. This class $\mathscr{V}_\Omega(\sum)$ is called the *variety* generated by $\sum$. For example, groups form the variety of (μ, ν, ε)-algebras generated by (1.3.7)–(1.3.9). Likewise rings form a variety, but fields do not. In the other direction we can from any set $\mathscr{C}$ of Ω-algebras form the set $\mathfrak{q}(\mathscr{C})$ of all laws holding in all algebras of $\mathscr{C}$. Now our Galois connexion relates each variety of Ω-algebras to a correspondence on $W_\Omega(X)$ of the form $\mathfrak{q}(\mathscr{C})$.

For any class $\mathscr{C}$ of Ω-algebras its members will be called *$\mathscr{C}$-algebras*. Our next task will be to determine the precise form of the set $\mathfrak{q}(\mathscr{C})$. A subalgebra of A is called *fully invariant* in A if it is mapped into itself by all endomorphisms of A; this definition also extends to congruences on A, as subalgebras of A^2.

Theorem 1.3.5. *Let $W = W_\Omega(X)$ be the Ω-word algebra on an infinite alphabet X. Then the Galois connexion between Ω-algebras and laws establishes a natural bijection between varieties of Ω-algebras and fully invariant congruences on W.*

Proof. For any class $\mathscr{C}$ of Ω-algebras let $\mathscr{C}^* = \mathfrak{q}(\mathscr{C})$ be the set of all laws holding in all $\mathscr{C}$-algebras, and for any set $\sum$ of laws let $\sum^* = \mathscr{V}_\Omega(\sum)$ be the variety defined by $\sum$. We first show that $\mathscr{C}^*$ is a fully invariant congruence on W. The congruence properties are clear: in every $\mathscr{C}$-algebra we have $u = u$ for any $u \in W$; if $u = v$ holds, then so does $v = u$, and if $u = v$, $v = w$ hold, then $u = w$ holds too. Further, if $u_i = v_i (i = 1, \ldots, n)$ are laws holding in $A \in \mathscr{C}$ and $\omega \in \Omega(n)$, then $u_1 \ldots u_n\omega = v_1 \ldots v_n\omega$ holds in A. Now let $(u, v) \in \mathscr{C}^*$ and let θ be any endomorphism of W. If $\alpha : W \to A$, where $A \in \mathscr{C}$, is any homomorphism, then so is $\theta\alpha$, whence $u\theta\alpha = v\theta\alpha$. Thus the law $u\theta = v\theta$ holds in A, so $(u\theta, v\theta) \in \mathscr{C}^*$ and this shows $\mathscr{C}^*$ to be a fully invariant congruence.

To complete the proof we have to show that

$$\mathscr{V}^{**} = \mathscr{V} \tag{1.3.11}$$

for any variety $\mathscr{V}$ and

$$\mathfrak{q}^{**} = \mathfrak{q} \tag{1.3.12}$$

for any fully invariant congruence $\mathfrak{q}$ on W.

By the definition of a variety, $\mathscr{V} = \sum^*$ for some $\sum \subseteq W^2$, hence $\mathscr{V}^{**} = \sum^{***} = \sum^* = \mathscr{V}$, which proves (1.3.11). To establish (1.3.12), we take a fully invariant congruence $\mathfrak{q}$ on W and first show that

$$W/\mathfrak{q} \in \mathfrak{q}^*. \tag{1.3.13}$$

This will follow if we can show that all the laws corresponding to the elements of $\mathfrak{q}$ hold in $W/\mathfrak{q}$. Let $(u, v) \in \mathfrak{q}$, let $\alpha : W \to W/\mathfrak{q}$ be any homomorphism and denote the natural homomorphism $W \to W/\mathfrak{q}$ by ν. We shall define an endomorphism α' of W such that

$$w\alpha'\nu = w\alpha \quad \text{for all} \quad w \in W; \tag{1.3.14}$$

to do so we pick for each $x \in X$ an element $x_0 \in W$ such that $x_0\nu = x\alpha$ and define $x\alpha' = x_0$. By Theorem 1.3.2 the mapping $\alpha' : X \to W$ so defined extends to a homomorphism and (1.3.14) holds for all $w \in X$; hence by Proposition 1.1.1 it holds generally. Now $\mathfrak{q}$ is fully invariant, hence $(u\alpha', v\alpha') \in \mathfrak{q}$ and so $u\alpha = u\alpha'\nu = v\alpha'\nu = v\alpha$, and this establishes (1.3.13).

To prove (1.3.12), we note that in any case $\mathfrak{q}^{**} \supseteq \mathfrak{q}$. If $(u, v) \notin \mathfrak{q}$, then $u = v$ is not a law in $W/\mathfrak{q}$, but $W/\mathfrak{q} \in \mathfrak{q}^*$ by (1.3.13), and so $(u, v) \notin \mathfrak{q}^{**}$. Therefore equality holds in (1.3.12). ■

Let $\mathscr{C}$ be a class of Ω-algebras. By a *free* $\mathscr{C}$-algebra on a set X we understand an algebra F in $\mathscr{C}$ with the following universal property: there is a mapping $\mu : X \to F$ such that every mapping f from X into a $\mathscr{C}$-algebra A can be factored uniquely by μ to give a homomorphism from F to A, i.e. there exists a unique homomorphism $f' : F \to A$ such that

$$\mu f' = f. \tag{1.3.15}$$

Remarks

1. If $\mathscr{C}$ contains non-trivial algebras, then μ is an embedding. For, given $a, b \in X$, $a \neq b$, we can map X to a $\mathscr{C}$-algebra by a mapping f such that $af \neq bf$; hence by (1.3.15), $a\mu \neq b\mu$.
2. If $\mathscr{C}$ admits subalgebras, then the free $\mathscr{C}$-algebra F is generated by the image $X\mu$. For otherwise we could replace F by the subalgebra generated by $X\mu$; since F is unique up to isomorphism, it must itself be generated by $X\mu$. Thus $X\mu$ generates F; it is called a *free generating set*.
3. If $\mathscr{C}$ admits subalgebras, F is a free $\mathscr{C}$-algebra on X and X' is a subset of X, then the subalgebra of F generated by X' is the free $\mathscr{C}$-algebra on X'. For this subalgebra is easily seen to possess the universal property.

Not every class has free algebras, but they exist in varieties, by our next result.

Proposition 1.3.6. *Let $\mathscr{V}$ be any variety of Ω-algebras and $\mathfrak{q}$ the congruence on $W = W_\Omega(X)$ consisting of all the laws holding in $\mathscr{V}$. Then $W/\mathfrak{q}$ is the free $\mathscr{V}$-algebra on X.*

Proof. By (1.3.13), $W/\mathfrak{q}$ is a $\mathscr{V}$-algebra, so it only remains to verify the universal property. Let us write $\nu : W \to W/\mathfrak{q}$ for the natural mapping. Given any mapping $f : X \to A$ to a $\mathscr{V}$-algebra, by Theorem 1.3.2 this extends to a homomorphism $\bar{f} : W \to A$. Given $u, v \in W$, if $u \equiv v \pmod{\mathfrak{q}}$, then (u, v) is a law in $\mathscr{V}$ and so holds in A, hence $u\bar{f} = v\bar{f}$. Thus $\mathfrak{q} \subseteq \ker \bar{f}$, and by the factor theorem there is a homomorphism $f' : W/\mathfrak{q} \to A$ such that $\bar{f} = \nu f'$. If $\mu : X \to W$ is the injection, we have $f = \mu\bar{f} = \mu\nu f'$, and f' is unique, since it is given on a generating set of $W/\mathfrak{q}$. Thus $W/\mathfrak{q}$ satisfies all the conditions for a free $\mathscr{V}$-algebra. ■

There is another way of forming free algebras, which leads to a useful criterion, due to Garrett Birkhoff, for a class of algebras to be a variety.

Theorem 1.3.7. *Let $\mathscr{C}$ be a class of Ω-algebras; $\mathscr{C}$ is a variety if and only if it is closed under the operations of taking subalgebras, homomorphic images and direct products.*

Proof. The necessity of the conditions is easy to check; given any Ω-algebra A, it is clear that any subalgebra and any homomorphic image of A satisfy all the laws holding in A. Moreover, if a law holds in every member of a family of Ω-algebras, then it also holds in their direct product. This shows that every variety satisfies the given conditions.

Conversely, let $\mathscr{C}$ be a class of Ω-algebras closed under subalgebras, homomorphic images and direct products. Then $\mathscr{C}$ contains the trivial algebra (as the direct product of the empty family). If there are no other algebras in $\mathscr{C}$, then we have the variety defined by the law $x_1 = x_2$. So we may now assume that $\mathscr{C}$ contains a non-trivial algebra. We can form a free $\mathscr{C}$-algebra on a given set X as follows. Consider the set of all $\mathscr{C}$-algebras with a generating set of cardinal not exceeding that of X. Take all mappings $f_\alpha : X \to A_\alpha$, where A_α is a $\mathscr{C}$-algebra and Xf_α a generating set of A_α, and in the direct product $P = \prod A_\alpha$ consider the subalgebra F generated by the elements (xf_α), $x \in X$. As a subalgebra of the direct product, F is again in $\mathscr{C}$. We claim that F satisfies the universal property relative to the mapping $\mu : x \mapsto (xf_\alpha)$. For if $f : X \to A$ is any mapping to a $\mathscr{C}$-algebra A and A' is the subalgebra generated by Xf, then the restriction $f|A'$ coincides with some f_α and so A' is a homomorphic image of F, the mapping $F \to A'$ being the projection on the appropriate factor. Hence we have a homomorphism $f' : F \to A$ such that $f = \mu f'$ and f' is unique since it is prescribed on a generating set of F. Thus F is the free $\mathscr{C}$-algebra on X.

Clearly we have

$$\mathscr{C} \subseteq \mathscr{C}^{**}, \tag{1.3.16}$$

and it remains to prove equality. Let $\mathfrak{q} = \mathscr{C}^*$ be the set of all laws holding in $\mathscr{C}$. By Proposition 1.3.6, the free $\mathscr{C}$-algebra is $W/\mathfrak{q}$. If $A \in \mathscr{C}^{**}$, we can write A as a homomorphic image of W, for an appropriate X, say $f : W \to A$. By the definition of $\mathscr{C}^{**} = \mathfrak{q}^*$, A satisfies all the laws of $\mathfrak{q}$, hence f can be factored by $\mathfrak{q}$; thus A is a homomorphic image of $W/\mathfrak{q}$, the free $\mathscr{C}$-algebra on X, and A is therefore itself a $\mathscr{C}$-algebra. Hence equality in (1.3.16) is established. ∎

We have already remarked that rings and groups are examples of varieties. We now see that fields (commutative or not) do not form a variety, since they do not admit direct products; for if E, F are any fields, their direct product $E \times F$ as a ring has zero-divisors and so cannot be a field.

Exercises

1. Show that if some operation symbols are written on the left and others on the right of their arguments, then ambiguities can arise.
2. Let $w = c_1 \ldots c_N$ be an Ω-word of the form $u_1 \ldots u_n\omega\,(u_i \in W, \omega \in \Omega(n))$. Show that any proper subsequence $c_ic_{i+1} \ldots c_j$, where $j - i < N - 1$, which is itself an Ω-word, can only occur within a single factor u_k.
3. Assuming a bijection $\mu : A^2 \leftrightarrow A$ between a set A and its Cartesian square A^2, show that every n-ary operation ω on A can be expressed in terms of the binary operation μ and a suitable unary operation.
4. Verify that the set of all essentially unary operations on a set is a clone. Deduce that any operation derived from essentially unary operations is again unary.
5. Let A be any Ω-algebra, X a set and for each $x \in X$, let $\delta_x : A^X \to A$ be the projection on the x-th factor. Show that the subalgebra of the direct power A^{A^X} generated by all the $\delta_x (x \in X)$ is the free algebra on X for the variety generated by the algebra A (i.e. the least variety containing A).
6. Show that modular lattices form a variety. Similarly for distributive lattices, and Boolean algebras.
7. Show that groups may be defined in terms of the operation $xy\alpha = xy^{-1}$ as non-empty algebras satisfying $xz\alpha yz\alpha\alpha = xy\alpha, xx\alpha yy\alpha y\alpha\alpha = y$. Show that abelian groups may be defined by $xxy\alpha\alpha = y, xy\alpha z\alpha = xz\alpha y\alpha$.
8. Show that any variety of groups defined by a finite set of laws can also be defined by a single law.
9. Show that the automorphism group of $W_\Omega(X)$ is isomorphic to the group of all permutations of X.
10. Let $\mathscr{V}$ be a variety of Ω-algebras and F the free $\mathscr{V}$-algebra on a set X. Given a homomorphism $f : A \to B$ between $\mathscr{V}$-algebras which is surjective, and a homomorphism $\theta : F \to B$, find a homomorphism $\theta' : F \to A$ such that $\theta = \theta' f$. (Hint. See the proof of Theorem 1.3.5.)

1.4 The diamond lemma

In many algebraic problems the elements of a set are defined as equivalence classes of formal expressions, where two expressions are considered as equivalent if one can pass from one to the other by a series of 'moves'. The problem is to decide when two expressions are equivalent. For example, the elements of a particular Ω-algebra A are given by Ω-words in a generating set; the defining relations in A allow the passage between certain words and we have to decide when two given words represent the same element of A (the *word problem* for A).

The situation may be represented by a graph as follows: the vertices of our graph are the different formal expressions, and each move, from u to v say, is represented by an edge from u to v. Now the equivalence classes are the connected components of our graph. Frequently the moves are of two sorts: direct moves (e.g. in a group, removing a factor xx^{-1}) and their inverses (inserting a factor xx^{-1} in a certain place); this means that we have a directed graph. An expression is *reduced* if it admits no direct moves and the main result of this section, the *diamond lemma*, gives conditions under which each equivalence class contains a single reduced expression. The conditions are of a form that frequently applies, and it leads to a simple solution of our problem: To test if two expressions are equivalent we apply direct moves until each is in reduced form; if these reduced forms are equal, then and only then are the two expressions equivalent.

Lemma 1.4.1 (Diamond lemma, M. H. A. Newman [1942]). *Let A be a set with an equivalence relation defined on it by moves as above, such that the following conditions are satisfied:*

(i) *Finiteness condition. For each $u \in A$ there exists an integer $r = r(u)$ such that no chain of direct moves applied to u has more than r terms.*
(ii) *Confluence condition. If $u \in A$ can be transformed to x by one direct move and to y by another, then there exists $v \in A$ which can be reached from each of x, y by an appropriate series of direct moves.*

Then each equivalence class of A contains exactly one reduced element.

Proof. By (i) we can from each element of A reach a reduced element by a finite series of direct moves, so each equivalence class contains a reduced element. Given $u \in A$, suppose that we reach a reduced element a in m moves, passing through the elements $u = a_0, a_1, \ldots, a_m = a$ and that $u = b_0, b_1, \ldots, b_n = b$ is another such chain leading to a reduced element b in n steps; we have to show that $a = b$. By (ii) we can reach a common element c_0 by direct moves applied to a_1 or to b_1, and by direct moves applied to c_0 we reach a reduced element c. We shall use induction on the least value of $r(u)$. Clearly $r(a_1) < r(u)$, hence by induction we have $c = a$, and similarly $c = b$, therefore $a = b$, as claimed. ∎

A typical application is the existence proof of a normal form for the elements of a free group (see Exercise 4 and Chapter 3). For a discussion of the applications to rings, with many illuminating examples, see Bergman [1978].

Exercises

1. In Lemma 1.4.1(ii) assume that if u is transformed to x by one direct move and to y by another, where $x \neq y$, then there exists $v \in A$ which can be reached from each of x, y by just one direct move. Show that all reduction chains from a given element to a reduced element have the same length. Show that the extra condition cannot be omitted.
2. (M. H. A. Newman) Show that the conclusion of Lemma 1.4.1 still applies if (ii) holds but instead of (i) we have merely the minimum condition: no element admits an infinite succession of direct moves. (Hint. Repeat the construction in the proof of Lemma 1.4.1 and use the minimum condition.)
3. Let A be a ring with an endomorphism α. Show that in the ring R generated by A and a symbol x satisfying $ax = xa^{\alpha}$ for all $a \in A$, every element can be uniquely

written as a polynomial in x : $\sum x^i a_i$ ($a_i \in A$), and hence that A is embedded in R (see Section 7.3 below).

4. Let X, X' be disjoint sets with a bijection $x \leftrightarrow x'$ between them. Write $Z = X \cup X'$ and on the set Z^* of all strings of letters from Z (including the empty string 1) define a product by juxtaposition (this is just the free monoid on Z, see Section 11.1 below). Define direct moves as the replacement of $fxx'g$ or $fx'xg$ by fg and their inverses as inverse moves. Apply the diamond lemma to deduce the existence of a normal form for the elements of a free group (see Section 3.5 below).
5. Let S be a semigroup (system with an associative multiplication) without idempotent (i.e. $x^2 \neq x$ for all $x \in S$) and satisfying

$$ua = ub \Rightarrow va = vb \quad \text{for all} \quad u, v, a, b \in S. \tag{1.4.1}$$

By adjoining formal solutions of the equations

$$xa = b \quad a, b \in S, \tag{1.4.2}$$

show that S can be embedded in a semigroup in which (1.4.2) has a solution for all a, b. (Hint. Adjoin a new symbol p to S and consider all words in $S \cup \{p\}$ with direct move $pa \to b$. Verify the conditions of Lemma 1.4.1 and show that distinct elements of S cannot be equivalent. Now show that the resulting semigroup again satisfies (1.4.1) and repeat the process (see Cohn [1956]).)

1.5 Ultraproducts

Let us again consider the direct product construction. Given a direct product $P = \prod A_i$ of Ω-algebras, we have seen that if a law $u = v$ holds in each factor A_i then it holds in the product. On the other hand, consider the statement occurring in the definition of a field:

$$\text{for all } a \neq 0 \text{ there exists } a' \text{ such that } aa' = 1. \tag{1.5.1}$$

This may well hold in each factor A_i and yet fail to hold in the direct product, as we see by taking the direct product of two fields; the element $(1, 0)$ is different from 0 but does not have an inverse.

In order to remedy the situation we introduce certain homomorphic images of direct products, called ultraproducts, which have the property that every sentence of first-order logic which holds in all the factors, also holds in the product. For a complete proof we would need a detailed description of what constitutes a sentence in first-order logic, i.e. a sentence without free variables, and in which all quantifications are over object variables (an 'elementary sentence'), and this would take us rather far afield. However, the construction itself is easily explained and has many uses in algebra. We describe it below and refer for further details to Bell and Slomson (1971), Barwise (1977) and Cohn (1981).

We shall need the concept of an ultrafilter. Let I be a non-empty set. By a *filter* on I one understands a collection $\mathscr{F}$ of subsets of I such that

F.1 $I \in \mathcal{F}, \emptyset \notin \mathcal{F}$,
F.2 if $X, Y \in \mathcal{F}$, then $X \cap Y \in \mathcal{F}$,
F.3 if $X \in \mathcal{F}$ and $X \subseteq X' \subseteq I$, then $X' \in \mathcal{F}$.

For example, given a subset A of I, if $A \neq \emptyset$, then the set of all subsets of I containing A is a filter, called the *principal filter* generated by A. More generally, if (A_λ) is any family of subsets of I, then the collection of all subsets containing a set of the form

$$A_{\lambda_1} \cap \ldots \cap A_{\lambda_\nu} \tag{1.5.2}$$

forms a filter, provided that none of the sets (1.5.2) is empty. This condition on the A_λ is called the *finite intersection property*. Thus any family of subsets of I with the finite intersection property is contained in a filter on I. Such a family is also called a *filter base*.

An *ultrafilter* on I is a filter which is maximal among all the filters on I. An alternative characterization is given by

Lemma 1.5.1. *A filter $\mathcal{F}$ on I is an ultrafilter if and only if for each subset A of I, either A or its complement A' belongs to $\mathcal{F}$.*

Of course A, A' cannot both belong to $\mathcal{F}$, because then $\mathcal{F}$ would contain $\emptyset = A \cap A'$.

Proof. Let $\mathcal{F}$ be an ultrafilter. If $A \notin \mathcal{F}$, then by **F**.3, no member of $\mathcal{F}$ can be contained in A and so each member of $\mathcal{F}$ meets A'. It follows that the family of all sets containing some $F \cap A'(F \in \mathcal{F})$ is a filter containing $\mathcal{F}$, but then it must equal $\mathcal{F}$ by maximality of the latter, so $A' \in \mathcal{F}$. Conversely, if for each $A \subseteq I$, either A or A' belongs to $\mathcal{F}$, consider a filter $\mathcal{F}_1 \supset \mathcal{F}$ and $B \in \mathcal{F}_1 \backslash \mathcal{F}$. By assumption, $B' \in \mathcal{F}$ and so $\emptyset = B \cap B' \in \mathcal{F}_1$ which is a contradiction. ■

The existence of ultrafilters is clear from Zorn's lemma:

Theorem 1.5.2. *Every filter on a set I is contained in an ultrafilter.*

Proof. Let $\mathcal{F}$ be a filter on I and consider the set of all filters containing $\mathcal{F}$. This set is easily seen to be inductive, hence it has a maximal member, which is the required ultrafilter. ■

For example, the principal filter generated by a one-element subset is an ultrafilter. When I is finite, every ultrafilter is of this form, but for infinite sets we can always find non-principal ultrafilters as follows. Let I be an infinite set and call a subset *cofinite* if it has a finite complement in I. The collection of all cofinite subsets of I clearly has the finite intersection property and so is contained in an ultrafilter, and the latter is easily seen to be non-principal.

We shall use filters to construct certain homomorphic images of direct products. Let $A_i (i \in I)$ be a family of Ω-algebras and let $\mathcal{F}$ be any filter on the index set I. Then the *reduced product*

$$\prod A_i/\mathcal{F} \tag{1.5.3}$$

is the homomorphic image of the direct product $P = \prod A_i$, defined by the rule:

$$\text{for any } x, y \in P, x \equiv y \Leftrightarrow \{i \in I | x\pi_i = y\pi_i\} \in \mathcal{F}, \tag{1.5.4}$$

where π_i is the projection on A_i. Let us call a subset of I *$\mathcal{F}$-large* if it belongs to $\mathcal{F}$. Then the definition states that $x \equiv y$ iff x and y agree on an *$\mathcal{F}$-large set. We have to verify that we obtain an Ω-algebra in this way, i.e. that the correspondence defined on P* by (1.5.4) is a congruence. Reflexivity and symmetry are clear and transitivity follows by **F**.2. Now take $\omega \in \Omega(n)$ and let $x_\nu \equiv y_\nu (\nu = 1, \ldots, n)$, say x_ν and y_ν agree on $A_\nu \in \mathcal{F}$. Then $A_1 \cap \ldots \cap A_n \in \mathcal{F}$ and on this set $x_1 \ldots x_n\omega$ and $y_1 \ldots y_n\omega$ agree. Thus we have

Theorem 1.5.3. *A reduced product of Ω-algebras is an Ω-algebra.* ■

More generally this holds for any $\mathcal{V}$-algebra, where $\mathcal{V}$ is a variety, because the reduced product is a homomorphic image of a direct product.

A reduced product formed with an ultrafilter is called an *ultraproduct*, or *ultrapower* if all factors are the same. Now the ultraproduct theorem for Ω-algebras asserts that an ultraproduct $\prod A_i/\mathcal{F}$ of Ω-algebras formed with an ultrafilter $\mathcal{F}$ is again an Ω-algebra; moreover, any elementary sentence holds in the ultraproduct precisely if it holds in each factor A_i for i in an $\mathcal{F}$-large set. As already indicated, we shall not prove the full form, but merely a special case, which illustrates the method and which is sufficient for our purposes.

Theorem 1.5.4. *Any ultraproduct of skew fields is a skew field.*

Proof. Let $D_i (i \in I)$ be a family of skew fields and $K = \prod D_i/\mathcal{F}$ their ultraproduct, formed with an ultrafilter $\mathcal{F}$ on I. By Theorem 1.5.3 and the remark following it, K is a ring. Let $a \in K$ and suppose that $a \neq 0$. Taking a representative (a_i) for a in $\prod D_i$, we have

$$J = \{i \in I | a_i = 0\} \notin \mathcal{F}, \tag{1.5.5}$$

because $a \neq 0$. Therefore its complement J' belongs to $\mathcal{F}$ by Lemma 1.5.1 and we can define b_i by

$$b_i = \begin{cases} a_i^{-1} & \text{if } i \in J', \\ 1 & \text{if } i \in J. \end{cases}$$

Let us denote the image of (b_i) in K by b. Since $a_ib_i = b_ia_i = 1$ for $i \in J'$ and $J' \in \mathcal{F}$, we find that $ab = ba = 1$. Hence every non-zero element of K has an inverse and so K is a skew field, as claimed. ■

It is instructive to take a reduced product and see where the proof fails; it was for (1.5.5) that we needed the property of ultrafilters singled out in Lemma 1.5.1.

To illustrate Theorem 1.5.4 and the ultraproduct theorem mentioned earlier, let us take a sentence Ψ in the language of fields and suppose that we can find fields of arbitrarily large characteristic for which Ψ holds. Then Ψ also holds in their ultraproduct, and this will be of characteristic 0, if it was formed with a non-principal ultrafilter. For suppose that k_i is a field of characteristic p_i, where $p_1 \leq p_2 \leq \ldots$ and $p_i \to \infty$ as $i \to \infty$. Then the sentence

$$\Phi_n : \underbrace{1 + 1 + \ldots + 1}_{n} = 0$$

holds in only finitely many of the k_i for each n, and hence its negation $\neg\Phi_n$ holds in their ultraproduct; this shows the latter to be of characteristic 0. Since Ψ holds in each k_i, it also holds in the ultraproduct. Thus we have

Proposition 1.5.5. *An elementary sentence which holds in a field k_i of finite characteristic p_i, where the p_i are unbounded, also holds in certain fields of characteristic 0.* ■

For example, consider the statement: every non-degenerate binary quadratic form is universal. This may be stated as

$$\forall a, b, c \, \exists x, y [a \neq 0 \wedge b \neq 0 \Rightarrow ax^2 + by^2 = c].$$

It holds for all finite fields of characteristic not two (see BA, Section 8.2); hence it also holds in certain fields of characteristic 0.

As a second illustration we observe that a field of characteristic p may be defined by the sentence '$\neg\Phi_1 \wedge \Phi_p$'. Hence a field of finite characteristic is defined by the 'infinite disjunction'

$$\neg\Phi_1 \wedge [\Phi_2 \vee \Phi_3 \vee \Phi_5 \vee \ldots].$$

This is not an elementary sentence as it stands. But we can assert further that it is not equivalent to any set of elementary sentences. For if it were, it would hold in all fields of finite characteristic and hence also in some fields of characteristic 0, which is clearly not the case.

Exercises

1. Show that any ultrafilter which includes a finite set must be principal.
2. Let A be an infinite set. Show that for every non-empty subset B of A there is an ultrafilter $\mathscr{F}_B$ including B. What is the condition on B for $\mathscr{F}_B$ to include all cofinite subsets?
3. Let I be a set and $\mathscr{P}(I)$ the Boolean algebra of all subsets of I (see BA, Section 3.4). Defining ideals of Boolean algebras as inverse images of 0 in homomorphisms, show that a filter on I is just the complement of a non-zero ideal in $\mathscr{P}(I)$. Which ideals correspond to ultrafilters?
4. Show that any formula $\Phi(x)$ holds in an ultraproduct $\prod A_i/\mathscr{F}$ iff it holds in all the factors A_i for an $\mathscr{F}$-large set of indices. (Hint. Verify that the formulae for which this is true include all atomic formulae and are closed under $\vee, \wedge, \neg, \forall, \exists$. Hence the result holds for sentences, i.e. formulae without free variables.)
5. (Compactness theorem of model theory) Let $\mathscr{T}$ be a set of elementary sentences about Ω-algebras. Show that if each finite subset P of $\mathscr{T}$ has a model (i.e. there is

an algebra in which each sentence of P holds), then $\mathcal{T}$ has a model. (Hint. For each $P \subseteq \mathcal{T}$ take a model A_P and form a suitable ultraproduct of the A_P)

6. Show that an integral domain R which is embeddable in a direct product of skew fields is embeddable in a skew field. (Hint. Let K_i $(i \in I)$ be the family of skew fields and for $c \in R^{\times}$ let I_c be the set of indices i for which c is inverted in K_i. Verify that the I_c form a filter base.)

1.6 The natural numbers

In a first approach to mathematics one usually takes the natural numbers for granted, but for a rigorous development it is necessary either to provide an axiomatic foundation for the natural numbers, or to deduce their properties from some other domain such as set theory. The latter alternative would of course make it necessary to include an axiomatic foundation of set theory and this would involve us far deeper in foundational questions than is appropriate here. Such a study would occupy a whole volume by itself and would not greatly help us in our understanding of algebra. We shall therefore confine ourselves to a derivation of the properties of the natural numbers from the Peano axioms and a brief discussion of their significance, as well as their relevance to algebra. As we shall see, the framework of universal algebra is particularly appropriate for this purpose.

We begin by writing down a system of axioms for the natural numbers. This is not so much to give a rigorous foundation as to make explicit the properties of numbers we are using. The notions of set theory will be used freely, in the intuitive form introduced in BA. We shall also use individual numbers (e.g. to label the axioms) without hesitation; no axioms are needed for the numbers up to 12, or for that matter up to 10^{100}. The purpose of the axioms is to allow us to deal with the set of *all* numbers.

The axioms, as stated essentially by Giuseppe Peano in 1889, are:

N.1 *1 is a natural number.*

N.2 *Every natural number n has a successor n', which is again a natural number.*

N.3 *1 is not the successor of any number.*

N.4 *Distinct numbers have distinct successors: $m \neq n \Rightarrow m' \neq n'$.*

N.5 *(Principle of induction) A set of numbers containing 1 and with each number its successor contains all numbers.*

The set of all natural numbers will be denoted by **N**. We can think of **N** as an algebra with a single unary operation, the *successor function* $x \mapsto x'$. Let us call an algebra with a single unary operation an *induction algebra*. By **N**.2, **N** is an induction algebra; moreover it is generated by the single element 1, by **N**.5. To elucidate the structure of **N** we begin with a general lemma on induction algebras.

Lemma 1.6.1. *Let A be an induction algebra. Then the subalgebra B generated by an element b of A consists of b and successors of elements of B.*

Proof. The set B_1 consisting of b and successors of elements of B is contained in B; it contains b and the successor of any element of B_1 and so is a subalgebra. Since B is the least subalgebra containing b, it follows that $B_1 = B$, as claimed. ∎

For example, if we take $A = \mathbf{N}$, $b = 1$ and remember **N.5**, we see from the lemma that every number different from 1 is the successor of a number. Thus if $n \neq 1$, then there is a number which we shall denote by $n - 1$ such that $(n-1)' = n$. By **N.4**, $n - 1$ is uniquely determined by n; it is called the *predecessor of n*.

For any $n \in \mathbf{N}$ we denote by $[n|$ the subalgebra generated by n.

Lemma 1.6.2. *For all* $n \in \mathbf{N}$, *we have* $n \notin [n'|$.

Proof. Let us first show that

$$n' \neq n. \tag{1.6.1}$$

For $n = 1$ this holds by **N.3**; if it holds for any $n \neq 1$, then it holds for n' by **N.4**, hence it holds for all n, by induction (i.e. **N.5**).

Now by Lemma 1.6.1, $[1'|$ consists entirely of successors of numbers, whereas 1 is not a successor, hence $1 \notin [1'|$. Suppose now that $n \notin [n'|$ but that $n' \in [n''|$. By (1.6.1), $n' \neq n''$, so n' must be the successor of an element in $[n''|$; but this can only be n (by **N.4**), so $n \in [n''|$ and $[n''| \subset [n'|$, therefore $n \in [n'|$, which contradicts the hypothesis. By induction we conclude that $n \notin [n'|$ for all n. ∎

Let us write $|n]$ for the complement of $[n'|$ in $\mathbf{N}$. By Lemma 1.6.2, $n \in |n]$; the elements of $|n]$ other than n will be called the *antecedents* of n. When $n \neq 1$, they clearly include the predecessor $n - 1$ of n. With these preparations we can prove a result on which the box principle is based.

Theorem 1.6.3. *Let* $m, n \in \mathbf{N}$. *There is an injective mapping from* $|m]$ *to* $|n]$ *if and only if* $|m] \subseteq |n]$. *Further there is a bijection between* $|m]$ *and* $|n]$ *if and only if* $m = n$.

Proof. If $|m] \subseteq |n]$, then the inclusion mapping is the required injection, and for $m = n$ this is a bijection. Conversely, assume that $f : |m] \to |n]$ is an injective mapping; we must show that $|m] \subseteq |n]$. When $m = 1$, then since $1 \notin [n'|$, we have $1 \in |n]$ and so $|1] \subseteq |n]$. We may therefore assume that $m \neq 1$ and use induction on m. Since $m \neq 1$, there is a predecessor $m - 1$; we define a mapping $g : |m-1] \to |n]$ by the rule

$$kg = \begin{cases} kf & \text{if } kf \neq n, \\ mf & \text{if } kf = n. \end{cases}$$

To check that this is a well-defined mapping we note that there is at most one number k such that $kf = n$, because f is injective. Denote this number by k_0; if $k_0 = m$ or k_0 is not defined (because the image of f does not include n), then g is just f restricted to $|m-1]$. Otherwise g differs from f only at k_0 and there it has the value mf which it assumes nowhere else, for the domain of g does not include m. Thus g is well-defined; moreover g is injective and it does not assume the value n. It follows that $n \neq 1$ and that g is an injective mapping from $|m-1]$ to

$|n-1]$. By the induction hypothesis, $|m-1] \subseteq |n-1]$; taking successors on both sides, we obtain $|m] \subseteq |n]$.

If there is a bijection between $|m]$ and $|n]$, we conclude that $|m] = |n]$ and therefore $m = n$, because n is the unique member of $|n]$ without a successor. ∎

A set S is called *finite* if there is a bijection between S and $|n]$, for some $n \in \mathbf{N}$. By what has been said, there can be at most one such n and this is called the *cardinal* of S. Thus for any finite set there is a natural number which is its cardinal.

Theorem 1.6.3 leads to the familiar ordering of the natural numbers: we write $m \leq n$ to mean that m is an antecedent of n, or equivalently, $|m] \subseteq |n]$. It is clear that this relation is reflexive and transitive, and by the last part of Theorem 1.6.3 we see that $m \leq n, n \leq m$ implies $m = n$. Thus we have a partial ordering. From the definition it is clear that $m \leq n$ implies $m' \leq n'$ and it is easy to show that '$\leq$' is a total ordering. Given $m, n \in \mathbf{N}$, if $m = 1$, then clearly $m \leq n$; similarly if $n = 1$, then $n \leq m$. Now if $m, n \neq 1$, we can form $m-1$, $n-1$ and by induction either $m-1 \leq n-1$ or $n-1 \leq m-1$. Taking successors we find that $m \leq n$ or $n \leq m$. We shall also adopt the usual notation of writing $m < n$ or $n > m$ to mean '$m \leq n$ but $m \neq n$' and $m \geq n$ to mean $n \leq m$.

In contrapositive form Theorem 1.6.3 shows that if $m \notin |n]$, so that $m > n$, then there can be no injective mapping from $|m]$ to $|n]$. In particular, taking $m = n'(= n+1)$, we see that when n' objects are distributed over n boxes, at least one box contains more than one element. This is just Dirichlet's Box Principle, already encountered in BA, *p*.2, where it was stated without formal proof.

The natural numbers have another property not shared by all ordered sets; they are *well-ordered*, i.e. every non-empty subset of $\mathbf{N}$ has a least element. Given $\emptyset \subset S \subseteq \mathbf{N}$, let T be the set of numbers m such that $m \leq n$ for all $n \in S$. Clearly $1 \in S$; we claim that there is a number $a \in T$ such that $a' \notin T$. For if $a' \in T$ for all $a \in T$, then by induction $T = \mathbf{N}$ and S must be empty, a contradiction. Hence there exists $a \in T$ such that $a' \notin T$ and it follows that a is the least number in S, since $a \leq n$ for all $n \in S$, and $a \in S$ since otherwise $a' \in T$. This proves

Theorem 1.6.4. *The set* $\mathbf{N}$ *of natural numbers is well-ordered.* ∎

Our next task is to establish a universal property of $\mathbf{N}$, which forms the basis of the process of definition by recursion.

Theorem 1.6.5. $\mathbf{N}$ *is the free induction algebra on 1. Thus if A is any induction algebra and $a \in A$, there is a unique homomorphism $\alpha : \mathbf{N} \to A$ such that $1\alpha = a$.*

Proof. In detail the assertion states that A is a set with a single unary operation $x \mapsto x'$, and given $a \in A$, there is a unique mapping $x \mapsto x\alpha$ from $\mathbf{N}$ to A such that

$$1\alpha = a, \quad x'\alpha = (x\alpha)' \quad \text{for all } x \in \mathbf{N}. \tag{1.6.2}$$

By Proposition 1.1.1 there can be at most one such mapping. To prove its existence we form the direct product $\mathbf{N} \times A$; this is again an induction algebra, with the operation $(x, y)' = (x', y')$. Let H be the subalgebra of $\mathbf{N} \times \mathbf{A}$ generated by $(1, a)$; further denote by p, q the projection mappings of $\mathbf{N} \times A$ on the factors $\mathbf{N}$ and A respectively, and by p_1, q_1 their restrictions to H.

The image of H under p_1 is a subalgebra of $\mathbf{N}$ containing 1, and hence is $\mathbf{N}$ itself, by **N**.5. Thus for each $x \in \mathbf{N}$,

$$\textit{there exists } y \in A \textit{ such that } (x, y) \in H. \tag{1.6.3}$$

We claim that for each $x \in \mathbf{N}$ there is exactly one such y. For by Lemma 1.6.1, H consists of $(1, a)$ together with successors of members of H. Any successor is of the form (x', y'); here the first component is different from 1, by **N**.3, so the y determined in (1.6.3) by $x = 1$ is unique. Let $\mathbf{N}_1$ be the subset of all $x \in \mathbf{N}$ occurring as the first component of just *one* member of H, i.e. for which the $y \in A$ obtained in (1.6.3) is unique. As we have seen, $1 \in \mathbf{N}_1$; if we can show that $\mathbf{N}_1$ contains with each element its successor, it will follow from **N**.5 that $\mathbf{N}_1 = \mathbf{N}$.

Let $x \in \mathbf{N}_1$ and suppose that $y_1, y_2 \in A$ are such that $(x', y_i) \in H (i = 1, 2)$; we have to show that $y_1 = y_2$. Since $x' \neq 1$, (x', y_i) is a successor in H, say $(x', y_i) = (u_i, v_i)' = (u_i', v_i')$, where $(u_i, v_i) \in H$. Equating first components, we find $x' = u_1' = u_2'$, and by **N**.4, $x = u_1 = u_2$, i.e. $(x, v_i) \in H$ for $i = 1, 2$. But $x \in \mathbf{N}_1$ and this means that $v_1 = v_2$; hence $v_1' = v_2'$ and so $y_1 = y_2$, as we had to show. Therefore $x' \in \mathbf{N}_1$ and by **N**.5 we conclude that $\mathbf{N}_1 = \mathbf{N}$.

We have now shown that for each $x \in \mathbf{N}$ there is a unique $y \in A$ such that $(x, y) \in H$. Writing $x\alpha$ for y, we find that $1\alpha = a$, $x'\alpha = (x\alpha)'$, so (1.6.2) holds and the proof is complete. ■

Functions on $\mathbf{N}$ are frequently defined recursively, for example the sum of the squares of the first n natural numbers may be defined as the function $g : \mathbf{N} \to \mathbf{N}$ such that

$$g(1) = 1, \; g(n + 1) = g(n) + (n + 1)^2.$$

It may appear intuitively obvious that this defines a function, but this needs to be proved. The basic reason is that $\mathbf{N}$ is the *free* induction algebra on 1, as we shall now see. The statement in Theorem 1.6.6 is a little more general, but the proof begins with the case exemplified above.

Theorem 1.6.6. *Given $a \in \mathbf{N}$ and any function from $\mathbf{N}$ to $\mathbf{N}$, there exists a unique function φ from $\mathbf{N}$ to itself, satisfying the equations:*

$$\varphi(1) = a, \; \varphi(n') = f(n, \varphi(n)). \tag{1.6.4}$$

Proof. Suppose first that f is independent of its first argument. Then we have to find $\varphi : \mathbf{N} \to \mathbf{N}$ to satisfy

$$\varphi(1) = a, \; \varphi(n') = f(\varphi(n)). \tag{1.6.5}$$

In this case the result follows immediately from Theorem 1.6.5, taking A there to be the set $\mathbf{N}$ with f as its successor function.

In the general case we take $A = \mathbf{N}^2$ with successor function

$$(x, y) \mapsto (x', f(x, y)),$$

and apply Theorem 1.6.5 with the element $(1, a)$ in place of a. We obtain a mapping $\varphi : \mathbf{N} \to \mathbf{N}^2$; if its projections on the factors are φ_1 and φ_2, then $\varphi_1(1) = 1$, $\varphi_1(n') = n'$, hence $\varphi_1(x) = x$ for all $x \in \mathbf{N}$, and now

$$\varphi_2(1) = a, \; \varphi_2(n') = f(n, \varphi_2(n)).$$

Thus φ_2 is the required function. ■

This result emphasizes the difference between 'proof by induction' and 'definition by recursion'. Whereas the former embodied **N.5**, the latter also relies on **N.3**, **N.4**, and a further argument (such as that leading to Theorem 1.6.6) is needed to prove it. From the algebraist's point of view the situation can be summed up by saying that a *proof* by induction depends on the fact that **N**, as induction algebra, is *generated* by 1, whereas the method of *definition* by recursion depends on the fact that N is generated *freely* by 1.

As an application let us see how Theorem 1.6.6 may be used to define addition and multiplication on **N**. Given $a \in \mathbf{N}$, there is a mapping $\alpha_a : \mathbf{N} \to \mathbf{N}$ such that

$$1\alpha_a = a', \quad x'\alpha_a = (x\alpha_a)'.$$

If we write $a + x$ in place of $x\alpha_a$, these equations take on a more familiar form:

$$a + 1 = a', \quad a + x' = (a + x)'. \tag{1.6.6}$$

From this definition it is easy to prove the associative and commutative laws of addition:

$$(a + b) + c = a + (b + c), \tag{1.6.7}$$

$$a + b = b + a. \tag{1.6.8}$$

We shall prove (1.6.7) as an example and leave (1.6.8) to the reader. For $c = 1$, (1.6.7) reduces to $(a + b) + 1 = a + (b + 1)$, i.e. $(a + b)' = a + b'$, which is true by the definition (1.6.6). If we assume that (1.6.7) holds for $c = n$, then $(a + b) + n' = [(a + b) + n]' = [a + (b + n)]' = a + (b + n)' = a + (b + n')$, by (1.6.6) and the case $c = n$. Hence (1.6.7) holds for $c = n'$, and so by induction it holds for all n.

Similarly we can define multiplication by constructing for each $a \in \mathbf{N}$, a mapping $\mu_a : \mathbf{N} \to \mathbf{N}$ such that $1\mu_a = a$, $x'\mu_a = x\mu_a + a$. The existence and uniqueness follow again from Theorem 1.6.6, and we write as usual $x\mu_a = ax$, so that the definition takes the form

$$a1 = a, \; a(x + 1) = ax + a. \tag{1.6.9}$$

The associative and commutative laws can again be proved without difficulty, as well as the distributive law. If we adjoin a new element, denoted by 0, to **N**, to satisfy $a + 0 = a$, $a0 = 0$, we have a monoid under addition and the usual procedure for embedding a commutative cancellation monoid in a group enables us to embed **N** in a group, which is the additive group of integers **Z**. All that is needed is a proof of the cancellation law: $a + c = b + c \Rightarrow a = b$; this presents no difficulty and may be left to the reader. Now the multiplication on **N** can easily be extended to the set **Z** of all integers and the distributive law verified; thus we have obtained a ring structure on **Z**. It is also not difficult to extend the ordering of **N** to **Z**; it is still preserved under addition but only under multiplication by a positive number (i.e. an element of **N**).

Let us consider the Peano axioms from the point of view of first-order logic. We remark that **N.1–N.4** are elementary sentences, whereas **N.5** is not. Without going into detail this can be explained by saying that when expressed formally, **N.5** involves quantification over *sets* of numbers and not merely numbers. It is easy to give examples of structures other than **N** which satisfy **N.1–N.4**. For example, take the set T consisting of the disjoint union of two induction algebras, T_1 isomorphic to **N** and T_2 isomorphic to **Z**. Then T satisfies **N.1–N.4**, though of course not **N.5**.

This leaves open the question whether it may not be possible to replace **N.5** by elementary sentences, so as to characterize the natural numbers by elementary sentences alone. This question can be answered negatively by forming an ultrapower I of **N** with a non-principal ultrafilter. By the ultraproduct theorem (see Theorem 1.5.3 and the remarks following it), I is again an induction algebra and satisfies all the elementary sentences holding in **N**, but I is not isomorphic to **N**, because it is again totally ordered, by the rule: $x \leq y$ iff $x_n \leq y_n$ for all components in a large set, but unlike **N** it is not well-ordered. To see this, we consider the sequence $a_r = (a_{r,n})$ of elements of I, where $a_{1,n} = n$ and for $r \geq 1$,

$$a_{r+1,n} = [\tfrac{1}{2}(a_{r,n} + 1)],$$

where $[x]$ denotes the greatest integer $\leq x$. Thus $a_1 = (1, 2, 3, 4, \ldots)$, $a_2 = (1, 1, 2, 2, 3, 3, 4, 4, \ldots)$, $a_3 = (1, 1, 1, 2, 2, 2, 3, 3, 3, \ldots)$ etc. This is an infinite strictly descending sequence, because the set $\{n \in \mathbf{N} | a_{r+1,n} \geq a_{r,n}\}$ is finite, for $r = 1, 2, \ldots$, and it shows incidentally that being well-ordered is not an elementary property.

An induction algebra A satisfying the same elementary sentences as **N** is said to be *elementarily equivalent* to **N**, or also a *model* of **N**; if A is not isomorphic to **N**, it is called a *non-standard model*. Such non-standard models of **N** allow one to introduce 'infinite' numbers, just as a non-standard model of the real numbers **R** may contain 'infinitesimal' numbers. Non-standard analysis is a powerful tool with many applications, but it lies outside the scope of this work (see e.g. Robinson (1963), Stroyan and Luxemburg (1976), Barwise (1977)).

Exercises

1. Prove the commutative law of addition in **N**.
2. Prove the associative and commutative laws of multiplication, the distributive law and the cancellation law in **N**.
3. Give a direct proof that an induction algebra generated by a single element 1 satisfies either **N.3** or **N.4**.

4. Show that in any induction algebra the union of two subalgebras is again a subalgebra.
5. Let f be the function on the non-negative integers, defined by $f(0) = 0, f(x') = x$. Describe the function $g(x, y)$ defined by $g(x, 0) = x$, $g(x, y') = f(g(x, y))$.
6. Define the natural ordering on $\mathbf{Z}$ in terms of the ordering on $\mathbf{N}$ and prove its compatibility with addition and multiplication by positive numbers.
7. Show that there is no total ordering on $\mathbf{Z}/m$, the set of integers mod m, preserving addition and multiplication.
8. Use Theorem 1.6.3 to give a proof that $\mathbf{N}$ is not finite.
9. Give a direct proof by induction that there exists no surjective mapping from $|m]$ to $|n]$ if $m < n$.

Further exercises for Chapter 1

1. Let $\mathscr{V}$ be a variety of Ω-algebras and assume that there is a $\mathscr{V}$-algebra C whose carrier is finite with more than one element. Let F_n be the free $\mathscr{V}$-algebra on an n-element generating set; by establishing a correspondence between C^n and homomorphisms from F_n to C, show that any generating set of F_n has at least n elements. Deduce that F_m and F_n are not isomorphic when $m \neq n$.
2. Show that the free distributive lattice on three generators has 18 elements. (Hint. Form the different expressions in x_1, x_2, x_3; try cases of one and two generators first.)
3. Show that the free distributive lattice with 0, 1 on n free generators is 2^{2^n}.
4. Define $\mathbf{N}$ as an algebra with the single unary operation λ by the rule

$$n\lambda = \begin{cases} n-1 & \text{if } n > 1, \\ 1 & \text{if } n = 1. \end{cases}$$

Show that any non-trivial homomorphic image of $\mathbf{N}$ is isomorphic to $\mathbf{N}$, but that $\mathbf{N}$ is not simple.
5. Let p_n be the number of equivalence relations on a set of n elements. Obtain the following recursion formula for p_n:

$$p_{n+1} = \sum_i \binom{n}{i} p_i, \quad p_0 = 1.$$

Show also that $\sum p_n x^n/n! = \exp[(\exp x) - 1]$.
6. (G. M. Bergman) On an infinite-dimensional vector space V, define a *filter of subspaces* as a set of subspaces of V satisfying **F**.1–**F**.3, (where **F**.1 now reads: $V \in \mathscr{F}, 0 \notin \mathscr{F}$), and define an ultrafilter again as a maximal filter. Show that if $\mathscr{F}$ is an ultrafilter, then every subspace of $\mathscr{V}$ either lies in $\mathscr{F}$ or has a complement in $\mathscr{F}$. Let $\mathscr{A}$ be the set of linear maps $V \to V$ which are 'continuous', i.e. the inverse image of an $\mathscr{F}$-space is an $\mathscr{F}$-space, and let $\mathscr{N}$ be the set of maps with kernel in $\mathscr{F}$. Show that $\mathscr{N}$ is an ideal in $\mathscr{A}$ and that $\mathscr{A}/\mathscr{N}$ is a skew field.
7. A ring R is called *prime* if $R \neq 0$ and $aRb = 0$ implies $a = 0$ or $b = 0$ (see Chapter 8 below). Show that an ultraproduct of prime rings is prime, but an

ultraproduct of simple rings need not be simple. Is an ultrapower of a simple ring necessarily simple? What about $\mathfrak{M}_n(K)$, where K is a skew field?

8. Let k be a field and $\mathcal{D}$ an ultrafilter on **N**. Show that there is a monoid homomorphism $\prod \mathfrak{M}_n(k)/\mathcal{D} \to k^{\mathbf{N}}/\mathcal{D}$, whose kernel is an ideal.
9. For any set S denote by $\{S\}$ the set whose single member is S (as usual). Let M be the induction algebra generated by $\emptyset$ in this way, with successor operation $S' = S \cup \{S\}$. Show that M satisfies **N**.1–**N**.5 with $\emptyset$ in place of 1.
10. Use Theorem 1.6.5 to define exponentiation on **N** by $a^1 = a$, $a^{b'} = a.a^b$. What goes wrong if we try to define this operation on $\mathbf{Z}/m$?

2 Homological algebra

The present chapter serves as a concise introduction to homological algebra. Only the basic notions of category theory (treated in BA) are assumed. The definition of abelian categories (Section 2.1) and of functors between them (Section 2.2) is followed by an abstract description of module categories in Section 2.3. A study of resolutions leads to the notion of homological dimension in Section 2.4; derived functors are then defined in Section 2.5 and exemplified in Section 2.6 by the instances that are basic for rings, Ext and Tor. Universal derivations are used in Section 2.7 to prove a form of Hilbert's syzygy theorem.

2.1 Additive and abelian categories

We have met general categories in BA, Section 3.3, but most of the instances have been categories of modules or at least categories with similar properties. For a general study we shall therefore postulate the requisite properties; this will lead to additive categories, and more particularly, abelian categories. Later, in Section 2.3, we shall see what further assumptions are needed to reach module categories.

We recall that an object I in a category $\mathscr{A}$ is called *initial* if for each $\mathscr{A}$-object X there is a unique morphism $I \to X$; dually, if there is a unique morphism $X \to I$ for each object X, then I is called a *final* object. As we saw in BA, Section 3.3, an initial (or final) object, when it exists, is unique up to a unique isomorphism. For example, the category Rg of rings and homomorphisms has the trivial ring, consisting of 0 alone, as final object, and $\mathbf{Z}$, the ring of integers, as initial object. Initial objects arise in the solution of universal problems. Thus let $\mathscr{A}$ be a concrete category, i.e. a category with a forgetful functor U to Ens, the category of sets and mappings, which is faithful, and denote by UX the underlying set of an $\mathscr{A}$-object X. We fix a set S and consider the category (S, U) whose objects are mappings $S \to UX$ ($X \in \mathrm{Ob}\mathscr{A}$) and whose morphisms are commutative triangles arising from an $\mathscr{A}$-map $f : X \to Y$ by applying U. This is the *comma category* based on S and U. An initial object in this category is said to have the *universal property* for the set S. For example, in the category of groups, the free group on S has the universal property for S.

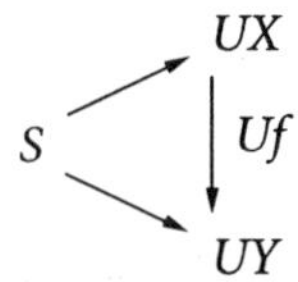

Let $\mathscr{A}$ be a category and (X_i) any family of $\mathscr{A}$-objects. Given $\mathscr{A}$-objects P, Y, each family of maps $\pi_i : P \to X_i$ gives rise to a natural mapping

$$\mathscr{A}(Y, P) \to \prod \mathscr{A}(Y, X_i), \tag{2.1.1}$$

where $g \mapsto g\pi_i$ and $\prod$ is the usual Cartesian product. When (2.1.1) is bijective for each Y we call P a *product* of the X_i with natural projections π_i. The product P with its maps π_i can also be described as the solution of a universal problem, for it is the final object in the category whose objects (A, f_i) are families of maps $f_i : A \to X_i$ and whose morphisms $\varphi : (A, f_i) \to (B, g_i)$ are families of commutative triangles, thus $\varphi : A \to B$ satisfies $f_i = \varphi g_i$. It follows that the product, when it exists, is unique up to isomorphism. We shall denote it by $\prod X_i$; thus we have an isomorphism, natural in Y:

$$\mathscr{A}\left(Y, \prod X_i\right) \cong \prod \mathscr{A}(Y, X_i). \tag{2.1.2}$$

Here $\prod$ on the left denotes the product just defined, and on the right the usual Cartesian product of sets.

Examples

1. In Ens, $\prod$ reduces to the Cartesian product. Likewise in Ab, the category of abelian groups, or more generally, in Mod_R, the category of right R-modules, $\prod$ is the direct product introduced in BA, Section 4.2.
2. In the category of all abelian torsion groups, $\prod X_i$ is the torsion subgroup of the ordinary direct product of the X_i.
3. In the category of all finite abelian groups the product does not exist; this is easily seen by taking an infinite family of non-trivial groups.
4. The product of the empty family (in any category where it exists) is the final object in the category. For here we have on the right of (2.1.1) the empty product, which by convention is a 1-element set.

There is a dual construction, the coproduct: given any family (X_i) of $\mathscr{A}$-objects, their *coproduct*, also called *sum*, is an $\mathscr{A}$-object S with maps $\mu_i : X_i \to S$, called the *natural injections*, such that (S, μ_i) is a product of the X_i in the dual category $\mathscr{A}^0$. For an explicit definition we need only reverse the arrows in the definition of the product. Thus (S, μ_i) is a coproduct if for any $\mathscr{A}$-object Y the natural mapping

$$\mathscr{A}(S, Y) \to \prod \mathscr{A}(X_i, Y)$$

given by $f \mapsto \mu_i f$ is a bijection. As before, the coproduct, when it exists, is unique up to isomorphism; denoting it by $\coprod X_i$, we have the isomorphism

$$\mathscr{A}\left(\coprod X_i, Y\right) \cong \prod \mathscr{A}(X_i, Y).$$

For example, in Ens the coproduct is the disjoint union, while in Ab or Mod_R it is the direct sum (see BA, Section 4.2). As usual, a *power* of M is a product of copies of M; similarly a *copower* is a coproduct (or sum) of copies of M. In particular, for a finite family of modules the product and coproduct are the same, except for the associated mappings (which go in opposite directions). The connexion between these two concepts becomes clearer in additive categories.

Definition. A category $\mathscr{A}$ is said to be *additive* if

Ad.1 *each* $\mathscr{A}(X, Y)$ *is an abelian group for an operation written* $+$,
Ad.2 *the composition* $\alpha, \beta \mapsto \alpha\beta$ *is biadditive, i.e.*

$$(\alpha + \alpha')\beta = \alpha\beta + \alpha'\beta, \alpha(\beta + \beta') = \alpha\beta + \alpha\beta', \tag{2.1.3}$$

Ad.3 *each finite family has a product and a coproduct.*

By applying **Ad.**3 to the empty family of objects we see that each additive category has an initial and a final object. In **Ad.**3 it is enough to assume the existence for pairs of objects and the empty family; the full strength can then be recovered by an easy induction argument. We also observe that axioms **Ad.**1–**Ad.**3 are self-dual. Further we remark that in **Ad.**3 it is enough to demand the existence of products (or only coproducts); the existence of the other sort results from the remarks following Theorem 2.1.1 below.

For example, the category Mod_R of R-modules is additive, since $\mathrm{Hom}_R(M, N)$ has an abelian group structure for which (2.1.3) holds. On the other hand, Ens, Rg and Gp are not additive, for there is no way of defining an abelian group structure on the hom sets to satisfy (2.1.3).

It is clear that a category with a single object is just a monoid; if the category also satisfies **Ad.**1 and **Ad.**2, it is a ring. Similarly in any additive category $\mathscr{A}$, the group $\mathscr{A}(X, X)$ is a ring for each $X \in \mathrm{Ob}\mathscr{A}$.

For finite families of objects in an additive category we can define a further type of product, which helps to clarify the connexion between products and coproducts. Let $\mathscr{A}$ be any additive category and $X_1, \dots, X_n$ any $\mathscr{A}$-objects. The *biproduct* of the family $(X_1, \dots, X_n)$, written $\coprod X_i$, is an object B with $2n$ maps $p_i : B \to X_i$, $q_i : X_i \to B$, such that

$$q_i p_j = \delta_{ij} 1_{X_i}, \quad \sum p_i q_i = 1_B \tag{2.1.4}$$

The three kinds of product are related by

Theorem 2.1.1. *In an additive category* $\mathscr{A}$ *let* (X_i) *be a finite family of objects. Given an* $\mathscr{A}$*-object* B *and maps* $p_i : B \to X_i$, $q_i : X_i \to B$, *such that* $q_i p_j = \delta_{ij}$, *the following conditions are equivalent:*

(a) (B, p_i) *is a product of the* X_i,
(b) (B, q_i) *is a coproduct of the* X_i,
(c) (B, p_i, q_i) *is a biproduct of the* X_i.

Proof. (a) $\Rightarrow$ (c). Let (B, p_i) be a product and write $\varphi = \sum p_i q_i$; then $\varphi p_i : B \to X_i$ satisfies $\varphi p_i = p_i$, for $\varphi p_i = \sum_j p_j q_j p_i = p_i$. By uniqueness, $\varphi = 1_B$ and so B is a biproduct.

(c) $\Rightarrow$(a). Given $f_i : A \to X_i$, we can define $f : A \to B$ by $f = \sum f_i q_i$. Then $fp_j = \sum_i f_i q_i p_j = f_j$, therefore (B, p_i) is a product. Thus we have shown that (a) $\Leftrightarrow$ (c) and by duality, (b) $\Leftrightarrow$ (c). ∎

We observe that any finite product (or coproduct) in any category satisfying **Ad.**1 and **Ad.**2 can be completed to a biproduct in a unique way. Given a product (B, p_i) of $X_1, \ldots, X_n$, fix j and define $f_i : X_j \to X_i$ by $f_i = \delta_{ij}$. Then there exists $q_j : X_j \to B$ such that $q_j p_i = \delta_{ij}$. This holds for $j = 1, \ldots, n$, and now (B, p_i, q_i) is a biproduct, by Theorem 2.1.1. Thus every finite product can be completed to a biproduct in a unique way, and by duality the same holds for finite coproducts. So we find

Corollary 2.1.2. *Let $\mathscr{A}$ be a category satisfying* **Ad.***1 and* **Ad.***2. Then any finite family of $\mathscr{A}$-objects has a coproduct if and only if it has a product, and the two are isomorphic. In particular, the product and coproduct of any finite family are isomorphic in any additive category.* ∎

Taking the empty family, we see that the initial and final object in any additive category are isomorphic. An object that is both initial and final is called a *zero object*. Thus every additive category has a zero object. By a *zero morphism* we understand a morphism which can be factored via a zero object. With this definition it is easily seen that in an additive category the neutral element in each hom group is the zero morphism.

We also note that on writing $p = (p_1, \ldots, p_n), q = (q_1, \ldots, q_n)^T$, we can express (2.1.4) in the form

$$pq = 1_B, \quad qp = \begin{pmatrix} 1_{X_1} & & 0 \\ & \cdots & \\ 0 & & 1_{X_n} \end{pmatrix}$$

Our next task is a categorical description of kernels; we begin with monomorphisms and subobjects. In any category (not necessarily additive) a map $\alpha : X \to Y$ is said to be *monic* or a *monomorphism* if whenever $\lambda\alpha$, $\mu\alpha$ are both defined, then

$$\lambda\alpha = \mu\alpha \quad \text{implies} \quad \lambda = \mu.$$

In an additive category this condition can of course be simplified to

$$\lambda\alpha = 0 \quad \text{implies} \quad \lambda = 0,$$

whenever $\lambda\alpha$ is defined. In Ens the monomorphisms are just the injective mappings. More generally, in any concrete category injective morphisms are monic; the converse holds frequently but not always. By a *subobject* of an object A we understand a pair (X, α) such that $\alpha : X \to A$ is monic. Two subobjects (X, α) and (X', α') of a given object A are said to be *equivalent* if there is an isomorphism $\lambda : X \to X'$ such that $\alpha = \lambda\alpha'$. It is clear that this leads to the expected notion of subset in the category Ens, or subgroup in Gp.

Dually a morphism $\alpha : X \to Y$ is said to be *epic* or an *epimorphism* if it is left cancellable:

$$\alpha\lambda = \alpha\mu \quad \text{implies} \quad \lambda = \mu.$$

In an additive category this can again be shortened to $\alpha\lambda = 0 \Rightarrow \lambda = 0$. If $f : A \to X$ is epic, the pair (X, f) is called a *quotient object* of A, and equivalence is defined as before. A quotient object in Ens is just a quotient set; in Gp it is a quotient group (by a normal subgroup).

Let $\mathscr{A}$ be an additive category; given a map $\alpha : X \to Y$, we shall define the kernel of α as a certain subobject of X. Consider all maps $\lambda : A \to X$ such that $\lambda\alpha = 0$; for fixed α we obtain a category by taking these maps λ as objects and as morphisms from λ to λ' maps φ from the source of λ to that of λ' such that $\lambda = \varphi\lambda'$, with the obvious composition rule (obtained by composing maps in $\mathscr{A}$). A final object in this category, if one exists, is called a *kernel* of α. Thus a kernel of α is a map $\lambda : A \to X$ such that $\lambda\alpha = 0$ and any other map $\lambda' : A' \to X$ satisfying $\lambda'\alpha = 0$ can be factored uniquely by λ, i.e. we have $\lambda' = \varphi\lambda$ for a unique map φ. This can be expressed more briefly by saying that the kernel is the largest subobject 'killed' (i.e. mapped to 0) by α. The kernel need not exist, but if it does, it is unique up to equivalence, and in fact is a subobject of X. For let (A, λ) be the kernel and assume that $f\lambda = 0$; then by the uniqueness of the factorization, $f = 0$. The kernel A of α or also the map λ to X, will be denoted by ker α.

Dually the *cokernel* of $\alpha : X \to Y$ is an initial object in the category of all maps $\mu : Y \to C$ such that $\alpha\mu = 0$. This is a quotient object of Y, unique up to equivalence if it exists; it (or also the map from Y) will be denoted by coker α.

Given any map $\alpha : X \to Y$, assume that ker α, coker α exist. Then we can define two further objects, the *image* of α, a subobject of Y, and the *coimage* of α, a quotient object of X:

$$\text{im}\ \alpha = \text{ker coker}\ \alpha, \ \text{coim}\ \alpha = \text{coker ker}\ \alpha.$$

Again they need not exist, but if they do, they are unique up to equivalence. Further we have the following diagram:

$$\begin{array}{ccc} \ker\alpha \to X & \xrightarrow{\alpha} & Y \to \operatorname{coker}\alpha \\ \downarrow & & \uparrow \\ \operatorname{coim}\alpha & \xrightarrow{\alpha'} & \operatorname{im}\alpha \end{array} \tag{2.1.5}$$

Here coim α is the largest quotient of X killing ker α, hence there is a map $\kappa : \text{coim}\ \alpha \to Y$ such that $\alpha = (\text{coim}\ \alpha)\kappa$. It follows that $(\text{coim}\ \alpha)\kappa(\text{coker}\ \alpha) = 0$; but coim is epic, so $\kappa(\text{coker}\ \alpha) = 0$, and since im α is the largest subobject of Y killed by coker α, there is a unique map $\alpha' : \text{coim}\ \alpha \to \text{im}\ \alpha$ to make the diagram commute. If we had proceeded in dual fashion, starting from im α and going via a map $X \to \text{im}\,\alpha$, we would have obtained another map $\alpha'' : \text{coim}\,\alpha \to \text{im}\,\alpha$ to make the square commute. Now $\alpha = (\text{coim}\,\alpha)\alpha'(\text{im}\,\alpha) = (\text{coim}\,\alpha)\alpha''(\text{im}\,\alpha)$; since coim α is epic and im α is monic, it follows that $\alpha' = \alpha''$, so the maps coincide and there is complete symmetry. In important cases α' is an isomorphism and this suggests the

Definition. An *abelian category* is an additive category $\mathscr{A}$ such that

Ab.1 *every map in* $\mathscr{A}$ *has a kernel and a cokernel,*
Ab.2 *the induced map* coim $\alpha \to$ im α *is an isomorphism.*

This definition can be applied to any category with a zero object, not necessarily additive; such categories are called *exact.*

These axioms, like the others, are self-dual. An example of an abelian category is Ab, the category of abelian groups, or more generally, the category Mod_R of right R-modules, for any ring R (BA, Section 4.2). By contrast the category Gp is not abelian (it is not even additive), and there are additive categories that are not abelian, such as the category of topological abelian groups and continuous homomorphisms; here **Ab**.2 need not hold, because there are continuous homomorphisms that are bijective but have no continuous inverse.

In an abelian category monic and epic maps have a simple description:

Proposition 2.1.3. *In any abelian category a map* α *is monic if and only if* $\ker\alpha = 0$ *and epic if and only if* $\text{coker}\,\alpha = 0$. *If* $\ker\alpha = \text{coker}\,\alpha = 0$, *then* α *is an isomorphism.*

Proof. By definition of ker α we have $\lambda\alpha = 0$ iff $\lambda = \lambda'(\ker\alpha)$ for some λ'. Hence $\lambda = 0$ holds for all such maps iff $\ker\alpha = 0$. This proves the first assertion; the second follows by duality. If $\alpha : X \to Y$ is such that $\ker\alpha = \text{coker}\,\alpha = 0$, then $\text{im}\,\alpha = Y$, $\text{coim}\,\alpha = X$ and $\alpha = \alpha'$ is an isomorphism. ∎

We observe that this result often fails to hold in more general categories, e.g. in Rg the inclusion map $\mathbf{Z} \to \mathbf{Q}$ is both epic and monic but is clearly not an isomorphism.

A sequence of objects and maps in an abelian category

$$\ldots \to A_{n-1} \xrightarrow{\alpha_n} A_n \xrightarrow{\alpha_{n+1}} A_{n+1} \to \ldots$$

is called a *complex* if $\alpha_n\alpha_{n+1} = 0$ for all n. This means that im α_n is a subobject of ker α_{n+1} for all n. If we have equality at A_n : $\text{im}\,\alpha_n = \ker\alpha_{n+1}$, the sequence is said to be *exact* at A_n. If the sequence is exact at each object, we speak of an *exact sequence.* It is clear that this generalizes the usage introduced for modules in BA, Section 4.2.

The simplest cases of exact sequences are $0 \to A \to 0$, which means that $A = 0$ (A is the zero object) and $0 \to A \to B \to 0$, which means that $A \cong B$, by Proposition 2.1.3. The first non-trivial case is that of a *short exact sequence*

$$0 \to A' \xrightarrow{\lambda} A \xrightarrow{\mu} A'' \to 0. \qquad (2.1.6)$$

In the case of modules this indicates that A' is isomorphic to a submodule of A, with quotient isomorphic to A''; thus A is an extension of A' by A''. In an abelian category we take (2.1.6) as the definition of an extension; thus we call A an *extension* of A' by

A'' when there is an exact sequence (2.1.6). A monomorphism λ clearly satisfies $\lambda = \ker \operatorname{coker} \lambda$, hence in the short exact sequence (2.1.6), $\lambda = \ker \mu$ and dually, $\mu = \operatorname{coker} \lambda$. The next case is that of an exact sequence with four non-zero terms; this arises for example when we analyse a general map $\alpha : X \to Y$ and obtain the exact sequence in the top line of diagram (2.1.5).

As for modules, the case of split exact sequences is important:

Proposition 2.1.4. *Given maps $\alpha : X \to Y, \beta : Y \to X$ in an abelian category such that $\alpha\beta = 1$, the canonical composite* $\ker \beta \to Y \to \operatorname{coker} \alpha$ *is an isomorphism.*

Proof. Write $\alpha' = \operatorname{coker} \alpha$, $\beta' = \ker \beta$, so that $\alpha\alpha' = 0 = \beta'\beta$; we have to show that $\beta'\alpha'$ is an isomorphism. Since β is epic, $\beta = \operatorname{coker} \beta'$; if $\beta'\alpha' f = 0$ for some f, then there exists g such that $\alpha' f = \beta g$, by the definition of β' as $\ker \beta$. Hence $g = \alpha\beta g = \alpha\alpha' f = 0$, so $\alpha' f = \beta g = 0$, but α' is epic and hence $f = 0$. This shows that $\beta'\alpha'$ is epic; by duality it is monic and so is an isomorphism. ∎

When maps α, β are related as in Proposition 2.1.4, α is called a *section* and β a *retraction.*

Corollary 2.1.5. *For a short exact sequence* (2.1.6) *in an abelian category the following conditions are equivalent:*

(a) *λ is a section,*
(b) *μ is a retraction,*
(c) *$A \cong A' \prod A''$ for suitable maps $\alpha : A \to A', \beta : A'' \to A$.*

Proof. (a) $\Rightarrow$ (c). By hypothesis there is a map α such that $\lambda\alpha = 1$. Put $\nu : \ker \alpha \to A$ for the canonical inclusion. Since $\mu = \operatorname{coker} \lambda$, it follows from Proposition 2.1.4 that $\nu\mu$ is an isomorphism and on writing $\beta = (\nu\mu)^{-1}\nu A'' \to A$, we find that $\beta\mu = 1$. We claim that A is a biproduct of A', A'' relative to the maps $\lambda, \beta; \alpha, \mu$. Clearly $\lambda\mu = 0$, $\beta\alpha = 0$, so it only remains to show that $\alpha\lambda + \mu\beta = 1$. Write $f = \alpha\lambda + \mu\beta - 1_A$; then $f\mu = \mu\beta\mu - \mu = 0$, hence $f = f'\lambda$ where $f' = f'\lambda\alpha = f\alpha = \alpha\lambda\alpha + \mu\beta\alpha - \alpha = \alpha - \alpha = 0$; it follows that $f = 0$ as claimed. Thus $A \cong A' \prod A''$; the converse is clear, hence (a) $\Leftrightarrow$ (c), and now (b) $\Leftrightarrow$ (c) follows by duality. ∎

A short exact sequence satisfying the equivalent conditions of this corollary is said to be *split exact.*

We recall from BA, Section 4.2 that in a category of modules, for any pair of maps with a common target, $\alpha : A \to C, \beta : B \to C$, there is a 'least common left multiple' P with maps $\alpha' : P \to B$, $\beta' : P \to A$ such that $\alpha'\beta = \beta'\alpha$ and for any pair α'', β'' such that $\alpha''\beta = \beta''\alpha$ there exists γ such that $\alpha'' = \gamma\alpha', \beta'' = \gamma\beta''$. It is called the *pullback* of the triple (α, β, C). This pullback exists in any abelian category, for, given $\alpha : A \to C$, $\beta : B \to C$, form the product $A \prod B$ with projections p, q on A, B respectively; now it is easily verified that $\ker(p\alpha - q\beta)$ is a pullback of α, β. A dual construction can be carried out for the pushout of a triple (C, α, β) as $\operatorname{coker}(\alpha i, \beta j)$, where i, j are the injections of A, B (the targets of α, β) into the coproduct $A \coprod B$.

The following property of pullbacks was proved in BA for the module case (Proposition 4.2.1); we now see that it holds quite generally:

Proposition 2.1.6. *Let A be an additive category. Given a pullback diagram (A, B, C; P) as shown below, if* $\ker \alpha'$ *exists, then* $\ker\alpha$ *exists and* $\ker\alpha \cong \ker\alpha'$. *A dual result holds for pushouts.*

$$\begin{array}{ccccccc} 0 \to & K' & \xrightarrow{\nu'} & P & \xrightarrow{\alpha'} & B \\ & \uparrow \mu & \nearrow & \downarrow \beta' & & \downarrow \beta \\ & K & \xrightarrow{\nu} & A & \xrightarrow{\alpha} & C \end{array}$$

Proof. Write $\ker \alpha' = (K', \nu')$; we shall show that $(K', \nu'\beta')$ is a kernel of α. In the first place $\nu'\beta'\alpha = \nu'\alpha'\beta = 0$; secondly, if $\nu : K \to A$ is such that $\nu\alpha = 0$, then the triangle ABC can be completed by K to a commutative square, with $0 : K \to B$, hence there is a unique map $\lambda : K \to P$ such that $\lambda\beta' = \nu$, $\lambda\alpha' = 0$. Since $\nu' = \ker \alpha'$, there is a unique map $\mu : K \to K'$ such that $\mu\nu' = \lambda$, hence $\mu\nu'\beta' = \nu$ and this shows that ν can be factored uniquely by $\nu'\beta'$; therefore ker $\alpha = (K', \nu'\beta')$, as claimed. ∎

In particular we see that in a pullback in an abelian category, α' is monic iff α is monic and dually for pushouts. Consider a commutative square, as in the above diagram. This corresponds to a complex

$$0 \to P \xrightarrow{\lambda} A \textstyle\coprod B \xrightarrow{\mu} C \to 0, \qquad (2.1.7)$$

where $\lambda = (\beta' i, \alpha' j)$, $\mu = p\alpha - q\beta$ and i, j, p, q are the natural injections and projections of the biproduct $A \coprod B$. The square is a pullback iff $P = \ker(p\alpha - q\beta)$, i.e. (2.1.7) is exact at P and $A \coprod B$; it is a pushout iff $C = \text{coker}(\beta' i, \alpha' j)$, i.e. (2.1.7) is exact at $A \coprod B$ and C. It follows that a pullback is also a pushout whenever μ is epic. Suppose now that α is epic and let ν be such that $\mu\nu = 0$. Then $\alpha\nu = ip\alpha\nu = i(p\alpha - q\beta)\nu = i\mu\nu = 0$, and hence $\nu = 0$; this means that μ is epic. Thus if in a pullback α is epic, then we have a pushout and so by Proposition 2.1.6, α' is also epic. This proves

Corollary 2.1.7. *Given a pullback diagram as in Proposition 2.1.6 in an abelian category, if* α *is epic, then so is* α'. *Dually, if in a pushout diagram* α *is monic, then so is* α'. ∎

Exercises

1. Show that Ens has an initial and a final object, but no zero object.
2. Show that in Rg the inclusion $\mathbf{Z} \to \mathbf{Q}$ is monic and epic but not an isomorphism. Is the inclusion $\mathbf{Z} \to \mathbf{R}$ an epimorphism?
3. Show that in a concrete category every monomorphism is injective.

4. Show that any epimorphism of groups is surjective. (Hint. If α with target G is not surjective, examine the maps from G to the group of all permutations of G.)
5. Show that the pushout of two maps $\alpha : C \to A$, $\beta : C \to B$, one of which is the zero map, is $A \oplus B$.
6. Show that $A' \xrightarrow{\lambda} A \xrightarrow{\mu} A''$ is exact iff the compositions im $\lambda \to A \to$ coim μ and ker $\mu \to A \to$ coker λ are both zero.
7. Show that in an abelian category a map α is monic iff $\alpha =$ ker coker α, and epic iff $\alpha =$ coker ker α.
8. Let $\mathscr{A}$ be an abelian category; a subcategory $\mathscr{A}_1$ is said to be *abelian* if with any morphism α, ker α and coker α (formed in $\mathscr{A}$) lie in $\mathscr{A}_1$. Verify that $\mathscr{A}_1$ is again an abelian category.
9. (3×3 lemma in abelian categories) Given three short exact sequences whose second and third terms, topped and tailed by 0's, form columns of exact sequences, so as to form a commutative diagram, show that there is just one way to fill in arrows between the first terms so as to make the diagram commutative, and the first column is then exact.
10. (Windmill lemma) Given two rows of short exact sequences with a common middle term, written as a row and column, say, form the pullback of the NW-square, the pushout of the SE-square and factorize the SW- and NE-squares through their images. Show that the resulting diagram is commutative, with exact rows and columns. Deduce the second isomorphism theorem for abelian categories: $A_1/(A_1 \cap A_2) \cong (A_1 + A_2)/A_2$.
11. Let R be any ring. Given R-modules and homomorphisms $\alpha : A \to C$, $\beta : B \to C$, show that the pullback of α and β is the submodule of $A \oplus B$ given by $\{(x, y) | x\alpha = y\beta\}$. Given $\alpha : C \to A$, $\beta : C \to B$, show that their pushout is $(A \oplus B)/K$, where $K = \{(z\alpha, z\beta) | z \in C\}$.
12. Show that the pushout of two K-algebras A, B relative to the natural maps $K \to A$, $K \to B$ is their tensor product over K.

2.2 Functors on abelian categories

Whenever we consider functors between additive categories we shall assume that they are additive; here $F : \mathscr{A} \to \mathscr{B}$ is called *additive* if $(\alpha + \beta)^F = \alpha^F + \beta^F$, whenever $\alpha + \beta$ is defined; thus the mapping $\mathscr{A}(X, Y) \to \mathscr{B}(X^F, Y^F)$ is a group homomorphism. For example, in any additive category $\mathscr{A}$ the hom functors $h^A : X \mapsto \mathscr{A}(A, X)$ and $h_A : X \mapsto \mathscr{A}(X, A)$ are additive functors from $\mathscr{A}$ to Ab; on the other hand, the functor $X \mapsto \mathrm{Hom}(X^*, X)$ between vector spaces, where X^* is the dual of X, is not additive. To give another example, an additive category with a single object is just a ring; now an additive functor between one-object categories is nothing other than a ring homomorphism, or an antihomomorphism in the case of a contravariant functor. Henceforth all functors are assumed to be additive unless otherwise stated.

Let $\mathscr{A}, \mathscr{B}$ be any categories. We recall (from BA, Section 3.3) that between two functors F, G from $\mathscr{A}$ to $\mathscr{B}$ a *natural transformation* is a family of morphisms $\varphi_X : X^F \to X^G$ such that for any $\mathscr{A}$-morphism $f : X \to Y$ we have $f\varphi_Y = \varphi_X f$;

a natural transformation with a natural transformation as inverse is a *natural isomorphism.*

If we apply an additive functor to a biproduct (B, p_i, q_i), the defining equations between the p's and q's are preserved, hence the result is again a biproduct. By Theorem 2.1.1 we obtain

Proposition 2.2.1. *Any additive functor acting on an abelian category preserves finite products, coproducts and biproducts.* ■

Clearly any functor takes zero maps to zero maps and hence transforms a complex into a complex. We shall be particularly interested in functors that preserve exactness. A functor T is said to be *exact* if it transforms each exact sequence

$$A \xrightarrow{\lambda} B \xrightarrow{\mu} C \tag{2.2.1}$$

into an exact sequence

$$A^T \xrightarrow{\lambda^T} B^T \xrightarrow{\mu^T} C^T. \tag{2.2.2}$$

For example, an equivalence between categories is an exact functor. We recall from BA, Section 3.3 that two categories $\mathscr{A}, \mathscr{B}$ are *equivalent* if there are two functors $T : \mathscr{A} \to \mathscr{B}, S : \mathscr{B} \to \mathscr{A}$ such that TS is naturally isomorphic to the identity functor on $\mathscr{A}$, and similarly ST is naturally isomorphic to the identity on $\mathscr{B}$. Any functor $T : \mathscr{A} \to \mathscr{B}$ defines for each pair X, Y of $\mathscr{A}$-objects a mapping

$$\mathscr{A}(X, Y) \to \mathscr{B}(X^T, Y^T). \tag{2.2.3}$$

The functor T is called *faithful* if (2.2.3) is injective and *full* if (2.2.3) is surjective. For an equivalence functor T, (2.2.3) is a bijection, so in this case T is full and faithful. Moreover, an equivalence functor T is *dense* in the sense that every $\mathscr{A}$-object is isomorphic to one of the form X^T, for some $\mathscr{A}$-object X. As we saw in BA, Proposition 3.3.1, a functor T is an equivalence iff it is full, faithful and dense.

All this holds in quite arbitrary categories; when $\mathscr{A}, \mathscr{B}$ are additive (and by assumption T is an additive functor), (2.2.3) is clearly a group homomorphism, and it follows easily from this that any equivalence functor is again exact.

However, exact functors are rare; most functors only satisfy a weaker condition. We define a functor to be *left exact* if it preserves kernels and *right exact* if it preserves cokernels. First we have a restatement of this condition.

Proposition 2.2.2. *A functor between abelian categories $T : A \to B$ is left exact if and only if the exactness of*

$$0 \to A \xrightarrow{\lambda} B \xrightarrow{\mu} C \tag{2.2.4}$$

implies the exactness of

$$A^T \xrightarrow{\lambda^T} B^T \xrightarrow{\mu^T} C^T. \tag{2.2.5}$$

Similarly T is right exact if and only if it preserves exactness when the 0 in (2.2.4) is at the other end (i.e. when coker $\mu = 0$*).*

Proof. The exactness of (2.2.4) is expressed by the equation $\lambda = \ker \mu$. If T preserves kernels, it follows that $\lambda^T = \ker \mu^T$ and so (2.2.5) is exact. Conversely, if (2.2.5) is exact, then by applying T to the exact sequence $0 \to 0 \to A \to B$, we find that the sequence $0 \to A^T \to B^T$ is exact, as well as (2.2.5), so $\lambda^T = \ker \mu^T$, and this shows T to be left exact. Similarly for right exactness. ■

Corollary 2.2.3. *A functor between abelian categories is exact if and only if it is left and right exact.*

Proof. Clearly an exact functor is left and right exact; conversely, if a functor T is left and right exact, it preserves kernels and cokernels, hence images and coimages. By hypothesis $\text{im}\ \lambda = \ker \mu$ in (2.2.1), hence $\text{im}\ \lambda^T = (\text{im}\ \lambda)^T = (\ker \mu)^T = \ker \mu^T$, so T is indeed exact. ■

We note that (2.2.1) is exact iff the sequence

$$0 \to \text{im}\ \lambda \to B \to \text{coim}\ \mu \to 0$$

is exact. Thus if T transforms short exact sequences into short exact sequences, then it is exact. The converse is clear, so we have

Corollary 2.2.4. *A functor between abelian categories is exact if and only if it preserves the exactness of short exact sequences.* ■

So far all functors were tacitly assumed to be covariant. If $T : \mathscr{A} \to \mathscr{B}$ is a contravariant functor, we shall call T *left exact* if the covariant functor $\text{op}.T : \mathscr{A}^0 \to \mathscr{B}$ is left exact. Right exact contravariant functors are defined correspondingly, by the right exactness of $\text{op}.T$. The reason for this form of the definition (rather than using $T.\text{op} : \mathscr{A} \to \mathscr{B}^0$) is to be found in

Theorem 2.2.5. *For any abelian category $\mathscr{A}$, the bifunctor $\mathscr{A}(X, Y)$ is left exact in each argument, i.e. h^X, h_Y are each left exact.*

Proof. For any $\mu : Y \to Y''$ in $\mathscr{A}$ the kernel of the induced mapping $\mathscr{A}(X, \mu) : \mathscr{A}(X, Y) \to \mathscr{A}(X, Y'')$ is the set of morphisms killed by μ, i.e. the maps that factor uniquely through $\ker \mu$. Thus

$$\ker \mathscr{A}(X, \mu) = \mathscr{A}(X, \ker \mu),$$

hence h^X is left exact, as claimed. Similarly, for any $\lambda : X' \to X$ in $\mathscr{A}$,

$$\ker \mathscr{A}(\lambda, Y) = \mathscr{A}(\text{coker}\ \lambda, Y),$$

therefore h_Y is left exact. ■

Let $\mathscr{A}$ be any category. A functor F from A to Ens is said to be *representable* if there is an $\mathscr{A}$-object P such that $X^F = \mathscr{A}(P, X)$; in other words, F is then naturally isomorphic to h^P and one also says that F is *represented by P*. When the category $\mathscr{A}$ is abelian, we can similarly define the representability of a functor from $\mathscr{A}$ to abelian groups. A contravariant functor G from $\mathscr{A}$ is called *representable* if there is an $\mathscr{A}$-object Q such that $Y^G = \mathscr{A}(Y, Q)$, thus G is naturally isomorphic to h_Q. For example, the dual of a vector space is representable, almost by definition: $V^* = \mathrm{Hom}_k(V, k)$. To give another example, consider $\mathbf{U}(R)$, the group of units of a ring R. It can be shown that this functor is representable by the infinite cyclic group $\mathbf{Z}$, thus $\mathbf{U}(R) \cong \mathrm{Mon}(\mathbf{Z}, R)$, where Mon is the category of monoids and R is considered as multiplicative monoid.

Sometimes we shall need a criterion for a functor to preserve inexact sequences; a sequence $\xrightarrow{\lambda}\xrightarrow{\mu}$ is called *inexact* if it is not exact, i.e. if im $\lambda \neq \ker \mu$.

Proposition 2.2.6. *A functor T between abelian categories preserves inexact sequences if and only if it is faithful.*

Proof. Suppose first T preserves inexact sequences; we must show that T is faithful, i.e. $\alpha \neq 0$ implies $\alpha^T \neq 0$. Given $\alpha : A \to B$, where $\alpha \neq 0$, the sequence $A \xrightarrow{1} A \xrightarrow{\alpha} B$ is inexact, hence it remains so on applying T, i.e. $\ker \alpha^T \neq A^T$ and so $\alpha^T \neq 0$.

Conversely, assume that T is faithful and consider the sequence (2.2.2). If this is exact, then $(\lambda\mu)^T = \lambda^T\mu^T = 0$, hence $\lambda\mu = 0$. Now let $\ker \mu = (B', i)$ and consider the composition $B' \xrightarrow{i} B \xrightarrow{\mu} C$. This is zero, hence so is the result of applying T and it gives rise to a map $(\ker \mu)^T \to \ker \mu^T$. Likewise there is a map $\mathrm{coker}\, \lambda^T \to (\mathrm{coker}\lambda)^T$, and the sequence

$$(\ker \mu)^T \to \ker \mu^T \to B^T \to \mathrm{coker}\, \lambda^T \to (\mathrm{coker}\, \lambda)^T$$

is exact at B^T; hence the composition $(\ker \mu)^T \to B^T \to (\mathrm{coker}\, \lambda)^T$ is zero, and it follows that the sequence (2.2.1) is exact at B. ■

There is a useful test for exactness in the case of adjoint functors. Given two functors $T : \mathscr{A} \to \mathscr{B}, S : \mathscr{B} \to \mathscr{A}$, we call $\mathscr{A}, \mathscr{B}$ an *adjoint pair*, or more precisely, S a *left adjoint* and T a *right adjoint* if for any $\mathscr{A}$-object X and $\mathscr{B}$-object Y,

$$\mathscr{A}(Y^S, X) \cong \mathscr{B}(Y, X^T), \tag{2.2.6}$$

where in the case of additive categories $\cong$ is an isomorphism of abelian groups which is natural in X and Y. The notion of an adjoint pair can of course be defined in quite general categories; then (2.2.6) is merely a bijection of sets (still natural in X and Y). For example, if U : Gp $\to$ Ens is the forgetful functor from groups, associating with each group its underlying set, then

$$\mathrm{Gp}(F_X, G) \cong \mathrm{Ens}(X, G^U),$$

where F_X is the free group on X. Generally nearly every universal construction arises as the left adjoint of a forgetful functor. To give another example, if i : Ab $\to$ Gp is the inclusion functor and $ab : \mathrm{Gp} \to \mathrm{Ab}$ is abelianization, i.e. passing from a group G to $G^{ab} = G/G'$, the universal abelian image (see BA, Section 3.3), then

$$\mathrm{Ab}(G^{ab}, A) \cong \mathrm{Gp}(G, iA).$$

The typical construction described by a right adjoint singles out a subset by some closure operation; see for example Exercise 8.

Returning to the general case of an adjoint pair (2.2.6), we observe that each of S, T determines the other up to natural isomorphism, for if we had

$$\mathscr{B}(Y, X^T) \cong \mathscr{B}(Y, X^{T'}), \tag{2.2.7}$$

let us first take $Y = X^T$ and denote by $\alpha : X^T \to X^{T'}$ the map on the right of (2.2.7) corresponding to the identity map on the left; next take $Y = X^{T'}$ and let $\beta : X^{T'} \to X^T$ be the map on the left corresponding to the identity map on the right. Then $\alpha\beta = 1_{X^T}$, $\beta\alpha = 1_{X^{T'}}$, so α is a natural isomorphism.

We also note that the hom functor as a bifunctor is faithful. Taking for example, $h^A : X \mapsto \mathscr{A}(A, X)$, we have for $\alpha : X \to Y$, $h^\alpha : \lambda \mapsto \lambda\alpha$, thus h^α is right multiplication by α, and choosing $\lambda = 1_X$ we find that $\lambda\alpha = 0$ for all λ implies $\alpha = 0$; similarly for h_β. With these preparations we have

Theorem 2.2.7. *Let S and T be a pair of adjoint functors between abelian categories $\mathscr{A}$ and $\mathscr{B}$. Then the left adjoint S is right exact and the right adjoint T is left exact. Moreover, if $\mathscr{A}$, $\mathscr{B}$ have arbitrary products and coproducts, then T preserves products and S preserves coproducts.*

Proof. Let us apply (2.2.6) to a short exact sequence

$$0 \to X' \xrightarrow{\lambda} X \xrightarrow{\mu} X''.$$

We obtain a commutative diagram of complexes

$$\begin{array}{ccccccc} 0 \to & \mathscr{A}(Y^S, X') & \to & \mathscr{A}(Y^S, X) & \to & \mathscr{A}(Y^S, X'') \\ & \downarrow\cong & & \downarrow\cong & & \downarrow\cong \\ 0 \to & \mathscr{B}(Y, X'^T) & \to & \mathscr{B}(Y, X^T) & \to & \mathscr{B}(Y, X''^T) \end{array} \tag{2.2.8}$$

By Theorem 2.2.5 the top row is exact, hence so is the bottom row, and this arises by applying the functor h^Y to the sequence

$$0 \to X'^T \to X^T \to X''^T. \tag{2.2.9}$$

But h^Y, when Y is allowed to vary, is faithful and so preserves inexact sequences; since the bottom row in (2.2.8) is exact, so is (2.2.9). This proves T to be left exact, and it preserves products, by (2.2.6). A dual argument shows S to be right exact and to preserve coproducts. ■

Let $\mathscr{C}$ be an abelian category and I a partially ordered set, regarded as a small category. We denote by $\mathscr{C}^I$ the functor category whose objects are functors from I to $\mathscr{C}$ with natural transformations as morphisms; thus the objects are families of $\mathscr{C}$-objects indexed by I, with families of $\mathscr{C}$-maps as morphisms. Explicitly a $\mathscr{C}$-object (A_i, α_{ij}) consists of $\mathscr{C}$-objects A_i with $\mathscr{C}$-maps $\alpha_{ij} : A_i \to A_j$ for $i \le j$, such that

$$\alpha_{ii} = 1, \quad \alpha_{ij}\alpha_{jk} = \alpha_{ik} \quad (i \le j \le k). \tag{2.2.10}$$

Conditions (2.2.10) are called the *coherence conditions* and a family satisfying them is said to be *coherent*. A morphism $f : (A_i, \alpha_{ij}) \to (B_i, \beta_{ij})$ is a family of maps $f_i : A_i \to B_i$ such that $\alpha_{ij}f_j = f_i\beta_{ij}$ for $i \le j$.

We have the *diagonal functor*

$$\Delta : \mathscr{C} \to \mathscr{C}^I, \tag{2.2.11}$$

which with each $\mathscr{C}$-object A associates the constant family (A_i, α_{ij}) with $A_i = A$, $\alpha_{ij} = 1$. The adjoint functors of Δ play an important role. The *direct limit* (also called the *inductive limit* or *colimit*) $\lim_\to$ is defined as the left adjoint of Δ:

$$\mathscr{C}(\lim_\to(A_i, \alpha_{ij}), B) = \mathscr{C}^I((A_i, \alpha_{ij}), \Delta(B)). \tag{2.2.12}$$

If $L = \lim(A_i)$ exists, there are maps $u_i : A_i \to L$ satisfying $u_i = \alpha_{ij}u_j$ for $i \le j$ such that any map (f_i) from (A_i, α_{ij}) to $\Delta(B)$ can be factored uniquely by (u_i), thus $f_i = u_i\lambda$ for all $i \in I$, for a unique $\lambda : L \to B$.

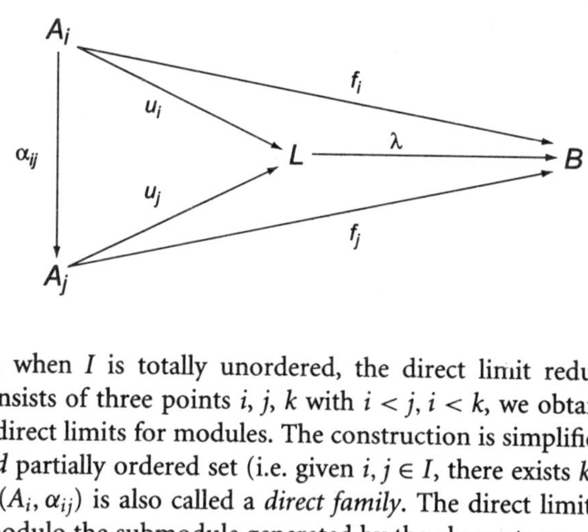

For example, when I is totally unordered, the direct limit reduces to the coproduct. If I consists of three points i, j, k with $i < j$, $i < k$, we obtain the pushout. Let us describe direct limits for modules. The construction is simplified if we assume I to be a *directed* partially ordered set (i.e. given $i, j \in I$, there exists $k \ge i, j$); in that case the family (A_i, α_{ij}) is also called a *direct family*. The direct limit L is the direct sum of the A_i modulo the submodule generated by the elements $x - x\alpha_{ij}(x \in A_i)$ for $i \le j$. For an example from field theory, let F be a field and (E_i) the family of all extensions of F of finite degree, with inclusions as mappings; then it is clear that we have a direct family. Its direct limit is a field Ω containing F, which is algebraic over F and algebraically closed, and hence is the algebraic closure of F (see BA, Section 7.3 and Section 11.8).

Similarly the *inverse limit* (also the *projective limit* or simply *limit*) $\lim_\leftarrow$ is defined as the right adjoint of Δ:

$$\mathscr{C}(B, \lim_\leftarrow (A_i, \alpha_{ij})) = \mathscr{C}^I(\Delta(B), (A_i, \alpha_{ij})).$$

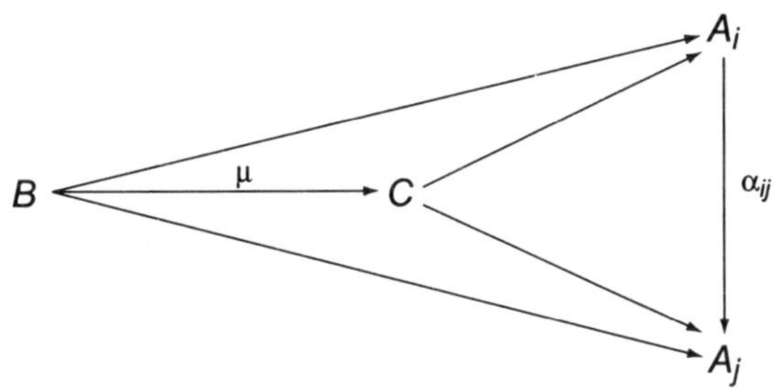

The inverse limit C has maps $\nu_i : C \to A_i$ such that $\nu_i\alpha_{ij} = \nu_j$ and any map (f_i) from $\Delta(B)$ to (A_i, α_{ij}) can be factored uniquely by (ν_i), thus $f_i = \mu\nu_i$ for all $i \in I$, for a unique $\mu : B \to C$.

When I is totally unordered, this reduces to the product of the A_i. For a triple i, j, k with $i < k$, $j < k$ it becomes the pullback. To describe the construction for modules we take I inversely directed and refer to (A_i, α_{ij}) as an *inverse family*. The inverse limit of such a family of modules is obtained by forming the product $\prod A_i$ and taking the submodule of all elements (x_i) such that $x_i\alpha_{ij} = x_j$.

To illustrate the notion of inverse limit consider a free group F of rank > 1. Let us write (N_i) for the family of all normal subgroups of finite index in F. Then $G_i = F/N_i$ is a finite group and for $N_i \subseteq N_j$ we have a natural homomorphism $\varphi_{ij} : G_i \to G_j$, and these homomorphisms are coherent. Since the intersection of any two of the N_i is again of finite index, we have an inverse family and we can form the inverse limit $G = \lim_{\leftarrow}(G_i)$. This group G is called a *profinite* group (as *pro*jective limit of *finite* groups). Since the natural homomorphisms $F \to G_i$ are compatible with the φ_{ij}, we have a canonical homomorphism $\gamma : F \to G$. As we shall see in Section 3.4, $\cap N_i = 1$, and it follows easily from this fact that γ is injective. However, γ is not surjective, for G, as inverse limit, is uncountable, whereas F is countable whenever its rank is at most countable. A similar construction is possible for abelian groups, thus for example $\mathbf{Z}$ can be embedded in a profinite group, or even in a pro-p-group (see Exercise 10).

The last example illustrates another important point, namely the lack of duality in general module categories. As we have seen, the notion of an abelian category can be developed in an entirely self-dual manner. However, the category of all modules over a given ring is not self-dual, except for very special rings, so in order to describe module categories axiomatically one will need axioms whose duals may not hold. We shall not carry out the full axiomatization (which can be found in most books on category theory) but merely list one axiom holding in all module categories but not always for their duals. This is Grothendieck's

AB5 Axiom. *Given a chain of subobjects* (A_i) *and any subobject* B *of an object, we have*

$$(\cup A_i) \cap B = \cup(A_i \cap B). \tag{2.2.13}$$

The following equivalent form will be more convenient for us:

I. The functor $\lim_{\rightarrow}$ *is exact.*

Since $\lim_{\rightarrow}$ is in any case right exact, as left adjoint, this requires that for any families of exact sequences $0 \rightarrow A_i \rightarrow B_i$ the sequence

$$0 \rightarrow \lim_{\rightarrow} (A_i) \rightarrow \lim_{\rightarrow} (B_i)$$

should again be exact. Like (2.2.13) this condition is easily verified in any category of modules. The dual states

I^0. *The functor* $\lim_{\leftarrow}$ is exact.

By duality we need only verify that the exactness of $A_i \rightarrow B_i \rightarrow 0$ entails that of $\lim_{\leftarrow} (A_i) \rightarrow \lim_{\leftarrow} (B_i) \rightarrow 0$. In most module categories this does not hold. For example, taking $A_i = \mathbf{Z}$ and B_i the family of finite images, we have $\lim_{\leftarrow} (A_i) = \mathbf{Z}$, but the mapping $\mathbf{Z} \rightarrow \hat{\mathbf{Z}} = \lim_{\leftarrow} (B_i)$ is far from surjective, since the limit $\hat{\mathbf{Z}}$ is uncountable.

In BA, Section 4.7 we have already met projective and injective modules. Their counterpart in abelian categories is of importance, because it can be used to remedy the lack of exactness of the hom functor (Theorem 2.2.5).

Definition. Let $\mathscr{A}$ be an abelian category. An $\mathscr{A}$-object P is called *projective* if the covariant hom functor $h^P = \mathscr{A}(P, -)$ is exact; an $\mathscr{A}$-object I is called *injective* if the contravariant hom functor $h_I = \mathscr{A}(-, I)$ is exact.

An alternative description of projective objects is given in

Theorem 2.2.8. *Let P be an object in an abelian category $\mathscr{A}$. Then the following conditions are equivalent:*

(a) *P is projective,*
(b) *every short exact sequence*

$$0 \rightarrow A \xrightarrow{\lambda} B \xrightarrow{\mu} P \rightarrow 0 \tag{2.2.14}$$

with P in third place splits,
(c) *given a diagram with exact row as shown, there exists a map $P \rightarrow B$ to make the triangle commutative.*

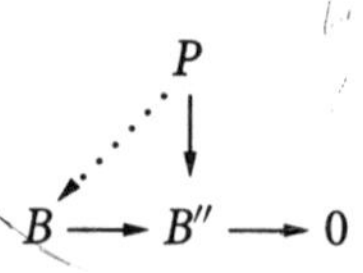

Condition (c) may be expressed by saying: every map from P to a quotient of B may be lifted to B. We note that the statement of this theorem is quite similar to that of Theorem 4.7.4 of BA for modules, but we shall not be able to use the proof given there, which depended on the existence of free modules. On the other hand, the proof given below provides another proof of Theorem 4.7.4 of BA.

Proof. (a) $\Rightarrow$ (b). Given a short exact sequence (2.2.14), we find by (a) that the sequence of abelian groups

$$0 \to \mathscr{A}(P, A) \to \mathscr{A}(P, B) \to \mathscr{A}(P, P) \to 0$$

is exact. Now $1_P \in \mathscr{A}(P, P)$ and by exactness there exists $\beta \in \mathscr{A}(P, B)$ such that $\beta\mu = 1_P$, hence (2.2.14) splits.

(b) $\Rightarrow$ (c). By forming the pullback of the given diagram we obtain

$$\begin{array}{ccccc} \ker \alpha & \longrightarrow & C & \xrightarrow{\alpha} & P & & \\ & & {\scriptstyle\beta}\downarrow & & \downarrow & & \\ & & B & \longrightarrow & B'' & \longrightarrow & 0 \end{array}$$

By Corollary 2.1.7, α is epic, hence by (b) the top row splits, so there is a map $f : P \to C$ and $f\beta : P \to B$ is the required map.

(c) $\Rightarrow$ (a). Given a short exact sequence, we apply $\mathscr{A}(P, -)$ and obtain

$$0 \to \mathscr{A}(P, B') \to \mathscr{A}(P, B) \to \mathscr{A}(P, B'') \to 0. \qquad (2.2.15)$$

By the left exactness of hom this can fail to be exact only at $\mathscr{A}(P, B'')$. But by (c) every map $P \to B''$ lifts to a map $P \to B$, and this means that (2.2.15) is also exact at $\mathscr{A}(P, B'')$. ■

Of course there is a dual characterization of injectives:

Theorem 2.2.9. *Let I be an object in an abelian category $\mathscr{A}$. Then the following conditions are equivalent:*

(a) *I is injective,*
(b) *every short exact sequence*

$$0 \to I \to B \to C \to 0$$

with I in first place splits,
(c) *given a diagram with exact row as shown, there is a map $A \to I$ such that the resulting triangle is commutative.*

$$\begin{array}{ccccc} 0 & \longrightarrow & A' & \longrightarrow & A \\ & & \downarrow & \swarrow & \\ & & I & & \end{array}$$

Here (c) may be expressed by saying: every map from a subobject of A to I can be extended to A.

The proof is dual to that of Theorem 2.2.8 and so may be left to the reader. ■

Although the notions of projective and injective module are dual, they can have very different appearance in actual categories and we shall return to this question for module categories in Section 2.3 and Section 4.6.

Exercises

1. Show that a functor between additive categories is additive iff it preserves finite products.
2. Use Exercise 1 to show that a functor between additive categories forming part of an adjoint pair is necessarily additive.
3. Show that a subcategory of an abelian category is abelian iff the inclusion functor is exact.
4. Show that for a faithful functor T in an abelian category, $C \neq 0$ implies $C^T \neq 0$. Show that for an exact functor this condition is sufficient as well as necessary.
5. Let $T : \mathscr{A} \to \mathscr{B}$, $S : \mathscr{B} \to \mathscr{A}$ be a pair of functors giving an equivalence of categories. Show that S, T is an adjoint pair as well as T, S.
6. Show that for any abelian category the following are equivalent: (a) every object is projective, (b) every object is injective, (c) every short exact sequence splits.
7. Let S, T be a pair of adjoint functors between abelian categories. Show that if S is left exact, then T preserves injectives; if T is right exact, then S preserves projectives.
8. For any group G denote by $\mathbf{Z}G$ the group ring of G over $\mathbf{Z}$. Show that the correspondence $G \mapsto \mathbf{Z}G$ is a functor from Gp to Rg whose right adjoint is the functor $R \mapsto \mathbf{U}(R)$, where $\mathbf{U}(R)$ is the group of units of R.
9. Show that a functor $\mathscr{A} \to \mathscr{B}$ is full and faithful iff $\mathscr{A}$ is equivalent to a full subcategory of $\mathscr{B}$.
10. Let p be a prime number. Verify that $\cap p^n\mathbf{Z} = 0$ and deduce that there is a natural injection $\mathbf{Z} \to \lim_{\leftarrow} (\mathbf{Z}/p^n)$.

2.3 The category $\mathbf{Mod}_R$

We have already seen that the category Mod_R of all right R-modules, for any ring R, is abelian. Frequently R will be a K-algebra (associative, with 1) , where K is some commutative ring; in that case the hom sets $\text{Hom}_R(M, N)$ are K-modules and not merely abelian groups. We shall say that we have a *K-linear* category in that case. A functor F between K-linear categories is required to be not merely additive but also *K-linear*:

$$(\alpha + \beta)^F = \alpha^F + \beta^F, \quad (\lambda\alpha)^F = \lambda.\alpha^F \quad (\lambda \in K).$$

As a rule K will be an arbitrary commutative ring, fixed in any given context, and all rings will be K-algebras. The case of abstract rings is included by taking $K = \mathbf{Z}$.

We recall that a right R-module structure on M can be described by saying that we have a homomorphism

$$f : R \to \text{End}_K(M). \tag{2.3.1}$$

Similarly, a left R-module structure on M corresponds to an antihomomorphism (2.3.1), i.e. a homomorphism $R^0 \to \text{End}_K(M)$ from the opposite ring; in detail this is a K-linear mapping f such that $(xy)f = yf.xf$. This remark is often used to avoid having to pass to the opposite ring. Thus if we have a homomorphism $R^0 \to \mathrm{End}_K(M)$, we shall regard M as a left R-module rather than a right R^0-module.

Let R, T be any rings and ${}_T\mathrm{Mod}_R$ the category of (T, R)-bimodules; clearly this is a subcategory of Mod_R. The following lemma on the transport of ring action is often useful.

Lemma 2.3.1. *Let R, S, T be any rings (or K-algebras) and $F : \mathrm{Mod}_R \to \mathrm{Mod}_S$ a covariant functor. Then F induces a functor $F' : {}_T\mathrm{Mod}_R \to {}_T\mathrm{Mod}_S$. Similarly a contravariant functor $G : \mathrm{Mod}_R \to {}_S\mathrm{Mod}$ induces a functor $G' : {}_T\mathrm{Mod}_R \to {}_S\mathrm{Mod}_T$.*

Proof. Given a (T, R)-bimodule M, we know that M^F is an S-module; further, for any $t \in T$ we can define the action of t on M^F as t^F. Since t defines an endomorphism of M_R, t^F defines an endomorphism of $(M^F)_S$, i.e. an element of $\mathrm{End}_S(M^F)$, and so

$$(xa)t^F = (xt^F)a \quad \text{for any } x \in M^F,\ a \in S.$$

We claim that the rule $t \mapsto t^F$ defines an antihomomorphism of T into $\mathrm{End}_S(M)$. For if $t, t' \in T$, then $(tt')^F = t'^F.t^F$, because M is a *left* T-module; hence M^F is indeed a (T, S)-bimodule. Moreover, a homomorphism a between (T, R)-bimodules may be characterized as an R-homomorphism centralizing T, hence α^F is an S-homomorphism centralizing T, i.e. a homomorphism between (T, S)-bimodules. The second part, referring to G, is proved similarly; since G is contravariant, it defines a *homomorphism* of T this time, which means that M^G is an (S, T)-bimodule. ∎

As an example, important for what follows, consider the hom functor. Let M be an (S, R)-bimodule and N a (T, R)-bimodule; we shall express this briefly by saying that we are in the situation $({}_SM_R, {}_TN_R)$. Consider $H = \mathrm{Hom}_R(M, N)$; when we regard M, N as right R-modules, H is just an abelian group (or a K-module). But the left T-module structure on N induces a left T-module structure on H, while the left S-module structure on M induces a right S-module structure on H; here the side is reversed because $\mathrm{Hom}(M, N)$ is contravariant in M. Thus we see that H is a left T-, right S-module; in fact it is a (T, S)-bimodule. To show this let us write (f, x) for the effect of $f \in H$ on $x \in M$. Then by definition we have for any $r \in R$, $s \in S$, $t \in T$, $(f, x)r = (f, xr)$, $(fs, x) = (f, sx)$, $(tf, x) = t(f, x)$. Hence $((tf)s, x) = (tf, sx) = t(f, sx) = t(fs, x) = (t(fs), x)$, i.e. $(tf)s = t(fs)$, as claimed.

A second functor of great importance is the tensor product. We recall from BA, Section 4.8 that for a K-algebra R and modules $(U_R, {}_RV)$ there is a K-module $U \otimes_R V$ with a mapping

$$\lambda : U \times V \to U \otimes_R V,$$

which is universal for K-bilinear mappings f from $U \times V$ to K-modules that are *R-balanced*, i.e. such that

$$(xr, y)f = (x, ry)f \quad \text{for all} \quad x \in U, y \in V, r \in R.$$

We remark that (assuming the tensor product over the commutative ring K as known), $U \otimes_R V$ may also be obtained as the homomorphic image of $U \otimes_K V$ by adding the relations $xr \otimes y = x \otimes ry (r \in R)$. Further we recall the equations of adjoint associativity which follow from the definition of the tensor product. For the situation $({}_QU_R, {}_RV_S, {}_TW_S)$ we have the natural isomorphism of (T, Q)-bimodules (adjoint associativity)

$$\mathrm{Hom}_S(U \otimes_R V, W) \cong \mathrm{Hom}_R(U, \mathrm{Hom}_S(V, W)). \tag{2.3.2}$$

This may be expressed by saying that $- \otimes_R V$ is the left adjoint of the functor $h^V = \mathrm{Hom}_S(V, -)$. By symmetry the same holds for $U \otimes_R -$ in the situation $({}_SU_R, {}_RV_T, {}_SW_Q)$, using the isomorphism

$$\mathrm{Hom}_S(U \otimes_R V, W) \cong \mathrm{Hom}_R(V, \mathrm{Hom}_S(U, W)). \tag{2.3.3}$$

By Theorem 2.2.7 we conclude

Proposition 2.3.2. *For a left R-module V over any ring R, the tensor product functor* $- \otimes_R V$ *is right exact and preserves direct sums; similarly for right R-modules.* ■

Further we recall the associative law for tensor products; in the situation $(U_R, {}_RV_S, {}_SW)$ we have

$$U \otimes_R (V \otimes_S W) \cong (U \otimes_R V) \otimes_S W. \tag{2.3.4}$$

We also recall the identity

$$U \otimes_R R \cong U \quad \text{for } U; \tag{2.3.5}$$

it corresponds to the well-known identity for the hom functor

$$\mathrm{Hom}_R(R, U) \cong U. \tag{2.3.6}$$

We have already met projective and injective objects in the category of modules in BA, Section 4.7. In particular, we see from the characterization given there that the projective R-modules are precisely the direct summands of free R-modules. There is no such explicit description of injective modules (but see Section 4.6 below); for the moment we note that by Theorem 2.2.9 a module M is injective iff every short exact sequence with M as first term splits, i.e. iff M is a direct summand in every module containing it as a submodule. This leads to the following criterion. An extension of modules $M \subseteq N$ is called *essential* and M is said to be a *large* submodule of N if M has a non-zero intersection with every non-zero submodule of N.

Proposition 2.3.3. *An R-module is injective if and only if it has no proper essential extension.*

Proof. Suppose that M is injective. If M is contained as a submodule in N, then it is a direct summand, so if $N \neq M$, the extension is not essential; thus M has no proper essential extension. Conversely, assume that M has no proper essential extension, and let L be any module containing M as a submodule. The family of all submodules of L meeting M in 0 is clearly inductive and so, by Zorn's lemma, has a maximal member, L_0 say. Consider $\bar{L} = L/L_0$; since $M \cap L_0 = 0$, M maps isomorphically to a submodule $\bar{M}$ of $\bar{L}$, and by the maximality of L, $\bar{L}$ is an essential extension of $\bar{M} \cong M$. So it cannot be proper, hence $\bar{L} = \bar{M}$, i.e. $M + L_0 = L$ and $M \cap L_0 = 0$, so M is a direct summand in L. It follows that M is injective, as claimed. ∎

We have already met Reinhold Baer's injectivity criterion in BA, Theorem 4.7.7; here is another very short proof, due to Peter Freyd.

Theorem 2.3.4 (Baer's criterion). *For any ring R, a left R-module M is injective if and only if every homomorphism from a left ideal of R into M can be extended to a homomorphism from R to M.*

Proof. The necessity is clear; to prove the sufficiency of the condition we show that when it holds, M has no proper essential extension. Let $M \subset L$ be a proper extension, fix $u \in L \backslash M$ and consider the pullback diagram shown, where $R \to L$ is the map $r \mapsto ru$.

$$\begin{array}{ccc} P & \longrightarrow & R \\ \phi \downarrow & & \downarrow \\ M & \longrightarrow & L \end{array}$$

By Proposition 2.1.6, the map $P \to R$ is monic, so P is isomorphic to a left ideal of R and by hypothesis the map $P \to M$ extends to a homomorphism $R \to M$. If $1 \mapsto v$ in this homomorphism, we have $x \mapsto xv = xu$ for all $x \in P$, and $x \in P$ if $xu \in M$, hence $R(u - v)$ is a submodule of L such that $R(u - v) \cap M = 0$, but $R(u - v) \neq 0$, because $v \in M$, $u \notin M$, so $u \neq v$. Thus L is not an essential extension of M and since L was arbitrary, it follows by Proposition 2.3.3 that M is injective. ∎

With every homomorphism of rings there are several transfer functors associated, which are often useful. Given any rings R, S and a homomorphism $f : R \to S$, any right S-module U may be defined as an R-module by putting

$$x.a = x(af) \quad \text{for } x \in U,\ a \in R.$$

This R-action on U is said to be defined by *pullback* along f (not to be confused with the pullback diagram in Section 2.1), and the resulting R-module is written ${}^fU$. The correspondence $U \mapsto {}^fU$ is a functor from $\mathbf{Mod}_S$ to $\mathbf{Mod}_R$ rather like the forgetful functor. We shall want to go in the opposite direction and construct an adjoint; thus we are given an R-module A and we ask for an associated S-module. There are two constructions, arising as the left adjoint and the right adjoint of the functor ${}^fU$; they are known as the *change-of-rings* constructions.

Proposition 2.3.5. *Let R, S be rings and* $f : R \to S$ *a homomorphism. Given a right R-module A, there is a right S-module* $A_f = A \otimes_R S$ *left adjoint to* ${}^fU$:

$$\mathrm{Hom}_S(A_f, U) \cong \mathrm{Hom}_R(A, {}^fU), \qquad (2.3.7)$$

and a right S-module $A^f = \mathrm{Hom}_R(S, A)$ *right adjoint to* ${}^fU$:

$$\mathrm{Hom}_S(U, A^f) \cong \mathrm{Hom}_R({}^fU, A). \tag{2.3.8}$$

Moreover, there is a map $\alpha : A \to A_f$ *which induces a homomorphism of R-modules (from A to* ${}^f(A_f)$*) and a map* $\beta : A^f \to A$ *inducing a homomorphism of R-modules (from* ${}^f(A^f)$ *to A). If the R-module structure on A arose by pullback along f from an S-module V, then V is a direct summand of* A_f *and of* A^f*, as R-modules.*

Proof. The proof is a simple verification, using (2.3.2), (2.3.5) and (2.3.6):

(i) $\mathrm{Hom}_S(A \otimes_R S, U) \cong \mathrm{Hom}_R(A, \mathrm{Hom}_S(S, U)) \cong \mathrm{Hom}_R(A, U)$,
(ii) $\mathrm{Hom}_S(U, \mathrm{Hom}_R(S, A)) \cong \mathrm{Hom}_R(U \otimes_S S, A) \cong \mathrm{Hom}_R(U, A)$.

Now $\alpha : A \to A \otimes_R S$ is just the map $a \mapsto a \otimes 1$ and $\beta : \mathrm{Hom}_R(S, A) \to A$ is the map $\varphi \mapsto 1\varphi$.

If $A = {}^fV$, we put $U = V$ in (2.3.7), (2.3.8) and consider the image of the identity map on the right of (2.3.7), (2.3.8); this provides maps $A_f \to V$, $V \to A^f$ which together with α, β respectively define a splitting of A_f, A^f respectively. ■

The module A_f is called the *induced* and A^f the *coinduced extension* of A along f. Here the variance refers to S, not A; in fact, both are covariant in A. They are sometimes called *relatively projective* and *relatively injective*, on account of the following property:

Corollary 2.3.6. *If A is projective as R-module, then* A_f *is projective as S-module; if A is injective as R-module, then* A^f *is injective as S-module.*

Proof. The first part follows because $\mathrm{Hom}_S(A_f, -)$ is exact whenever $\mathrm{Hom}_R(A, -)$ is exact, by (2.3.7); similarly the second part follows by (2.3.8). ■

An abelian category is said to possess *enough projectives* if every object can be written as a quotient of a projective object. For example, the category Mod_R of right modules over any ring R has enough projectives, because every module is a homomorphic image of a free (hence projective) module, by BA, Theorem 4.6.3. Dually, an abelian category is said to have *enough injectives* if every object can be embedded as a subobject of an injective object. For example, $\mathbf{Z}$ as $\mathbf{Z}$-module is contained in $\mathbf{Q}$ which is injective. Let us show that Mod_R has enough injectives.

Proposition 2.3.7. *Let R be any ring. Then* Mod_R *(as well as* ${}_R\mathrm{Mod}$*) has enough injectives, i.e. every R-module can be embedded in an injective R-module.*

Proof. We first take the special case $R = \mathbf{Z}$. Every abelian group A can be written as a quotient of a free abelian group: $A \cong F/N$. Now F is a direct sum of copies of $\mathbf{Z}$ and by embedding $\mathbf{Z}$ in $\mathbf{Q}$ we can embed the abelian group F in a vector space over $\mathbf{Q}$, G say. Clearly G is divisible as $\mathbf{Z}$-module and hence so is G/N, and it contains $F/N \cong A$ as a submodule. But for a $\mathbf{Z}$-module 'divisible' is the same as 'injective' (by BA, Proposition 4.7.8), so the $\mathbf{Z}$-module A has been embedded in an injective $\mathbf{Z}$-module.

Consider now the general case. Given any ring R, there is a natural homomorphism $f : \mathbf{Z} \to R$, obtained by mapping $n \mapsto n.1$, and we can consider any R-module M as $\mathbf{Z}$-module by pullback along f. By what has been proved, M can be embedded in an injective $\mathbf{Z}$-module I, hence the coinduced extension $M^f = \text{Hom}_{\mathbf{Z}}(R, M)$ is a submodule of I^f, by the left exactness of Hom. By Proposition 2.3.5, M is a direct summand of M^f, hence it is an R-submodule of I^f, and I^f is injective as R-module by Corollary 2.3.6. ■

It is possible to go beyond Proposition 2.3.7 and describe the 'least' injective module containing a given module:

Theorem 2.3.8. *Let R be any ring. Given R-modules M, E, where $M \subseteq E$, the following conditions are equivalent:*

(a) *E is a maximal essential extension of M,*
(b) *E is a minimal injective module containing M.*

Such an extension E exists for any R-module M, and if E' is another extension of M satisfying (a) and (b), then there is an isomorphism from E to E' leaving M elementwise fixed.

Proof. (a) $\Rightarrow$ (b). If E is a maximal essential extension of M, then any essential extension F of E is an essential extension of M, for any non-zero submodule of F meets E and hence M non-trivially. By maximality we have $F = E$, so E has no proper essential extensions and is therefore injective, by Proposition 2.3.3, but any submodule of E containing M has E as essential extension and so cannot be injective unless it is the whole of E, again by Proposition 2.3.3. Hence E is a minimal injective module containing M.

(b) $\Rightarrow$ (a). Assume that E is a minimal injective module containing M and let F be any essential extension of M; we claim that F can be embedded in E. For the inclusion of M in E extends to a homomorphism $f : F \to E$ because E is injective, and $M \cap \ker f = 0$, hence $\ker f = 0$, because F is an essential extension of M. Thus F is embedded in E. If F is a maximal essential extension of M, then as we have just seen, we can take F to be a submodule of E and by the first part of the proof F is injective. It follows that F is a direct summand of E, and so, by the minimality of E, we have $E = F$, as we had to show.

We can always construct such an E by taking an injective module I containing M (Proposition 2.3.7) and inside I taking a maximal essential extension of M, using Zorn's lemma. Finally, if E, E' are two modules both satisfying (a) and (b), then the identity mapping on M extends to a homomorphism $\alpha : E \to E'$ by the injectivity of E'. The kernel of α meets M in 0, hence $\ker \alpha = 0$ by (a), so $\text{im}\,\alpha$ is an injective submodule of E' and hence $\text{im}\,\alpha = E'$ by (b). This shows α to be an isomorphism. ■

The module E in Theorem 2.3.8, first constructed by Eckmann and Schopf in 1953, is called the *injective hull* of M. Although E is determined up to isomorphism by M, this isomorphism is not unique and the correspondence of E to M is not a functor. Later, in Chapter 4, we shall find that over certain rings there is a dual notion of projective cover for every finitely generated module.

Exercises

1. For any ring R show that the finitely generated right R-modules and homomorphisms form a full subcategory of Mod_R. Is it abelian? What about the full subcategory of cyclic right R-modules?
2. Prove the rules (2.3.4)–(2.3.6) in detail.
3. Let (A_i) be a family of objects in an abelian category. Show that if $\prod A_i$ exists then it is injective iff each A_i is injective; likewise, if $\coprod A_i$ exists, then it is projective iff each A_i is projective. (Warning. Exact sequences need not be preserved under products or coproducts in general abelian categories.)
4. An object P in a category $\mathscr{A}$ is called a *generator* of $\mathscr{A}$ if $h^P = \mathscr{A}(P, -)$ is faithful. Show that a generator in Mod_R is faithful as R-module, i.e. any non-zero element of R defines a non-zero action.
5. Show that in an abelian category with arbitrary coproducts an object P is a generator iff every object is a quotient of a copower of P. Deduce that an abelian category with arbitrary coproducts and a projective generator has enough projectives.
6. Dualize Exercises 4 and 5 to show that an abelian category with arbitrary products and an injective cogenerator (i.e. $h_I = A(-, I)$ is exact and faithful) has enough injectives.
7. Show that R is a generator of Mod_R. More generally, show that M is a generator of Mod_R iff R is a direct summand of ${}^nM$ for some $n \geq 1$.
8. Let $f : R \to S$ be a ring homomorphism. Show that for any modules U_R, ${}_SV$ we have $U_f \otimes_S V \cong U \otimes_R {}^fV$.
9. Show that for $f : R \to S$ as in Exercise 8 and modules U_R, ${}_RV$ we have ${}^f(U_f) \otimes_R V \cong U \otimes_R {}^f(V_f)$. Show further that when R, S are commutative, then $U_f \otimes_S V_f \cong (U \otimes_R V)_f$.
10. Show that if M is a finitely generated module over a Noetherian ring R, then $M^* = \mathrm{Hom}_R(M, R)$ is again finitely generated.
11. Show that in Baer's criterion (Theorem 2.3.4) it is enough to test all large left ideals.

2.4 Homological dimension

For any abelian group A we can write down a presentation $A \cong F/K$, where F is a free abelian group and K as subgroup of F is also free. For modules over a ring we still have such a presentation, but now K need no longer be free, nor even projective. If we take the view that projective modules are particularly simple (justified in the sequel), we may next take a presentation of K and hope that the process terminates, i.e. that K is in some sense closer to being projective than A. Our first objective is to assign a numerical value to this lack of projectivity.

Given any R-module A, we take a projective module P_0 mapping onto A. The kernel K_0 need not be projective, but we can again take a projective P_1 mapping onto K_0. This map has kernel K_1 and we can continue the process, giving rise to a commutative diagram with exact row as follows:

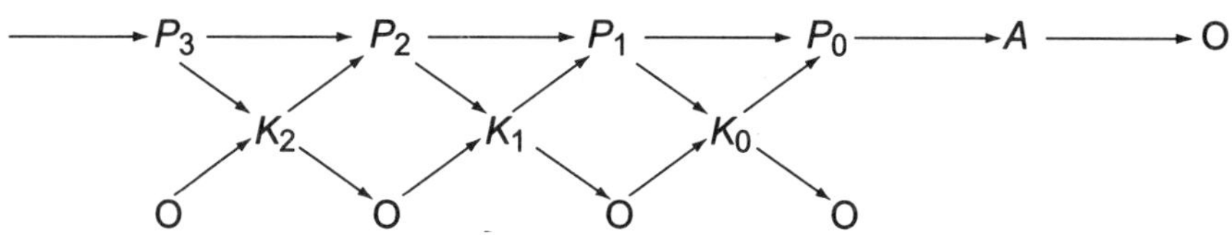

As a rule one omits the kernels and just writes the exact sequence

$$\ldots \to P_3 \to P_2 \to P_1 \to P_0 \to A \to 0. \tag{2.4.1}$$

This is called a *projective resolution* of A. For example, let $R = k[x, y]$ be the polynomial ring in x, y over a field k and consider k as R-module, by pullback along the natural homomorphism $R \to R/(x, y) \cong k$. We resolve k by mapping R to $R/(x, y)$; the kernel is the ideal (x, y) and we next take the map $R \to (x, y)$ defined by $(a, b) \mapsto ax - by$. Its kernel is the set (cy, cx), $c \in R$, and this is isomorphic to R. Thus we have obtained a resolution

$$0 \to R \to R^2 \to R \to k \to 0. \tag{2.4.2}$$

As a second example, take $R = \mathbf{Z}/4$, $A = \mathbf{Z}/2$. Clearly A is not projective, but we have a homomorphism $R \to A$ consisting of multiplication by 2, with kernel $2\mathbf{Z}/4 \cong A$, hence we obtain an infinite resolution

$$\ldots \to R \to R \to A \to 0. \tag{2.4.3}$$

It is clear that a projective resolution, possibly infinite, exists for any module, because Mod_R has enough projectives, and the resolution terminates when we reach a projective kernel. In order to compare different resolutions of a given module we need Schanuel's lemma. It is useful to have this in an extended form.

Proposition 2.4.1. *Let R be any ring and M an R-module. Given two short exact sequences $0 \to A \to P \to M \to 0$ and $0 \to B \to Q \to M \to 0$, where P is projective, we have an exact sequence*

$$0 \to A \to P \oplus B \to Q \to 0. \tag{2.4.4}$$

Proof. If we form the pullback of $P \to M$, $Q \to M$ and recall Proposition 2.1.6 and Corollary 2.1.7, we obtain an exact commutative diagram, where $A' \cong A$, $B' \cong B$:

$$\begin{array}{ccccccccc} & & 0 & & 0 & & \\ & & \downarrow & & \downarrow & & \\ & & A' & \rightarrow & A & & \\ & & \downarrow & & \downarrow & & \\ 0 \rightarrow & B' & \rightarrow C & \rightarrow & P & \rightarrow 0 \\ & \downarrow & \downarrow & & \downarrow & & \\ 0 \rightarrow & B & \rightarrow Q & \rightarrow & M & \rightarrow 0 \\ & & \downarrow & & \downarrow & & \\ & & 0 & & 0 & & \end{array}$$

Since P is projective, the middle horizontal sequence splits and $C \cong P \oplus B' \cong P \oplus B$. Now the middle vertical (with A' replaced by its isomorph A) is the desired exact sequence (2.4.4). ■

If Q as well as P is projective, (2.4.4) splits and we obtain

Lemma 2.4.2 (Schanuel's lemma). *Given two short exact sequences as in Proposition 2.4.1, if P and Q are projective, then $P \oplus B \cong Q \oplus A$.* ■

This result suggests the following definition. Two modules M, N are called *projectively equivalent* if there exist projectives P, Q such that

$$P \oplus M \cong Q \oplus N.$$

It is clear that this is in fact an equivalence relation; we shall denote the class of M by $[M]$ and note that $[M] = 0$ iff M is projective.

On the set of all equivalence classes we can define an operation as follows. Given M, we resolve it by a projective:

$$0 \rightarrow A \rightarrow P \rightarrow M \rightarrow 0,$$

and write $\pi(M) = [A]$. By Schanuel's lemma the class $[A]$ depends only on M, not on the resolution chosen. If we replace M by $M \oplus Q$, where Q is projective, we have a resolution

$$0 \rightarrow A \rightarrow P \oplus Q \rightarrow M \oplus Q \rightarrow 0,$$

and this shows that $\pi(M)$ depends in fact only on the class $[M]$ and not on M itself; π is sometimes called the *loop functor*.

We can now define the *homological* (or *projective*) dimension of a module M as

$$\mathrm{hd}(M) = \min\{n | \pi^{n+1}(M) = 0\}, \quad \text{where } \pi^0(M) = [M].$$

This depends only on the class of M; the definition shows that $\mathrm{hd}(M) \leq n$ iff M has a projective resolution of length $\leq n$, i.e. of the form (2.4.1) with $P_i = 0$ for $i > n$.

The *global dimension* of a ring R is defined as

$$\text{gl.dim.}(R) = \sup\{\text{hd}(M) | \text{all } M_R\}.$$

If necessary, we shall distinguish the *right* and *left* global dimensions, formed from right or left modules. In general these two numbers may be distinct, although they coincide for Noetherian rings (see Section 2.6 below). The rings of global dimension 0 are just the semisimple rings, for they are the rings for which every module is projective (BA, Theorem 5.2.7). As an example of a ring of infinite global dimension we have the ring $\mathbf{Z}/4$, as the resolution (2.4.3) shows.

There is an analogous development using injective resolutions, based on the fact that Mod_R also has enough injectives (Proposition 2.3.7). Given an R-module M, we form an injective resolution

$$0 \to M \to I_0 \to I_1 \to I_2 \ldots, \tag{2.4.5}$$

by embedding M in an injective module I_0, then embedding the cokernel in an injective I_1 and so on. The dual of Schanuel's lemma shows that in a short resolution $0 \to M \to I \to L \to 0$, the 'injective class' of L (defined like the projective equivalence class) depends only on that of M, and so may be written $\iota(M)$. The least integer n such that $\iota^{n+1}(M) = 0$ is called the *cohomological* or *injective dimension* of M, written $\text{cd}(M)$. As before we can define the corresponding global dimension of R, but as we shall see in Section 2.6 below, this agrees with the global dimension defined in terms of homological dimension. In the special case of global dimension 0 this is already clear from the definition of a semisimple ring as a ring over which any short exact sequence splits.

To illustrate these ideas we shall examine the class of rings of global dimension 1. We shall need a lemma which may be regarded as the dual of Baer's injectivity criterion (Theorem 2.3.4).

Lemma 2.4.3. *An R-module P is projective if and only if every homomorphism from P to a quotient of an injective module I can be lifted to I itself.*

Proof. The condition is necessary by Theorem 2.2.8. Conversely, assume that it holds. Given a short exact sequence $0 \to A \to B \to P \to 0$, we embed B in an injective module I and form the pushout:

$$\begin{array}{ccccccccc}
0 & \to & A & \to & B & \xrightarrow{\alpha} & P & \to & 0 \\
 & & & & \downarrow{\scriptstyle\beta} & \swarrow{\scriptstyle\alpha'} & \downarrow{\scriptstyle\beta'} & & \\
0 & \to & A & \to & I & \longrightarrow & C & \to & 0
\end{array}$$

Since β is monic, the pushout is a pullback, by the dual of the argument following Proposition 2.1.6. Now by hypothesis there is a map $\theta : P \to I$ such that $\beta' = \theta\alpha'$, therefore by the pullback property there is a map $\lambda : P \to B$ such that $\lambda\alpha = 1$, $\lambda\beta = \theta$. Thus the given sequence splits and this shows P to be projective. ∎

Theorem 2.4.4. *For any ring R the following conditions are equivalent:*

(a) *every quotient of an injective right R-module is injective,*
(b) *every submodule of a projective right R-module is projective,*
(c) *every right ideal of R is projective.*

Proof. (a) $\Rightarrow$ (b). Given the diagram

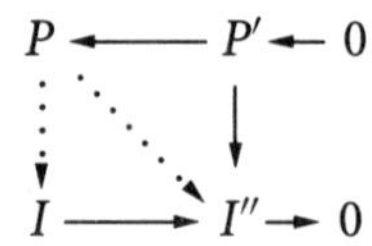

where P is projective, we have to fill in the map $P' \to I$ to produce a commutative triangle. By Lemma 2.4.3 we may assume that I is injective, and then I'' is injective, by (a). We can therefore fill in $P \to I''$, and then $P \to I$ (because P is projective). Now the composition $P' \to P \to I$ provides the required map.

(b) $\Rightarrow$ (c) is trivial and the proof of (c) $\Rightarrow$ (a) is dual to the first part, using the diagram below.

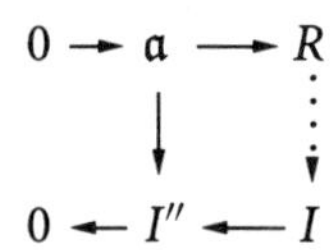

■

A ring satisfying the conditions of this theorem is said to be *right hereditary*. For example, any principal ideal domain (commutative or not) is right (and also left) hereditary. From (b) it is clear that the right hereditary rings are just the rings of global dimension at most 1, and (a) shows that the same class is obtained by computing the global dimension from injective resolutions.

In the commutative case hereditary integral domains have another more illuminating description. We recall from BA, Section 10.5 that an ideal $\mathfrak{a}$ in a commutative integral domain R is called *invertible* if there is an R-submodule $\mathfrak{b}$ of its field of fractions K such that $\mathfrak{ab} = R$. In BA, Proposition 10.5.1 we saw that an ideal is invertible iff it is non-zero projective; in particular such an ideal must be finitely generated. In fact a commutative hereditary domain is precisely a Dedekind domain, as the description of the latter in BA, Section 10.5 shows.

Exercises

1. Given two projective resolutions (P_i), (P'_i) of finite length of a module M, show that $P_0 \oplus P'_1 \oplus P_2 \oplus \ldots \cong P'_0 \oplus P_1 \oplus P'_2 \oplus \ldots$ (extended Schanuel lemma).
2. Find the global dimension of $\mathbf{Z}/n$. (Hint. Take first the case of a prime power.)
3. Let R be a right hereditary Noetherian ring and M a finitely generated submodule of R^I, as right R-module, for some set I. Show that M is a direct sum of modules isomorphic to right ideals of R, and hence is projective. (Hint. Among the direct summands of M isomorphic to a direct sum of right ideals, pick a maximal one.)

4. Show that if R is as in Exercise 3 and M a finitely generated right R-module, then $M = M_0 \oplus \mathfrak{a}_1 \oplus \ldots \oplus \mathfrak{a}_r$, where the $\mathfrak{a}_i$ are right ideals of R and $M_0 = \cap\{\ker f \mid f : M \to R\}$.
5. Let R be a ring such that every right ideal is free as right R-module. Show that every submodule of a free right R-module is free (a ring with IBN in which every right ideal is free is called a *right fir*, see Section 8.7 below). Show that if R is also commutative, it must be a principal ideal domain.
6. Show that a left fir (defined by symmetry as in Exercise 5) which is right Noetherian is also a right fir, in fact right principal (see Section 8.7 below).
7. Show that if R and S are right hereditary rings and U is any (R, S)-bimodule, projective as right S-module, then the triangular matrix ring $\begin{pmatrix} R & U \\ 0 & S \end{pmatrix}$ is again right hereditary.
8. Show that $\mathbf{Q}$ is not projective, as $\mathbf{Z}$-module. Deduce that the triangular matrix ring $\begin{pmatrix} \mathbf{Z} & \mathbf{Q} \\ 0 & \mathbf{Q} \end{pmatrix}$ is right but not left hereditary.
9. Show that over a commutative integral domain R, every quotient of a divisible module is divisible. Deduce that R is a Dedekind domain iff every divisible R-module is injective.
10. In the ring $\mathbf{R}[\sin\theta, \cos\theta]$ show that the ideal generated by $\sin\theta$ and $1 - \cos\theta$ is projective but not principal, and hence not a direct summand (this ring is a Dedekind domain, see BA, Section 10.5).

2.5 Derived functors

We have seen that the hom functor and tensor product are only left and right exact respectively, and we shall now describe a way of measuring this lack of exactness. The method is quite general and applies to any functor which is left or right exact. The basic idea of the construction is as follows. Let F be a functor which is right exact covariant, say. Given any module A, we take a projective resolution

$$\ldots \to X_n \to \ldots \to X_1 \to X_0 \to A \to 0,$$

and apply F:

$$\ldots \to FX_n \to \ldots \to FX_1 \to FX_0 \to FA \to 0.$$

In general this will no longer be exact, but it still is a complex. From any such complex one can form homology groups $H_n(A)$ described below, which measure the lack of exactness of F; taking the X_i projective ensures that these groups depend only on A and F, but not on the choice of the resolution X.

Before we enter on the actual construction, we need some properties of commutative diagrams. These are true in any abelian category, but we shall only consider the case of modules, where they can be verified by diagram-chasing. As a matter of fact, most of the results then follow for general abelian categories, because every small abelian category has an exact embedding into a module category (see Mitchell (1965) and Further Exercise 12 of Chapter 4), but we shall not make use of this fact.

Given any commutative square I:

$$\begin{array}{ccc} & \xrightarrow{\alpha} & \\ \downarrow \beta & I & \downarrow \gamma \\ & \xrightarrow{\delta} & \end{array}$$

we define the *image ratio* of I as $i(I) = (\text{im } \gamma \cap \text{im } \delta)/\text{im}(\beta\delta)$ and the *kernel ratio* of I as $k(I) = \ker(\beta\delta)/(\ker \alpha + \ker \beta)$. We note that if γ or δ is monic, then $i(I) = 0$, and if α or β is epic, then $k(I) = 0$. The ratios of two adjacent squares are related by

Lemma 2.5.1 (Two-square lemma). *Given a commutative diagram with exact rows:*

$$\begin{array}{ccccc} & \xrightarrow{\lambda} & & \xrightarrow{\mu} & \\ \downarrow \alpha & I & \downarrow \beta & II & \downarrow \gamma \\ & \xrightarrow{\lambda'} & & \xrightarrow{\mu'} & \end{array}$$

we have $i(I) \cong k(II)$.

Proof. We must show that

$$\frac{\text{im } \lambda' \cap \text{im } \beta}{\text{im } \lambda\beta} \cong \frac{\ker(\beta\mu')}{\ker \beta + \ker \mu}.$$

Clearly $\text{im } \beta \cap \text{im } \lambda' = \text{im } \beta \cap \ker \mu' = \{x\beta | x\beta\mu' = 0\} = (\ker \beta\mu')\beta$, and $\text{im } \lambda\beta = (\text{im } \lambda)\beta = (\ker \mu)\beta = (\ker \beta + \ker \mu)\beta$. Now both $\ker \beta\mu'$ and $\ker \beta + \ker \mu$ contain $\ker \beta$, hence by the third isomorphism theorem,

$$k(II) \cong \frac{\ker \beta\mu'}{\ker \beta + \ker \mu} \cong \frac{\ker(\beta\mu')\beta}{(\ker \beta + \ker \mu)\beta} \cong i(I).$$

■

Lemma 2.5.2. *Given the commutative diagram with exact rows,*

$$\begin{array}{ccccccc} & & A & \xrightarrow{\lambda} & B & \xrightarrow{\mu'} & C \to 0 \\ & & \downarrow \alpha & & \downarrow \beta & & \downarrow \gamma \\ 0 \to & & A' & \xrightarrow{\lambda'} & B' & \xrightarrow{\mu'} & C' \end{array}$$

the following diagram is commutative with exact rows and columns:

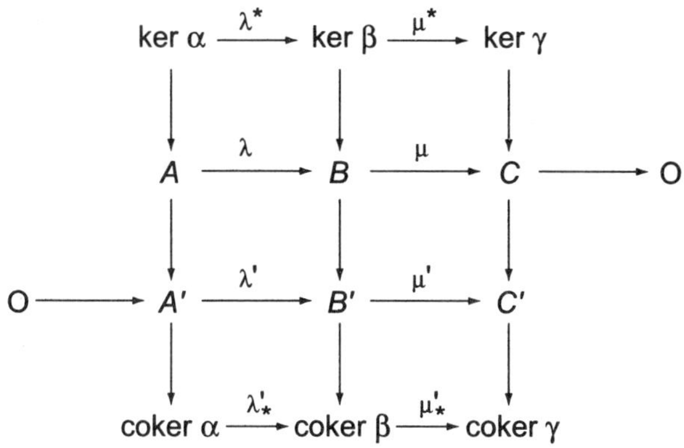

Here λ^*, μ^* are the maps induced between the kernels and λ'_*, μ'_* are the maps induced between the cokernels. Moreover, if λ is monic, then so is λ^* and if μ' is epic, so is μ'_*.

The proof, by diagram chasing, is straightforward and may be left to the reader. ■

Lemma 2.5.3 (Snake lemma). *Given the diagram in the hypothesis of Lemma 2.5.2, there exists a homomorphism $\Delta : \ker\gamma \to \operatorname{coker}\alpha$ such that the sequence*

$$\ker\alpha \xrightarrow{\lambda^*} \ker\beta \xrightarrow{\mu^*} \ker\gamma \xrightarrow{\Delta} \operatorname{coker}\alpha \xrightarrow{\lambda'^*} \operatorname{coker}\beta \xrightarrow{\mu'^*} \operatorname{coker}\gamma$$

is exact.

Proof. (J. Lambek) We have to prove exactness at $\ker\gamma$ and at $\operatorname{coker}\alpha$, for a suitable Δ, and for this it is enough to show that $\operatorname{coker}\mu^* \cong \ker\lambda'_*$. Writing $X = \operatorname{coker}\mu^*$, $Y = \ker\lambda'$, we have the following commutative diagram with exact rows and columns:

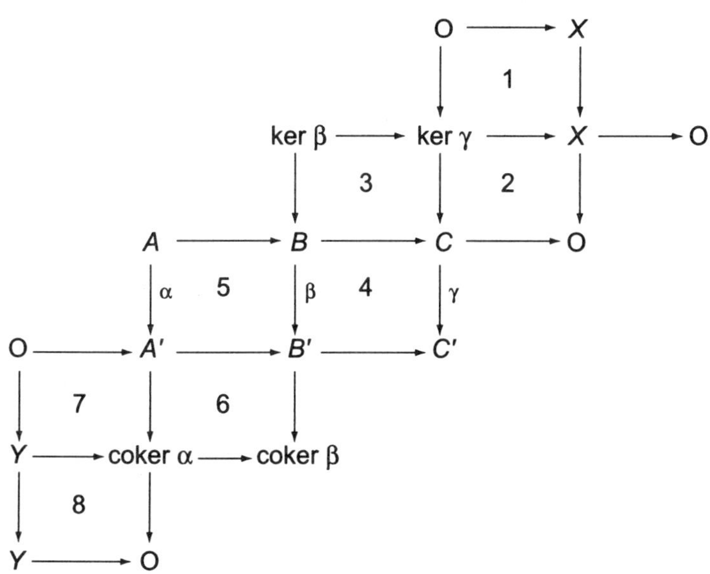

By the 2-square lemma, $X \cong i(1) \cong k(2) \cong i(3) \cong k(4) \cong i(5) \cong k(6) \cong i(7) \cong k(8) \cong Y$. ■

We can now return to the task of constructing the homology of a complex. It will be convenient to treat a special case first, that of a differential module. By a *differential module* X we understand an R-module X with an endomorphism d of square zero: $d^2 = 0$. If we regard these modules as objects of a category Diff_R, the maps in the category are taken to be homomorphisms preserving the structure, i.e. the homomorphisms $f : X \to Y$ such that the square shown commutes:

$$\begin{array}{ccc} X & \xrightarrow{d} & X \\ \downarrow f & & \downarrow f \\ Y & \xrightarrow{d} & Y \end{array}$$

These maps are traditionally called *chain-maps*. The condition $d^2 = 0$ means that $\mathrm{im}\, d \subseteq \ker d$. We shall write $\mathrm{im}\, d = B$, $\ker d = C$ and call the elements of B *boundaries* and those of C *cycles*. Finally $H = C/B$ is called the *homology group* of X. (For an excellent concise indication of the geometrical background, see Mac Lane (1963).)

In general B and C and hence $H = H(X)$ are merely abelian groups, but when R is a K-algebra, they are K-modules. We have

Theorem 2.5.4. *For any K-algebra R, $H : \mathrm{Diff}_R \to \mathrm{Mod}_K$ is a covariant K-linear functor from differential modules to K-modules.*

The proof is a straightforward verification, which may be left to the reader. ■

We observe that a complex may be regarded as a special case of a differential module. To see this we need only replace modules by graded modules (regarding R as a graded ring concentrated in degree 0). The type of complex we encounter will generally be a graded module with an antiderivation, also called a *differential*, of degree $r = 1$ or -1. In the case $r = -1$ one speaks of a *chain complex*, in the case $r = 1$ of a *cochain complex*. Thus given a chain complex

$$\ldots \to X_n \xrightarrow{d_n} X_{n-1} \xrightarrow{d_{n-1}} \ldots \to X_1 \xrightarrow{d_1} X_0 \to 0, \tag{2.5.1}$$

we can regard this as a graded differential module and the homology group is again a graded module; in detail we have

$$H_n(X) = \ker d_n / \mathrm{im}\, d_{n+1} \quad (n = 0, 1, \ldots).$$

Here d_0 is taken to be the zero map, thus $H_0(X) = X_0/\mathrm{im}\, d_1$. It is clear that the complex given by (2.5.1) is exact precisely when $H(X) = 0$, so that $H(X)$ may be taken as a measure of the lack of exactness of X. We note further that if in (2.5.1) each X_n for $n \geq 1$ is projective, then (2.5.1) is a projective resolution of the R-module A iff

$$H_n(X) = \begin{cases} A & \text{for } n = 0, \\ 0 & \text{for } n \neq 0. \end{cases} \tag{2.5.2}$$

A complex (X_n) satisfying (2.5.2) is said to be *acyclic* over A.

We shall state the next few results for differential modules rather than complexes (= graded differential modules), for the sake of simplicity. The change to complexes can easily be made by the reader.

By Theorem 2.5.4 an isomorphism of differential modules induces an isomorphism of homology groups, but since structure is lost in passing to homology, one would expect a much wider class of mappings to induce isomorphisms. The appropriate notion is suggested by the topological background.

Definition. Two chain maps $f, g : X \to Y$ of differential modules are said to be *homotopic*: $f \approx g$, if there is an R-homomorphism $s : X \to Y$, called a *homotopy*, such that

$$s.d_Y + d_X.s = f - g. \tag{2.5.3}$$

It is clear that this relation between chain maps is an equivalence; moreover it has the desired property:

Proposition 2.5.5. *Homotopic chain maps induce the same homology map, i.e. if $f \approx g$, then $H(f) = H(g)$.*

Proof. By linearity it is enough to show that if $f \approx 0$, then $H(f) = 0$. If $c \in C(X)$, then $cd = 0$ and since $f = sd + ds$, it follows that $cf = csd + cds = csd \in B(Y)$. Thus $C(X)f \subseteq B(Y)$ and so $H(f) = 0$. ■

It is clear how this definition and proposition have to be modified if X, Y are graded, say both are chain complexes. For f, g to be homotopic they must be of the same degree r say, and then the homotopy s will have degree $r + 1$.

A chain map between differential modules, $f : X \to Y$ is said to be a *chain equivalence* or *homotopy equivalence* if there is a second chain map $f' : Y \to X$ such that $ff' \approx 1_X$, $f'f \approx 1_Y$. This is easily seen to be an equivalence relation between chain maps and now Proposition 2.5.5 yields

Corollary 2.5.6. *If $f : X \to Y$ is a chain equivalence, then $H(f)$ is an isomorphism.* ■

We now come to a basic property of short exact sequences of differential modules:

Theorem 2.5.7. *Given a short exact sequence of differential modules:*

$$\mathbf{X}: \quad 0 \to X' \xrightarrow{\alpha} X \xrightarrow{\beta} X'' \to 0,$$

there exists a homomorphism $\Delta : H(X'') \to H(X')$ natural in X, such that the triangle

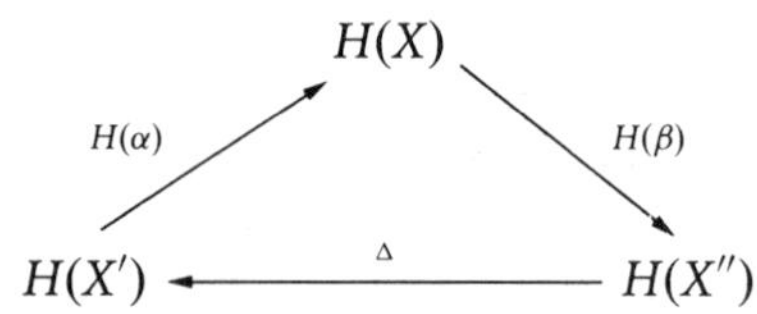

is exact.

Δ is known as the *connecting homomorphism.*

Proof. The exact sequence X may be written

$$
\begin{array}{ccccc}
0 \to C' & \longrightarrow & C & \longrightarrow & C'' \\
\downarrow & & \downarrow & & \downarrow \\
0 \to X' & \longrightarrow & X & \longrightarrow & X'' \to 0 \\
\downarrow d & & \downarrow d & & \downarrow d \\
0 \to X' & \longrightarrow & X & \longrightarrow & X'' \to 0 \\
\downarrow & & \downarrow & & \downarrow \\
X'/B' & \longrightarrow & X/B & \longrightarrow & X''/B'' \to 0
\end{array}
$$

where C and X/B are the kernel and cokernel of d respectively. By Lemma 2.5.2 the whole diagram is commutative, with exact rows and columns. Now the map $d : X \to X$ induces a map $X/B \to C$, because $Xd \subseteq C$ and $Bd = 0$, and this map has both kernel and cokernel equal to $H = C/B$. Hence we have a commutative diagram

$$
\begin{array}{ccccc}
H' & \longrightarrow & H & \longrightarrow & H'' \\
\downarrow & & \downarrow & & \downarrow \\
X'/B' & \longrightarrow & X/B & \longrightarrow & X''/B'' \to 0 \\
\downarrow & & \downarrow & & \downarrow \\
0 \to C' & \longrightarrow & C & \longrightarrow & C'' \\
\downarrow & & \downarrow & & \downarrow \\
H' & \longrightarrow & H & \longrightarrow & H''
\end{array}
$$

which has exact rows and columns, by Lemma 2.5.2. Further, by the snake lemma, there is a homomorphism $\Delta : H'' \to H'$ which makes the homology triangle exact, and which from its derivation is natural in X. ■

The important case of this theorem is that where X', X, X'' are in fact graded modules, usually chain or cochain complexes, with maps of degree zero between them, and with d of degree -1 or 1 respectively. In the case of chain complexes, say, the exact triangle takes on the form of an infinite sequence

$$\ldots \to H_n(X') \to H_n(X) \to H_n(X'') \to H_{n-1}(X') \to \ldots \to H_0(X'') \to 0,$$

which is called the *exact homology sequence* associated with the short exact sequence **X**.

Any R-module may be trivially regarded as a chain complex concentrated in degree 0, i.e. $M_0 = M, M_n = 0$ for $n \neq 0$, and $d = 0$. In the sequel all chain complexes will be zero in negative dimension, i.e. $X_n = 0$ for $n < 0$. The complex X is said to be *over* M if there is an exact sequence $X \to M \to 0$ (regarding M as a trivial chain complex in the way described above). In full this sequence reads

$$\ldots \to X_n \to X_{n-1} \to \ldots \to X_1 \to X_0 \to M \to 0. \tag{2.5.4}$$

If this is exact, it is called a *resolution*, or also an *acyclic complex*; then $H_n(X) = 0$ for $n > 0$, $H_0(X) \cong M$. If each X is projective, we have a *projective resolution*. An important property of projective resolutions is that they are universal among resolutions of M. This follows from the more general

Theorem 2.5.8 (Comparison theorem). *Given two complexes*

$$\begin{array}{ccc} X & \xrightarrow{\varepsilon} & M \to 0 \\ \vdots & & \downarrow \varphi \\ \vee & & \\ X' & \xrightarrow{\varepsilon'} & M' \to 0 \end{array}$$

where X is projective, X' is a resolution of M' and φ is a homomorphism, then there exists a chain map $f : X \to X'$ such that the resulting diagram commutes, and f is unique up to homotopy.

We shall also say that f is *over* φ or that f *lifts* φ.

Proof. We have to construct $f_n : X_n \to X'_n$ such that $f_n d' = d f_{n-1}$ $(n \geq 1)$ and $f_0 \varepsilon' = \varepsilon\varphi$. We construct these maps recursively, using the fact that X_n is projective and $\mathrm{im}(df_{n-1}) \subseteq \mathrm{im}\, d'$ (by the exactness of X'). At the n-th stage (when f_{n-1} has been constructed) we have the diagram

$$\begin{array}{ccccc} & \swarrow\!\cdots & X_n & & \\ & & \downarrow df_{n-1} & & \\ X'_n & \xrightarrow{\varepsilon'} & \mathrm{im}\, d' & \longrightarrow & 0 \end{array}$$

and this can be completed by a map $f_n : X_n \to X'_n$ because X_n is projective; this still applies for $n = 0$, taking ε, φ in place of d, f_{n-1}. To show that the map f so constructed is unique, suppose that $f + h$ is another map lifting φ; then h lifts 0 and we have to find $s_n : X_n \to X'_{n+1}$ such that

$$s_n d' + d s_{n-1} = h_n, \quad s_0 d' = h_0.$$

The construction needed is quite similar to the one just carried out and may be left to the reader. ■

In particular, if we have two projective resolutions of M, we can lift the identity on M to a map of complexes which is unique up to homotopy, hence we obtain

Corollary 2.5.9. *Any two projective resolutions of a module M are chain equivalent and hence give rise to isomorphic homology groups, for any functor.* ■

We now have all the means at our disposal for constructing derived functors.

Theorem 2.5.10. *Let F be a covariant right exact functor on* Mod_R. *Then there exist functors $F_n (n = 0, 1, \ldots)$ such that*

(i) $F_0 M \cong FM$,
(ii) $F_n P = 0$ *for* $n > 0$ *if P is projective,*
(iii) *to each short exact sequence of R-modules*

$$\mathbf{A}: \quad 0' \to A \to A'' \to 0.$$

there corresponds a long exact sequence

$$\ldots \to F_n A' \to F_n A \to F_n A'' \xrightarrow{\Delta} F_{n-1} A' \to$$

$$F_{n-1} A \to \ldots \to F_0 A' \to F_0 A \to F_0 A'' \to 0, \tag{2.5.5}$$

where the connecting homomorphism Δ is natural in $\mathbf{A}$. Moreover, F_n is determined up to natural isomorphism by (i)–(iii).

Proof. We begin by proving the uniqueness. Given any short exact sequence

$$0 \to K \to P \to A \to 0, \tag{2.5.6}$$

where P is projective, we have for any F_n satisfying (i)–(iii), the exact sequence

$$0 \to F_1 A \to FK \to FP \to FA \to 0;$$

thus any exact sequence (2.5.6) with P projective determines $F_1 A$. If $F_1 A$ is obtained from (2.5.6) and $F'_1 A$ from

$$0 \to K' \to P' \to A \to 0,$$

where P' is again projective, we form the pullback Q of $P \to A$ and $P' \to A$ and apply F. We thus obtain the commutative diagram:

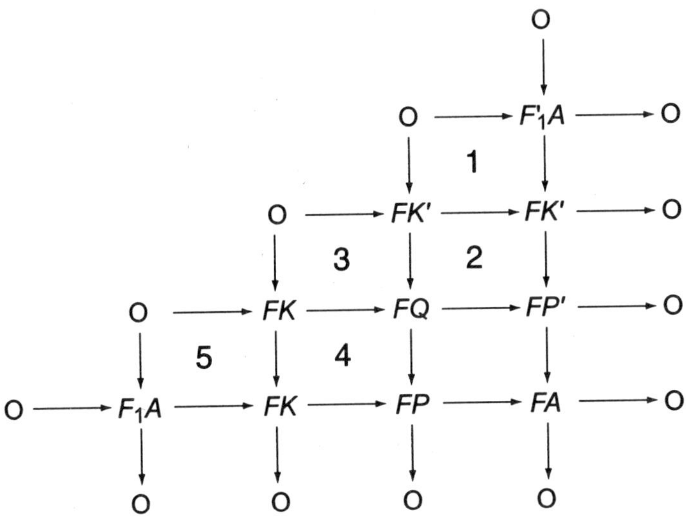

The row and column meeting in FQ are split exact, because they arose by applying F to split exact rows and columns (split because P and P' are projective). The remaining rows and columns are also exact, and by the 2-square lemma we have $F'_1A \cong i(1) \cong k(2) \cong i(3) \cong k(4) \cong i(5) \cong F_1A$. This shows that F_1A is determined up to isomorphism by (i)–(iii). For $n > 1$ the short exact sequence (2.5.6) yields the long exact sequence

$$\ldots \to F_nP \to F_nA \to F_{n-1}K \to F_{n-1}P \to \ldots$$

and $F_nP = F_{n-1}P = 0$ by (ii). Thus in terms of the loop operator π introduced in Section 2.4 we have the formula

$$F_nA \cong F_{n-1}(\pi(A)); \tag{2.5.7}$$

this makes sense since F is constant on projective equivalence classes for $n > 0$, by (ii). Now the uniqueness follows by induction on n.

It remains to prove the existence of F_n; here we may use any projective resolution for A, say $X \to A \to 0$. Applying F, we get a complex $FX \to FA \to 0$. In detail this reads

$$\ldots \to FX_n \to FX_{n-1} \to \ldots \to FX_1 \to FX_0 \to FA \to 0.$$

We put $F_nA = H_n(FX)$ and assert that this satisfies (i)–(iii).

(i) We have $X_1 \to X_0 \to A \to 0$ and by the right exactness of F obtain the exact sequence

$$FX_1 \to FX_0 \to FA \to 0;$$

hence $FA \cong FX_0/\mathrm{im}\, FX_1 \cong H_0(FX)$.

(ii) If A is projective, we can take as our resolution

$$0 \to A \to A \to 0;$$

applying F, we find $0 \to FA \to FA \to 0$, hence $F_nA = 0$ for $n > 0$.

(iii) Given a short exact sequence **A** as in (iii), we take projective resolutions X', X'' of A' and A'' and construct a projective resolution X of A by induction on n. If the kernels at the $(n-1)$-th stage are K'_n, K_n, K''_n, we have

$$\begin{array}{ccccccccc} 0 & \longrightarrow & X'_n & \longrightarrow & X_n & \longrightarrow & X''_n & \longrightarrow & 0 \\ & & \downarrow & & \vdots & & \downarrow & & \\ 0 & \longrightarrow & K'_n & \longrightarrow & K_n & \longrightarrow & K''_n & \longrightarrow & 0 \end{array}$$

where $X_n = X'_n \oplus X''_n$. Since X''_n is projective, we have a map $X''_n \to K_n$, while the map $X'_n \to K_n$ arises by composition (via K'_n). By definition of X_n as a direct sum (i.e. product) we obtain a map $X_n \to K_n$ to make the squares commute, and a simple diagram chase shows that this map is epic (this also follows from the 5-lemma). By induction on n we thus have a resolution X of A. Now the row

$$0 \to X' \to X \to X'' \to 0$$

is split exact, by definition; applying F, we obtain the exact sequence of complexes

$$0 \to FX' \to FX \to FX'' \to 0,$$

and Theorem 2.5.7 provides us with the exact homology sequence, which is the required exact sequence. ■

The functors F_n constructed in Theorem 2.5.10 are called the *(left) derived functors* of F. The same result, appropriately modified, gives a construction of right derived functors of a left exact covariant functor, using injective resolutions. Any given module A can be embedded in an injective module I_0 by Proposition 2.3.7; by embedding the cokernel in an injective module I_1 and continuing in this fashion, we obtain an injective resolution

$$0 \to A \to I_0 \to I_1 \to \dots . \qquad (2.5.8)$$

For any left exact covariant functor F we have a series of functors F^n such that

(i) $F^0A \cong FA$,
(ii) $F^nI = 0$ for $n > 0$ if I is injective,
(iii) for each short exact sequence **A** as in Theorem 2.5.10 there is a corresponding long exact sequence

$$0 \to F^0A' \to F^0A \to F^0A'' \to \dots \to F^{n-1}A' \xrightarrow{\Delta} F^nA' \to F^nA \to F^n \qquad (2.5.9)$$

where the connecting homomorphism Δ is natural in **A**, and F^n is determined up to isomorphism by (i)–(iii).

The proof is exactly analogous to that of Theorem 2.5.10. We note that here the index appears as a superscript, and its value increases along the sequence, whereas in (2.5.5) the index is a subscript, which decreases as we go along the sequence.

For a contravariant functor the roles are reversed; if F is left exact, we define F by means of a projective resolution and obtain a long exact sequence (2.5.9), while for a right exact contravariant functor F we define F by an injective resolution and obtain the long exact sequence (2.5.5). To sum up, a projective resolution is needed for a right exact covariant and a left exact contravariant functor, and an injective resolution for a left exact covariant or a right exact contravariant functor.

Of course the construction of Theorem 2.5.10 can be carried out for any functor, not necessarily right (or left) exact. In general we obtain in this way the left derived functor of F; similarly we can form the right derived functor (using an injective resolution) and together they form a long exact sequence extending in both directions (see Exercise 9).

Exercises

1. Show that if I is a pullback or pushout square, then $i(I) = 0$ and $k(I) = 0$.
2. Show that the category of chain complexes and chain maps is an abelian category.
3. Let M be a finitely presented R-module, i.e. with a resolution

$$0 \to G \to F \to M \to 0,$$

where F is free and both F and G are finitely generated. Given any short exact sequence

$$0 \to A \to B \to M \to 0,$$

where B is finitely generated, show that A is finitely generated. (Hint. Use Proposition 2.4.1.)
4. Verify that homotopy is an equivalence between chain maps.
5. Show that if $f \approx g$ is a homotopy of chain maps, where $f, g : X \to Y$ and $h : Y \to Z$, then $fh \approx gh$; likewise $ef \approx eg$ for $e : U \to X$.
6. Prove Theorem 2.5.4.
7. (5-lemma) Given a commutative diagram with exact rows

$$\begin{array}{ccccccccc} & \longrightarrow & & \longrightarrow & & \longrightarrow & & \longrightarrow & \\ f_1\downarrow & & f_2\downarrow & & f_3\downarrow & & f_4\downarrow & & f_5\downarrow \\ & \longrightarrow & & \longrightarrow & & \longrightarrow & & \longrightarrow & \end{array}$$

show that if f_1, f_2, f_4, f_5 are isomorphisms, then so is f_3. More precisely, if f_1 is epic and f_4, f_5 are monic, show that f_3 is monic, and dually. Deduce that f_3 is an isomorphism whenever f_1, f_5 are isomorphisms, f_2 is epic and f_4 is monic.
8. Verify that the isomorphisms $F_n \to F'_n$ between two functors satisfying (i)–(iii) of Theorem 2.5.10 are compatible with Δ.
9. For any covariant functor F on modules, F_nM $(n \geq 0)$ is defined as in the proof of Theorem 2.5.10 in terms of projective resolutions of M. Show that there is a

natural transformation $F_0M \to FM$ and that F_0 is right exact; if now F^nM is defined similarly in terms of an injective resolution, there is a natural transformation $FM \to F^0M$ and F^0 is left exact. Hence obtain an exact sequence of derived functors (F_n, F^n are called the *left* and *right derived* functors of F, respectively).

10. Let R be a ring with IBN and M an R-module with a finite resolution (F_i) by free modules of finite rank. Show that the integer $\chi(M) = \sum(-1)^i rkF_i$ depends only on M and not on the resolution; it is called the *Euler characteristic* of M. Given a short exact sequence $\mathbf{A}$ of modules with finite resolutions by free modules of finite rank, show that $\chi(A) = \chi(A') + \chi(A'')$. Define χ for a complex C and show that $\chi(H(C)) = \chi(C)$.

2.6 Ext, Tor and global dimension

The most important functors to which the construction of Section 2.5 has been applied are $A \otimes B$ and $\mathrm{Hom}(A, B)$. These are bifunctors, i.e. functors of two arguments, and it is possible to form derived functors in two ways, by resolving either the first or the second argument. We shall find that the results in these two cases are the same; this is based on a general criterion which we shall now derive.

Let $F(A, B)$ be any bifunctor, covariant right exact in each argument, say. Then F is said to be *R-balanced* or simply *balanced* if $F(A, -)$ is an exact functor whenever A is projective and $F(-, B)$ is exact whenever B is projective. The same definition applies if F is contravariant left exact in either argument, while for a covariant left exact or a contravariant right exact argument, 'projective' is replaced by 'injective'.

Theorem 2.6.1. *Let F be a bifunctor, covariant right exact in each argument, and denote by F'_n, F''_n the derived functors obtained by resolving the first and second argument of F respectively. If F is balanced, then*

$$F'_n(A, B) \cong F''_n(A, B), \tag{2.6.1}$$

by an isomorphism which is natural in A and B.

Proof. By the uniqueness of derived functors (Theorem 2.5.10) we need only take $F'_n(A, B)$ for fixed A and verify that it satisfies the conditions of Theorem 2.5.10 as a functor in B.

(i) $F'_0(A, B) \cong F(A, B)$ by definition,

(ii) If B is projective, then $F(-, B)$ is exact. We apply this functor to a projective resolution of A:

$$X \to A \to 0, \tag{2.6.2}$$

and obtain

$$F(X, B) \to F(A, B) \to 0.$$

This is still a resolution, by the exactness of $F(-, B)$. Hence $F'_n(A, B) = 0$ for $n > 0$, by the definition of $F'_n(-, B)$.

(iii) Given a short exact sequence

$$0 \to B' \to B \to B'' \to 0, \tag{2.6.3}$$

and a projective resolution (2.6.2) of A, we apply $F(X, -)$ to (2.6.3). Since $F(X, -)$ is exact, we obtain an exact sequence of complexes:

$$0 \to F(X, B') \to F(X, B) \to F(X, B'') \to 0.$$

From this the long exact sequence is obtained by applying Theorem 2.5.7. ■

A similar argument applies when there is a change of side or variance.

Let us apply these results to $\mathrm{Hom}(A, B)$. This is left exact covariant in B and left exact contravariant in A; moreover it is balanced. By Theorem 2.6.1 the derived functor may be obtained either by a projective resolution of A or by an injective resolution of B. It is written

$$\mathrm{Ext}^n_R(A, B) \quad \text{or} \quad \mathrm{Ext}^n(A, B).$$

To account for the name we shall briefly indicate an interpretation of Ext. Given two R-modules A, B over a ring R, an *extension* of A by B is a module E together with a short exact sequence

$$0 \to A \to E \to B \to 0. \tag{2.6.4}$$

We can form the category Ex, whose objects are short exact sequences, with the obvious maps between them: a morphism is a triple of homomorphisms making the diagram

$$\begin{array}{ccccccccc} 0 & \to & A & \longrightarrow & E & \longrightarrow & B & \to & 0 \\ & & \downarrow & & \downarrow & & \downarrow & & \\ 0 & \to & A' & \longrightarrow & E' & \longrightarrow & B' & \to & 0 \end{array}$$

commutative. We observe that if the maps $A \to A'$ and $B \to B'$ are isomorphisms, then so is $E \to E'$, by the 5-lemma, and we then have an isomorphism of extensions.

From the extension (2.6.4) we form the exact homology sequence

$$0 \to \mathrm{Hom}(B, A) \to \mathrm{Hom}(B, E) \to \mathrm{Hom}(B, B) \xrightarrow{\Delta} \mathrm{Ext}^1(B, A) \to \dots$$

Consider the image in $\mathrm{Ext}^1(B, A)$ of the identity map j on $B : j\Delta$. This is called the *obstruction* or the *characteristic class* of the extension. Clearly it depends only on the isomorphism type of the extension (2.6.4). Moreover, it is zero iff (2.6.4) splits, for $j\Delta = 0$ iff j is induced by a homomorphism $B \to E$, which is just the condition for (2.6.4) to split, by Corollary 2.1.5.

We could also apply $\mathrm{Hom}(-, A)$ to the sequence (2.6.4) and get

$$0 \to \mathrm{Hom}(B, A) \to \mathrm{Hom}(E, A) \to \mathrm{Hom}(A, A) \xrightarrow{\Delta} \mathrm{Ext}^1(B, A) \to \dots$$

This will give the same obstruction; in fact we have a bijection from the set of isomorphism classes of extensions of A by B to $\mathrm{Ext}^1(B, A)$. We shall return to this topic in Section 3.1.

The homological dimension of a module was defined in Section 2.4 in terms of the loop functor π; we now show how to express it in terms of Ext. For simplicity we shall not distinguish between the class $[\pi A]$ and a module in it.

Proposition 2.6.2. *For any R-module A over any ring R the following conditions are equivalent:*

(a) $\mathrm{hd}\, A \leq n$,
(b) $\mathrm{Ext}^k(A, -) = 0$ *for all* $k > n$,
(c) $\mathrm{Ext}^{n+1}(A, -) = 0$.

Proof. For any $k > 0$ and any R-module B we have, by (2.5.7) and its dual,

$$\mathrm{Ext}^k(\pi A, B) = \mathrm{Ext}^{k+1}(A, B) = \mathrm{Ext}^k(A, \iota B) \quad \text{for} \quad k > 0. \tag{2.6.5}$$

Now (a) states that $\pi^n A$ is projective; so in that case we have for any $k > n$,

$$\mathrm{Ext}^k(A, -) = \mathrm{Ext}^{k-1}(\pi A, -) = \ldots = \mathrm{Ext}^{k-n}(\pi^n A, -) = 0$$

and (b) follows. Clearly (b) $\Rightarrow$ (c), so assume (c). Then

$$\mathrm{Ext}^1(\pi^n A, -) = \mathrm{Ext}^2(\pi^{n-1} A, -) = \ldots = \mathrm{Ext}^{n+1}(A, -) = 0;$$

this means that $\mathrm{Hom}(\pi^n A, -)$ is exact, i.e. $\pi^n A$ is projective, so (a) holds. ■

In the dual situation we can assert a little more:

Proposition 2.6.3. *For any R-module B over any ring R the following are equivalent:*

(a) $\mathrm{cd}\, B \leq n$,
(b) $\mathrm{Ext}^k(-, B) = 0$ *for all* $k > n$,
(c) $\mathrm{Ext}^{n+1}(-, B) = 0$,
(d) $\mathrm{Ext}^{n+1}(C, B) = 0$ *for all cyclic modules C.*

The proof that (a)–(c) are equivalent is entirely analogous to that of Proposition 2.6.2 and it is clear that (c) implies (d). Conversely, assume (d) for right modules say; then for any right ideal $\mathfrak{a}$ of R,

$$\mathrm{Ext}^1(R/\mathfrak{a}, \iota^n B) = \mathrm{Ext}^{n+1}(R/\mathfrak{a}, B) = 0,$$

hence the short exact sequence

$$0 \to \mathfrak{a} \to R \to R/\mathfrak{a} \to 0$$

leads to the exact sequence

$$0 \to \mathrm{Hom}(R/\mathfrak{a}, \iota^n B) \to \mathrm{Hom}(R, \iota^n B) \to \mathrm{Hom}(\mathfrak{a}, \iota^n B) \to 0.$$

This shows that any homomorphism $\mathfrak{a} \to \iota^n B$ is obtained by restriction from a homomorphism $R \to \iota^n B$. By Baer's criterion (Theorem 2.3.4) it follows that $\iota^n B$ is injective, i.e. (a). ■

From Proposition 2.6.2 we see that the global homological dimension of R may be defined as $\sup\{n | \mathrm{Ext}^n \neq 0\}$, while Proposition 2.6.3 shows that this determines the global cohomological dimension. Hence we have

Corollary 2.6.4. *For any ring R, the (right) global homological and cohomological dimensions are equal, and may be defined as*

$$\mathrm{r.gl.dim}(R) = \sup\{n | \mathrm{Ext}^n(C, B) \neq 0 \text{ for cyclic } C \text{ and any } B\}$$

i.e. $\sup(\mathrm{hd}\, C)$ *for cyclic right R-modules C.* ■

Of course it must be borne in mind that the global dimension defined here refers to right R-modules, and in general it will be necessary to distinguish this from the left global dimension, $\mathrm{l.gl.dim}(R)$. As we shall soon see, for Noetherian rings these numbers coincide, but we shall also meet more general examples where they differ (see Exercise 8 of Section 2.4).

We now turn to the tensor product. The functor $A \otimes B$ is covariant right exact in each argument and is balanced. We therefore have a unique derived functor, sometimes called the *torsion product*, written

$$\mathrm{Tor}_n^R(A, B) \quad \text{or} \quad \mathrm{Tor}_n(A, B).$$

As an example consider the case $R = \mathbf{Z}$. Here $\mathrm{Tor}_n = 0$ for $n > 1$, because $\mathbf{Z}$ is hereditary and so all projective resolutions have length at most 1. Writing $\mathbf{C}_k$ for the cyclic group of order k, we have an exact sequence

$$0 \to \mathbf{Z} \xrightarrow{k} \mathbf{Z} \to \mathbf{C}_k \to 0,$$

where k indicates multiplication by k. Tensoring up with $\mathbf{C}_k$ we get

$$0 \to \mathrm{Tor}_1(\mathbf{C}_k, \mathbf{C}_k) \to \mathbf{C}_k \otimes \mathbf{Z} \to \mathbf{C}_k \otimes \mathbf{Z} \to \mathbf{C}_k \otimes \mathbf{C}_k \to 0.$$

Denote a generator of $\mathbf{C}_k$ by c; then under the induced map $1 \otimes k$ we have $c \otimes 1 \mapsto c \otimes k \mapsto ck \otimes 1 = 0$. Therefore $\ker(1 \otimes k) = \mathbf{C}_k$ and we find

$$\mathrm{Tor}_1^{\mathbf{Z}}(\mathbf{C}_k, \mathbf{C}_k) \cong \mathbf{C}_k. \tag{2.6.6}$$

Since Tor, like $\otimes$, preserves direct sums, (2.6.6) together with the equation $\mathrm{Tor}_1(\mathbf{C}_h, \mathbf{C}_k) = 0$ for coprime h, k is enough to determine Tor for any finitely generated abelian group.

A right R-module A is said to be *flat* if $A \otimes -$ is an exact functor, or equivalently, if $\mathrm{Tor}_1^R(A, -) = 0$. For example, any projective module is flat. A corresponding definition applies to left R-modules.

Using Tor, we can define another dimension for modules, the *weak dimension*. This is defined, for a right R-module A, as

$$\mathrm{wd}\, A = \sup\{n | \mathrm{Tor}_n^R(A, -) \neq 0\}.$$

In terms of the loop functor π we can also define wd A as the least integer k such that $\pi^k A$ is flat. Now the *weak global dimension* of R is defined as

$$\text{w.gl.dim}(R) = \sup\{n | \text{Tor}_n^R \neq 0\}.$$

From the definition (and the symmetry of Tor) it is clear that this function is left–right symmetric. To compare it with the global dimension, let us consider the situation $(A_{R,\mathbf{Z}}B_{R,\mathbf{Z}}C)$ and define a natural transformation

$$A \otimes_R \text{Hom}_{\mathbf{Z}}(B, C) \to \text{Hom}_{\mathbf{Z}}(\text{Hom}_R(A, B), C) \tag{2.6.7}$$

by mapping the element $a \otimes f$, where $a \in A$, $f \in \text{Hom}_{\mathbf{Z}}(B, C)$, to

$$\varphi(a, f) : \theta \mapsto (\theta a)f, \quad \text{where } \theta \in \text{Hom}_R(A, B).$$

This map is clearly R-balanced and biadditive, and hence defines a homomorphism (2.6.7), which is easily seen to be natural. Moreover, for $A = R$ it is an isomorphism, hence it is an isomorphism for any finitely generated projective R-module A. We now fix C to be $\mathbf{Z}$-injective, i.e. divisible and consider the two sides of (2.6.7) as a functor in A. Both sides are covariant right exact; if we apply them to a projective resolution (2.6.2) of A we obtain a natural transformation

$$\text{Tor}_n^R(A, \text{Hom}_{\mathbf{Z}}(B, C)) \to \text{Hom}_{\mathbf{Z}}(\text{Ext}_R^n(A, B), C). \tag{2.6.8}$$

Suppose now that R is right Noetherian and A is finitely generated. Then each term X in the resolution (2.6.2) may be taken to be finitely generated and so (2.6.8) will be an isomorphism. Thus we have proved

Proposition 2.6.5. *Let R be a right Noetherian ring and C a divisible abelian group. Then for any right R-modules A, B such that A is finitely generated, we have*

$$\text{Tor}_n^R(A, \text{Hom}_{\mathbf{Z}}(B, C)) \cong \text{Hom}_{\mathbf{Z}}(\text{Ext}_R^n(A, B), C) \quad \text{for all } n \geq 0, \tag{2.6.9}$$

by a natural isomorphism. ■

We shall use this result to compare the weak and homological dimensions. In the first place, any R-module A satisfies

$$\text{hd}\, A \geq \text{wd}\, A, \tag{2.6.10}$$

because every projective module is flat. It follows that for any ring R,

$$\text{w.gl.dim}(R) \leq \text{r.gl.dim}(R),\ \text{l.gl.dim}(R). \tag{2.6.11}$$

Now assume that R is right Noetherian and A is finitely generated over R. Choose n such that $n \leq \text{hd}\, A$; then $\text{Ext}_R^n(A, B) \neq 0$ for some R-module B, and moreover, we can find a divisible group C into which $\text{Ext}_R^n(A, B)$ has a non-zero homomorphism (indeed an embedding, by Proposition 2.3.7). Thus for suitable B, C the right-hand side of (2.6.9) is non-zero; looking at the left-hand side, we deduce that wd $A \geq n$. Hence equality must hold in (2.6.10) and we have proved

Theorem 2.6.6. *If R is a right Noetherian ring, then for any finitely generated right R-module A,* $\mathrm{wd}\, A = \mathrm{hd}\, A$. ■

Corollary 2.6.7. *For any right Noetherian ring R,*

$$\mathrm{r.gl.dim}(R) = \mathrm{w.gl.dim}(R) \leq \mathrm{l.gl.dim}(R). \tag{2.6.12}$$

Proof. By Corollary 2.6.4 the right global dimension is the supremum of the projective dimensions of the cyclic right R-modules, and by Theorem 2.6.6 this cannot exceed the weak global dimension of R; by (2.6.10) it cannot be less, and so the equality in (2.6.11) follows. The inequality follows similarly from (2.6.10), bearing in mind that the weak global dimension is left–right symmetric. ■

If R is left and right Noetherian, we obtain by symmetry,

Corollary 2.6.8. *In a Noetherian ring R,*

$$\mathrm{r.gl.dim}(R) = \mathrm{l.gl.dim}(R).$$ ■

Exercises

1. Show that if $\mathrm{Ext}^n_R(A, C) = 0$ for all cyclic modules C, then the same holds for all finitely generated modules C. Do the same for $\mathrm{Ext}^n_R(C, A)$ and $\mathrm{Tor}^R_n(A, C)$.
2. Verify that two extensions of R-modules A by B are isomorphic iff they correspond to the same element of $\mathrm{Ext}^1_R(B, A)$.
3. Show that Tor preserves direct sums in both arguments, while Ext preserves direct products in the second argument and converts direct sums in the first argument to direct products.
4. Verify that the two ways of defining the characteristic class of an extension agree.
5. Let F be a bifunctor from R-modules which is covariant right exact and preserves direct sums. Show that F is balanced iff $F(R, -)$ and $F(-, R)$ are exact. Similarly if F is contravariant and convert direct sums to direct products.
6. For any abelian group A denote by tA its torsion subgroup. Show that $\mathrm{Tor}^{\mathbf{Z}}_1(\mathbf{Q}/\mathbf{Z}, A) \cong tA$.
7. Show that $\mathrm{Tor}^{\mathbf{Z}}_1(\mathbf{C}_h, \mathbf{C}_k) \cong \mathbf{C}_d$, where d is the highest common factor of h and k.
8. Show that for any abelian groups A, B, $\mathrm{Tor}_1(A, B) \cong \mathrm{Tor}_1(tA, B)$; hence calculate $\mathrm{Tor}_1(A, B)$ for two finitely generated abelian groups.
9. For any abelian group A show that $\mathrm{Ext}^1(\mathbf{C}_n, A) \cong A/nA$, by applying $\mathrm{Hom}(-, A)$ to a suitable short exact sequence. Deduce that any extension of finite abelian groups A by B splits if A, B are of coprime orders.
10. Show that (2.6.7) is not always an isomorphism. (Hint. Use Theorem 2.6.6 and Exercise 8 of Section 2.4.)

2.7 Tensor algebras, universal derivations and syzygies

We recall from BA, Section 6.2 that for any K-module U over a commutative ring K we can form a tensor ring as a graded ring whose components are the tensor powers of U. More generally, we can replace K by a general ring A and take U to be an A-bimodule. We define the n-th tensor power of U as

$$U^n = U \otimes_A U \otimes_A \ldots \otimes_A U \quad (n \text{ factors}). \tag{2.7.1}$$

Now the *tensor A-ring* on U is defined as

$$\mathbf{T}_A(U) = \oplus_{n=0}^{\infty} U^n, \tag{2.7.2}$$

where the multiplication is defined componentwise by the isomorphism

$$U^r \otimes U^s \cong U^{r+s} \tag{2.7.3}$$

which follows from the associative law for tensor products. If A is commutative and the left and right actions on U agree, the ring defined by (2.7.2) is an A-algebra, but for general A the ring $\mathbf{T}_A(U)$ so obtained is an *A-ring*, i.e. a ring with a homomorphism $A \to \mathbf{T}_A(U)$. If U is the free K-module on a set X, we write $T_K(U)$ as $K\langle X\rangle$; this is just the *free K-algebra* on X (see BA, Section 6.2). The ring $\mathbf{T}_A(U)$ has the special property that any A-linear mapping of U into an A-ring R can be extended to a homomorphism of $\mathbf{T}_A(U)$ into R:

Theorem 2.7.1. *Let A be any ring and U an A-bimodule. Then $\mathbf{T}_A(U)$ is the universal A-ring for A-linear mappings of U into A-rings: there is a homomorphism $\lambda : U \to \mathbf{T}_A(U)$ such that for every A-linear map $f : U \to R$ into an A-ring R there is a homomorphism $f^* : \mathbf{T}_A(U) \to R$ such that*

$$f = \lambda f^*. \tag{2.7.4}$$

Proof. The map λ may be taken as the embedding which identifies U with U^1. Given an A-linear mapping $f : U \to R$, we extend f to $\mathbf{T}_A(U)$ by defining

$$(u_1, \ldots, u_n)f = (u_1 f)\ldots(u_n f).$$

By the properties of the tensor product this defines a mapping f^* from U to R, which is easily seen to be a homomorphism; f^* is unique since it is determined on the generating set U and (2.7.4) holds, almost by definition. ∎

It turns out that derivations have the same property. We recall that for any ring homomorphisms $\alpha : C \to A$, $\beta : C \to B$ and an (A, B)-bimodule M, a mapping $\delta : C \to M$ is called an *(α, β)-derivation* if δ is linear and satisfies

$$(xy)^\delta = x^\alpha y^\delta + x^\delta y^\beta \quad \text{for all} \quad x, y \in C. \tag{2.7.5}$$

By means of the triangular matrix ring $\begin{pmatrix} A & M \\ 0 & B \end{pmatrix}$ we can rephrase the definition by saying that $\delta : C \to M$ is an (α, β)-derivation precisely if the mapping

$$x \mapsto \begin{pmatrix} x^\alpha & x^\delta \\ 0 & x^\beta \end{pmatrix} \quad x \in C,$$

is a homomorphism. The proof is a simple verification which may be left to the reader. By applying Theorem 2.7.1 we thus obtain

Corollary 2.7.2. *Given an A-bimodule U over a ring A and linear mappings α, β of U into A, denote the extensions to $\mathbf{T}_A(U)$ (which exist by Theorem 2.7.1) again by α, β. Then any (α, β)-derivation δ of U into an A-bimodule M can be extended in just one way to an (α, β)-derivation of $\mathbf{T}_A(U)$ into M.* ■

Any derivation followed by a module homomorphism is again a derivation and we can ask whether there is a module with a derivation which is universal. For simplicity we shall assume that $\alpha = \beta = 1$. Thus R is an A-ring and we are dealing with derivations from R to R-bimodules. These derivations form the objects of a category in which the morphisms are commutative triangles and we are looking for an initial object in this category, i.e. an R-bimodule Ω with a derivation $\delta : R \to \Omega$ such that every derivation from R can be uniquely factored by δ. Such a bimodule indeed exists and can be described explicitly.

Proposition 2.7.3 (Eilenberg). *Let R be any A-ring with multiplication mapping $\mu : R \otimes_A R \to R$. This mapping gives rise to an exact sequence*

$$0 \to \Omega \longrightarrow R \otimes_A R \xrightarrow{\mu} R \to 0, \tag{2.7.6}$$

where $\Omega = \ker \mu$ is an R-bimodule, generated as left (or right) R-module by the elements $x \otimes 1 - 1 \otimes x$ $(x \in R)$ and split exact as left or right R-module sequence.

This R-bimodule Ω is the *universal derivation bimodule* for R, with the universal derivation

$$\delta : x \mapsto x \otimes 1 - 1 \otimes x \quad (x \in R). \tag{2.7.7}$$

Proof. It is clear that the multiplication $\mu : x \otimes y \mapsto xy$ is an R-bimodule homomorphism and its kernel Ω contains the elements $x \otimes 1 - 1 \otimes x$. Suppose that $\sum x_i \otimes y_i \in \Omega$; then $\sum x_i y_i = 0$ in R and so

$$\sum x_i \otimes y_i = \sum (x_i \otimes 1 - 1 \otimes x_i) y_i = \sum x_i (1 \otimes y_i - y_i \otimes 1).$$

This shows Ω to be generated by the elements $x \otimes 1 - 1 \otimes x$ as left or right R-module.

Now the mapping (2.7.7) is a derivation:

$$\begin{aligned}(xy)^\delta &= xy \otimes 1 - 1 \otimes xy = x(y \otimes 1 - 1 \otimes y) + (x \otimes 1 - 1 \otimes x)y \\ &= x.y^\delta + x^\delta.y.\end{aligned}$$

It is universal, for if $d : R \to M$ is any derivation, we can define a homomorphism $f : \Omega \to M$ as follows: if $\sum x_i \otimes y_i \in \Omega$, then $\sum x_i y_i = 0$, hence $\sum x_i^d.y_i + x_i.y_i^d = 0$ and so we may put

$$\left(\sum x_i \otimes y_i\right)f = \sum x_i^d.y_i = -\sum x_i.y_i^d. \tag{2.7.8}$$

This is an R-bimodule homomorphism, since $(\sum a(x_i \otimes y_i))f = -\sum ax_i.y_i^d$, $(\sum (x_i \otimes y_i)a)f = \sum x_i^d.y_ia$. Moreover, $\delta f : x \mapsto x \otimes 1 - 1 \otimes x \mapsto x^d$. Hence $d = \delta f$ and f is unique since it is prescribed on the generating set $\{x \otimes 1 - 1 \otimes x\}$ of Ω. Finally to show that (2.7.6) is split exact, we observe that the mapping $\phi : R \to R \otimes_A R$ given by $x \mapsto x \otimes 1$ is a left R-module mapping (and $x \mapsto 1 \otimes x$ a right R-module mapping) such that $\phi\mu = 1$. ■

The module Ω in (2.7.6), regarded as a universal derivation bimodule of R, is often denoted by $\Omega_A(R)$. We shall be particularly interested in this bimodule when R is the tensor A-ring on an A-bimodule U, $R = \mathbf{T}_A(U)$. In that case we can give an explicit description.

Proposition 2.7.4. *Let A be a K-algebra, U an A-bimodule and $R = \mathbf{T}_A(U)$ the tensor A-ring on U. Then the universal derivation bimodule of R is given by*

$$\Omega_A(R) = R \otimes U \otimes R \tag{2.7.9}$$

and the exact sequence

$$0 \to R \otimes U \otimes R \xrightarrow{\alpha} R \otimes R \xrightarrow{\mu} R \to 0 \tag{2.7.10}$$

is split exact as sequence of left (or right) R-modules.

Here and in the proof that follows, all tensor products are understood over A.

Proof. As before, we write $\Omega = \ker \mu$ and consider the canonical derivation $\delta : R \to \Omega$ given by $x^\delta = 1 \otimes x - x \otimes 1$. We saw in the proof of Proposition 2.7.3 that Ω is generated as left or right R-module by the x^δ as x ranges over R. But R is generated by U as A-ring, therefore Ω is generated as A-bimodule by the u^δ, where $u \in U$. Thus the restriction map $\delta|U : U \to \Omega$ gives rise to an R-bimodule map $\alpha : R \otimes U \otimes R \to R \otimes R$ such that $(1 \otimes u \otimes 1)^\alpha = u^\delta$ and clearly im $\alpha = \Omega$. Now (2.7.10) is split since this is the case for (2.7.6). ■

To construct free objects in other varieties of algebras it is often simplest to take free algebras and apply the factor theorem. We illustrate the method by the case of symmetric algebras, which is of some interest in itself. To begin with we describe the process of 'abelianizing' a ring, which is analogous to the corresponding notion for groups (see BA, Section 3.3).

Theorem 2.7.5. *To any ring R there corresponds a commutative ring R^{ab} with a homomorphism $\nu : R \to R^{ab}$ which is universal for homomorphisms of R into commutative* *rings. Thus each homomorphism f from R to a commutative ring can be factored uniquely by ν.*

Proof. Let $\mathfrak{c}$ be the commutator ideal of R, i.e. the ideal generated by all the commutators $xy - yx$, where $x, y \in R$, and write $R^{ab} = R/\mathfrak{c}$, with the natural homomorphism $\nu : R \to R^{ab}$. Then any homomorphism f from R to a commutative ring maps $xy - yx$ to 0, for all $x, y \in R$, hence $\ker f \supseteq \mathfrak{c}$, and so by the factor theorem (Theorem 1.2.4) f can be factored uniquely by ν. ■

We remark that if X is a generating set of R, then the commutator ideal of R is already generated by the elements $xy - yx$ for all $x, y \in X$. For let $\mathfrak{b}$ be the ideal generated by these commutators and write $\lambda : R \to R/\mathfrak{b}$ for the natural mapping. Then $R/\mathfrak{b}$ is generated by the elements $x\lambda$, $x \in X$. Now fix $x_1 \in X$; $x_1\lambda$ commutes with every $y\lambda$, $y \in X$, so the centralizer of $x_1\lambda$ contains a generating set of $R/\mathfrak{b}$ and hence is the whole ring, i.e. each $x_1\lambda$ lies in the centre of $R/\mathfrak{b}$. Since x_1 could be any element of X, the centre contains a generating set and so it is the whole ring, i.e. $R/\mathfrak{b}$ is commutative. This means that $\mathfrak{b} = \ker \lambda \supset \mathfrak{c}$, hence $\mathfrak{b} = \mathfrak{c}$ as claimed.

Given a K-module U over a commutative ring K, let $\mathbf{T}(U)$ be its tensor algebra. We define the *symmetric algebra* on U as the algebra

$$\mathbf{S}(U) = \mathbf{T}(U)^{ab}. \tag{2.7.11}$$

By what has just been said, we see that $\mathbf{S}(U)$ can be obtained from $\mathbf{T}(U)$ by imposing the relations

$$xy = yx \quad \text{for all} \quad x, y \in U, \tag{2.7.12}$$

because $\mathbf{T}(U)$ is generated by U. It is clear from the definition that $\mathbf{S}(U)$ is universal for K-linear mappings from U to commutative K-algebras, so we have

Theorem 2.7.6. *Let K be a commutative ring. For any K-module U there is a commutative K-algebra* $\mathbf{S}(U)$ *with a K-linear mapping* $\mu : U \to \mathbf{S}(U)$ *which is universal for K-linear mappings from U to commutative K-algebras.* $\mathbf{S}(U)$ *can be obtained from the tensor algebra by imposing the relations (2.7.12).* ■

Let $\delta : U \to \mathbf{S}(U)$ be any K-linear mapping. We ask: when can δ be extended to a derivation of $\mathbf{S}(U)$? The answer is 'always'. For what we require is a homomorphism from $\mathbf{S}(U)$ to $\begin{pmatrix} \mathbf{S}(U) & \mathbf{S}(U) \\ 0 & \mathbf{S}(U) \end{pmatrix}$ extending the mapping

$$f : u \mapsto \begin{pmatrix} u & u^{\delta} \\ 0 & u \end{pmatrix}.$$

Now a simple verification shows that uf and vf commute, for any $u, v \in U$, so by the universal property of $\mathbf{S}(U)$ there is a unique K-algebra homomorphism from $\mathbf{S}(U)$ extending f. This proves

Proposition 2.7.7. *Let U be a K-module (over a commutative ring K) and $\mathbf{S}(U)$ its symmetric algebra. Then any K-linear mapping $\delta : U \to \mathbf{S}(U)$ extends to a unique derivation of $\mathbf{S}(U)$.* ∎

We also note the following test for algebraic dependence in fields. If E/k is an extension field generated by $x_1, \ldots, x_n$, then it is easily verified that the universal derivation module $\Omega_k(E)$ is spanned by the dx_i. Let us write D_i or $\partial/\partial x_i$ for the derivation of the polynomial ring $k[x_1, \ldots, x_n]$ with respect to x_i; this is the derivation over k which maps x_i to 1 and x_j for $j \neq i$ to 0.

Theorem 2.7.8. *Let E/k be a field extension in characteristic 0, and let (x_i) be any family of elements of E. Then*

(i) *the x_i are algebraically independent if and only if the dx_i are linarly independent over E,*
(ii) *$E/k(x_i)$ is algebraic if and only if the dx_i span $\Omega_k(E)$ as E-space,*
(iii) *(x_i) is a transcendence basis for E if and only if the dx_i form a basis for $\Omega_k(E)$ over E.*

Proof. This follows from the fact that any polynomial relation $f(x_1, \ldots, x_n) = 0$ corresponds to a relation $\sum D_i f(x).dx_i = 0$. ∎

For our last result in this section we shall need a change-of-rings theorem which is also generally useful. Let $f : R \to S$ be a homomorphism of rings; we saw in Section 2.3 that every S-module U can be considered as R-module ${}^f U$ by pullback along f; in particular S itself becomes an R-bimodule in this way. Further, every R-module A gives rise to an induced extension $A_f = A \otimes_R S$ and a coinduced extension $A^f = \mathrm{Hom}_R(S, A)$.

Theorem 2.7.9. *Let R, S be any rings and $f : R \to S$ a homomorphism. If S is projective as right R-module, then there is a natural isomorphism*

$$\mathrm{Ext}^n_S(U, A^f) \cong \mathrm{Ext}^n_R({}^f U, A) \quad (U_S, A_R). \tag{2.7.13}$$

If S is flat as left R-module, then there is a natural isomorphism

$$\mathrm{Ext}^n_S(A_f, U) \cong \mathrm{Ext}^n_R(A, {}^f U) \quad (U_S, A_R) \tag{2.7.14}$$

and

$$\mathrm{Tor}^S_n(A_f, U) \cong \mathrm{Tor}^R_n(A, {}^f U) \quad (A_R, {}_S U). \tag{2.7.15}$$

Proof. Let us take an injective resolution $0 \to A \to I$ and apply the functor $\mathrm{Hom}_R(S, -)$, which is exact because S_R is projective. We obtain an exact sequence

$$0 \to A^f \to I^f,$$

and here the terms I_n^f are injective, as coinduced modules, so this is an injective resolution of A^f. If we now apply the hom functor to this resolution and bear in mind that by (2.3.3) and (2.3.4),

$$\mathrm{Hom}_S(U, I_n^f) \cong \mathrm{Hom}_R(U, I_n),$$

we obtain (2.7.13). Similarly, if $_RS$ is flat, then $- \otimes_R S$ is exact. Hence if $P \to A \to 0$ is a projective resolution, then so is $P_f \to A_f \to 0$. We now apply the hom functor and use the fact that

$$\mathrm{Hom}_S((P_n)_f, U) \cong \mathrm{Hom}_R(P_n, {}^fI)$$

to obtain (2.7.14); in the same way we apply the tensor product to the isomorphism

$$(P_n)_f \otimes_S U \cong P_n \otimes_R {}^fU$$

to obtain (2.7.15). ∎

We conclude this section by finding an estimate for the global dimension of a tensor ring:

Theorem 2.7.10 (Yu. V. Roganov [1975]). *Let K be any commutative ring, C a K-algebra with* r.gl.dim$(C) = n$, *U a C-bimodule and $R = \mathbf{T}_C(U)$ the tensor C-ring on U with canonical map $f : C \to R$.*

(i) *If $_CU$ is flat, then*

$$n \le \text{r.gl.dim}(R) \le n + 1, \qquad (2.7.16)$$

and r.gl.dim$(R) = n + 1$ *if and only if* hd$(A \otimes_C U) = n$ *for some right C-module A.*

(ii) *If U_C is projective, then (2.7.16) holds, and* r.gl.dim$(R) = n + 1$ *if and only if* cd$(\mathrm{Hom}_C(U, B)) = n$ *for some right C-module B.*

(iii) *If $_CU$ is flat and* w.gl.dim$(C) = m$, *then*

$$m \le \text{w.gl.dim}(R) \le m + 1, \qquad (2.7.17)$$

and w.gl.dim$(R) = m + 1$ *if and only if* wd$(A \otimes_C U) = m$ *for some A_C.*

Proof. (W. Dicks) Throughout this proof all tensor products are understood to be over C, unless otherwise stated.

Let $\Omega = R \otimes U \otimes R$ be the universal derivation bimodule for R and consider the sequence (2.7.10). Since it is split exact, it remains so on tensoring (over R) with any right R-module M. Recalling that M qua C-module is just ${}^fM$, we thus obtain the exact sequence

$$0 \to {}^fM \otimes U \otimes R \to {}^fM \otimes R \to M \to 0,$$

which can also be written

$$0 \to ({}^fM \otimes U)_f \to ({}^fM)_f \to M \to 0.$$

Hence we obtain the exact homology sequence

$$\ldots \to \mathrm{Ext}_R^{\nu}(M,N) \to \mathrm{Ext}_R^{\nu}(({}^fM)_f,N) \to \mathrm{Ext}_R^{\nu}(({}^fM\otimes U)_f,N)$$
$$\to \mathrm{Ext}_R^{\nu+1}(M,N) \to \ldots$$

for any right R-module N. Since ${}_CU$ is flat, so is ${}_CR$ and so by Theorem 2.7.8 this simplifies to

$$\ldots \to \mathrm{Ext}_R^{\nu}(M,N) \to \mathrm{Ext}_C^{\nu}({}^fM,{}^fN) \to \mathrm{Ext}_C^{\nu}({}^fM\otimes U,{}^fN) \to \mathrm{Ext}_R^{\nu+1}(M,N) \to \ldots \tag{2.7.18}$$

It follows that $r.\mathrm{gl.dim}(R) \le n+1$. Moreover, by the definition of n, $\mathrm{Ext}_C^{n+1}({}^fM,{}^fN)=0$, so we have a surjection

$$\mathrm{Ext}_C^{n}({}^fM\otimes U,{}^fN) \to \mathrm{Ext}_R^{n+1}(M,N) \to 0. \tag{2.7.19}$$

We next show that $r.\mathrm{gl.dim}(R) \ge n$. Choose right C-modules A, B such that $\mathrm{Ext}_C^{n+1}(A,B) \ne 0$ and consider A, B as right R-modules with trivial U-action, i.e. $AU = BU = 0$. The C-module structure is then recovered by pullback along f. Taking $M = A$, $N = B$ in (2.7.18) and observing that $\mathrm{Hom}_R(A,B) \to \mathrm{Hom}_C(A,B)$ is then an isomorphism, we conclude by exactness that $\mathrm{Hom}_C(A,B) \to \mathrm{Hom}_C(A\otimes U,B)$ is then the zero map. The same applies if we resolve B, hence by (2.7.18) we have the exact sequence

$$0 \to \mathrm{Ext}_C^{n-1}(A\otimes U,B) \to \mathrm{Ext}_R^{n}(A,B) \to \mathrm{Ext}_C^{n}(A,B) \to 0. \tag{2.7.20}$$

It follows that $\mathrm{Ext}_R^n(A,B) \ne 0$, and so $\mathrm{r.gl.dim}(R) \ge n$; this proves (2.7.16). Now if $\mathrm{hd}(A\otimes U) = n$ for some A_C, then by (2.7.20) with n replaced by $n+1$, we have $\mathrm{r.gl.dim}(R) = n+1$, while if $\mathrm{hd}(A\otimes U) < n$ for all A_C, then $\mathrm{hd}({}^fM\otimes U) < n$ and by (2.7.19), $\mathrm{r.gl.dim}(R) \le n$.

The proof of (ii) is similar. For any right R-module M we have

$$\mathrm{Hom}_R(R\otimes U\otimes R, M) \cong \mathrm{Hom}_C(R, \mathrm{Hom}_R(U\otimes R, M))$$
$$\cong \mathrm{Hom}_C(R, \mathrm{Hom}_C(U, {}^fM)) \cong \mathrm{Hom}_C(U,{}^fM)^f.$$

In particular, $\mathrm{Hom}_R(R\otimes R, M) = ({}^fM)^f$; hence if we apply $\mathrm{Hom}_R(-,M)$ to (2.7.10) we obtain the exact sequence

$$0 \to \mathrm{Hom}_C(U,{}^fM)^f \to ({}^fM)^f \to M \to 0,$$

and using Theorem 2.7.9 we thus find the exact sequence

$$\ldots \to \mathrm{Ext}_R^{\nu}(M,N) \to \mathrm{Ext}_C^{\nu}(M,N) \to \mathrm{Ext}_C^{\nu}(\mathrm{Hom}_C(U,{}^fM),{}^fN) \to \mathrm{Ext}_R^{\nu+1}(M,N)$$

We see again that $\mathrm{r.gl.dim}(R) \le n+1$, and the surjection (2.7.19) is replaced by

$$\mathrm{Ext}_C^{n}(\mathrm{Hom}_C(U,M),N) \to \mathrm{Ext}_R^{n+1}(M,N) \to 0.$$

The argument as before gives the analogue of (2.7.20):

$$0 \to \mathrm{Ext}_C^{n-1}(\mathrm{Hom}_C(U,A),B) \to \mathrm{Ext}_R^{n}(A,B) \to \mathrm{Ext}_C^{n}(A,B) \to 0$$

and the rest follows as before.

To prove (iii) we apply $\otimes N$ and obtain the exact homology sequence

$$\ldots \to \mathrm{Tor}^R_{\nu+1}(M,N) \to \mathrm{Tor}^C_\nu({}^fM \otimes U, {}^fN) \to \mathrm{Tor}^C_\nu({}^fM, {}^fN) \to \mathrm{Tor}^R_\nu(M,N) \to \ldots.$$

Hence $\mathrm{w.gl.dim}(R) \leq m+1$. We obtain the exact sequence

$$0 \to \mathrm{Tor}^R_{m+1}(M,N) \to \mathrm{Tor}^C_m({}^fM \otimes U, {}^fN), \tag{2.7.21}$$

and instead of (2.7.20) we find

$$0 \to \mathrm{Tor}^C_m(A,B) \to \mathrm{Tor}^R_m(A,B) \to \mathrm{Tor}^C_{m-1}(A \otimes U, B) \to 0. \tag{2.7.22}$$

It follows that $\mathrm{Tor}^R_m(A,B) \neq 0$, so $\mathrm{w.gl.dim}(R) \geq m$. If $\mathrm{wd}(A \otimes U) = m$ for some A_C, then replacing m by $m+1$ in (2.7.21) we find $\mathrm{w.gl.dim}(R) = m+1$; if $\mathrm{wd}(A \otimes U) < m$ for all A_C, then $\mathrm{wd}({}^fM \otimes U) < m$ and by (2.7.22) we have $\mathrm{w.gl.dim}(R) \leq m$. ∎

Taking $U = C$, we obtain what is in effect the Hilbert syzygy theorem:

Corollary 2.7.11. *For any ring C and a central indeterminate x,* $\mathrm{r.gl.dim}(C[x]) = \mathrm{r.gl.dim}(C) + 1$.

Proof. If $\mathrm{r.gl.dim}(C) = n$, then $U^n = C \otimes \ldots \otimes C \cong C$, as right C-module. If the generator of U is written x, then it is easily checked that $\mathbf{T}(U) \cong C[x]$. The rest is clear from Theorem 2.7.10 (i), since $A \otimes U \cong A$. ∎

In particular, since any field is semisimple, we have

$$\mathrm{gl.dim}(k[x_1, \ldots, x_n]) = n,$$

for any field k, even skew.

This means that when we resolve a module M over the polynomial ring $k[x_1, \ldots, x_n]$ by means of a free resolution $F = (F_i)$ say, then the submodule of relations in F_0 is not generally free and any generating set for these relations has further relations. These relations are known as syzygies, from the Greek 'συζυγοσ': yoked or paired (usually applied to the conjunction of heavenly bodies). By the above result, any free resolution of at most n steps leads to a projective module; Hilbert's theorem states slightly more than this, namely that the free resolution F can be chosen so as to terminate (in a free module) at the n-th step (see Eisenbud (1995)). This sharpening is also a consequence of the rather deeper theorem of Quillen and Suslin (see e.g. Lam (1978)), which states that every finitely generated projective module over the polynomial ring $k[x_1, \ldots, x_n]$ over a field k is free.

When K is any commutative ring and U is free as K-module, then by taking $C = K$ in Theorem 2.7.10 we obtain a formula for the free K-algebra $K\langle X\rangle$:

Corollary 2.7.12. *For any commutative ring K and any set X,*

$$\mathrm{r.gl.dim}(K\langle X\rangle) = \mathrm{l.gl.dim}(K\langle X\rangle) = \mathrm{gl.dim}(K) + 1.$$

In particular, a free algebra over a field is hereditary. ∎

Thus for any free algebra $k\langle X\rangle$ over a field k, every right ideal (and every left ideal) is projective. In fact these ideals are free, of unique rank (see Cohn (1985)); for the case of finitely generated right (or left) ideals this will be proved in Section 8.7 and the full result is in Section 11.5 below.

Exercises

1. Verify that $U \mapsto \mathbf{S}(U)$ is a functor.
2. Show that $\mathbf{S}(U \oplus V) \cong \mathbf{S}(U) \otimes \mathbf{S}(V)$.
3. Given a surjective homomorphism $\mu : U \to V$ of K-modules (where K is a commutative ring), show that $S(\mu)\cdot: S(U) \to S(V)$ is surjective.
4. Writing $[x, y] = xy - yx$, verify the identity $[xy, z] = x[y, z] + [x, z]y$, which expresses the fact that the mapping $u \mapsto [u, z]$ is a derivation. Use this result to give another proof of the remark following Theorem 2.7.5, that the commutator ideal of a ring R is generated by all $[x, y]$, where x, y range over a generating set of R.
5. Find extensions of Proposition 2.7.3 and Proposition 2.7.4 to (α, β)-derivations.
6. Given a situation $(A_R, {}_R B_S)$, where A_R and B_S are flat, show that $(A \otimes B)_S$ is flat. Deduce that if a C-bimodule U is flat as right C-module, then so is $T_C(U)$. Do the same for 'projective' in place of 'flat'.
7. Find an extension of Theorem 2.7.8 to the case of prime characteristic.
8. Apply Theorem 2.7.8 to prove that the transcendence degree (in characteristic 0) is an invariant of the dimension. What can be said in prime characteristic?
9. A finitely generated R-module M is called *stably free* if integers r, s exist such that $M \oplus R^r \cong R^s$. Given that every finitely generated projective over a polynomial ring is stably free, derive Hilbert's form of the syzygy theorem from Corollary 2.7.11.
10. Let k be a field and R the k-algebra generated by (disjoint) finite sets $X_1, \ldots, X_r$ with the defining relations $xy = yx$ precisely when x, y lie in different sets. Show that $\text{gl.dim}(R) = r$. Find the global dimension of R when the relation $xy = yx$ holds precisely for x, y in the same set.

Further exercises on Chapter 2

1. Let K be a commutative ring; describe products and coproducts in the category of K-algebras.
2. Let $\mathscr{C}$ be a small category admitting products. If for some objects X, Y, $\mathscr{C}(X, Y)$ has two distinct members, show that $\mathscr{C}(X, Y^I)$ has at least $2^{|I|}$ members, for any set I. Deduce that any $\mathscr{C}(X, Y)$ has at most one morphism (and so $\mathscr{C}$ is a preordered set).
3. Show that every map $\alpha : A \to B$ in an abelian category gives rise to an exact sequence

$$0 \to \ker \alpha \to A \xrightarrow{\alpha} B \to \operatorname{coker} \alpha \to 0. \tag{2.8.1}$$

Given $f : A \to B$ and $g : B \to A$ such that $fg = 1$, show that f is monic and g is epic. By applying the windmill lemma (Exercise 10 of Section 2.1) to the sequences obtained from (2.8.1) for f, g, deduce that A is isomorphic to a summand in a coproduct representation of B.

4. Prove Yoneda's lemma for additive categories: If $F : \mathscr{A} \to Ab$ is given and for $p \in X^F$ a natural transformation $p^{\blacklozenge} : h^X \to F$ is defined by the rule that for $a \in \mathscr{A}(X, Y)$, $\alpha \mapsto p\alpha^F$ maps Yh^X to Y^F, verify the naturality and show that the resulting map $X^F \to \text{Nat}(h^X, F)$ to the set of natural transformations is an isomorphism.
5. If $G : \mathscr{A} \to Ab$ is a contravariant functor, show that $X^G \cong \text{Nat}(h_X, G)$, where $h_X : Y \to A(X, Y)$ is defined by the rule
$$\mathscr{A}(X, Y) \cong \text{Nat}(h^Y, h^X) \cong \text{Nat}(h_X, h_Y).$$
6. Use Yoneda's lemma to show that the left adjoint of a functor, if it exists, is unique up to isomorphism.
7. Show that a functor is exact iff it is right exact and preserves monics or left exact and preserves epics.
8. Show that any left exact functor preserves pullbacks.
9. Show that in any category $\mathscr{A}$ the product of two $\mathscr{A}$-objects X, Y is the object representing the functor $A \mapsto \mathscr{A}(A, X) \times \mathscr{A}(A, Y)$ (if it exists). Similarly their coproduct is the object representing $A \mapsto \mathscr{A}(X, A) \times \mathscr{A}(Y, A)$.
10. Let X be an object in an abelian category. Assuming that the equivalence classes of subobjects of X form a set, partially ordered by inclusion, show that this set is a modular lattice.
11. In an additive category consider the sequence
$$P \xrightarrow{\lambda} A \prod B \xrightarrow{\mu} Q.$$
Show that if i, j, p, q are the natural injections and projections of the biproduct $A \prod B$, then the square formed by the maps $\lambda p, -\lambda q, i\mu, j\mu$ commutes if $\lambda\mu = 0$, is a pullback if $\lambda = \ker \mu$ and is a pushout if $\mu = \text{coker}\ \lambda$.
12. Given a pullback in an additive category (with notation as in Section 2.1) show that α' is monic iff α is (for abelian categories this follows from Proposition 2.1.6, but not in general).
13. Let P be a pullback of $\alpha : A \to C$, $\beta : B \to C$ in the category of rings. Show that if P is an integral domain, then one of α, β, say α is injective, and P is isomorphic to a subring of B.
14. Show that in an abelian category the intersection of two subobjects of a given object can be defined as a pullback and describe the dual concept.
15. Given a 3×3 'matrix' of short exact sequences between modules U_{ij} $(i, j = 1, 2, 3)$ forming a commutative diagram as in the 3×3 lemma, show that the kernel of the composite map $U_{22} \to U_{33}$ is $\text{im}(U_{12}) + \text{im}(U_{21})$. Deduce that for any two short exact sequences of modules U_i, V_i the kernel of the map $U_2 \otimes V_2 \to U_3 \otimes V_3$ is $\text{im}(U_1 \otimes V_2) + \text{im}(U_2 \otimes V_1)$.
16. Let $\mathscr{A}$, $\mathscr{B}$ be abelian categories with direct sums and F, G two right exact functors from $\mathscr{A}$ to $\mathscr{B}$, which preserve coproducts. Show that if there is a natural

transformation $t : F \to G$ such that for a generator P of $\mathscr{A}$, $t : P^F \cong P^G$ is an isomorphism, then t is a natural isomorphism. (Hint. Apply t to a 'presentation' of an object and use the 5-lemma.)

17. (Eilenberg–Watts) Show that for any functor $S : \mathrm{Mod}_A \to \mathrm{Mod}_B$ the following are equivalent: (a) S has a right adjoint $T : \mathrm{Mod}_B \to \mathrm{Mod}_A$, (b) S is right exact and preserves coproducts, (c) $S = - \otimes_A P$, for some (A, B)-bimodule P, unique up to isomorphism. Show that when (a)–(c) hold, the adjoint is $Y^T = \mathrm{Hom}_B(P, Y)$. (Hint. Use Exercise 16; see also Section 4.4.)
18. Show that for a finitely generated projective R-module P, $\mathrm{Hom}_R(P, M) \cong P^* \otimes M$, where $P^* = \mathrm{Hom}_R(M, R)$. (Hint. Use Exercise 16.)
19. A complex (C, d) is said to *split* if it is homotopic to $(H(C), 0)$. Show that (C, d) splits iff (a) C has a homotopy s of degree 1 such that $dsd = d$ or equivalently, (b) $B(C)$, $Z(C)$ are direct summands of C. (Hint. If s is a homotopy as in (a), then sd is a projection on $B(C)$ and $1 - ds$ is a projection on $Z(C)$.)
20. Given a chain map $f : X \to X'$ between complexes, to obtain an exact sequence which includes the maps $f^\bullet : H_nX \to H_nX'$, define a new complex $M(f)$, the *mapping cone* of f, as $M_n = X_{n+1} \oplus X'_n$ with $(x, x')d = (-xd, x'd + xf)$. Verify that there is a short exact sequence

$$0 \to X' \to M \to \sum X \to 0,$$

where $(\sum X)_n = X_{n-1}$, and obtain the associated exact homology sequence. Deduce that $M(f)$ is acyclic iff $f^\bullet$ is chain equivalence.

21. Given a chain map $f : X \to X'$ between complexes, define the *mapping cylinder* $N(f)$ on f as the complex $N_n = C_n \oplus C_{n-1} \oplus C'_n$ with $(x, y, x')d = (xd + y, -yd, x'd - yf)$. Verify that there is a short exact sequence

$$0 \to X \to N(f) \to M(f) \to 0,$$

where $M(f)$ is the mapping cone from Exercise 20. Using the associated homology sequence, show that $\alpha : X' \to N(f)$, $\beta : N(f) \to X'$ are mutually homotopy inverse, where $\alpha : x' \mapsto (0, 0, x')$ and $\beta : (x, y, x') \mapsto x' + xf$.

22. (Eilenberg trick) Let P be a projective module, say $F = P \oplus P'$ is free. By expressing the direct sum S of countably many copies of F in two ways, show that $P \oplus S \cong S$.
23. Show that the triangular matrix ring $\begin{pmatrix} k & k \\ 0 & k \end{pmatrix}$ over a field k is hereditary. Is it a principal ideal ring?
24. Given a short exact sequence of modules

$$0 \to A' \to A \to A'' \to 0,$$

show that $\mathrm{hd}(A) \leq \max \mathrm{hd}(A'), \mathrm{hd}(A'')$ with equality except possibly when $\mathrm{hd}(A'') = \mathrm{hd}(A') + 1$. Find similar bounds for $\mathrm{hd}(A')$, $\mathrm{hd}(A'')$ in terms of the other two.

25. Let R be a ring and M an R-module. Show that if $\mathrm{gl.dim}(R) = r$ and $\mathrm{hd}(M) = r - 1$, then every submodule of M has homological dimension at most $r - 1$.

26. A ring is said to be *right semihereditary* if every finitely generated right ideal is projective. Show that in a right semihereditary ring the finitely generated right ideals form a sublattice of the lattice of all right ideals.
27. A ring R is said to be *weakly semihereditary* if, given finitely generated projective modules P, P_0, P_1 and maps $\alpha : P_0 \to P$, $\beta : P \to P_1$ such that $\alpha\beta = 0$, there is a decomposition $P = P' \oplus P''$ such that $\operatorname{im} \alpha \subseteq P' \subseteq \ker \beta$. Show that R is weakly semihereditary iff for any matrices $A \in {}^rR^n$, $B \in {}^nR^s$ such that $AB = 0$ there exists an idempotent $n \times n$ matrix E over R such that $AE = A$, $EB = 0$. Deduce that the condition is left–right symmetric. Show also that every right (or left) semihereditary ring is weakly semihereditary.
28. Show that over a right semihereditary ring R every finitely generated submodule of a projective module is projective and that every finitely generated projective module is isomorphic to a direct sum of a finite number of finitely generated right ideals.
29. Show that an injective R-module E is a cogenerator iff $\mathrm{Hom}_R(S, E) \neq 0$ for every simple R-module S. Verify that $\mathbf{Q}/\mathbf{Z}$ is an injective cogenerator for $\mathbf{Z}$ (see also Section 4.6).
30. Show that for any commutative ring K, $\mathrm{Ext}_K^n(A, B)$ and $\mathrm{Tor}_n^K(A, B)$ have a natural K-module structure. Show that if, moreover, K is Noetherian and A, B are finitely generated, then $\mathrm{Ext}_K^n(A, B)$ and $\mathrm{Tor}_n^K(A, B)$ are finitely generated as K-modules.

3 Further group theory

Group theory has developed so much in recent years that a separate volume would be needed even for an introduction to all the main areas of research. The most a chapter can do is to give the reader a taste by a selection of topics; our choice was made on the basis of general importance or interest, and relevance in later applications. Thus ideas from extension theory (Section 3.1) are used in the study of simple algebras, while the notion of transfer (Section 3.3) has its counterpart in rings in the form of determinants. Hall subgroups (Section 3.2) are basic in the deeper study of finite groups, the ideas of universal algebra are exemplified by free groups (Section 3.4) and linear groups (Section 3.5) lead to an important class of simple groups, as do symplectic groups (Section 3.6) and orthogonal groups (Section 3.7).

We recall some standard notations from BA. If a group G is generated by a set X, we write $G = \text{gp}\{X\}$, and we put $\text{gp}\{X|R\}$ for a group with generating set X and set of defining relations R. For subsets X, Y of G, XY denotes the set of all products xy, where $x \in X, y \in Y$. We write $N \triangleleft G$ to indicate that N is a normal subgroup in G, i.e. mapped into itself by all inner automorphisms of G. If H, K are subgroups of G, then HK is a subgroup precisely when $HK = KH$; in particular this holds when H or K is normal in G. We also recall the *modular law*: given subgroups K, L, M of G, if $K \subseteq M$, then $K(L \cap M) = KL \cap M$.

3.1 Group extensions

In BA, Section 2.3 we saw that every finite group has a composition series; the factors of this series are simple groups and for the structure of G one has (i) to study these simple groups, and (ii) to determine how G is composed from them. For the moment we concentrate on (ii); in its simplest form this is the extension problem. It is of great general interest and some special cases will be of use to us later.

An extension of groups may be written as a short exact sequence

$$1 \to A \xrightarrow{\lambda} E \xrightarrow{\mu} G \to 1. \tag{3.1.1}$$

As in the case of abelian groups (**Z**-modules) this means that A is isomorphic to a normal subgroup A_1 of E and $E/A_1 \cong G$. We shall call E an *extension* of A by G.

Given any two groups A, G, their direct product $G \times A$ is such an extension, but in general there will be many others. Two extensions of A by G, say E_1 and E_2 are said to be *isomorphic* if there is an isomorphism $f : E_1 \to E_2$ that the diagram

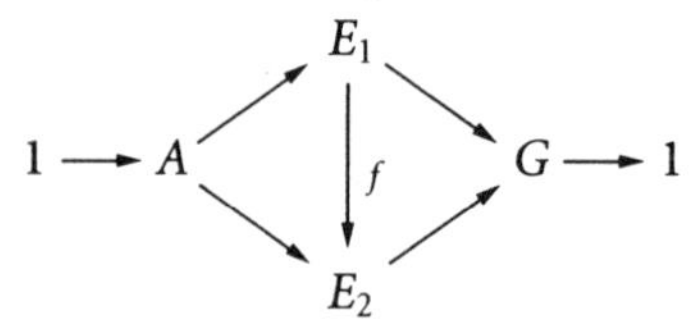

commutes. By the 5-lemma any such homomorphism f is an isomorphism. The aim of extension theory is to obtain a survey of all (isomorphism classes of) extensions of a prescribed pair of groups. In principle this is solved by Schreier's theorem (Theorem 3.1.2 below), but the solution is not very explicit and the description in terms of cohomology is more illuminating, although not exhaustive.

We begin by discussing an important special case, that of split extensions. The extension (3.1.1) is said to *split* if there is a homomorphism $\alpha : G \to E$ such that $\alpha\mu = 1$; this means that we can choose the transversal for the subgroup $A\lambda$ in E to be a subgroup (not necessarily normal in E). Consider any split extension (3.1.1) and denote the images of G, A in E by G_1, A_1 respectively. Any element x of E has the same image $x\mu$ as some element $g \in G_1$, thus $(g^{-1}x)\mu = 1$, and by exactness, $g^{-1}x \in A_1$. Hence each $x \in E$ has the form

$$x = ga, \quad \text{where } g \in G_1,\ a \in A_1. \tag{3.1.2}$$

This just means that $E = G_1A_1$. Next we observe that $G_1 \cap A_1 = 1$, for the restriction of μ to G_1 is injective and its kernel is $G_1 \cap A_1$. It follows that the expression (3.1.2) for x is unique: if $x = ga = g'a'$, where $g, g' \in G_1$, $a, a' \in A_1$, then $a'a^{-1} = g'^{-1}g \in G_1 \cap A_1 = 1$, hence $a' = a$, $g' = g$. Let us identify G with G_1 and A with A_1, so that we have $E = G \times A$, as *sets*. By hypothesis A is normal in E; if G is also normal in E, then E is just the direct product of G and A, as is easily checked, but in general G need not be normal in E; each element of G then defines an inner automorphism of E, which induces an automorphism of the normal subgroup A. If we write α_g for the automorphism of A induced by $g \in G$, then we have the commutation rule

$$ag = g.a\alpha_g \quad \text{for any } a \in A, g \in G. \tag{3.1.3}$$

It is easily verified that the mapping

$$g \mapsto \alpha_g \tag{3.1.4}$$

is a homomorphism from G to Aut A, the automorphism group of A, and (3.1.3) is enough to determine the multiplication in E:

$$(ga)(g'a') = gg'.(a\alpha_{g'})a'. \tag{3.1.5}$$

Thus the extension E is determined by G, A and the action of G on A; this is just the semidirect product of G and A with the action α, which we have met in BA,

Section 2.4, denoted by $G \ltimes A$ or $G \ltimes_\alpha A$. Given any groups G, A and a homomorphism $\alpha : G \to \mathrm{Aut}\, A$, we can define a group structure on the set $G \times A$ by the rule

$$(g, a)(g', a') = (gg', a\alpha_{g'}a').$$

We saw in BA, Section 2.4 (and can easily verify directly) that E is a group in which the elements $(g, 1)$ form a subgroup isomorphic to G and the elements $(1, a)$ form a normal subgroup isomorphic to A, with quotient isomorphic to G; thus E is a split extension of A by G. We state the result as

Theorem 3.1.1. *Let G, A be any groups and $\alpha : G \to \mathrm{Aut}\, A$ a homomorphism. Then the semidirect product $G \ltimes_\alpha A$ is a split extension of A by G with action α, and all split extensions arise in this way.* ■

For example, the dihedral group $\mathbf{D}_m$ of order 2m can be written as a semidirect product of cyclic groups:

$$\mathbf{D}_m \cong \mathbf{C}_2 \ltimes \mathbf{C}_m,$$

where $\mathbf{C}_2$ acts on $\mathbf{C}_m$ by the automorphism $x \mapsto x^{-1}$.

Let us now return to general extensions. Given any extension (3.1.1), where we take A to be identified with its image in E by means of λ, let us denote the elements of A by latin letters and those of G by greek letters. For each $\alpha \in G$ choose an element $g_\alpha \in E$ which maps to α, taking $g_1 = 1$ in order to simplify later formulae. In general it will no longer be possible to choose the g_α to form a subgroup (they form a transversal of G in E), but in any case we have $g_\alpha g_\beta \equiv g_{\alpha\beta} \pmod A$, hence

$$g_\alpha g_\beta = g_{\alpha\beta} m_{\alpha,\beta} \quad \text{where } m_{\alpha,\beta} \in A, \tag{3.1.6}$$

and since $g_1 = 1$, we have

$$m_{\alpha,1} = m_{1,\alpha} = 1. \tag{3.1.7}$$

Further, each g_α induces an automorphism θ_α of A:

$$g_\alpha^{-1} a g_\alpha = a\theta_\alpha \quad \text{for all } a \in A, \alpha \in G, \tag{3.1.8}$$

where $\theta_1 = 1$. It is no longer true that θ has to be a homomorphism, i.e. $\theta_\alpha\theta_\beta$ will not in general equal $\theta_{\alpha\beta}$, but will differ from it by an inner automorphism; by (3.1.6) we have

$$\theta_\alpha\theta_\beta = \theta_{\alpha\beta}\iota(m_{\alpha,\beta}), \tag{3.1.9}$$

where for any $x \in A$, $\iota(x) : u \mapsto x^{-1}ux$ is the inner automorphism defined by x. As is easily verified, the set Inn A of all these inner automorphisms is a normal subgroup of Aut A, the group of all automorphisms of A. The quotient $\mathrm{Aut}(A)/\mathrm{Inn}(A)$ is called the *automorphism class group* and (3.1.9) shows that we have a homomorphism

$$\bar{\theta} : G \to \mathrm{Aut}(A)/\mathrm{Inn}(A). \tag{3.1.10}$$

The set $m_{\alpha,\beta}$ is called a *factor set* of the extension, *normalized* because it satisfies (3.1.7). In E we have $g_\alpha(g_\beta g_\gamma) = g_\alpha g_{\beta\gamma} m_{\beta,\gamma} = g_{\alpha\beta\gamma} m_{\alpha,\beta\gamma} m_{\beta,\gamma}$ and $(g_\alpha g_\beta)g_\gamma = g_{\alpha\beta} m_{\alpha,\beta} g_\gamma = g_{\alpha\beta} g_\gamma (m_{\alpha,\beta}\theta_\gamma) = g_{\alpha\beta\gamma} m_{\alpha\beta,\gamma}(m_{\alpha,\beta}\theta_\gamma)$, hence by the associative law we obtain the factor set condition

$$m_{\alpha,\beta\gamma} m_{\beta,\gamma} = m_{\alpha\beta,\gamma}(m_{\alpha,\beta}\theta_\gamma) \quad \text{for all } \alpha, \beta, \gamma \in G. \tag{3.1.11}$$

Conversely, given any groups G and A and mappings $\theta : G \to \text{Aut}\, A$ and $m : G^2 \to A$ such that $\theta_1 = 1$ and (3.1.7), (3.1.9) and (3.1.11) hold, we can define a multiplication on the set $G \times A$ by putting

$$(\alpha, a)(\beta, b) = (\alpha\beta, m_{\alpha,\beta}(a\theta_\beta)b); \tag{3.1.12}$$

then it is straightforward to verify that we obtain a group in this way, which is an extension of A by G with factor set $\{m_{\alpha,\beta}\}$ and automorphisms θ_α. The associative law follows from (3.1.11), the neutral is $(1, 1)$ and the inverse of (α, a) is $(\alpha^{-1}, (m^{-1}_{\alpha^{-1},\alpha} a^{-1})\theta_\alpha^{-1})$; the verification may be left to the reader. Here the normalization (3.1.7) is not essential, but without it the formulae become a little more complicated.

Let E be an extension with factor set $\{m_{\alpha,\beta}\}$ arising from the transversal $\{g_\alpha\}$. If $\{g'_\alpha\}$ is a second transversal, then

$$g'_a = g_\alpha c_\alpha, \qquad \text{where } c_\alpha \in A,\ c_1 = 1,$$

and the new factor set $\{m'_{\alpha,\beta}\}$ obtained from the g'_α is related to the old by the equations $g_{\alpha\beta} c_{\alpha\beta} m'_{\alpha,\beta} = g'_{\alpha\beta} m'_{\alpha,\beta} = g'_\alpha g'_\beta = g_\alpha c_\alpha g_\beta c_\beta = g_\alpha g_\beta (c_\alpha\theta_\beta) c_\beta = g_{\alpha\beta} m_{\alpha,\beta}(c_\alpha\theta_\beta)c_\beta$. Hence

$$m'_{\alpha,\beta} = c^{-1}_{\alpha\beta} m_{\alpha,\beta}(c_\alpha\theta_\beta)c_\beta; \tag{3.1.13}$$

we shall express (3.1.13) by saying that $\{m_{\alpha,\beta}\}$ and $\{m'_{\alpha,\beta}\}$ are *associated*. Similarly, we obtain from (3.1.8)

$$\theta'_\alpha = \theta_\alpha \iota(c_\alpha). \tag{3.1.14}$$

Conversely, if $m'_{\alpha,\beta}$ and θ'_α are defined by (3.1.13) and (3.1.14), they lead to an extension isomorphic to the given one, as we see by retracing our steps. We sum up these results in

Theorem 3.1.2 (Schreier's extension theorem). *Given two groups G and A, a homomorphism $\bar{\theta} : G \to \text{Aut}(A)/\text{Inn}(A)$ and a map $m : G^2 \to A$ satisfying the factor set condition (3.1.11), we can define a multiplication on the set $G \times A$ by (3.1.12) and so obtain a group which is an extension of A by G. All extensions of A by G are obtained in this way and two extensions are isomorphic if and only if their factor sets are associated.* ∎

To prove this result in full, some verifications are needed, which are straightforward (though tedious) and will therefore be left to the reader. Instead we shall examine a special case, important for the applications, in more detail.

Let us assume that A is abelian; in that case Inn A is trivial and so the map $\theta : G \to \text{Aut } A$ is a homomorphism. Moreover, as (3.1.14) shows, the automorphism θ_α depends only on α and not on the choice of θ. So in this case we have an action of G on A (by automorphisms) and in place of $x\theta_\alpha$ we shall simply write x^α. This action is trivial precisely when A is contained in the centre of E; we call this a *central extension.*

In what follows we shall write A as an additive group, but G will still be multiplicative. The action of G then turns A into a G-module and the factor set condition (3.1.11) now reads

$$m_{\alpha,\beta\gamma} + m_{\beta,\gamma} = m_{\alpha\beta,\gamma} + m^{\gamma}_{\alpha,\beta}, \tag{3.1.15}$$

while associated factor sets are related by the equations

$$m'_{\alpha,\beta} = m_{\alpha,\beta} + c^{\beta}_{\alpha} + c_\beta - c_{\alpha\beta}. \tag{3.1.16}$$

Our aim is to obtain a homological description of all extensions with abelian kernel. We recall that G-modules may equally well be regarded as $\mathbf{Z}G$-modules, where $\mathbf{Z}G$ is the group algebra of G over $\mathbf{Z}$. Moreover, any left G-module A can be regarded as a right G-module by using the canonical antiautomorphism of $\mathbf{Z}G$. Explicitly we put

$$a.g = g^{-1}a \quad \text{for all } a \in A,\ g \in G.$$

Let A, B be any right G-modules and consider Hom(A, B), the group of all (abelian group) homomorphisms from A to B. We can define a G-module structure on this group as follows. If $f \in \text{Hom}(A, B)$, we put

$$f^s : a \mapsto (f(as^{-1}))s \quad \text{for } s \in G. \tag{3.1.17}$$

This is indeed a G-module structure, for (3.1.17) can be written

$$f^s(as) = f(a)s,$$

hence $f^{st}(ast) = (f(a))st = (f^s(as))t = (f^s)^t(ast)$, therefore $f^{st} = (f^s)^t$, and the remaining laws are clear.

For any G-module A we define

$$A^G = \{x \in A | xs = x \quad \text{for all } s \in G\};$$

thus A^G is the largest submodule left fixed by G. For example, if A, B are any G-modules, then $f \in \text{Hom}(A, B)$ is left fixed by G iff $f^s = f$, i.e. $f(as) = f(a)s$ for all $s \in G$. This just means that f is a G-homomorphism from A to B, hence we have

$$(\text{Hom}(A, B))^G = \text{Hom}_G(A, B). \tag{3.1.18}$$

In particular, regarding $\mathbf{Z}$ as a G-module by the trivial action: $ns = n$ for all $n \in \mathbf{Z}$ and $s \in G$, we have

$$\text{Hom}_G(\mathbf{Z}, A) = (\text{Hom}(\mathbf{Z}, A))^G \cong A^G. \tag{3.1.19}$$

It is clear from (3.1.19) that the functor $A \mapsto A^G$ is left exact; we can therefore form the right derived functor, as described in Sections 2.5 and 2.6. The n-th right derived functor of A^G is written $H^n(G, A)$ and is called the n-th *cohomology group* of the G-module A. By (3.1.19) we see that

$$H^n(G, A) = \mathrm{Ext}_G^n(\mathbf{Z}, A), \quad H^0(G, A) = A^G,$$

where for the subscript on the right we have written G rather than $\mathbf{Z}G$. We note that for any coinduced module A we have $H^n(G, A) = 0$. For $\mathbf{Z}G$ is free as $\mathbf{Z}$-module, hence by the change-of-rings formula (2.7.13), we have

$$\mathrm{Ext}_G^n(\mathbf{Z}, B^f) \cong \mathrm{Ext}_{\mathbf{Z}}^n(\mathbf{Z}, B) = 0 \quad \text{for all } n \geq 1.$$

This holds for any coinduced module $B^f = \mathrm{Hom}_{\mathbf{Z}}(\mathbf{Z}G, B)$, where $f : \mathbf{Z} \to \mathbf{Z}G$.

Similarly the n-th *homology group* of the G-module A is defined as

$$H_n(G, A) = \mathrm{Tor}_n^G(\mathbf{Z}, A).$$

If A is induced, say $A = B_f = \mathbf{Z}G \otimes_{\mathbf{Z}} B$, then $\mathrm{Tor}_n^G(\mathbf{Z}, B_f) \cong \mathrm{Tor}_n^{\mathbf{Z}}(\mathbf{Z}, B) = 0$ for all $n \geq 1$, hence we have $H_n(G, A) = 0$ for any induced G-module A.

Let $\varepsilon : \mathbf{Z}G \to \mathbf{Z}$ be the *augmentation map*, defined as $\varepsilon : \sum a_s s \mapsto \sum a_s$. Its kernel IG is called the *augmentation ideal* of $\mathbf{Z}G$, and from the split exact sequence

$$0 \to IG \to \mathbf{Z}G \to \mathbf{Z} \to 0 \tag{3.1.20}$$

we obtain, by tensoring with a G-module A, the exact sequence

$$0 \to (IG)A \to A \to \mathbf{Z} \otimes A \to 0.$$

Hence $\mathbf{Z} \otimes A \cong A/(IG)A$ and we see that $H_n(G, A)$ is the left derived functor of $A/(IG)A$. We note that $A/(IG)A$, sometimes written A_G, is the largest quotient of A with trivial G-action, as is easily seen. Hence we find that

$$H_0(G, A) \cong \mathbf{Z} \otimes A \cong A_G. \tag{3.1.21}$$

From the exact sequence (3.1.20) and the definitions of H_n, H^n we obtain by shifting dimensions,

$$H_n(G, A) = \mathrm{Tor}_{n-1}^G(A, IG), \tag{3.1.22}$$

$$H^n(G, A) = \mathrm{Ext}_G^{n-1}(IG, A). \tag{3.1.23}$$

We recall that to construct $H_n(G, A)$ or $H^n(G, A)$ we take a projective resolution X of $\mathbf{Z}$ (as trivial G-module) and form the complex $A \otimes X$ or $\mathrm{Hom}(X, A)$ respectively. Taking first $H_n(G, A)$ let us form $A \otimes X$, where the X_n are now left modules; thus X_n consists of all n-chains, i.e. $(n+1)$-tuples over G with the action

$$g(s_0, s_1, \ldots, s_n) = (gs_0, gs_1, \ldots, gs_n).$$

The differential $d : X_n \to X_{n-1}$ is given by the formula

$$(s_0, \ldots s_n)d = \sum_{i=0}^{n} (-1)^i (s_0, \ldots, \hat{s}_i, \ldots, s_n), \tag{3.1.24}$$

where the caret on $\hat{s}_i$ means that this term is to be omitted. The augmentation map $\varepsilon : X_0 \to \mathbf{Z}$ is defined by $s_0\varepsilon = 1$. With these definitions we have $d^2 = 0$, as we see by counting how often the term $(s_0, \ldots, \hat{s}_p, \ldots, \hat{s}_q, \ldots, s_n)$ where $p < q$, occurs in $(s_0, \ldots, s_n)d^2$.

To prove exactness, we define a homotopy $h : X_n \to X_{n+1}$ by the rule: $(s_0, \ldots, s_n)h = (1, s_0, \ldots, s_n)$ for $n \geq 0$, $1.h = 1$ for $n = -1$. Then it is easily verified that $hd + dh = 1$. This resolution is called the *standard* (or *bar*) *resolution* of $\mathbf{Z}$ over $\mathbf{Z}G$, in homogeneous form. The elements of ker d, im d are called *cycles* and *boundaries* respectively. Sometimes the inhomogeneous form is more convenient to use; this is defined as

$$[t_1|t_2|\ldots|t_n] = (1, t_1, t_1t_2, \ldots, t_1\ldots t_n) \quad n \geq 1, [\,] = (1). \tag{3.1.25}$$

Now the differential is given by

$$[t_1|t_2|\ldots|t_n]d = t_1[t_2|\ldots|t_n] + \sum(-1)^i[t_1|\ldots|t_{i-1}|t_it_{i+1}|t_{i+2}|\ldots|t_n] + (-1)^n[t_1|\ldots|t_{n-1}].$$

In low dimensions we have

$$(1)\varepsilon = (1), [g]d = g[\,] - [\,] = g - 1, [g|h]d = g[h] - [gh] + [g]. \tag{3.1.26}$$

As an illustration let us calculate $H_1(G, \mathbf{Z})$. We take the resolution X of $\mathbf{Z}$ and tensor with $\mathbf{Z}$:

$$\ldots \to \mathbf{Z} \otimes X_2 \to \mathbf{Z} \otimes X_1 \to \mathbf{Z} \otimes X_0 \to \mathbf{Z} \to 0.$$

Since $\mathbf{Z}$ is a trivial G-module, we see that at $\mathbf{Z} \otimes X_1$ the kernel is all of $\mathbf{Z} \otimes X_1$ while the image is generated by the elements $1 \otimes u$, where $u = [h] - [gh] + [g]$, by (3.1.26). Thus $H_1(G, \mathbf{Z})$ consists of all sums of terms $1 \otimes [g]$ subject to the relation

$$1 \otimes [gh] = 1 \otimes [g] + 1 \otimes [h].$$

This is just G made abelian, $G^{\text{ab}} = G/G'$, hence we have

Proposition 3.1.3. *For any group G we have $H_1(G, \mathbf{Z}) = \mathrm{Tor}_1^G(\mathbf{Z}, \mathbf{Z}) \cong G^{\text{ab}}$.* ∎

We next turn to the construction of $H^n(G, A)$ using the standard resolution. The elements of $\mathrm{Hom}_G(X_n, A)$ are *n-cochains*, those of ker d, im d the *cocycles* and *coboundaries* respectively. Taking X_n now as right G-module, we find that an n-cochain is a function $f : G^{n+1} \to A$ such that

$$f(s_0g, \ldots, s_ng) = f(s_0, \ldots, s_n)g \quad \text{for } g \in G.$$

Since g is arbitrary in G, f is completely determined by its values when its last argument is 1. Let us write

$$\varphi(t_1, \ldots, t_n) = f(t_1\ldots t_n, t_2\ldots t_n, \ldots, t_{n-1}t_n, t_n, 1).$$

φ is the inhomogeneous cochain corresponding to the homogeneous cochain f. In terms of φ the coboundary is given by the formula

$$(\varphi d)(t_1, \dots, t_{n+1}) = \varphi(t_2, \dots, t_{n+1}) + \sum (-1)^i \varphi(t_1, \dots, t_i t_{i+1}, \dots, t_{n+1}) \\ + (-1)^{n+1} \varphi(t_1, \dots, t_n) t_{n+1}. \tag{3.1.27}$$

Let us again write out the cases of low dimensions:

$n = 1$. A 1-cocycle is a map $\varphi : G \to A$ such that $\varphi d = 0$. Thus C^1 consists of all φ satisfying

$$\varphi(gg') = \varphi(g') + \varphi(g)g'. \tag{3.1.28}$$

It is a coboundary iff

$$\varphi(g) = c - cg \text{ for some fixed } c \in A. \tag{3.1.29}$$

A function φ satisfying (3.1.28) is sometimes called a *derivation* (it is an $(\varepsilon, 1)$-derivation in the sense of BA, Section 6.2) or a *crossed homomorphism*. If (3.1.29) holds, it is said to be *inner* or a *principal* crossed homomorphism. Writing $\mathrm{Der}(G, A)$ for the group of derivations and $\mathrm{IDer}(G, A)$ for the subgroup of inner derivations, we have

$$H^1(G, A) = C^1/B^1 \cong \mathrm{Der}(G, A)/\mathrm{IDer}(G, A).$$

If G acts trivially on A, the derivations are just the homomorphisms and the inner derivations are 0, so in this case $H^1(G, A) \cong \mathrm{Hom}(G, A)$. Since A is abelian, the homomorphisms $G \to A$ correspond to the homomorphisms $G^{\mathrm{ab}} \to A$, and so we obtain

Proposition 3.1.4. *For any group G and any module A with trivial G-action $ag = a$ for all $a \in A$, $g \in G$, we have*

$$H^1(G, A) \cong \mathrm{Hom}(G, A) \cong \mathrm{Hom}(G^{\mathrm{ab}}, A).$$ ■

$n = 2$. A 2-cocycle is a map $\varphi : G \to A$ such that

$$\varphi(a, bc) + \varphi(b, c) = \varphi(ab, c) + \varphi(a, b)c.$$

This is precisely the condition (3.1.15) for a factor set; moreover, a 2-coboundary is a factor set of the form $\varphi(b) - \varphi(ab) + \varphi(a)b$; now (3.1.16) shows that two factor sets are associated iff they differ by a coboundary. This establishes

Theorem 3.1.5. *Let G be any group and A a G-module. Then there is a natural bijection between the group $H^2(G, A)$ and the set of isomorphism classes of extensions of A by G with the given G-action on A.* ■

We observe that since $H^2(G, A)$ has a group structure, there is a multiplication of extensions of A by G. It is obtained by taking the product (resp. sum) of the corresponding factor sets (resp. cocycles) and is known as the *Baer product* (resp. *sum*).

Some simple calculations are suggested in the exercises; we add a general result which is often useful.

Proposition 3.1.6. *Let G be a finite group of order r and A any G-module. Then for any $n > 0$, each element of $H^n(G, A)$ has order dividing r.*

Proof. We define a 'homotopy mod r' by the equation

$$(\varphi h)(s_2, \ldots, s_n) = \sum_g \varphi(g, s_2, \ldots, s_n).$$

If in (3.1.27) we sum over t_1, we obtain

$$(\varphi dh)(t_2, \ldots, t_{n+1}) = r\varphi(t_2, \ldots, t_{n+1}) - (\varphi hd)(t_2, \ldots, t_{n+1}).$$

Hence we have

$$dh + hd = r.1.$$

If c is any cocycle, then $cd = 0$, hence $rc = cdh + chd = (ch)d$, therefore rc is a coboundary and so $rH(G, A) = 0$, as claimed. ∎

For finite A the order of A annihilates $H^n(G, A)$, so Proposition 3.1.6 yields

Corollary 3.1.7. *If G, A are both finite, of coprime orders, then $H^n(G, A) = 0$ for any $n > 0$.* ∎

It is clear from Proposition 3.1.6 that if A is uniquely divisible, as abelian group, then $H^n(G, A) = 0$ for any finite group G. This remark can be used to compute H^2 of a finite group over $\mathbf{Z}$:

Proposition 3.1.8. *Let G be any finite group and $\mathbf{K} = \mathbf{Q}/\mathbf{Z}$ the group of rational numbers* mod 1, *as trivial G-module. Then*

$$H^2(G, \mathbf{Z}) \cong \mathrm{Hom}(G, \mathbf{K}). \tag{3.1.30}$$

Proof. The exact sequence $0 \to \mathbf{Z} \to \mathbf{Q} \to \mathbf{K} \to 0$ leads to the derived sequence

$$H^1(G, \mathbf{Q}) \to H^1(G, \mathbf{K}) \to H^2(G, \mathbf{Z}) \to H^2(G, \mathbf{Q}).$$

Since $\mathbf{Q}$ is uniquely divisible, the extreme terms are 0 and so the other two are isomorphic: $H^2(G, \mathbf{Z}) \cong H^1(G, \mathbf{K})$. Since the G-action on $\mathbf{K}$ is trivial, $H^1(G, \mathbf{K}) \cong \mathrm{Hom}(G, \mathbf{K})$ and the result follows. ∎

A similar proof, replacing $\mathbf{Q}$ by $\mathbf{R}$, shows that $H^2(G, \mathbf{Z}) \cong \mathrm{Hom}(G, \mathbf{T})$, where $\mathbf{T} = \mathbf{R}/\mathbf{Z}$. More directly we see that $\mathrm{Hom}(G, \mathbf{T}) = \mathrm{Hom}(G, \mathbf{K})$ for any finite group G, because every homomorphism of G into $\mathbf{T}$ has its image in the torsion subgroup of $\mathbf{T}$, and this is just $\mathbf{K}$. We also note that $\mathrm{Hom}(G, \mathbf{T}) \cong \mathrm{Hom}(G, \mathbf{C}^\times)$, because $\mathbf{T} \cong \mathbf{C}^\times$ by the isomorphism $x \mapsto \exp(2\pi i x)$.

The group $H^2(G, \mathbf{K})$ is called the *multiplicator* of G. It arises when we consider representations of G by linear transformations of a projective space, briefly *projective* transformations. If $\alpha \in G$ is represented by a matrix $\rho(\alpha)$ over C say, we have

$$\rho(\alpha)\rho(\beta) = c_{\alpha,\beta}\rho(\alpha\beta),$$

where $c_{\alpha,\beta} \in \mathbf{C}^{\times}$ is a factor set, corresponding to an element of $H^2(G, \mathbf{K})$.

As we have seen, from a short exact sequence of G-modules we can derive a long exact sequence by applying hom or the tensor product. This argument cannot be applied directly to an exact sequence describing a group extension

$$1 \to N \xrightarrow{\lambda} G \xrightarrow{\mu} L \to 1. \tag{3.1.31}$$

Nevertheless a similar result can be obtained on going via a short exact sequence of L-modules. It is given by

Theorem 3.1.9. *Given an exact sequence (3.1.31) describing a group extension G, there is a 5-term exact sequence for any L-module A:*

$$0 \to H^1(L, A) \xrightarrow{\mu^*} H^1(G, A) \xrightarrow{\lambda^*} \mathrm{Hom}_L(N^{\mathrm{ab}}, A) \xrightarrow{t} H^2_L(L, A) \xrightarrow{\mu^*} H^2_G(G, A). \tag{3.1.32}$$

The map μ^*, arising from μ, is called the *inflation map* and λ^*, arising from the inclusion λ is called the *restriction map*. The connecting map t is called the *transgression*.

Proof. By tensoring (3.1.20) with $\mathbf{Z}L$ over G we obtain the exact sequence

$$0 \to \mathrm{Tor}_1^G(\mathbf{Z}L, \mathbf{Z}) \to \mathbf{Z}L \otimes_G IG \to \mathbf{Z}L \otimes_G \mathbf{Z}G \to \mathbf{Z}L \otimes_G \mathbf{Z} \to 0.$$

Here the last two terms just represent the augmentation of $\mathbf{Z}L$, so the kernel in the third term is IL. For the first term we have, by Theorem 2.7.9 and Proposition 3.1.3, $\mathrm{Tor}_1^G(\mathbf{Z}L, \mathbf{Z}) \cong \mathrm{Tor}_1^N(\mathbf{Z}, \mathbf{Z}) \cong N^{\mathrm{ab}}$. Hence we obtain the exact sequence of L-modules

$$0 \to N^{\mathrm{ab}} \to \mathbf{Z}L \otimes_G IG \to IL \to 0. \tag{3.1.33}$$

From the associativity of the tensor product we know that for any L-module A,

$$\mathrm{Hom}_L(\mathbf{Z}L \otimes_G IG, A) \cong \mathrm{Hom}_G(IG, A).$$

Let us write the two sides of this formula as $P(A) \cong Q(A)$ and write P^n, Q^n for the left derived functors. We take a short resolution of A with I injective:

$$0 \to A \to I \to C \to 0,$$

and apply P, Q, recalling the P^1 vanishes on injectives:

$$\begin{array}{ccccccccc}
0 \to & P^0(A) & \to & P^0(I) & \to & P^0(C) & \to & P^1(A) & \to 0 \\
& \downarrow\cong & & \downarrow\cong & & \downarrow\cong & & \downarrow & \\
0 \to & Q^0(A) & \to & Q^0(I) & \to & Q^0(C) & \to & Q^1(A) & \to Q^1(I)
\end{array}$$

It follows that there is an injection from $P(A)$ to $Q(A)$, i.e. from $\mathrm{Ext}(\mathbf{Z}L \otimes IG, A)$ to $\mathrm{Ext}(IG, A)$. If we now apply $\mathrm{Hom}(-, A)$ to (3.1.33), we find

$$0 \to \mathrm{Hom}_L(IL, A) \to \mathrm{Hom}_L(\mathbf{Z}L \otimes_G IG, A) \to \mathrm{Hom}_L(N^{\mathrm{ab}}, A)$$
$$\to \mathrm{Ext}^1_L(IL, A) \to \mathrm{Ext}^1_L(\mathbf{Z}L \otimes IG, A).$$

This reduces to (3.1.32) if we use (3.1.23) on the first two terms, replace the last term by $\mathrm{Ext}^1_G(IG, A)$ and again use (3.1.23). ■

Occasionally the cohomology groups are needed for a more general coefficient ring. If K is any commutative ring and KG is the group algebra of G over K, then the standard resolution $\mathbf{X} \to Z \to 0$ on tensoring with K over $\mathbf{Z}$ becomes

$$\mathbf{X} \otimes_{\mathbf{Z}} K \to K \to 0,$$

and this is still a projective resolution, because the original resolution was $\mathbf{Z}$-split, by the homotopy found earlier. Thus we obtain, for any KG-module A,

$$\mathrm{Ext}^n_{\mathbf{Z}G}(\mathbf{Z}A) \cong \mathrm{Ext}^n_{KG}(KG \otimes_G \mathbf{Z}) \cong \mathrm{Ext}^n_{KG}(K, A). \tag{3.1.34}$$

Similarly we have

$$\mathrm{Tor}^{\mathbf{Z}G}_n(\mathbf{Z}, A) \cong \mathrm{Tor}^{KG}_n(K, A). \tag{3.1.35}$$

We recall that the terms in (3.1.34) vanish for $n \geq 1$ if A is coinduced, and those in (3.1.35) vanish if A is induced. But for G-modules over a finite group G, induced and coinduced mean the same thing, for if $f : K \to KG$ is the inclusion map, then there is an isomorphism of KG-modules:

$$\mathrm{Hom}_K(KG, U) \cong U \otimes KG,$$

for any K-module U, given by $\alpha \mapsto \sum s\alpha \otimes s^{-1}$.

Exercises

1. Supply the details for the proof of Theorem 3.1.2.
2. Examine the case of unnormalized factor sets.
3. (O. Hölder) Let E be an extension of A by B, where A, B are cyclic groups of orders m, n respectively (such a group E is called *metacyclic*). Show that E has a presentation $\mathrm{gp}\{a, b \mid a^m = 1,\ b^n = 1,\ b^{-1}ab = a^s\}$, where $s \equiv 1 \pmod m$ and $r(s-1) \equiv 0 \pmod m$. Conversely, given m, n, r, s satisfying these relations, show that there is a metacyclic group with this presentation.
4. Show that an extension of A by G with factor set $\{m_{\alpha,\beta}\}$ splits iff there exist $c_\alpha \in A$ such that $m_{\alpha,\beta} = c_{\alpha\beta}^{-1}(c_\alpha \theta_\beta)c_\beta$. Let E be an extension of an abelian group A by G. By adjoining free abelian generators c_α to E and using the above relations to define $c_\alpha\theta_\beta$ show that E can be embedded in a group E^* which is a semidirect product of A^* and G, where $A^* \supseteq A$.
5. Show that the group of isometries of Euclidean n-space is a split extension of the normal subgroup of translations by the orthogonal group.

6. Show that $H^1(G, A)$ is in natural bijection with the set of isomorphism classes of *A-torsors*, i.e. left A-sets with regular A-action such that $(a.x)^\alpha = a^\alpha.x^\alpha$ $(\alpha \in G, a \in A)$.
7. Show that the augmentation ideal, defined as in (3.1.20), is $\mathbf{Z}$-free on the $s-1$ $(1 \neq s \in G)$. Verify that $\mathrm{Hom}_G(IG, A)$ is isomorphic to the group of 1-cocycles.
8. Using the mapping $s \mapsto s - 1 \pmod{IG^2}$ of G into $IG/(IG)^2$ show that $G^{ab} \cong IG/(IG)^2$. Deduce another proof of Proposition 3.1.3.
9. Let G be a finite group of order r. Show that every cocycle in $H^n(G, \mathbf{C}^\times)$ is cohomologous to a cocycle whose values are r-th roots of 1. Deduce that the multiplicator of G has order at most r^{r^2}.
10. Let $G = \mathbf{C}_m$, the cyclic group of order m, with generator s and write $D = s - 1$, $N = 1 + s + \ldots + s^{m-1}$ in $\mathbf{Z}G$. Verify that there is a free resolution $W \to \mathbf{Z} \to 0$, where $W_n = \mathbf{Z}G$ and $d_{2n} : W_{2n} \to W_{2n-1}$ is $d_{2n} = N$, while $d_{2n-1} = D$. With the notation ${}_NA = \{a \in A | aN = 0\}$ for any G-module A show that $H^{2n}(G, A) = A^G/AN$, $H^{2n-1}(G, A) = {}_NA/AD$ and $H_n(G, A) = H^{n-1}(G, A)$ for all $n > 1$.

3.2 Hall subgroups

One of the first results in group theory is Lagrange's theorem, which tells us that the order of a subgroup of G is a divisor of the order of G. One soon discovers that the converse is false: for example, Alt_4 has order 12 but contains no subgroup of order 6. The first positive result in this direction was Sylow's theorem, which showed that for prime powers the converse of Lagrange's theorem is true. A significant generalization was found by Philip Hall, who showed in 1928 that in a soluble group G, for every factorization of the order of G into two coprime integers, $|G| = mn$, $(m, n) = 1$, there is a subgroup of order m, and in 1937 he showed that solubility is necessary for this to happen.

It is convenient to begin with some notation. Let π be a set of prime numbers. By the *π-part* of an integer $n = \prod p^{\alpha_p}$ we understand the number $n_\pi = \prod_{p \in \pi} p^{\alpha_p}$. The complementary set of primes is written π', so for any positive integer n we have

$$n = n_\pi . n_{\pi'} .$$

For any finite group G, $\pi(G)$ denotes the set of primes dividing $|G|$. If $\pi(G) \subseteq \pi$, the group G is called a π-group. A *Hall subgroup* is a subgroup of G whose order is prime to its index. A π-subgroup of G whose index is prime to π (i.e. not divisible by any prime in π) is called a *Hall π-subgroup*; e.g. when $\pi = \{p\}$, a Hall p-subgroup is just a Sylow p-subgroup. A Hall p'-subgroup is also called a *p-complement*. To establish the existence of Hall subgroups in soluble groups we shall need some preliminary results.

Lemma 3.2.1. *Let G be a finite group and H, K any subgroups. Then*

$$(HK : H) = (K : H \cap K). \tag{3.2.1}$$

In particular, if $(G : H) = (K : H \cap K)$, then $HK = G$.

Proof. Clearly HK is a union of cosets of H, say $HK = Hk_1 \cup \ldots \cup Hk_r$, where $k_i \in K$ and $r = (HK : H)$. We claim that, writing $D = H \cap K$, we have the coset decomposition

$$K = Dk_1 \cup \ldots \cup Dk_r; \tag{3.2.2}$$

this will establish (3.2.1). Any $x \in K$ can by hypothesis be written as $x = hk_i$ where $h \in H$. Here $h = xk_i^{-1} \in H \cap K = D$; if $hk_i = h'k_j$, then $k_ik_j^{-1} \in H$ and it follows that $i = j$. This proves the coset decomposition (3.2.2), and (3.2.1) follows. Now if $(G : H) = (K : H \cap K)$, then $(G : H) = (HK : H)$, hence G and HK contain the same number of cosets of H and so $HK = G$. ■

We note that in the special case where H or K is normal in G, the lemma follows from the second isomorphism theorem (Theorem 1.2.6).

Lemma 3.2.2. *If H, K are subgroups of a finite group G whose indices in G are coprime, then $HK = G$ and $(G : H \cap K) = (G : H)(G : K)$.*

Proof. Put $(G : H) = m$, $(G : K) = n$; by Lemma 3.2.1 we have

$$(G : H \cap K) = (G : K)(K : H \cap K) = (G : K)(HK : H).$$

Clearly $(HK : H)$ is a factor of $m = (G : H)$, hence

$$(G : H \cap K) = nm_1, \quad \text{where } m_1 | m.$$

Similarly,

$$(G : H \cap K) = mn_1, \quad \text{where } n_1 | n.$$

Hence $m_1n = n_1m$, so $n/n_1 = m/m_1 = 1$, because m and n are coprime, and it follows that

$$(G : H \cap K) = mn.$$

Moreover, $(G : H) = m = m_1 = (HK : H)$, hence $HK = G$, by Lemma 3.2.1. ■

We shall need the result that a normal Hall subgroup always has a complement. When the normal subgroup is abelian, this follows from Corollary 3.1.7; the general case can be deduced from this by an induction on the order:

Theorem 3.2.3 (Schur–Zassenhaus). *Let G, H be two finite groups. If the orders of G, H are coprime, then every extension of H by G splits. Moreover, if either G or H is soluble, then any two complements of H in an extension are conjugate.*

Proof. Write $|G| = n.|H| = r$, so that $(n, r) = 1$, and consider an extension

$$1 \to H \to E \xrightarrow{\lambda} G \to 1.$$

If H is abelian, the extension is represented by an element in $H^2(G, H)$ and this group is 0 by Corollary 3.1.7, so the extension splits. In the general case we use induction on r. Clearly we need only find a subgroup K of order n in E; this must be a complement to H, because $H \cap K = 1$ by comparing orders, and so $\lambda|K$ maps onto G.

Take a prime divisor p of r and let P be a Sylow p-subgroup of E; if its normalizer is $N = N_E(P)$, then $N_H(P) = H \cap N$ and the number of conjugates of P in N is $(H : H \cap N) = (E : N)$.

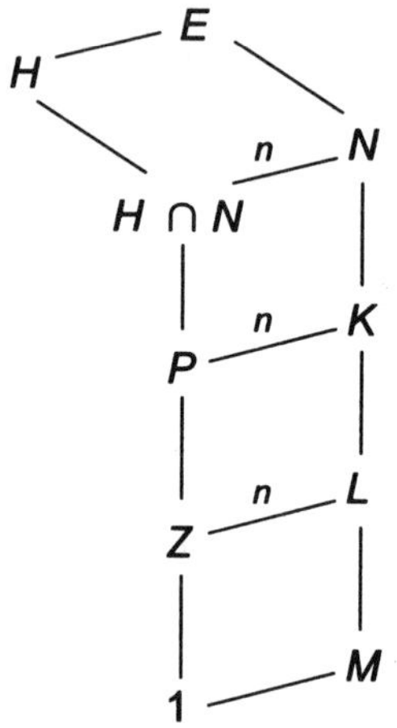

It follows that $n = (E : H) = (N : N \cap H)$. Now P and $H \to N$ are normal in N and N/P is an extension of $(H \cap N)/P$ by $N/(H \cap N)$, whose orders divide r and n respectively, and so are coprime. Since $P \neq 1$, we can apply the induction hypothesis and find a subgroup K of N such that $P \subset K \subset N$ and $(K : P) = n$.

Let Z be the centre of P; this is a non-trivial characteristic subgroup of P, hence $Z \triangleleft N$ and so $Z \triangleleft K$. By induction we obtain a subgroup L of K such that $Z \subset L \subset K$ and $(L : Z) = n$. Now L is an extension of the abelian group Z and the orders are again coprime, so we have a subgroup M of L of order n, and this is the required complement.

If an extension E has two complements G_1, G_2 of H, suppose first that H is abelian. An element x of G corresponds to $xf_1, xf_2 = xf_1.c_x$ of G_1, G_2, where c_x is a cocycle of G in H. Since $H^1(G, H) = 0$, we have $c_x = u^{-x}u$ for some $u \in H$, hence $xf_2 = xf_1.u^{-x}u = u^{-1}xf_1u$ and so G_2 is conjugate to G_1, as claimed.

Next assume that H is soluble and let G_1, G_2 be two complements. Applying the previous case to E/H' we find that $G_2H' = (G_1H')^u$ for some $u \in H$, i.e. $G_1^uH' = G_2H'$. By induction on the derived length of H there exists $v \in H'$ such that $G_1^{uv} = G_2$ and the result follows.

Finally assume that G is soluble and let G_1, G_2 again be complements of H in E. If G is a p-group, then G_1, G_2 are Sylow p-subgroups of E and hence are conjugate, by Sylow's theorem. In the general case let M_1 be a minimal normal subgroup of G_1 and M_2 the corresponding subgroup of G_2. Then $M_1H = M_2H$ and by induction there exists $x \in H$ such that $M_1^x = M_2$. Replacing G_1 by G_1^x we may thus assume that $M_1 = M_2$. Now $G_1H/M_1 = G_2H/M_2$ and by the induction hypothesis there exists $y \in H$ such that $G_1^y/M_1 = G_2/M_2$; hence $G_1^y = G_2$ as we had to show. ■

We remark that by the Feit–Thompson theorem every group of odd order is soluble. This shows that the hypothesis of Theorem 3.2.3 is always satisfied, since of two coprime integers at least one must be odd.

Let π be any set of primes and let G be a finite group. It is clear that if H is a Hall π-subgroup of G, then H is also a Hall π-subgroup of any subgroup of G containing H, and under any homomorphism f from G, the image Hf is a Hall subgroup of Gf. Further, if p, q are distinct primes, H is a Hall p'-subgroup and K a Hall q'-subgroup of G, then $|HK|$ is divisible by $|G|_{p'}$ and $|G|_{q'}$, hence by $|G|$ and so $HK = G$. Moreover $H \cap K$ is a Hall p'-subgroup of K, a q'-subgroup of H and a $\{p, q\}'$-subgroup of G.

We shall also need an auxiliary result. For any finite group G, a minimal normal subgroup is a direct product of simple groups; we shall only need the case where G is soluble, when a minimal normal subgroup is a direct power of a cyclic group of prime order p, i.e. an elementary abelian p-group.

Lemma 3.2.4. *Let G be a finite soluble group. Then any minimal normal subgroup is elementary abelian.*

Proof. Any minimal normal subgroup H of G is abelian, because its derived group is a proper subgroup and so must be the trivial group. Now H can have no characteristic subgroups and so must be a p-group for some prime p, and moreover all elements satisfy $x^p = 1$, i.e. it is elementary abelian, as claimed. ■

With these preparations we can establish the existence of Hall subgroups.

Theorem 3.2.5 (P. Hall [1928]). *Let G be a finite soluble group and π a finite set of primes. Then any π-subgroup of G is contained in a Hall π-subgroup, hence Hall π-subgroups exist for any π, and any two Hall π-subgroups are conjugate in G.*

Proof. We shall use induction on $|G|$. Let G be a finite soluble group and M a minimal normal subgroup of G. By Lemma 3.2.4, M is an elementary abelian p-group, for some prime p. Write $\bar{G} = G/M$; by the induction hypothesis $\bar{G}$ contains a Hall π-subgroup $\bar{H}$, where $H \supseteq M$. Moreover, if A is a π-subgroup of G, its image $\bar{A}$ in $\bar{G}$ is a π-subgroup and so is contained in some conjugate of $\bar{H}$, say $\bar{A} \subseteq \bar{H}^x$; it follows that $A \subseteq H^x$. If $p \in \pi$, then $(G : H)_\pi = 1$ and H is a π-subgroup, hence a Hall π-subgroup of G, so is H^x and A has been embedded in a Hall π-subgroup. Since 1 is always a π-subgroup, this shows that Hall π-subgroups exist. Moreover, all Hall π-subgroups are conjugate, for if A is a Hall π-subgroup and $A \subseteq H^x$, then $A = H^x$ because $|A| = |H|$.

There remains the case $p \notin \pi$. By Theorem 3.2.3 there is a complement K of M in H and all complements are conjugate. Since $|K| = |\bar{H}| = |G|_\pi$, K is a Hall π-subgroup of G, and all the Hall π-subgroups are conjugate. If A is any π-subgroup of G, we have as before $A \subseteq H^x$ for some $x \in G$. Either $H \subset G$, then by induction, A is contained in a Hall π-subgroup of H, which is also a Hall π-subgroup of G; or $H = G$. Then $A \subset G = KM$ and hence

$$AM = AM \cap KM = (AM \cap K)M.$$

Now A and $AM \cap K$ are Hall subgroups of AM and so are conjugate: $A^y = AM \cap K \subseteq K$ $(y \in M)$. It follows that A is contained in a Hall π-subgroup of G. ■

For the converse we shall use an interesting solubility criterion due to Helmut Wielandt:

Theorem 3.2.6. *Let G be a finite group. If G has three soluble subgroups whose indices are pairwise coprime, then G is soluble.*

Proof. Let the subgroups be H_1, H_2, H_3; if $H_1 = 1$, then $|G| = (G : H_1)$ is prime to $(G : H_2)$, so the latter is 1 and $H_2 = G$ is soluble. Hence we may assume that $H_1 \neq 1$. Let M be a minimal normal subgroup of H_1; since H_1 is soluble, M is an elementary abelian p-group, for some prime p. Now p cannot divide the indices of both H_2 and H_3, say it is prime to $(G : H_2)$. Then $p||H_2|$, hence H_2 contains a non-trivial Sylow p-subgroup P, which is also a Sylow p-subgroup of G. Let P_1 be a Sylow p-subgroup of H_1; then $P_1 \subseteq P^x$ for some $x \in G$. We may replace H_2 by H_2^x without affecting the hypothesis; then $P_1 \subseteq P$ and $M \subseteq P$, because $M \lhd H_1$. Thus we have $M \subseteq H_1 \cap H_2$. Now by Lemma 3.2.2, $G = H_1 H_2$; hence any $x \in G$ has the form $x = x_1 x_2, x_i \in H_i$; therefore $M^x = M^{x_2} \subseteq H_2$. Hence H_2 contains the normal closure of M in G: $K = M^G = \text{gp}\{M^y | y \in G\} \subseteq H_2$. Since H_2 is soluble, so is K. Further, the subgroups KH_i/K of G/K satisfy the hypothesis, so G/K is soluble, hence so is G. ■

We can now complete the proof of Hall's criterion; we recall that a p-complement is a subgroup of p-power index and order prime to p; in particular, if p does not divide the order of G, then G itself is the only p-complement.

Theorem 3.2.7(P. Hall [1937]). *Any finite group is soluble if and only if it contains a p-complement for each prime p.*

Proof. For soluble groups the result follows by Theorem 3.2.5. Now assume that $|G| = p_1^{\alpha_1} \dots p_r^{\alpha_r}$, where the p_i are distinct primes, and that G has a p_i-complement H_i for $i = 1, \dots, r$. If $r = 1$ or 2, G is soluble by Burnside's $p^\alpha q^\beta$-theorem (Theorem 6.8.3 below), hence we may assume $r \geq 3$. We claim that each H_i is soluble. Clearly $(G : H_i) = p_i^{\alpha_i}$, hence by Lemma 3.2.2, $H_1 \cap H_i$ is a Hall π_i-subgroup of G, where $\pi_i = \pi(G) \backslash \{p_1, p_i\}$, and it follows that $H_1 \cap H_i$ is a p_i-complement of H_1. By induction H_1 is soluble, similarly for $H_2, \dots, H_r$; now we apply Theorem 3.2.6 to complete the proof. ■

Exercises

1. By a *Hall system* in a finite group G is meant a family $\sum$ of Hall subgroups of G, one of order d for each divisor d of $|G|$ prime to $|G|/d$, such that each $H \in \sum$ is the intersection of all subgroups in $\sum$ whose orders are divisible by $|H|$. Verify that every family of p-complements, where p runs over the prime divisors of $|G|$, gives rise to a Hall system, and that for any $H, K \in \sum$, $HK = KH \in \sum$.

2. Find all Hall systems in Sym_4.
3. Show that Lemma 3.2.1 holds for any group G such that the subgroup H is of finite index. Extend Lemma 3.2.2 similarly.
4. Let G be a finite group and P_i a Sylow p_i-subgroup, for each prime divisor p_i of $|G|$ $(i = 1, \ldots, r)$. Show that if $P_iP_j = P_jP_i$ for $i, j = 1, \ldots, r$, then G is soluble and the different products of the P_i constitute a Hall system. Show that G is nilpotent iff it has a single Hall system.
5. Show that if G is a finite soluble (non-trivial) group, then for some prime p there is a non-trivial abelian normal p-subgroup of G. (Hint. Examine the last link in the derived series for G.)
6. Let G be a finite soluble group and p a prime divisor of $|G|$. Show that the number of p-complements in G is a power of p. Use Lemma 3.2.2 to deduce that all Hall systems in G are conjugate.
7. Let G be a finite soluble group. Show that every subgroup containing the normalizer of a Hall subgroup is its own normalizer.
8. By examining Sym_5 show that Theorem 3.2.6 does not remain true when the number of subgroups is reduced to 2.
9. Show that in any finite group each minimal normal subgroup is a direct product of isomorphic simple groups.

3.3 The transfer

Unlike rings, groups have only one binary operation, which makes it harder to form constructions like the determinant. Even the latter depends on the commutativity of the ring for its definition; later, in Section 9.2, we shall see that determinants can be defined over a skew field, to take values in an abelian group, and it turns out that there is an analogous construction for groups, the transfer, which we shall describe now, with some applications.

To define it, take any group G and a subgroup H of finite index, n say, and consider a fixed decomposition of G:

$$G = Hs_1 \cup Hs_2 \cup \ldots \cup Hs_n. \tag{3.3.1}$$

We can represent G by matrices as follows: Each $a \in G$ permutes the cosets in (3.3.1) by right multiplication, thus there is a permutation σ_a of $1, 2, \ldots, n$ such that $Hs_ia = Hs_{i\sigma_a}$. Hence $s_ias_{i\sigma_a}^{-1} \in H$ and we can define an $n \times n$ matrix $\mu(a) = (\mu_{ij}(a))$ with entries in H by the equation

$$\mu_{ij}(a) = \begin{cases} s_ias_{i\sigma_a}^{-1} & \text{if } j = i\sigma_a, \\ 0 & \text{otherwise.} \end{cases}$$

Thus $\mu(a)$ has a single non-zero entry in each row and one in each column; it is called a *monomial matrix* over H. To show that we have indeed a representation, take $a, b \in G$; we have $\sigma_{ab} = \sigma_a\sigma_b$ because σ is a permutation representation of G

on the cosets (3.3.1). Hence we have $s_ias_{i\sigma_a}^{-1}s_{i\sigma_a}bs_{i\sigma_a\sigma_b}^{-1} = s_i(ab)s_{i\sigma_{ab}}^{-1} \in H$ and it follows that

$$\mu_{ik}(ab) = \begin{cases} \mu_{ii\sigma_a}(a)\mu_{i\sigma_a i\sigma_{ab}}(b) & \text{if } k = i\sigma_{ab}, \\ 0 & \text{otherwise.} \end{cases}$$

Thus $\mu_{ik}(ab) = \sum_j \mu_{ij}(a)\mu_{jk}(b)$, and we have indeed a representation. Of course the representation still depends on the choice of transversal (s_i); to free ourselves of this dependence, consider a homomorphism $f : H \to A$ to an abelian group A (written multiplicatively) and put $\alpha_{ij}(a) = \mu_{ij}(a)f$. We claim that the mapping $V : G \to A$ defined by

$$aV = |\det \alpha(a)| = \prod_{i=1}^{n} (s_ias_{i\sigma_a}^{-1})f \tag{3.3.2}$$

is a homomorphism independent of the choice of transversal. For if $t_i = h_is_i$ $(h_i \in H)$ is another transversal and we form the corresponding expression, we obtain

$$\prod_{i=1}^{n} (h_is_ias_{i\sigma_a}^{-1}h_{i\sigma_a}^{-1})f = \prod_{i=1}^{n} (h_if)\prod_{i=1}^{n} (s_ias_{i\sigma_a}^{-1})f\,(\prod_{i=1}^{n} h_if)^{-1} = aV,$$

because σ is a permutation of $1, \ldots, n$. Now the homomorphism property follows from the multiplication law for determinants. This mapping V is called the *transfer* of G into A (German: Verlagerung). We remark that $\ker V \supseteq G'$, so that we can factor V by the natural map from G to $G^{ab} = G/G'$, to obtain a homomorphism from G^{ab} to A.

Let us examine the action of $a \in G$ on G/H more closely. Suppose that G/H consists of r orbits of n_i points $(i = 1, \ldots, r)$ under the action of $\mathrm{gp}\{a\}$, so that $\sum n_i = n = (G : H)$. If we take our transversal in the form t_ia^j $(i = 1, \ldots, r,$ $j = 0, 1, \ldots, n-1)$, then the action on the i-th orbit is represented by the cycle $(t_i, t_ia, \ldots, t_ia^{n_i-1})$, and its contribution to the transfer is $t_ia^{n_i}t_i^{-1}$. So we have

$$aV = \prod_{i=1}^{r} (t_ia^{n_i}t_i^{-1})f, \quad \text{where } t_ia^{n_i}t_i^{-1} \in H. \tag{3.3.3}$$

We shall apply the transfer to prove the existence of normal p-complements under suitable conditions. Here we need

Lemma 3.3.1. *Let G be a finite group, P a Sylow p-subgroup of G and X, Y two subsets of G normalized by P and conjugate in G. Then X and Y are conjugate in* $N_G(P)$.

Proof. By hypothesis $Y = X^b$ for some $b \in G$, and X, Y are normalized by P, hence $Y = X^b$ is normalized by P^b. Thus $N = N_G(Y)$ contains P and P^b; clearly they are Sylow subgroups of N and by Sylow's theorem they are conjugate in N, say $P^{bc} = P$ for $c \in N$. Writing $a = bc$, we have $a \in N_G(P)$, hence $X^a = X^{bc} = Y^c = Y$, because $c \in N$. ∎

Theorem 3.3.2 (Burnside). *Let G be a finite group and suppose that the Sylow p-subgroup P of G is contained in the centre of its normalizer. Then P has a normal complement in G.*

Proof. By hypothesis P is abelian, so we can take the transfer $V : G \to P$. For any $u \in P$ and n_i, t_i as in (3.3.3) (for $H = P$), u^{n_i} and $t_i u^{n_i} t_i^{-1}$ lie in P, are normalized by P and are conjugate in G. By the above lemma they are conjugate in $N_G(P)$, hence equal (because P lies in the centre of $N_G(P)$), so we obtain

$$t_i u^{n_i} t_i^{-1} = u^{n_i}.$$

It follows that $uV = u^n$, where $n = (G : P)$. Since p is prime to $(G : P)$ and P is an abelian p-group, it follows that $V : G \to P$ is surjective and the kernel is of index $|P|$, so it is the desired complement. ■

Corollary 3.3.3. *Let G be a finite non-abelian group with a non-trivial cyclic Sylow 2-subgroup. Then G has a normal 2-complement and so cannot be simple.*

Proof. Let P, of order 2^α, be a Sylow 2-subgroup of G. Since it is cyclic, its automorphism group has order $\varphi(2^\alpha) = 2^{\alpha-1}$, therefore $(N_G(P) : P)$ is a factor of 2^α, but it also divides $(G : P)$, which is odd, hence $N_G(P) = P$, so the hypothesis of Theorem 3.3.2 is satisfied and we obtain a normal complement of P. ■

Exercises

1. In the definition (3.3.2) the sign of the determinant was ignored. Show that taking the sign into account only amounts to taking the sign of the permutation representation of G on G/H.
2. Show that the corestriction mapping $H_1(G, \mathbf{Z}) \to H_1(H, \mathbf{Z})$ (induced by the restriction from G to H) is just the transfer (see also Section 5.6).
3. Show that if G has an abelian Sylow p-subgroup P with normalizer N, then $P \cap G' = P \cap N'$ and $P = (P \cap N') \times (P \cap Z(N))$. Show also that the maximal p-factor group of G (i.e. the maximal quotient group which is a p-group) is $\cong P \cap Z(N)$.
4. Show that if a Sylow p-subgroup P of G has trivial intersection with any of its distinct conjugates, then any two elements of P which are conjugate in G are conjugate in $N_G(P)$.
5. Let P, Q be two distinct Sylow p-subgroups of G, chosen so that (for fixed p) $P \cap Q$ has maximal order. Show that the only conjugates of $P \cap Q$ in G that are contained in P are conjugate to $P \cap Q$ in $N_G(P)$.

3.4 Free groups

Since groups form a class of algebras defined by laws (Section 1.3), it is clear what is to be understood by a free group on a given set. But a peculiarity of the laws defining groups allows the elements of a free group to be written in an easily recognized form, that we shall now describe.

Let X be any non-empty set. By a *group word* in X we understand an expression

$$u_1 u_2 \dots u_n, n \geq 0, \tag{3.4.1}$$

where each u is either an element x or x^{-1} for some $x \in X$. Formally we can regard the expressions (3.4.1) as the elements of the free monoid on $X \cup X^{-1}$, where we have put $X^{-1} = \{x^{-1} | x \in X\}$. For $n = 0$, (3.4.1) reduces to the empty word, written 1. We define an *elementary reduction* of a word $u_1 \dots u_n$ as the process of omitting a pair of neighbouring factors of the form xx^{-1} or $x^{-1}x$, for some $x \in X$. The inverse process, inserting a factor xx^{-1} or $x^{-1}x$ at some point, is an *elementary expansion*. A word is said to be *reduced* if it contains no pairs of neighbouring factors xx^{-1} or $x^{-1}x$; thus a word admits an elementary reduction precisely if it is not reduced. Two group words in X are said to be *equivalent* if we can transform one into the other by a series of elementary reductions and expansions.

It is clear that in any group G generated by a set X, two equivalent group words represent the same element. When there are relations in G between the group words in X, there will be inequivalent group words representing the same group element, but as we shall see, in a free group inequivalent groups words represent distinct group elements.

Theorem 3.4.1. *Let X be a set and F the free group on X. Then every element of F is represented by exactly one reduced group word in X and two group words represent the same element of F if and only if they are equivalent.*

Proof. If $X = \emptyset$, F is the trivial group consisting of 1 alone, so we may assume that X is not empty. It is clear from the definitions that every element of F is represented by a group word in X, and that equivalent group words represent the same element. The multiplication of group words by juxtaposition is associative, and it is easily checked that the equivalence class of a product depends only on those of the factors, not on the factors themselves. Further, the empty word 1 acts as neutral under multiplication. Hence the set of equivalence classes forms a monoid under multiplication, and this is in fact a group, since $u_1 u_2 \dots u_n$ has the inverse $u_n^{-1} \dots u_1^1$, where $u_i^{-1} = x^{-1}$ if $u_i = x$, $u_i^{-1} = x$ if $u_i = x^{-1}$:

$$u_1 \dots u_n u_n^{-1} \dots u_1^{-1} \sim u_1 \dots u_{n-1} u_{n-1}^{-1} \dots u_1^{-1} \sim \dots \sim 1,$$

by elementary reductions. It only remains to show that each group word is equivalent to exactly one reduced group word. Given a group word (3.4.1), we apply elementary reductions as often as possible; each such reduction reduces the length, so we arrive at a reduced form after a finite number of steps. In order to show that this form is independent of the order in which the reductions are made, we shall use the diamond lemma (Lemma 1.4.1). We have to show that if a word f is reduced to g_1, g_2 by different elementary reductions, then there is a word h which can be obtained by reduction from g_1 as well as g_2. There are two cases: (i) The terms being reduced do not overlap, say $u_i u_{i+1} = xx^{-1}$, $u_j u_{j+1} = yy^{-1}$, where $j > i + 1$ and $x, y \in X \cup X^{-1}$, and of course $(x^{-1})^{-1} = x$. Clearly these reductions can be performed in either order and the outcome will be the same. (ii) The terms overlap; then a subword $uu^{-1}u$ or $u^{-1}uu^{-1}$ occurs. Taking $uu^{-1}u$, we can either reduce uu^{-1} to 1 or $u^{-1}u$ to 1, and we are left with u in each case, so again the outcome is the same, and similar reasoning applies to $u^{-1}uu^{-1}$. Now it follows by Lemma 1.4.1 that we have a unique reduced form. ■

Here is another proof, not using the diamond lemma. The essential step is to show that distinct reduced words represent distinct group elements. We define F as a permutation group on the set W of all reduced group words as follows. Given $w = u_1 \dots u_n \in W$ and $x \in X$, we put

$$w\alpha_x = \begin{cases} u_1 \dots u_n x & \text{if } u_n \neq x^{-1} \text{ or } n = 0, \\ u_1 \dots u_{n-1} & \text{if } u_n = x^{-1}. \end{cases}$$

$$w\beta_x = \begin{cases} u_1 \dots u_n x^{-1} & \text{if } u_n \neq x \text{ or } n = 0, \\ u_1 \dots u_{n-1} & \text{if } u_n = x. \end{cases}$$

It is easily checked that α_x is a permutation of W with inverse β_x. Thus F has been defined as a permutation group on W. Now, given any reduced word $u_1 \dots u_n$, if we apply $\alpha_{u_1} \dots \alpha_{u_n}$ (where $\alpha_{x^{-1}} = \beta_x$), to the empty word, we find $u_1 \dots u_n$, hence distinct words define distinct permutations of W and so represent different elements of F. ■

A free generating set in a free group will also be called a *basis*. Let F be the free group with basis $X = \{x_1, \dots, x_d\}$ and denote by F^2 the subgroup generated by all squares. Then F/F^2 is the elementary abelian 2-group generated by the images of $x_1, \dots, x_d$, and hence is of order 2^d. In particular this shows that the number d of elements in a basis is independent of the choice of basis. It is called the *rank* of F, written rk F, and the above argument shows that free groups of different ranks cannot be isomorphic (see also Further Exercise 1 of Chapter 1).

We remark that by the results of Section 1.3 every group G can be written as a homomorphic image of a free group; more precisely, if G can be generated by d elements, then it can be expressed as a quotient of a free group of rank d.

We note that Theorem 3.4.1 solves the word problem for free groups, for it provides a means of deciding when two group words (in a given free generating set) represent the same group element. Let us now look at the conjugacy problem and show that in free groups this can be solved in a similar manner.

Two group words f, g are said to be *cyclic conjugates* if $f = uv$, $g = vu$ for suitable words u, v. Thus $xy^{-1}z^{-1}xy$ and $xyxy^{-1}z^{-1}$ are cyclic conjugates. We remark that even when a word is in reduced form, it may have a cyclic conjugate which is not reduced, e.g. $x^{-1}yx$. A group word is said to be *cyclically reduced* if all its cyclic conjugates are reduced. Now we have

Proposition 3.4.2. *Let F be a free group on a set X. Then every element of F has a cyclically reduced conjugate and two elements of F are conjugate if and only if their cyclically reduced forms are cyclic conjugates.*

It is clear that one can check in a finite number of steps whether two reduced words are cyclic conjugates; for words of length n we need only compare n different words.

Proof. Let $f, g \in F$; if their reduced forms are cyclic conjugates, then $f = uv$, $g = vu$ and hence $g = u^{-1}fu$. Conversely, suppose that f, g are conjugate. By passing to appropriate conjugates we may suppose that both f and g are in cyclically reduced form. Let

$$g = u^{-1}fu, \tag{3.4.2}$$

where u is in reduced form. Since g is cyclically reduced, the right-hand side cannot be in reduced form, say u and f begin with the same letter x. Then f cannot end in x^{-1} (because f is cyclically reduced), so no cancellation can take place between f and u, and the only way to reduce the right-hand side of (3.4.2) is by cancelling an initial portion of u against that of f. So we have either $u = u_1u_2$, $f = u_1f_1$, hence $u^{-1}fu = u_2^{-1}f_2u_1u_2$; this is reduced and so must equal g; but the latter is cyclically reduced, so $u_2 = 1$, $u_1 = u$ and $f = uf_1$, $g = f_1u$, showing f, g to be cyclic conjugates. Or we have $u = fu_2$; then $u^{-1}fu = u_2^{-1}fu_2$ and the argument shows as before that $u_2 = 1$; now it follows again that f, g are cyclically conjugate. ■

We conclude this section by proving the Schreier subgroup theorem, first proved by Otto Schreier in 1927; this is an important result in its own right, while the proof illustrates some of the methods used in the study of free groups.

Given any group G with generating set X and a G-set M, we define its *diagram* (G, X, M) *as the graph whose vertices are the points of* M, with an edge from p to q whenever $q = px$ for some $x \in X$. More precisely, the diagram may be regarded as a directed graph (digraph), in which each $x \in X$ is represented by a directed edge and x^{-1} by its opposite.

For example, the symmetric group Sym_3 with generating set $\{(12), (23)\}$ acting on the set $\{1, 2, 3\}$ has the diagram shown below; the second diagram represents Sym_3 acting on the set of all arrangements of 1, 2, 3. In both cases continuous lines represent (1 2) and broken lines (2 3).

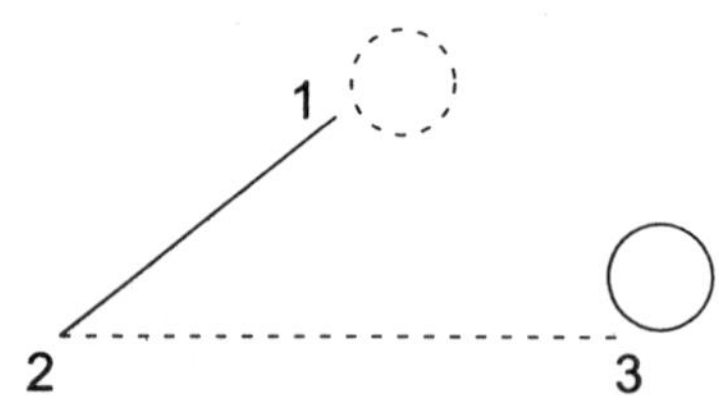

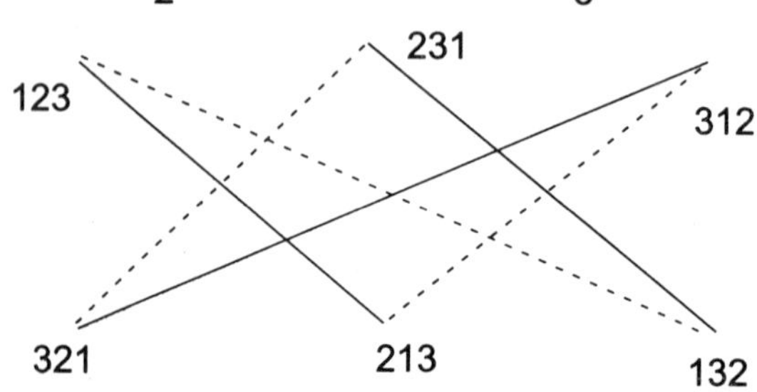

We note that the second diagram may be regarded as the diagram of the regular representation (G, X, G); such a diagram contains all the information contained in the multiplication table, but in a more accessible form. For example, we can from the diagram read off the relation $[(1\,2)(2\,3)]^3 = 1$, corresponding to a circuit based at 123.

It is easily seen that the diagram of (G, X, M) has a connected graph iff G acts transitively on M. When this is so, we can take M to consist of the cosets in G of a stabilizer of a point, and from this point of view a diagram may be called a *coset diagram*. Any path in such a coset diagram $(G, X, G/H)$ represents an element of G applied to a certain coset; in particular any element of H may be represented by a circuit beginning and ending at the coset H. Since the graph is connected, it contains a subgraph on all the vertices which is a tree (BA, Theorem 1.3.3), i.e. a connected graph without circuits. Such a tree including all the vertices of our graph is also called a *spanning tree* for the graph. Let us express this in terms of our group. Given any coset Hu, there is a path in the coset diagram from H to Hu (where the orientation in forming the path is disregarded). Each such path corresponds to a group word w on X such that $Hw = Hu$. Now the choice of spanning tree in our graph amounts to the choice of a particular representative $w = w_1 \dots w_r$ ($w_i \in X \cup X^{-1}$) for our coset. It has the property that any left factor $w_1 \dots w_i$ of w is the chosen representative of this coset. A transversal with this property is called a *Schreier transversal*. So the choice of a spanning tree amounts to choosing a Schreier transversal, and it is not hard to verify that any Schreier transversal in turn leads to a spanning tree. Therefore the existence of a spanning tree assures us that every subgroup of G has a Schreier transversal, relative to the given generating set of G. We note that neither X nor $(G : H)$ need be finite for the spanning tree to exist.

In any group G with generating set X and a subgroup H consider a coset diagram $(G, X, G/H)$ and choose a spanning tree Γ. Any edge α in our diagram, from p to q say, gives rise to a circuit as follows. Denote by p_0 the vertex corresponding to the coset H. There is a unique path w_1 within Γ from p_0 to p, and a unique path w_2 from q to p_0; hence $w_1\alpha w_2$ is a circuit on p_0. Here we have $w_2 = (w_1\alpha)^{-1}$ precisely if $\alpha \in \Gamma$, so our circuit is trivial (reduced to p_0) precisely if $\alpha \in \Gamma$. If $\alpha \notin \Gamma$, so that the circuit is non-trivial, it will be called a *basic* circuit, corresponding to α. As we saw earlier, any circuit on p_0 corresponds to an element of H. We now observe that any circuit on p_0 can be written as a product of basic circuits. For, given edges $\alpha_1, \dots, \alpha_r$ forming a circuit on p_0, suppose that α_i goes from p_{i-1} to p_i, where $p_r = p_0$, and let w_i be the path within Γ from p_0 to p_i; in particular, w_0 is the empty path.

Then we have

$$\alpha_1 \dots \alpha_r = \alpha_1 w_1^{-1} w_1 \alpha_2 w_2^{-1} \dots w_{r-1}^{-1} w_{r-1} \alpha_r. \tag{3.4.3}$$

Here $w_{i-1}\alpha_i w_i^{-1}$ either is trivial and so can be omitted, or it is basic. This expresses our circuit as a product of basic circuits, and it shows that H is generated by the elements corresponding to basic circuits. Thus we have

Proposition 3.4.3. *Let G be a finitely generated group. Then any subgroup of finite index is finitely generated; more precisely, if G has a generating set of d elements and $(G : H) = m$, then H can be generated by $m(d-1)+1$ elements.*

It only remains to prove the last part. The coset diagram $(G, X, G/H)$ has m vertices and dm edges; by BA, Theorem 1.3.3, a spanning tree has $m-1$ edges. We saw that H has a generating set whose elements correspond to the non-tree edges, and their number is $dm - (m-1) = m(d-1)+1$. ∎

By a slight refinement of the argument we obtain Schreier's subgroup theorem:

Theorem 3.4.4. *Any subgroup of a free group is free. If F is free of finite rank d and H is a subgroup of finite index m in F, then H has rank given by*

$$\text{rk}\, H = (F : H)(\text{rk}\, F - 1) + 1. \tag{3.4.4}$$

Equation (3.4.4) is known as *Schreier's formula* (see also Section 11.5 below).

Proof. We take the coset diagram of $(F, X, F/H)$ and choose a spanning tree Γ. As we saw, H is generated by the elements corresponding to basic circuits; our aim is to show that these elements generate H freely. For each non-tree edge α we have a basic circuit $\bar{\alpha}$ and it will be enough to show that if $\alpha_1 \dots \alpha_r$ is a reduced word $\neq 1$ in the non-tree edges, then

$$\bar{\alpha}_1 \dots \bar{\alpha}_r \neq 1, \tag{3.4.5}$$

where $\bar{\alpha}_i = w_{i-1}\alpha_i w_i^{-1}$ as before. If we write out the left-hand side of (3.4.5) as a word in X, we find that although there may be cancellation, this cannot involve the α_i; they cannot cancel against any part of any w_j because the α_i are not in Γ, and they cannot cancel against each other because the original word $\alpha_1 \dots \alpha_r$ was reduced. Hence (3.4.5) follows and this shows the basic circuits to be a free generating set of H. Now (3.4.4) follows as before by counting the non-tree edges. ∎

Finally we shall prove that free groups are residually finite p-groups, for any prime p. We recall from Section 1.2 that this means: given any element $c \neq 1$ in a free group F, there exists a normal subgroup N of F not containing c, such that F/N is a finite p-group. This will also show that for any prime p, F can be expressed as a subdirect product of finite p-groups.

Let F be the free group on $x_1, \dots, x_d$ and consider a non-trivial word

$$x_{i_1}^{\alpha_1} \dots x_{i_r}^{\alpha_r}, \quad i_\rho \neq i_{\rho+1}, \quad \alpha_\rho \neq 0, \quad r \geq 1. \tag{3.4.6}$$

Let m be a positive integer so large that p^m does not divide $\alpha_1 \dots \alpha_r$ and take G to be the group of all upper unitriangular matrices of order $r+1$ over $\mathbf{Z}/p^m$, i.e. matrices of the form $I + N$, where $N = (n_{ij})$, $n_{ij} = 0$ for $i \geq j$. It follows that $N^{r+1} = 0$, hence $(I+N)^{p^t} = I + N^{p^t} = I$ for $p^t \geq r+1$; this shows G to be a finite p-group. Our object will be to find a homomorphism $F \to G$ such that (3.4.6) is not mapped to 1. We map x_i to A_i, where

$$A_i = \prod_k (I + e_{kk+1}),$$

where the product is taken over all k such that $i_k = i$. Then we have $e_k A_{i_k} = e_k + e_{k+1}$, where the e are unit row vectors; hence

$$e_1 A_{i_1}^{\alpha_1} \dots A_{i_r}^{\alpha_r} = e_1 + \alpha_1 \dots \alpha_r e_{r+1} + \dots,$$

where the final dots indicate terms in $e_2, \dots, e_r$. We have thus found a homomorphism of the required form; we remark that G does not depend on the rank d of F, and this may be taken to be infinite. We thus obtain

Theorem 3.4.5. *The free group (of any rank) is residually a finite p-group, for any prime p.* ∎

An interesting consequence is due to Wilhelm Magnus. We recall that for any group G, the *lower central series* is defined recursively as $\gamma_i(G) = (G, \gamma_{i-1}(G))$, $\gamma_1(G) = G$, where (G, H) denotes the subgroup generated by all commutators $(x, y) = x^{-1}y^{-1}xy$ $(x \in G, y \in H)$.

Corollary 3.4.6 (W. Magnus). *In a free group F,* $\cap_n \gamma_n(F) = 1$.

Proof. Suppose that $\cap_n \gamma_n(F) \neq 1$ and take a word $c \neq 1$ in the intersection. By Theorem 3.4.5 there is a normal subgroup N not containing c such that F/N is a finite p-group (for some p). Since F/N is nilpotent, of class h say, we have $\gamma_{h+1}(F) \subseteq N$ and so $c \notin \gamma_{h+1}(F)$, which is a contradiction. ∎

For another proof of this corollary see Further Exercise 15.

Exercises

1. Give a direct proof that every abelian subgroup of a free group is cyclic.
2. In a free group, if $w = u^n$, where $n \geq 1$, u is called a *root* of w, *primitive* if n is maximal for a given w. Show that every element of a free group has a unique primitive root. Show that two elements $u, v \neq 1$ have a common root or inverse roots iff they commute.
3. Let F be a free group. Show that every subgroup of finite index meets every subgroup $\neq 1$ non-trivially.
4. Define a group G to be *projective* (for this exercise only) if any homomorphism $G \to H_1$, where H_1 is a quotient of a group H, can be lifted to a homomorphism $G \to H$. Show that a group is projective iff it is free. (Note that this allows one to define free groups without reference to a basis.)
5. Let G be any group and X a subset such that $X \cap X^{-1} = \varnothing$ and any non-empty product of elements of $X \cup X^{-1}$ with no factors xx^{-1} or $x^{-1}x$ is $\neq 1$. Show that X is a free generating set of the subgroup generated by it.
6. Show that a free group of rank d cannot be generated by fewer than d elements.
7. A group is called *Hopfian* (after Heinz Hopf) if every surjective endomorphism of G is an automorphism. Show that every free group of finite rank is Hopfian.
8. Show that in a free group of rank 3 any subgroup of finite index has odd rank.

9. In the free group on x, y, z find a basis for the subgroup generated by all the squares.
10. Show that in the free group on x, y, the elements $y^{-n}xy^n$ form a free generating set. Deduce that any non-cyclic free group contains a free subgroup of countable rank.
11. Show that in any free group F of rank > 1, the derived group F' has countable rank.
12. Let G be a group with generating set X and H a subgroup. Show how to construct a Schreier transversal and verify that it corresponds to a spanning tree of the graph $(G, X, G/H)$.

3.5 Linear groups

In Section 3.1 we saw how in principle at least every finite group can be built up from simple groups. This leaves the problem of determining the finite simple groups, and it was only around 1980 that the classification of the simple finite groups was completed. The full list contains 18 families (including the cyclic groups of prime order) and another 26 'sporadic' groups. This is not the place to give a detailed account (see Gorenstein (1982)), but we shall present some of the families that have been known since the work of Leonard Eugene Dickson (1901). It is well known that the alternating group Alt_n is simple for $n \geq 5$. In this section we shall describe the linear groups and in the next two sections we deal with the symplectic and orthogonal groups, collectively usually known under the name of *classical groups* since the time of Hermann Weyl (1939). These families of groups can be shown to correspond to certain infinite families of simple Lie algebras, though we shall not do so here. It was Chevalley's paper [1955], describing how these families of Lie algebras could also be defined over finite fields, that provided the impetus for research in the 1960s and 1970s that led to the classification of the simple finite groups.

Let k be any field and V an n-dimensional vector space over k. The automorphisms of V form a group whose members may be described, after the choice of a basis in V, by invertible $n \times n$ matrices over k. This is the *general linear group* of degree n over k, written $\mathbf{GL}(V)$ or $\mathbf{GL}_n(k)$. The determinant function provides a homomorphism from $\mathbf{GL}_n(k)$ to $k^\times$ whose kernel $\mathbf{SL}_n(k)$ is the *special linear group*, consisting of all $n \times n$ matrices over k with determinant 1. We remark that the general linear group may be defined more generally over any ring R as $\mathbf{GL}_n(R)$, the set of all invertible $n \times n$ matrices over R. When R is commutative we also find the subgroup $\mathbf{SL}_n(R)$ as before, but this definition breaks down for more general rings. Later, in Chapter 9, we shall find a way of defining $\mathbf{SL}_n(K)$ for a skew field K, but for the present we shall mainly be concerned with a commutative field k as ground ring.

We begin by finding generating sets for $\mathbf{GL}_n(k)$ and $\mathbf{SL}_n(k)$. We denote by E_{ij} the matrix with (i, j)-entry 1 and all others zero. By an *elementary matrix* we understand a matrix of the form

$$B_{ij}(a) = I + aE_{ij}, \quad \text{where } a \in k, i \neq j.$$

It is clear that this matrix lies in $\mathbf{SL}_n(k)$, with inverse $B_{ij}(-a)$. A diagonal matrix $\sum a_i E_{ii}$ lies in $\mathbf{GL}_n(k)$ iff $\prod a_i \neq 0$, while $\prod a_i = 1$ is the condition for it to belong to $\mathbf{SL}_n(k)$.

Taking V again as our vector space over k, let $u \neq 0$ be a vector in V. By a *transvection* along u we understand a linear mapping T of V keeping u fixed and such that $xT - x$ is in the direction of u. Thus we can express T as

$$xT = x + \lambda(x)u,$$

where λ is a linear functional on V such that $\lambda(u) = 0$. For example, with the standard basis $e_1, \ldots, e_n$ in k^n, $B_{1n}(a)$ maps $x = \sum x_i e_i$ to $x + ax_1 e_n$ and so is a transvection along e_n (for $n > 1$). Given a transvection T along u, let us choose a basis $u_1, \ldots, u_n$ of T such that $u_1 = u$ while $u_2, \ldots, u_n$ forms a basis for the kernel of $T - I$. Then T takes the form $xT = x + \lambda(x)u_1$, so T is represented by an elementary matrix relative to this basis.

For our first result we take our ring to be a Euclidean domain; we recall (from BA, Section 10.2) that a *Euclidean domain* is an integral domain in which the Euclidean algorithm holds (relative to a norm function).

Theorem 3.5.1. *For any (commutative) Euclidean domain R and any $n \geq 1$, the special linear group $\mathbf{SL}_n(R)$ is generated by all elementary matrices and the general linear group $\mathbf{GL}_n(R)$ by all elementary and diagonal matrices with units along the main diagonal. In particular, this holds for any field.*

Proof. The result is clear for $n = 1$, so assume $n > 1$. Let $A \in \mathbf{SL}_n(R)$; it is clear that right multiplication by $B_{ij}(c)$ corresponds to the operation of adding the i-th column, multiplied by c, to the j-th column. Now the Euclidean algorithm shows that multiplication on the right alternatively by $B_{12}(p)$ and $B_{21}(q)$ for suitable $p, q \in R$ reduces a_{11}, a_{12} to d, 0 respectively, where d is an HCF of a_{11} and a_{12}. By operating similarly on other columns we reduce all elements of the first row after the first one to zero. We next repeat this process on the rows, to reduce all elements in the first column after the first one to zero, by multiplying by suitable elementary matrices on the left. This either leaves the top row unchanged or it replaces the $(1, 1)$-entry by a proper factor. By an induction on the number of factors we reduce A to the form $a \oplus A_1$, where A_1 is $(n-1) \times (n-1)$, and another induction, this time on n, reduces A to diagonal form. Using the factorization

$$\begin{pmatrix} c & 0 \\ 0 & c^{-1} \end{pmatrix} = \begin{pmatrix} 1 & 0 \\ c^{-1} & 1 \end{pmatrix} \begin{pmatrix} 1 & -c \\ 0 & 1 \end{pmatrix} \begin{pmatrix} 1 & 0 \\ c^{-1} & 1 \end{pmatrix} \begin{pmatrix} 1 & 0 \\ -1 & 1 \end{pmatrix} \begin{pmatrix} 1 & 1 \\ 0 & 1 \end{pmatrix} \begin{pmatrix} 1 & 0 \\ -1 & 1 \end{pmatrix},$$

we can express $\mathrm{diag}(a_1, a_2, \ldots, a_n)$ by $\mathrm{diag}(1, a_1 a_2, a_3, \ldots, a_n)$ and a product of elementary matrices, and hence by another induction express A as a product of elementary matrices.

The same process applied to a matrix of $\mathbf{GL}_n(R)$ reduces it to a diagonal matrix and this proves the second assertion. ∎

For any commutative ring R the determinant may be regarded as a homomorphism from $\mathbf{GL}_n(R)$ to $\mathbf{U}(R)$ and it yields the exact sequence

$$1 \to \mathbf{SL}_n(R) \to \mathbf{GL}_n(R) \xrightarrow{\det} \mathbf{U}(R) \to 1. \tag{3.5.1}$$

Let us recall that a group G is called *perfect* if it coincides with its derived group G'. Since R is commutative, it follows that $\mathbf{GL}_n(R)' \subseteq \mathbf{SL}_n(R)$; for fields we have the following more precise relation:

Proposition 3.5.2. *Let k be any field and $n \geq 2$. Then*

$$\mathbf{SL}_n(k)' = \mathbf{GL}_n(k)' = \mathbf{SL}_n(k), \tag{3.5.2}$$

except when $n = 2$ and k consists of 2 or 3 elements. Thus $\mathbf{SL}_n(k)$ is perfect (with the exceptions listed).

Proof. As we have remarked, we have $\mathbf{SL}_n(k)' \subseteq \mathbf{GL}_n(k)' \subseteq \mathbf{SL}_n(k)$. To establish equality it is enough, by Theorem 3.5.1, to show that every elementary matrix is a product of commutators. It is easily checked that for any distinct indices i, j, k,

$$(B_{ik}(a), B_{kj}(1)) = B_{ik}(-a)B_{kj}(-1)B_{ik}(a)B_{kj}(1) = B_{ij}(a).$$

This expresses every elementary matrix as a commutator when $n \geq 3$. For $n = 2$ we have

$$\left(\begin{pmatrix} b^{-1} & 0 \\ 0 & b \end{pmatrix}, \begin{pmatrix} 1 & c \\ 0 & 1 \end{pmatrix}\right) = \begin{pmatrix} 1 & (1-b^2)c \\ 0 & 1 \end{pmatrix}; \tag{3.5.3}$$

hence if k contains an element b such that $b \neq 0$, $b^2 \neq 1$, then we can express any elementary matrix $B_{12}(a)$ as a commutator by taking $c = (1-b^2)^{-1}a$ in (3.5.3), and similarly for $B_{21}(a)$. This shows that $\mathbf{SL}_n(k)' = \mathbf{SL}_n(k)$ except when $n = 2$ and $b^3 = b$ for all $b \in k$, and this happens only when $|k| = 2$ or 3. ■

We shall see that $\mathbf{SL}_2(\mathbf{F}_2)$ and $\mathbf{SL}_2(\mathbf{F}_3)$ are soluble (see Exercise 1), so the exceptions made in Proposition 3.5.2 do in fact occur.

Our next aim is to show that $\mathbf{SL}_n(k)$ modulo its centre is simple (except when $n = 2$ and $|k| \leq 3$). By taking k to be finite we thus obtain a family of finite simple groups. The centre of $\mathbf{GL}_n(k)$ or $\mathbf{SL}_n(k)$ clearly consists of all scalar matrices and the quotients are known as the *projective* linear groups, written

$$\mathbf{PGL}_n(k) = \mathbf{GL}_n(k)/Z, \quad \mathbf{PSL}_n(k) = \mathbf{SL}_n(k)/(Z \cap \mathbf{SL}_n(k)),$$

where Z is the centre of $\mathbf{GL}_n(k)$. Let us define *projective n-space* $\mathbf{P}^n(k)$ as the set of all $(n+1)$-tuples $x = (x_0, x_1, \ldots, x_n) \in k^{n+1}$, where the x_i are not all 0 and $x = y$ iff $x_i = \lambda y_i$ for some $\lambda \in k^\times$. Thus the points of $\mathbf{P}^n(k)$ are given by the ratios of $n+1$ coordinates. It is clear that $\mathbf{PGL}_{n+1}(k)$ and $\mathbf{PSL}_{n+1}(k)$ act in a natural way on the points of $\mathbf{P}^n(k)$.

To prove the simplicity of $\mathbf{PSL}_n$ (following Iwasawa) we shall need to recall the notion of primitivity and derive some of its properties. We recall that a permutation group G acting on a set S is *transitive* if for any $p, q \in S$ there exists $g \in G$

such that $pg = q$, thus S consists of a single G-orbit. The *stabilizer* of $p \in S$ is $\mathrm{St}_p = \{x \in G | px = p\}$, a subgroup of G. When G acts on S, it acts in a natural way on S^n for $n \geq 1$; if this action is transitive, G is said to be *n-fold* transitive. The G-action on S is called *primitive* if G acts transitively and there is no partition of S (other than the trivial ones consisting of S alone and of the 1-element subsets) which is compatible with the G-action. If T is another G-set with a map $f : S \to T$ such that $(pg)f = (pf)g$ $(p \in S, g \in G)$, i.e. f is compatible with G, then it is easily seen that the action on S is primitive iff there is no compatible map from S onto a G-set with more than one element, which is not injective. It is also not hard to show that G is imprimitive precisely when S contains a proper subset S_0 with more than one element such that for each $g \in G$ either $S_0 g = S_0$ or $S_0 g \cap S_0 = \emptyset$. For example, any 2-fold transitive group is primitive, for if $p, q \in S_0$ then there exists $g \in G$ with $pg \in S_0$, $qg \notin S_0$, hence $S_0 g$ meets S_0 but is not equal to it.

It follows that for $n \geq 2$ the action of $\mathbf{PSL}_n(k)$ on $\mathbf{P}^{n-1}(k)$ is primitive, since it is doubly transitive. To establish this result, take two distinct pairs of points in $P^{n-1}(k), p_i, q_i$ $(i = 1, 2)$ with coordinate vectors x_i, y_i respectively, so that x_1, x_2 are linearly independent, as well as y_1, y_2; then there exists $A \in \mathbf{SL}_n(k)$ such that $x_1 A = y_1, x_2 A = cy_2$ for some $c \in k^\times$, hence A maps p_i to q_i for $i = 1, 2$. By the earlier remark it follows that $\mathbf{PSL}_n(k)$ is primitive.

We shall need a couple of lemmas giving conditions for primitivity:

Lemma 3.5.3. *Let G be a permutation group acting on a set S and denote the stabilizer of $p \in S$ by St_p. If G is transitive, then it is primitive if and only if St_p is a maximal proper subgroup of G.*

Proof. Assume G to be imprimitive and let $f : S \to T$ be a compatible mapping onto another G-set which is neither injective nor a constant mapping. Take $p, q \in S$ such that $pf = qf$ and denote the stabilizer of pf by L. If $pg = p$, then $(pf)g = (pg)f = pf$, hence $\mathrm{St}_p \subseteq L \subseteq G$. By transitivity there exists $h \in G$ such that $ph = q$, hence $(pf)h = (ph)f = qf = pf$. This shows that $h \in L$, but $h \notin \mathrm{St}_p$, so $\mathrm{St}_p \subset L$ and L is a proper subgroup of G, because f is non-constant. Thus St_p is not maximal.

Conversely, if $\mathrm{St}_p \subset K \subset G$ for some subgroup K of G, we may regard S as the coset space $\cup \mathrm{St}_p x$ and the mapping $\mathrm{St}_p x \mapsto Kx$ is compatible and neither constant nor injective, hence G cannot be primitive. ∎

Lemma 3.5.4. *Let G be a primitive permutation group acting on a set S. Then any non-trivial normal subgroup N acts transitively and $G = \mathrm{St}_p.N$ for the stabilizer St_p of any point p of S.*

Proof. Consider the orbits under the action of N. For any $p \in S$, $g \in G$, we have $pNg = pgN$, and this equals pN if $pg = p$ and otherwise is disjoint. By primitivity it follows that $pN = S$. Now fix $p \in S$ and take any $g \in G$; then $pg = pu$ for some $u \in N$, hence $gu \in \mathrm{St}_p$ and so $g \in \mathrm{St}_p.N$ as claimed. ∎

This leads to a criterion for simplicity:

Proposition 3.5.5. *Let G be a primitive permutation group acting on S. If for some $p \in S$ the stabilizer St_p contains an abelian subgroup A normal in St_p such that G is generated by all the conjugates $g^{-1}Ag$ $(g \in G)$, then any non-trivial normal subgroup of G contains the derived group G'. In particular, if G is perfect, then it must be simple.*

Proof. Let $N \neq 1$ be a normal subgroup of G. By Lemma 3.5.4, N is transitive and $G = \mathrm{St}_p.N$, where $p \in S$. We claim that AN is normal in G. Let $a \in A$, $n \in N$; any $g \in G$ has the form bm, where $b \in \mathrm{St}_p$, $m \in N$, and $(bm)^{-1}anbm = m^{-1}b^{-1}anbm = m^{-1}a_1b^{-1}nbm = ma_1n_1m = a_1m_1n_1m$ for some $a_1 \in A, m_1$, $n_1 \in N$. Hence AN is normal in G and so contains all the conjugates $g^{-1}Ag$. It follows that $AN = G$; now $G/N \cong A/(A \cap N)$ is abelian, therefore $N \supseteq G'$ and the conclusion follows. ■

We shall use these results to prove that $\mathbf{PSL}_n(k)$ is simple, taking the action of $\mathbf{PSL}_n(k)$ on $\mathbf{P}^{n-1}(k)$.

Theorem 3.5.6. *Let k be a field and $n \geq 2$. Then $\mathbf{PSL}_n(k)$ is simple except when $n = 2$ and $|k| \leq 3$.*

Proof. As we have seen, the action of $G = \mathbf{PSL}_n(k)$ on $\mathbf{P}^{n-1}(k)$ is primitive; moreover, G is perfect, for if $H = \mathbf{SL}_n(k)$ and Z denotes the centre of H, then $G = H/Z$, hence by Proposition 3.5.2, $G' = H'Z/Z = H/Z = G$. To complete the proof we shall verify the conditions of Proposition 3.5.5. Let p be the point with coordinates $(1, 0, \dots, 0)$. This is left fixed by any matrix

$$\begin{pmatrix} a & 0 \\ b & A \end{pmatrix} \text{ with inverse } \begin{pmatrix} a^{-1} & 0 \\ -A^{-1}ba^{-1} & A^{-1} \end{pmatrix},$$

$$\text{where } a \in k^{\times}, b \in {}^{n-1}k, A \in \mathbf{GL}_{n-1}(k). \tag{3.5.4}$$

Consider the subgroup L consisting of all matrices

$$\begin{pmatrix} 1 & 0 \\ c & I \end{pmatrix}, \quad \text{where } c \in {}^{n-1}k. \tag{3.5.5}$$

Clearly this is an abelian subgroup, and transforming by (3.5.4), we find

$$\begin{pmatrix} a^{-1} & 0 \\ -A^{-1}ba^{-1} & A^{-1} \end{pmatrix}\begin{pmatrix} 1 & 0 \\ c & I \end{pmatrix}\begin{pmatrix} a & 0 \\ b & A \end{pmatrix} = \begin{pmatrix} 1 & 0 \\ A^{-1}ca & I \end{pmatrix}.$$

Thus L is a normal abelian subgroup of St_p. Clearly every elementary matrix is conjugate to a matrix in L, so by Theorem 3.5.1, L and its conjugates generate G. Thus G satisfies the conditions of Proposition 3.5.5, and it is perfect, therefore it is simple. ■

Let us determine the order of $\mathbf{PSL}_n(k)$ when k is finite, with q elements say. We begin with $\mathbf{GL}_n(k)$; this group acts on $V = k^n$, a vector space with q^n elements. Any element of $\mathbf{GL}_n(k)$ is completely determined by the image of the standard basis $e_1, e_2, \dots, e_n$ of V, and this image can be any basis of V. Thus e_1 can map to any non-zero vector, giving $q^n - 1$ choices, e_2 can map to any vector independent of the image of e_1, giving $q^n - q$ choices, e_3 can map to any vector not a linear combination of the images of e_1, e_2, giving $q^n - q^2$ choices, and so on. In this way we find that

$$|\mathbf{GL}_n(k)| = (q^n - 1)(q^n - q)\ldots(q^n - q^{n-1}), \quad \text{where } |k| = q.$$

To determine $|\mathbf{SL}_n(k)|$ we use the exact sequence (3.5.1) and find

$$|\mathbf{SL}_n(k)| = (q^n - 1)(q^n - q)\ldots(q^n - q^{n-2})q^{n-1}.$$

To find the order of $\mathbf{PSL}_n(k)$ we need to calculate the order of its centre Z. We recall that Z consists of all matrices cI such that $c^n = 1$, so we need to find the number of solutions of $c^n = 1$ in k. But k is a cyclic group of order $q - 1$, so the number we want is $d = (n, q - 1)$. Thus

$$|\mathbf{PSL}_n(k)| = (q^n - 1)(q^n - q)\ldots(q^n - q^{n-2})q^{n-1}/d, \quad \text{where } d = (n, q - 1).$$

Exercises

1. By examining the action of $\mathbf{PSL}_2(k)$ on $\mathbf{P}^1(k)$ show that $\mathbf{PSL}_2(\mathbf{F}_2) \cong \mathrm{Sym}_3$, $\mathbf{PSL}_2(\mathbf{F}_3) \cong \mathrm{Alt}_3$. Show also that $\mathbf{GL}_2(\mathbf{F}_2) \cong \mathrm{Sym}_3$.
2. Show that $\mathbf{PSL}_2(\mathbf{F}_4) \cong \mathbf{PSL}_2(\mathbf{F}_5) \cong \mathrm{Alt}_5$.
3. Show that $\mathbf{PSL}_2(\mathbf{F}_9) \cong \mathrm{Alt}_6$, $\mathbf{PSL}_4(\mathbf{F}_2) \cong \mathrm{Alt}_8$.
4. Apply Proposition 3.5.5 to show that Alt_5 is simple.
5. How much remains true of Theorem 3.5.1 when commutativity is dropped?
6. Show that $\mathbf{PSL}_3(\mathbf{F}_4)$ and $\mathbf{PSL}_4(\mathbf{F}_2)$ have the same order but are not isomorphic. (Hint. Compare the Sylow 2-subgroups.)

3.6 The symplectic group

Another class of simple groups is formed by the symplectic groups. We recall that a symplectic space is a vector space V over a field k with a regular bilinear form b which is *alternating*, i.e. $b(x, x) = 0$, and hence $b(x, y) = -b(y, x)$. Thus every vector is isotropic. Relative to a basis this form is described by an *alternating* matrix (also called *skew-symmetric*): $A^T = -A$. Since the form is regular, its matrix A is non-singular, and it follows that a symplectic space is always even-dimensional. The *symplectic group* of V, $\mathbf{Sp}(V)$ or $\mathbf{Sp}_{2m}(k)$, where $2m = \dim V$, is the group of all symplectic transformations, i.e. all isometries of V. Any symplectic space V of dimension $2m$ has a basis of the form $u_1, \ldots, u_m, v_1, \ldots, v_m$ where $b(u_i, v_j) = \delta_{ij}$, $b(u_i, u_j) = b(v_i, v_j) = 0$. A basis of this form is called a *symplectic basis*. Relative to this basis the matrix of the form b becomes

$$J = \begin{pmatrix} 0 & I \\ -I & 0 \end{pmatrix} \tag{3.6.1}$$

Now the symplectic transformations may be described by the matrices P such that

$$PJP^T = J. \tag{3.6.2}$$

To establish the existence of a symplectic basis we recall a result from BA. By a *hyperbolic pair* of vectors we understand a pair $u, v \in V$ such that $b(u, v) = 1$; clearly a two-dimensional symplectic space has a basis consisting of a hyperbolic pair; such a space is called a *hyperbolic plane*. We recall that a subspace N is called *totally isotropic* if the form restricted to N is identically zero.

Lemma 3.6.1. *Let V be a symplectic space, U any subspace and U_0 a maximal totally isotropic subspace of U. Then $\dim U \leq 2 \dim U_0$ and any basis $u_1, \ldots, u_r$ of U_0 can after renumbering form part of a basis $u_1, \ldots, u_r, v_1, \ldots, v_s$ of U such that the (u_i, v_i) $(i = 1, \ldots, s \leq r)$ are mutually orthogonal hyperbolic pairs. Moreover, this basis of U can be completed to a symplectic basis of V.*

Proof. Given U and U_0 as stated, if $U_0 \neq U$, then no vector of U_0 is orthogonal to all of U_0, so there is $v_1 \in U$ such that $b(u_i, v_1) = 0$ for all i except one, which may be taken as 1 by renumbering the u's; thus $b(u_1, v_1) \neq 0$ and replacing v_1 by $v_1/b(u_1, v_1)$ we have $b(u_1, v_1) = 1$. If $\langle U_0, v_1 \rangle \neq U$, we can repeat the process and after a finite number of steps we reach a basis of U of the required form. By the maximality of U_0 we have $s \leq r$, hence $\dim U = r + s \leq 2r$.

Now if $s < r$, we can in V find v_r such that $b(u_i, v_r) = \delta_{ir}$ (because V is regular); continuing in this way we find $u_1, \ldots, u_r, v_1, \ldots, v_r$, a symplectic basis for a subspace W of V. If $W \neq V$, we can have an orthogonal sum $V = W \perp W^\perp$; by induction on $\dim V$ we can find a symplectic basis for $W^\perp$ which together with the basis found for W forms a symplectic basis for V. ∎

For any $u \neq 0$ in V and any $c \in k$ the linear mapping defined by

$$\tau_{u,c} : x \longmapsto x + cb(x, u)u \tag{3.6.3}$$

is a transvection along u, which is easily verified to be symplectic; its properties are given by

Lemma 3.6.2. *Let V be a symplectic space of dimension $2m$ and u any non-zero vector in V. Then any transvection which is symplectic, with kernel $u^\perp$, has the form $\tau_{u,c}$ for some $c \in k$. Moreover,*

(i) *for fixed u the mapping $c \longmapsto \tau_{u,c}$ is an injective homomorphism of the additive group of k into $\mathbf{Sp}_{2m}(k)$,*
(ii) *for any $\sigma \in \mathbf{Sp}_{2m}(k)$, $\sigma^{-1}\tau_{u,c}\sigma = \tau_{u\sigma,c}$ and*
(iii) *for $a \in k^\times$, $\tau_{au,c} = \tau_{u,a^2c}$.*

Proof. Any linear functional on V with kernel $u^\perp$ is a multiple of $b(-, u)$, hence the symplectic transformation with kernel $u^\perp$ has the form $\tau_{u,c}$ for some $c \in k$. The other properties are verified without difficulty. ∎

We shall use transvections to find a generating set for $\mathbf{Sp}_{2m}(k)$:

Theorem 3.6.3. *For any field k, and any $m \geq 1$, $\mathbf{Sp}_{2m}(k)$ is transitive on the hyperbolic pairs, and is generated by the set of all symplectic transvections.*

Proof. We denote by T the subgroup generated by all symplectic transvections and divide the proof into three parts:

(i) T is transitive on the non-zero vectors of V. For if $u_1, u_2 \neq 0$ and $b(u_1, u_2) \neq 0$, let $c \in k$ be such that $cb(u_1, u_2) = 1$ and put $u = u_1 - u_2$. Then $\tau_{u,c}$ maps u_1 to $u_1 + cb(u_1, u_1 - u_2)(u_1 - u_2) = u_1 - (u_1 - u_2) = u_2$. If $b(u_1, u_2) = 0$ and $v \in V$ is such that $b(u_i, v) \neq 0$ $(i = 1, 2)$, then the result just proved yields transvections to map u_1 to v and v to u_2, so it only remains to find v. If u_1, u_2 are linearly dependent, we can take any v not in $u_1^{\perp}$. Otherwise u_1, u_2 are linearly independent and such that $b(u_1, u_2) = 0$; then we can by Lemma 3.6.1, find v_1, v_2 such that (u_1, v_1) and (u_2, v_2) are orthogonal hyperbolic pairs, and now $v = v_1 + v_2$ is the required vector. Thus T has been shown to be transitive on the non-zero vectors of V.

(ii) Next we show that T is transitive on the hyperbolic pairs in V. Let (u_i, v_i) $(i = 1, 2)$ be two such pairs. By what has been proved we may take $u_1 = u_2 = u$. If $b(v_1, v_2) \neq 0$, then (as we saw in (i)) there is a symplectic transvection τ along $v_2 - v_1$ with $v_1\tau = v_2$; since $b(u, v_1 - v_2) = b(u, v_1) - b(u, v_2) = 0$, we have $u\tau = u$, so τ maps (u, v_1) to (u, v_2). If $b(v_1, v_2) = 0$, we use the hyperbolic pair $(u, v_1 + u)$; since $-b(v_1, v_1 + u) = b(v_1 + u, v_1) = 1$, we can find symplectic transvections to map (u, v_1) to $(u, v_1 + u)$ and $(u, v_1 + u)$ to (u, v_2).

(iii) We now use induction on m to show that $T = \mathbf{Sp}_{2m}(k)$. Suppose that $m = 1$, and that u, v is a symplectic basis. Any linear transformation has the form

$$u \longmapsto u' = au + bv,$$

$$v \longmapsto v' = cu + dv,$$

and this will be symplectic iff $b(u', v') = 1$, i.e. $ad - bc = 1$. Thus $\mathbf{Sp}_2(k)$ consists precisely of all matrices with determinant 1. By Theorem 3.5.1 this group is generated by all elementary matrices, and these are easily seen to be transvections. Now assume that $m > 1$, let $\sigma \in \mathbf{Sp}_{2m}(k)$ and take any hyperbolic pair (u, v) in V. By (ii) there exists $\tau \in T$ such that $(u, v)\sigma = (u, v)\tau$, hence $\sigma\tau^{-1}$ leaves u, v fixed. Therefore it maps $W = (u, v)^{\perp}$ into itself and defines an isometry there. By the induction hypothesis $\sigma\tau^{-1}|W = \prod \tau'_i$, where τ'_i is a symplectic transvection on W. We can extend τ'_i to a symplectic transvection τ_i on V by defining it as the identity on $\langle u, v\rangle$. Then $\sigma = \prod \tau_i.\tau$ and this shows that $\sigma \in T$. ■

Since a symplectic transvection clearly has determinant 1, we obtain

Corollary 3.6.4. *Every symplectic transformation has determinant 1.* ■

Theorem 3.6.3 also allows us to determine the centre of $\mathbf{Sp}_{2m}(k)$:

Corollary 3.6.5. *The centre of* $\mathbf{Sp}_{2m}(k)$ *consists of the transformations* I *and* $-I$.

Proof. If τ is a symplectic transvection along u, then $x\tau - x$ is proportional to u, hence $x\tau.\sigma - x\sigma$ is proportional to $u\sigma$, for any $\sigma \in \mathbf{Sp}_{2m}(k)$. Now $x\sigma.\tau - x\sigma$ is proportional to u and not always 0, so if σ and τ commute, then $u\sigma$ must be proportional to u. For a given σ this can happen for all u only if σ is a scalar, say $\sigma = c.1$. Now the condition (3.6.2) shows that $c^2 = 1$, hence $c = \pm 1$. ■

We next determine the commutator structure of the symplectic group.

Theorem 3.6.6. $\mathbf{Sp}_{2m}(k)$ *is perfect except when* $m = 1$ *and* $|k| \leq 3$ *or* $m = 2$ *and* $|k| = 2$.

Proof. In the proof of Theorem 3.6.3 we saw that $\mathbf{Sp}_2(k) \cong \mathbf{SL}_2(k)$ and by Proposition 3.5.2 this is perfect when $|k| > 2$, so we may assume that $m \geq 2$. Suppose first that $|k| > 3$; we shall show that every transvection $\tau_{u,a}$ is a commutator. In k there exists $c \neq 0$ such that $c^2 \neq 1$. Put $b = (1 - c^2)^{-1}a$, $d = -c^2 b$; then $b + d = a$, hence $\tau_{u,a} = \tau_{u,b}\tau_{u,d}$. If σ is any symplectic mapping such that $u\sigma = cu$ (Theorem 3.6.3), then

$$\sigma^{-1}\tau_{u,b}^{-1}\sigma = \sigma^{-1}\tau_{u,-b}\sigma = \tau_{u\sigma,-b} = \tau_{cu,-b} = \tau_{u,-c^2 b} = \tau_{u,d}.$$

Hence

$$\tau_{u,a} = \tau_{u,d}\tau_{u,b} = \sigma^{-1}\tau_{u,b}^{-1}\sigma\tau_{u,b}, \tag{3.6.4}$$

and this is the required expression.

There remains the case where k has two or three elements. It will be enough to find a transvection $\tau_{u,a} \neq 1$ in the derived group; since $\mathbf{Sp}_{2m}(k)$ is transitive on the non-zero vectors of V, we then have $\sigma^{-1}\tau_{u,a}\sigma = \tau_{u\sigma,a}$ in the derived group, and in case $|k| = 3$, $\tau_{u,2a} = \tau_{u,a}^2$, so it follows that the derived group contains all transvections and so by Theorem 3.6.3, coincides with $\mathbf{Sp}_{2m}(k)$. We shall write our transformations as matrices relative to a symplectic basis. Let $A, B \in k_m$; if A is invertible and B is symmetric, then

$$S_A = \begin{pmatrix} A & 0 \\ 0 & (A^T)^{-1} \end{pmatrix} \quad \text{and } R_B = \begin{pmatrix} I & B \\ 0 & I \end{pmatrix}$$

are symplectic, as is easily verified. With this notation we have

$$(S_A, R_B) = R_C, \quad \text{where } C = B - A^{-1}B(A^T)^{-1}, \tag{3.6.5}$$

and for suitable choice of A, B this is a symplectic transvection.

Suppose first that $|k| = 3$, $m = 2$. Taking $A = \begin{pmatrix} 1 & 1 \\ 1 & 0 \end{pmatrix}$, $B = \begin{pmatrix} 2 & 1 \\ 1 & 0 \end{pmatrix}$, we find that $B - A^{-1}B(A^T)^{-1} = 2E_{11}$, so we have obtained a transvection which is a commutator. The same argument applies for $m > 2$, taking A as I and B as 0 on the remaining coordinates.

There remains the case $|k| = 2$, $m \geq 3$. When $m = 3$, we take

$$A = \begin{pmatrix} -1 & 0 & 1 \\ 0 & 0 & 1 \\ 0 & 1 & 0 \end{pmatrix}, \quad B = \begin{pmatrix} 0 & 1 & 0 \\ 1 & 1 & 1 \\ 0 & 1 & 1 \end{pmatrix}.$$

It is easily verified that $B - A^{-1}B(A^T)^{-1} = E_{11}$, so we have again a symplectic transformation which is a commutator; for $m > 3$ we again take A, B to be I, 0 respectively on the new coordinates. So we have in all cases expressed a transvection as a commutator, and the result follows. ∎

Finally we come to the simplicity proof, which runs along similar lines to that for the general linear group (see Section 3.5).

Theorem 3.6.7. *The projective symplectic group* $\mathbf{PSp}_{2m}(k)$ *is simple for all fields k and all integers* $m \geq 1$, *except when* $m = 1$ *and* $|k| \leq 3$ *or* $m = 2$ *and* $|k| = 2$.

Proof. Since $\mathbf{Sp}_2(k) \cong \mathbf{SL}_2(k)$, we may assume that $m > 1$. We consider the action of $G = \mathbf{PSp}_{2m}(k)$ on the space $P = \mathbf{P}^{2m-1}(k)$ and begin by showing that G is primitive. Let Q be a set in a partition of P compatible with the G-action and containing more than one point. Suppose first that Q contains a pair of points $\langle x \rangle$, $\langle y \rangle$ defined by a hyperbolic pair of vectors (x, y). Given any other point $\langle z \rangle$ in P, if $b(x, z) \neq 0$, we may assume that $b(x, z) = 1$; since G is transitive on the hyperbolic pairs, by Theorem 3.6.3, there exists $\sigma \in G$ mapping (x, y) to (x, z), hence $Q\sigma \cap Q \neq 1$, so $Q\sigma = Q$, but $\langle z \rangle \in Q\sigma = Q$, so $Q = P$. If $b(x, z) = 0$, we may assume that $\langle z \rangle \neq \langle x \rangle$, because otherwise $\langle z \rangle = \langle x \rangle \in Q$. Then there exists $w \in V$ such that $b(x, w) = b(z, w) = 1$, and by what has been shown, $\langle z \rangle \in Q$. Further there exists $\sigma \in G$ mapping (x, w) to (z, w), hence $Q\sigma \cap Q \neq \emptyset$, so again $\langle z \rangle \in Q\sigma = Q$ and it follows again that $Q = P$. The alternative is that the subspace defining Q is totally isotropic. Let $\langle x, y \rangle$ be a plane in Q and choose $w \in V$ such that $b(x, w) = 1$, $b(y, w) = 0$. Writing $H = \langle x, w \rangle$, we have $V = H \perp H^{\perp}$. Further, $y \in H^{\perp}$ and for $0 \neq z \in H^{\perp}$ there exists a symplectic transformation σ leaving H fixed and mapping y to z. Since $\langle x \rangle \in Q$, it follows that $Q\sigma = Q$ and since $\langle y \rangle \in Q$, we have $\langle z \rangle \in Q\sigma = Q$. Hence Q contains all points defined by vectors in $H^{\perp}$. Since $m > 1$, $H^{\perp}$ contains a hyperbolic pair, so the first part of the argument can be applied to show that $Q = P$, and this shows G to be primitive.

In order to apply Proposition 3.5.5, we need to find a normal abelian subgroup of a stabilizer whose conjugates generate G. Take a point $\langle x \rangle$, let S be its stabilizer and let A be the group of transvections $\tau_{x,a}$ $(a \in k)$. Then (3.6.4) shows that $A \triangleleft S$ and by Theorem 3.6.3, A and its conjugates generate G. Hence G is indeed simple, with the exceptions listed. ∎

Again the exceptions actually occur (see Exercises 2 and 3).

It still remains to compute the order of $\mathbf{Sp}_{2m}(k)$ when k is finite.

Proposition 3.6.8. *If k is a field of q elements and $m \geq 1$, then*

$$|\mathbf{Sp}_{2m}(k)| = q^{2m-1}(q^{2m}-1).|\mathbf{Sp}_{2m-2}(k)|, \tag{3.6.6}$$

hence

$$|\mathbf{Sp}_{2m}(k)| = q^{2m-1}(q^{2m}-1)q^{2m-3}(q^{2m-2}-1)\ldots q(q^2-1). \tag{3.6.7}$$

For $\mathbf{PSp}_{2m}(k)$ the order is the same when q is even and has one half this value when q is odd.

Proof. Let V be a two-dimensional space over k; we first determine the number of hyperbolic pairs in V. For a hyperbolic pair (x, y), x may be any non-zero vector in V, so there are $q^{2m} - 1$ choices. Given a particular hyperbolic pair (x, y_0), any other hyperbolic pair with first vector x has the form (x, y), where $y = y_0 + z$ for a vector $z \in x^{\perp}$, and here we have q^{2m-1} choices for z. Hence there are $q^{2m-1}(q^{2m} - 1)$ pairs in all. By Theorem 3.6.3, $\mathbf{Sp}_{2m}(k)$ is transitive on the set of these pairs, and the stabilizer of a particular pair (x, y) is isomorphic to the symplectic group of $\langle x, y\rangle^{\perp}$, which is $\mathbf{Sp}_{2m-2}(k)$. Hence we obtain (3.6.6) and now (3.6.7) is an easy consequence. The final remark follows because the centre is ± 1, as we saw in Corollary 3.6.5 and $-1 = 1$ when q is even. ■

Exercises

1. Show that in the action of $\mathbf{PSp}_{2m}(k)$ on $\mathbf{P}^{2m-1}(k)$, the stabilizer of a point has exactly three orbits.
2. Verify that $\mathbf{PSp}_2(\mathbf{F}_2)$ and $\mathbf{PSp}_2(\mathbf{F}_3)$ are both soluble, of orders 6 and 12 respectively, and express them as permutation groups.
3. Show that $\mathbf{Sp}_4(\mathbf{F}_2) \cong \mathrm{Sym}_6$ by considering its action on quintuples of vectors $u_1, \ldots, u_5$ such that $b(u_i, u_j) = 1$ for $i \neq j$ in a four-dimensional symplectic space over $\mathbf{F}_2$. (Hint. Find the number of hyperbolic pairs in a quintuple and the number of quintuples containing a given hyperbolic pair.)

3.7 The orthogonal group

The orthogonal group is the group of all orthogonal transformations of a quadratic space V, denoted by $\mathbf{O}(V)$ or $\mathbf{O}_n(k)$, where k is the ground field and n the dimension of the space. This group leads to a class of simple groups, just as the symplectic group does, but this time the result is dependent on the precise structure of the underlying quadratic space. Our treatment follows Iwasawa and Tamagawa, as in the account of Jacobson (1985). The field k is again assumed to have characteristic not equal to 2, and we shall denote the quadratic form by q and the associated bilinear form by b, so that $b(x, x) = q(x)$. In what follows we shall assume that dim $V \geq 3$; the case dim $V = 1$ or 2 is easily dealt with separately (see Exercises 4 and 5). We begin by determining the centre of $\mathbf{O}(V)$. This turns out to be the same as that of the symplectic group, but for the proof we use symmetries instead of transvections.

Proposition 3.7.1. *Let V be a regular quadratic space of dimension ≥ 3. Then the centre of $O(V)$ is ± 1.*

Proof. If u is any anisotropic vector in V and $\sigma_u : x \mapsto x - 2(b(x,u)/q(u))u$ the symmetry defined by u, then $\langle u \rangle = \{x \in V | x\sigma_u = -x\}$; hence any α in the centre of $\mathbf{O}(V)$ has the property that $u\alpha \in \langle u \rangle$ for every anisotropic vector u. Let $u_1, \ldots, u_n$ be an orthogonal basis of V; then $u_i\alpha = \pm u_i$, so by suitable numbering we may assume that $u_i\alpha = u_i$ for $i = 1, \ldots, s$ and $u_i\alpha = -u_i$ for $i > s$. If $u_i + u_j$ is anisotropic, we have $(u_i + u_j)\alpha = \pm(u_i + u_j)$ and it follows that either $i, j \leq s$ or $i, j > s$. Thus if $1 \leq s < n$, then $u_1 + u_n$ is isotropic; hence for $1 < i < n$, $q(u_1 + u_i + u_n) = q(u_i) \neq 0$, so $(u_1 + u_i + u_n)\alpha = \varepsilon(u_1 + u_i + u_n)$, where $\varepsilon = \pm 1$, but also $(u_1 + u_i + u_n)\alpha = u_1 + \varepsilon' u_i - u_n$ and this leads to a contradiction. Therefore s is 0 or n and $a = 1$ or -1. ■

The symmetry σ_u is defined only for anisotropic vectors u; in the isotropic case one has the following replacement, going back to Carl Ludwig Siegel.

Let u be an isotropic vector, choose v so that u, v is a hyperbolic pair and take $0 \neq w \in \langle u, v \rangle^{\perp}$. We have $V = \langle v \rangle \oplus u^{\perp}$, hence the equations

$$\begin{cases} x\rho_{u,w} = x + b(x,w)u & (x \in u^{\perp}) \\ v\rho_{u,w} = v - q(w)u - w, & \end{cases} \tag{3.7.1}$$

define the linear transformation $\rho_{u,w}$ completely and it is easily checked that $\rho_{u,w}$ is orthogonal. Moreover, it is proper (i.e. of determinant 1); in fact it is unipotent, i.e. $1 - \rho_{u,w}$ is nilpotent.

We remark that $\rho_{u,w}$ is uniquely determined as the orthogonal transformation which maps x to $x + b(x,w)u$ for all $x \in u^{\perp}$. For this mapping can be extended to an orthogonal transformation, by Witt's theorem (BA, Theorem 8.5.5) and if there were two such mappings, their quotient would leave $u^{\perp}$ fixed. Now $v\alpha$ has the form $\lambda u + \mu v + z$, where $z \in \langle u, v \rangle^{\perp}$. For any $x \in \langle u, v \rangle^{\perp}$ we have $0 = b(x, v) = b(x\alpha, v\alpha) = b(x, \lambda u + \mu v + z) = b(x, z)$; hence $z \in \langle u, v \rangle^{\perp\perp} = 0$, i.e. $z = 0$. Further, $1 = b(u,v) = b(u\alpha, v\alpha) = b(u, \lambda u + \mu v) = \mu$, thus $\mu = 1$ and finally $0 = q(v) = q(v\alpha) = q(\lambda u + v) = \lambda$, hence $\lambda = 0$. So $v\alpha = v$ and therefore $\alpha = 1$. This shows $\rho_{u,w}$ to be uniquely determined by its effect on $u^{\perp}$.

We shall use the transformation $\rho_{u,w}$ to construct an abelian normal subgroup of the stabilizer of u.

Lemma 3.7.2. *Let V be a regular quadratic space of dimension ≥ 3, containing a hyperbolic pair u, v and put $W = \langle u, v \rangle$. Then the set*

$$A_u = \{\rho_{u,w} | w \in W\}$$

is a subgroup of $\mathbf{O}(V)$ *which is abelian and normal in the stabilizer of* u. *Moreover, the mapping*

$$w \longmapsto \rho_{u,w} \tag{3.7.2}$$

is an isomorphism of W *with* A_u.

Proof. Given $w, w' \in W$, we see from (3.7.1) that $\rho_{u,w}\rho_{u,w'}$ and $\rho_{u,w+w'}$ both map x to $x + b(x, w + w')u$, for any $x \in u^{\perp}$, hence they agree on the whole of V. This shows (3.7.2) to be a homomorphism. If w lies in the kernel of (3.7.2), then by (3.7.1), $w \in u^{\perp\perp} = \langle u \rangle$ (see BA, Proposition 8.1.3), but the only multiple of u in W is 0; this shows (3.7.2) to be injective. It follows that A_u is an abelian subgroup of $\mathbf{O}(V)$. Clearly it is contained in the stabilizer of u. Moreover, for any $\sigma \in \mathbf{O}(V)$ we have $\sigma^{-1}\rho_{u,w}\sigma = \rho_{u\sigma,w}$, hence A_u is mapped into itself by any σ leaving u fixed. ∎

Our next aim will be to show that under some mild restrictions on V, the derived group $\mathbf{O}(V)'$ is generated by all the A. That some restriction is needed is clear since there are no A unless the Witt index of V is positive. It will be convenient to write Ω for the subgroup of $\mathbf{O}(V)$ generated by all the A_u, where u ranges over all isotropic vectors. We begin by establishing some transitivity properties.

Lemma 3.7.3. *Let* V *be a regular quadratic space,* $\dim V \geq 3$. *Then*

(i) *for any isotropic* $u \in V$, A_u *is transitive on the one-dimensional isotropic subspaces not orthogonal to* u,
(ii) *for any two linearly independent isotropic vectors* u_1, u_2 *there is a vector* v *and* $\lambda_1, \lambda_2 \in k$ *such that* $(\lambda_i u_i, v)$ *is a hyperbolic pair,*
(iii) *the subgroup* Ω *of* $\mathbf{O}(V)$ *generated by the* A_u *is transitive on the hyperbolic pairs.*

Proof. (i) Let z, y be isotropic vectors not orthogonal to u; we may assume that $b(u, x) = b(u, y) = 1$. Write $y = \lambda u + \mu x + z$, where $z \in \langle u, x\rangle^{\perp}$; since $b(u, y) = 1$, we have $\mu = 1$ and the relation $q(y) = 0$ shows that $\lambda + q(z) = 0$. Hence $y = x - q(z)u + z = x\rho_{u,-z}$ and (i) follows.

(ii) Let u_1, u_2 be as stated; if $b(u_1, u_2) \neq 0$, we may assume that $b(u_1, u_2) = 1$. The space $\langle u_1, u_2\rangle^{\perp}$ contains an anisotropic vector w, by the regularity of V. We put $v = u_1 - q(w)u_2 + w$; then $q(v) = -q(w) + q(w) = 0$, $b(u_1, v) = -q(w) \neq 0$, $b(u_2, v) = 1$, so with $\lambda_1 = -1/q(w)$, $\lambda_2 = 1$, v satisfies the required conditions. If $b(u_1, u_2) = 0$, then since u_1, u_2 are linearly independent, there is a linear functional equal to 1 on u_1, u_2, hence by the regularity of b there exists v such that $b(u_i, v) = 1$, and by subtracting suitable multiples of u_1 from v we can ensure that $q(v) = 0$. Then (u_i, v) are again two hyperbolic pairs.

(iii) Let (u_i, v) $(i = 1, 2)$ be any two hyperbolic pairs. If u_1, u_2 are linearly independent, we can by (ii) find v and λ_i such that $(\lambda_i u_i, v)$ are hyperbolic pairs; by (ii) there is then an element of A_v mapping $\langle u_1\rangle$ to $\langle u_2\rangle$. Thus for some $\sigma \in \Omega$, $u_1\sigma = cu_2$. We now have the hyperbolic pairs $(u_1\sigma, v_1\sigma) = (cu_2, v_1\sigma)$ and (u_2, v_2); applying (i) again, we find $\tau \in A_{u_2}$ such that $v_1\sigma\tau = v_2$, $u_2\tau = u_2$, so $\sigma\tau$ maps (u_1, v_1) to (cu_2, v_2), but $1 = b(u_1, v_1) = b(cu_2, v_2) = c$, hence $c = 1$. If u_1, u_2 are linearly dependent, we can by (i) apply an element of Ω to (u_1, v_1) to obtain a hyperbolic pair with first vector linearly independent of u_2 and now proceed as before. ■

We recall the Cartan–Dieudonné theorem, which shows that $\mathbf{O}(V)$ is generated by symmetries (see BA, Corollary 8.3.3). It follows that the square of every orthogonal transformation lies in the derived group $\mathbf{O}(V)'$. For if $\alpha = \sigma_1 \dots \alpha_r$, then $\alpha^2 = \sigma_1 \dots \sigma_r \sigma_1 \dots \sigma_r \equiv \sigma_1^2 \dots \sigma_r^2 \equiv 1 \pmod{\mathbf{O}(V)'}$. We also note that for $\dim V \geq 3$ we have $\mathbf{O}(V)' = \mathbf{SO}(V)'$ (BA, Theorem 8.3.4). In two dimensions this need not hold; as an example (which is used later) let us determine $\mathbf{O}(V)$ for a hyperbolic plane.

Let H be a hyperbolic plane with the hyperbolic pair u, v as basis. Any linear mapping has the form

$$u' = au + bv, \quad v' = cu + dv,$$

and this is an isometry iff $q(u') = q(v') = 0$, $b(u', v') = 1$. Thus $ab = cd = 0$, $ad + bc = 1$, so either $b = c = 0$ or $a = d = 0$, and the isometries of H have the following forms:

$$\lambda_a = \begin{pmatrix} a & 0 \\ 0 & a^{-1} \end{pmatrix} \text{ or } \lambda_a \tau = \begin{pmatrix} 0 & a \\ a^{-1} & 0 \end{pmatrix}, \quad \text{where } \tau = \begin{pmatrix} 0 & 1 \\ 1 & 0 \end{pmatrix}. \tag{3.7.3}$$

The λ_a form a group isomorphic to $k^\times$ and $\mathbf{O}(H)$ is an extension by a cyclic group of order 2, acting by inversion: $\tau\lambda_a\tau = \lambda_a^{-1}$. In particular,

$$\lambda_{a^2} = (\tau, \lambda_a). \tag{3.7.4}$$

We can now determine the subgroup Ω:

Theorem 3.7.4. *Let V be a regular quadratic space of dimension ≥ 3 and positive Witt index n. The $\mathbf{SO}(V) = \mathbf{O}(V)'$; more precisely, the subgroup Ω generated by the A_u (u isotropic) coincides with the derived group: $\Omega = \Omega' = \mathbf{O}(V)'$, except when $n = 4$, $\nu = 2$ and $n = 3$, $|k| = 3$.*

Proof. As a first step we show that $\Omega \supseteq \mathbf{O}(V)'$. We fix a hyperbolic pair (u, v) and write $O_1 = \mathbf{O}(\langle u, v\rangle)$, $W = \langle u, v\rangle^\perp$. We claim that every symmetry is conjugate under Ω to one in O_1. For if x is any anisotropic vector, then $x_1 = u + q(x)v$ satisfies $q(x_1) = q(x)$, hence there exists $\rho \in \mathbf{O}(V)$ such that $x_1\rho = x$. By Lemma 3.7.3(iii) there exists $\alpha \in \Omega$ mapping $(u\rho, v\rho)$ to (u, v); since $x_1 \in \langle u, v\rangle$, we have $x = x_1\rho \in \langle u\rho, v\rho\rangle$ and so $x\alpha = x_1\rho\alpha \in \langle u, v\rangle$. Hence $\alpha^{-1}\sigma_x\alpha = \sigma_{x\alpha}$, where $x\alpha \in \langle u, v\rangle$, as we wished to show.

Now $O_1 = \mathbf{O}(\langle u, v\rangle)$ is generated by symmetries $\sigma_x (x \in \langle u, v\rangle)$, and the restriction $\sigma_x|W$ is the identity; it follows that the mapping $\sigma_x \mapsto \sigma_x|\langle u, v\rangle$ defines an isomorphism $O_1 \cong \mathbf{O}(H)$, and clearly SO_1 corresponds to $\mathbf{SO}(H)$ in this isomorphism. Now any $\rho \in \mathbf{SO}(H)$ can be written as a product of an even number of symmetries:

$$\rho = \sigma_1 \dots \sigma_{2r},$$

and we have seen that $\sigma_i = \nu_i^{-1}\tau_i\nu_i$, where τ_i is a symmetry in O_1 and $\nu_i \in \Omega$. Thus we have

$$\rho = (\nu_1^{-1}\tau_1\nu_1)\dots(\nu_{2r}^{-1}\tau_{2r}\nu_{2r}).$$

Since Ω is normal in $\mathbf{O}(V)$, this relation can be written $\rho = \mu\tau_1\dots\tau_{2r}$, where $\mu \in \Omega$, so we have $\mathbf{SO}(V) \subseteq \Omega.SO_1$. But as we saw, $\Omega \subseteq \mathbf{SO}(V)$, and we conclude that

$$\mathbf{SO}(V) = \Omega.SO_1. \tag{3.7.5}$$

Therefore $\mathbf{SO}(V)/\Omega \cong SO_1/SO_1 \cap \Omega$, but $SO_1 \cong k^\times$ is abelian, by the earlier remark, so $\Omega \supseteq \mathbf{SO}(V)' = \mathbf{O}(V)'$ and we find that

$$\Omega \supseteq \mathbf{SO}(V)'. \tag{3.7.6}$$

To complete the proof we show that Ω is perfect, for then $\Omega = \Omega' = \mathbf{SO}(V)'$. It will be enough to show that $\rho_{u,w} \in \Omega'$ for all isotropic u and all w orthogonal to a hyperbolic plane containing u. So we may take u, v, W, O_1 as before. Let λ_a, τ be the transformations in O_1 defined as in (3.7.3), so that $\lambda_{a^2} \in \Omega$ by (3.7.4) and (3.7.6). For any $w \in W$ we have

$$\lambda_{a^2}^{-1}\rho_{u,w}^{-1}\lambda_{a^2}\rho_{u,w} = \rho_{a^2u,-w}\rho_{u,w} = \rho_{u,-a^2w}\rho_{u,w} = \rho_{u,(1-a^2)w}.$$

When $|k| > 3$, we can choose $a \in k^\times$ such that $a^2 \neq 1$ and replacing w by $(1-a^2)^{-1}w$ we see that $\rho_{u,w} \in \Omega'$, hence $\Omega = \Omega' = \mathbf{O}(V)'$ when $|k| > 3$.

Suppose now that $|k| = 3$; then $n \geq 4$ and for $n = 4$, $\nu = 1$. Thus $\dim W \geq 2$ and we have an orthogonal basis $w_1 = w, w_2, \dots, w_r$ for W. If $q(w_1) = q(w_2)$, then the map $\alpha : w_1, w_2, \dots, w_r \mapsto -w_2, w_1, w_3, \dots, w_r$ is an isometry, hence $\alpha^2 \in \mathbf{O}(V)' \subseteq \Omega$ and α^2 maps w to $-w$. Thus

$$\rho_{u,w}^{-1}\alpha^{-2}\rho_{u,w}\alpha^2 = \rho_{u,-w}\rho_{u,-w} = \rho_{u,-2w} = \rho_{u,w}$$

and this shows that $\rho_{u,w} \in \Omega'$.

It remains to justify the assumption $q(w_1) = q(w_2)$. Since $q(w_1) \in k^\times$ and $|k| = 3$, $q(w_1)$ is 1 or -1. For $n = 4$, W is isotropic, so $q(w_1)$, $q(w_2)$ have the same sign and hence must be equal. When $n \geq 5$, $W \cap w^\perp$ is regular and at least two-dimensional, so q restricted to this subspace is universal (see BA, Theorem 8.2.7), and it follows that W contains w' orthogonal to w such that $q(w') = q(w)$. Thus we can in all cases find a basis of the required form. ∎

We now have the means at our disposal to prove the main structure theorem for orthogonal groups. The result was first established, with a restriction on the index, by Dickson (1901), and in its full generality by Dieudonné in 1940.

Theorem 3.7.5. *Let V be a regular quadratic space of dimension $n \geq 3$ and of positive Witt index ν. Denote by C the centre of $\mathbf{SO}(V)$. Then $\mathbf{SO}(V)/C$ is simple, except when $n = 4$, $\nu = 2$ and $n = 3$, $|k| = 3$.*

Proof. We remark that C has order 2 when n is even and is trivial when n is odd. For by Proposition 3.7.1 the centre can only contain 1 and -1 and $\det(-1) = (-1)^n$.

Consider the quadric cone Q defined as the set of points $\langle x\rangle$ in $\mathbf{P}^{n-1}(k)$ satisfying $q(x)=0$. By Lemma 3.7.3(i), Ω acts transitively on Q. We first show that the action is primitive except when $n=4$, $\nu=2$. If $b(x,y)\neq 0$ for any two linearly independent isotropic vectors x, y, then by Lemma 3.7.3(iii), Ω is 2-fold transitive, and hence primitive. In particular, this always holds for $\nu=1$, so we may assume henceforth that $\nu\geq 2$ and hence $n\geq 5$. Let S be a subset with more than one point in a partition of Q stable under Ω; we have to show that $S=Q$. Given $\langle x_1\rangle, \langle x_2\rangle\in S$, if $b(x_1,x_2)=0$, then there exist $y_1,y_2\in Q$ such that $(x_1,y_1), (x_2,y_2)$ are orthogonal hyperbolic pairs. By Lemma 3.7.3(i) applied to $\langle x_2,y_2\rangle^{\perp}$ there exists $\alpha\in\Omega$ mapping x_1 to y_1 and leaving x_2,y_2 fixed. Since $\langle x_2\rangle\in S$, we have $S\alpha=S$ and $\langle x_1\rangle\in S$, therefore $\langle y_1\rangle\in S$. Now if $\langle z\rangle$ is any point of Q different from $\langle x_1\rangle$, we can by Lemma 3.7.3(ii) find v such that (x_1,v), (z,v) are hyperbolic pairs. By Lemma 3.7.3(iii) there exists $\beta\in\Omega$ mapping (x_1,y_1) to (x_1,v), hence $S\beta=S$ and $\langle v\rangle\in S\beta=S$. Similarly there is $\gamma\in\Omega$ mapping $\langle x_1,v\rangle$ to $\langle z,v\rangle$, hence $\langle v\rangle\in S\gamma=S$ and $z=x_1\gamma$, so $\langle z\rangle\in S$. Since z was arbitrary, this means that $S=Q$.

We may therefore assume that for any distinct points $\langle x_1\rangle, \langle x_2\rangle$ in S, $b(x_1,x_2)\neq 0$. Given $\langle z\rangle\neq\langle x_1\rangle$ in Q as before, we can find v such that (x_1,v), (z,v) are hyperbolic pairs, and $\beta\in\Omega$ maps (x_1,x_2) to (x_1,v). It follows that $S\beta=S$ and $\langle v\rangle\in S$. If now $\gamma\in\Omega$ is chosen so as to map (x_1,v) to (z,v), then since $v\gamma=v$, we have $S\gamma=S$ and $z=x_1\gamma$, hence $\langle z\rangle\in S$ and we again find that $S=Q$.

We thus see that Ω acts primitively on Q and Ω is perfect, by Theorem 3.7.4. Moreover, if u is isotropic, then A_u is a normal abelian subgroup of the stabilizer of $\langle u\rangle$ and the conjugates of A_u generate Ω, by the proof of Theorem 3.7.4. Hence by Proposition 3.5.5, Ω is simple and now the result follows by Theorem 3.7.4. ■

Some of the exceptions of Theorem 3.7.5 will be considered in the exercises. Let us now take up some particular cases to show that the hypotheses on the Witt index cannot be omitted, without striving for full generality. We take a quadratic form over $\mathbf{R}$; if its index is 0, the form must be definite, say positive definite, and in suitable coordinates it will have an orthonormal basis. For simplicity consider the case $n=3$, thus $\mathbf{SO}(V)$ is the group of rotations in 3-space. We claim that $\mathbf{SO}(V)$ acts primitively on the unit sphere S. For let T be a subset of S with more than one point, stable under all rotations. Given $p,q\in T$, T must include all points of the circle through q about p as axis. If the points at opposite ends of a diameter of this circle are a spherical distance d apart, then T will include points at any distance $\leq d$ from q, and by repetition, points at any finite (spherical) distance, hence $T=S$ and the action is primitive. The rotations about a point form an abelian subgroup whose conjugacy classes generate $\mathbf{SO}(V)$, and this shows $\mathbf{SO}(V)$ to be simple. The same argument applies for any odd dimension ≥ 3, while for even dimensions ≥ 6, $\mathbf{PSO}(V)$ is simple. When $\dim V=4$, we have $PO\cong G\times G$, where $G=\mathbf{PSL}_2(k)$ when V has index 2, and $G=\mathbf{SO}(\mathbf{R}^3)$ for a Euclidean 3-space when $V=\mathbf{R}^4$ is Euclidean. When $\mathbf{R}^4$ has index 1 (e.g. the Lorentz metric of relativity theory), then $PO\cong\mathbf{PSL}_2(\mathbf{C})$; in this case Ω consists of all rotations which do not reverse the time direction.

The argument just used to show that for a definite quadratic form $\mathbf{PSO}(\mathbf{R}^n)$ is simple depended essentially on the fact that R is Archimedean ordered. For an ordered field K which is non-Archimedean (i.e. there are elements greater than any integer) it can be shown that $\mathbf{PSO}(K^n)$ is not simple: the infinitesimal rotations generate a proper normal subgroup (see Exercise 6). This happens, for example, for the field of formal Laurent series $\mathbf{R}((x))$, ordered by the sign of its first coefficient.

For a finite field it is again possible to calculate the order of the orthogonal group, but this depends on the quadratic character of the determinant as well as the parity of the dimension (see Exercises 7–9).

Exercises

1. Give the details of the proof that for a real Euclidean space V of dimension ≥ 5, $\mathbf{PSO}(V)$ is simple.
2. Verify that $\rho_{u,w}$ defined by (3.7.1) satisfies $(1-\rho_{u,w})^3 = 0$. When is $(1-\rho_{u,w})^2 = 0$?
3. Let V be an n-dimensional quadratic space. Given two anisotropic vectors x, y, show that $\sigma_x\sigma_y$ is a rotation in the plane $\langle x, y\rangle$ leaving $\langle x, y\rangle^{\perp}$ fixed. Verify that for a Euclidean V the angle of rotation is twice the angle between x and y.
4. Use the method of proof of Proposition 3.7.1 to find the centre of $\mathbf{O}_1(k)$, $\mathbf{O}_2(k)$.
5. Examine the form Lemmas 3.7.2 and 3.7.3 take when dim $V = 1$ or 2.
6. Let V be a Euclidean space over an ordered field K which is non-Archimedean. Show that the rotation through an infinitesimal angle generates a proper normal subgroup (α is *infinitesimal* if $n\alpha < 1$ for all $n \in \mathbf{Z}$).
7. Show that over a finite field of odd characteristic every regular quadratic form has the form $\langle 1^{n-1}, d\rangle$ where d is the determinant and that $|k^\times/k^{\times 2}| = 2$, so that there are just two classes of forms in each dimension. (Hint. Recall from BA, Section 8.2 that a quadratic form of rank ≥ 2 over a finite field is universal.)
8. Let $|k| = q$ be odd. Show that the number of solutions of $\sum_1^m (x_i^2 - y_i^2) = b$ is $q^{2m-1} - q^{m-1} + \delta_{0b}q^m$, the number of solutions of $\sum_1^m (x_i^2 - y_i^2) - (d-1)y_m^2 = b$ is $q^{2m-1} + q^{m-1} - \delta_{0b}q^m$, and the number of solutions of $\sum_1^m (x_i^2 - y_i^2) - z^2 = b$ is $q^{2m} + (-b/q)q^m$, where $(-b/q)$ is a Legendre symbol, i.e. 0, 1 or -1 according as $-b$ is 0, a non-zero square or not a square in k.
9. Show that for a regular form over a finite field $\mathbf{F}_q$ (q odd), $|\mathbf{O}_{2m}(\mathbf{F}_q)| = (q^{2m-1} - q^{m-1})|\mathbf{O}_{2m-1}(\mathbf{F}_q)|$, $|\mathbf{O}_{2m+1}(\mathbf{F}_q)| = (q^{2m} + (-d/q)q^m)|\mathbf{O}_{2m}(\mathbf{F}_q)|$, where d is the determinant of the form. Hence calculate the order of the orthogonal group.
10. What form do the equations (3.7.1) take when u is replaced by $-u$? Show that $\rho_{-u,-w} = \rho_{u,w}$ and that the normalizer of A_u includes any σ mapping u to $-u$.

Further exercises on Chapter 3

1. Let A be a group and α an automorphism of A. Show that there is a split extension E of A by an infinite cyclic group such that α is induced by an inner automorphism of E. Show also that any extension of A by an infinite cyclic group splits.

2. Defining an automorphism of an extension E as an isomorphism of E with itself, show that the group of all automorphisms of an extension E of an abelian group A by a group G is isomorphic to the group of 1-cocycles of G in A. Show that if the extension is split and $H^1(G, A) = 0$, then any two complements of A in E are conjugate.
3. Let F be the free group on X and kF the group algebra over a field k. Obtain the following analogue of (2.7.10):
$$0 \to {}^X(kF \otimes_k kF) \to kF \otimes_k kF \to kF \to 0.$$
By applying $k \otimes_{kF}$ deduce that the augmentation ideal IF is free on $x - 1$ $(x \in X)$ as F-module. Hence show that $H^n(F, A) = H_n(F, A) = 0$ for $n > 1$.
4. Show that in a finite soluble group G, with a Hall subgroup of order r, the number h_r of subgroups of order r in G has the form $h_r = c_1 \dots c_n$, where each c_i is a prime power dividing the order of a chief factor of G and $c_i \equiv 1$ modulo a prime factor of r. (Hint. Put $|G| = rs$, where $(r, s) = 1$ and first treat the case when G has a normal subgroup of index $r's'$, where $r'|r, s'|s$ and $s' > 1$.)
5. Let E/k be a finite Galois extension of degree $m = q_1 \dots q_r$, where the q_i are powers of distinct primes. Show that if E contains a subfield E_i of degree q_i over k, for $i = 1, \dots, r$, then
$$E = E_1 \otimes \dots \otimes E_r, \ [E_i : k] = q_i.$$
Use Hall's theorem to show that such a decomposition of E exists iff $G = \mathrm{Gal}(E/k)$ is soluble and that any two such decompositions of E are conjugate by an element of G.
6. Show that for any finite Galois extension E/k, a decomposition as in Exercise 5, where the E_i/k are all Galois extensions, exists iff $\mathrm{Gal}(E/k)$ is nilpotent.
7. Show that the order of a finite simple group is divisible either by 12 or by the cube of the least prime dividing its order. (Hint. If a Sylow p-subgroup P has order p or p^2, it is abelian; now use Theorem 3.3.2 to describe the action of $N_G(P)$ on P.)
8. Show that $\mathbf{PSL}_3(\mathbf{F}_2) \cong \mathbf{PSL}_2(\mathbf{F}_7)$ and find its order. (Hint. This is the subgroup of Sym_7 in the action on the columns of the array
$$\begin{matrix} 1 & 2 & 3 & 4 & 5 & 6 & 7 \\ 2 & 3 & 4 & 5 & 6 & 7 & 1 \\ 4 & 5 & 6 & 7 & 1 & 2 & 3 \end{matrix}$$
which map each column into another. The columns may be interpreted as lines in the projective plane over $\mathbf{F}_2$, $\mathbf{P}^2(\mathbf{F}_2)$.)
9. Describe $\mathbf{O}(V)^{\mathrm{ab}}$ for a two-dimensional anisotropic space V.
10. Show that in the action of $\mathbf{PSp}_{2m}(k)$ on $\mathbf{P}^{2m-1}(k)$, the stabilizer of a point has exactly three orbits.
11. Let X be a subset of a free group F and define an *elementary transformation* of X as one of the following: (i) replace x by xy $(x, y \in X, x \neq y)$, (ii) replace x by x^{-1} $(x \in X)$, (iii) omit 1. A *Nielsen transformation* is a series of elementary trans-

formations. Show that if X' is obtained from X by a Nielsen transformation, then $\mathrm{gp}\{X'\} = \mathrm{gp}\{X\}$. Show also that any finite subset can be reduced by a Nielsen transformation, i.e. brought to the form where $1 \notin X$, $x, y \in X \cup X^{-1}$, $xy \neq 1$ implies $|xy| \geq |x|$, $|y|$ and $x, y, z \in X \cup X^{-1}$, $xy \neq 1$, $yz \neq 1$ implies $|xyz| \geq |x| - |y| + |z|$, where $|w|$ is the length of w in terms of a basis.

12. (H. Zieschang) Let U be a reduced set in a free group F (see Exercise 11). For each $u \in U \cup U^{-1}$ denote by $a(u)$ the longest prefix of u cancelled in any product $vu \neq 1$, $v \in U \cup U^{-1}$. Show that $u = a(u)m(u)a(u^{-1})^{-1}$ is reduced, where $m(u) \neq 1$, and if $w = u_1 \dots u_n$, then in the reduced form of w, $m(u_1), \dots, m(u_n)$ are uncancelled.
13. Show that a reduced subset X of a free group is a basis of $\mathrm{gp}\{X\}$. Deduce that every finitely generated subgroup of a free group is free.
14. Let F be a free group of finite rank on a basis X and let A be an automorphism of F. By applying Exercises 11 and 12 to the set X^α show that every automorphism of F can be obtained by a permutation and a Nielsen transformation of the generating set.
15. (M. Takahashi) Let F be a free group and $F = F_1 \supset F_2 \supset \dots$ a series of subgroups such that F_{i+1} contains no element of a basis of F_i. Show that relative to any basis of F_1 any element $w \neq 1$ of F_{i+1} satisfies $|w| > i$. Deduce that any infinite series of characteristic subgroups of F intersects in 1, and so obtain another proof of Magnus's theorem (Corollary 3.4.6).

4 Algebras

In Section 5.2 of BA we saw that semisimple Artinian rings can be described quite explicitly as direct products of full matrix rings over skew fields (Wedderburn's theorem). Later in Chapters 7 and 8 we shall see what can be said when the Artinian hypothesis is dropped, but in many cases, such as the study of group algebras in finite characteristic, it is important to find out more about the non-semisimple (but Artinian) case. There is now a substantial theory of such algebras which is still developing, and a full treatment is beyond the framework of this book, but some of the basic properties are described here.

One of the main results, the Krull–Schmidt theorem, asserts the uniqueness of decompositions of a module as a direct sum of indecomposables. This is established in Section 4.1 for finitely generated modules over Artinian rings, but as we shall see in Section 4.3, for projective modules it holds over the somewhat larger class of semi-perfect rings. These rings are also of interest because they allow the construction of a projective cover for each finitely generated module (Section 4.2).

The rest of the chapter is concerned with conditions for two rings to have equivalent module categories (Section 4.4), leading in Section 4.5 to Morita equivalence; Section 4.6 deals with flat modules and their relation to projective and injective modules, while Section 4.7 studies the homology of algebras and in particular separable algebras.

4.1 The Krull–Schmidt theorem

A finitely generated module over a general Artinian ring may not be semisimple, i.e. a direct sum of simple modules, but it can always be written as a direct sum of indecomposable modules. Moreover, these indecomposable summands do not depend on the decomposition chosen, but are unique up to isomorphism. This is the content of the Krull–Schmidt theorem; our aim in this section is to prove this result, but we shall do so in a slightly more general form.

We recall a *local ring* is a ring R in which the set of all non-units forms an ideal $\mathfrak{m}$; this is then the unique maximal ideal and it is easily seen that $R/\mathfrak{m}$ is a skew field, called the *residue class field* of R. From the definition it is clear that a ring is local precisely if the sum of any two non-units is a non-unit or equivalently, for any non-unit c, $1 - c$ is a unit. The maximal ideal of a local ring is its Jacobson radical;

since the latter is nilpotent in any Artinian ring (BA, Theorem 5.3.5), it follows that an Artinian ring is local iff every element is either nilpotent or a unit. Such rings arise naturally as endomorphism rings of indecomposable modules, as we shall now see. In what follows we shall write our modules as left modules and put module homomorphisms on the right, except when otherwise stated.

Lemma 4.1.1. (Fitting's lemma). *Let R be any ring and M an R-module of finite composition length. Given any endomorphism α of M, there exists a direct decomposition*

$$M = M_0 \oplus M_1, \tag{4.1.1}$$

such that M_0, M_1 both admit α, and α restricted to M_0 is nilpotent, while its restriction to M_1 is an automorphism.

Proof. We have $M \supseteq M\alpha \supseteq M\alpha^2 \supseteq \dots, 0 \subseteq \ker \alpha \subseteq \ker \alpha^2 \subseteq \dots$; since M has finite length, there exists n (at most equal to the length of M) such that $M\alpha^n = M\alpha^{n+1} = \dots$, $\ker \alpha^n = \ker \alpha^{n+1} = \dots$. We put $M_1 = M\alpha^n$, $M_0 = \ker \alpha^n$ and then have $M_1\alpha^n = M\alpha^{2n} = M\alpha^n = M_1$. Thus for any $x \in M, x\alpha^n = x_1\alpha^n$ for some $x_1 \in M_1$; hence $x - x_1 \in \ker \alpha^n = M_0$ and so $x \in M_0 + M_1$. This proves that $M = M_0 + M_1$, and this sum is direct, for if $y \in M_0 \cap M_1$, then $y = x\alpha^n$, hence $0 = y\alpha^n = x\alpha^{2n}$. Thus $x \in \ker \alpha^{2n} = \ker \alpha^n$ and $y = x\alpha^n = 0$. This shows the sum to be direct and it establishes the decomposition (4.1.1); clearly α is nilpotent on M_0 and bijective on M_1. ∎

If in this lemma M is indecomposable, then M_0 or M_1 is the zero module and we obtain

Corollary 4.1.2. *Let M be an indecomposable R-module of finite length. Then any endomorphism of M is either nilpotent or an automorphism, thus* $\mathrm{End}_R(M)$ *is a local ring.* ∎

A local ring in which every non-unit is nilpotent is said to be *completely primary.* Thus the endomorphism ring of an indecomposable module of finite length is completely primary. This result does not hold without some restriction on the module M beyond being indecomposable (see Exercise 5), but we remark that conversely, any module with local endomorphism ring is indecomposable. For if M is decomposable, say $M = M_1 \oplus M_2$, then the projection $e_1 : M \to M_1$ is an idempotent of $\mathrm{End}(M)$ which for a non-trivial decomposition is neither 0 nor 1. Hence $1 - e_1$ is not a unit and so $\mathrm{End}(M)$ cannot be local. Sometimes a module with local endomorphism ring is called 'strongly indecomposable'.

In what follows we shall state our conclusions for modules with local endomorphism rings. By Corollary 4.1.2 they will apply to any indecomposable modules of finite length, in particular to any finitely generated indecomposable modules over Artinian rings, but they will also apply in some other cases.

Lemma 4.1.3. *Let R be any ring and V an indecomposable R-module such that* $\mathrm{End}_R(V) = E$ *is a local ring with maximal ideal $\mathfrak{m}$. Given any R-module M and homo-*

morphisms $\alpha : V \to M$, $\beta : M \to V$, *we have* $\alpha\beta \in \mathfrak{m}$ *unless* V *is a direct summand of* M.

Proof. If $\alpha\beta \notin \mathfrak{m}$, it is a unit in E, so there exists $\gamma \in E$ such that $\alpha\beta\gamma = \gamma\alpha\beta = 1$. This shows that α is injective and we have an exact sequence

$$0 \to V \xrightarrow{\alpha} M \to \operatorname{coker} \alpha \to 0.$$

The sequence is split by $\beta\gamma$, hence $M \cong V \oplus \operatorname{coker} \alpha$. ■

In what follows we shall fix a left R-module V with local endomorphism ring $E = \operatorname{End}_R(V)$. The maximal ideal of E will be denoted by $\mathfrak{m}$ and we put $K = E/\mathfrak{m}$ and write $x \mapsto [x]$ for the natural homomorphism $E \to K$. Given any left R-module M, we can consider $\operatorname{Hom}_R(V, M)$ as left E-module in a natural way. We shall write $[V, M] = \operatorname{Hom}_R(V, M)/\mathfrak{m}\operatorname{Hom}_R(V, M)$; this is a left E-module which is annihilated by $\mathfrak{m}$, so it can be defined as a left vector space over K in a natural way. Similarly we can consider $\operatorname{Hom}_R(M, V)$ as a right E-module and hence define $[M, V] = \operatorname{Hom}_R(M, V)/\operatorname{Hom}_R(M, V)\mathfrak{m}$ as a right K-space. Next we define a bilinear mapping

$$b : [V, M] \times [M, V] \to K$$

by the following rule: Given $\alpha \in [V, M]$, $\beta \in [M, V]$, take homomorphisms f, g such that $[f] = \alpha$, $[g] = \beta$ and define

$$b(\alpha, \beta) = [fg]. \tag{4.1.2}$$

This is a well-defined operation on α, β, for if $[f] = [f']$, say $f' = f + \sum \lambda_i h_i$, where $\lambda_i \in \mathfrak{m}$, $h_i \in \operatorname{Hom}(V, M)$, then $[f'g] = [fg] + \sum \lambda_i [h_i g] = [fg]$, hence $[fg]$ depends only on $[f]$, not on f, and similarly for g. In this way (4.1.2) defines a pairing of the spaces $[V, M]$, $[M, V]$. We define the rank of b in the usual way as the rank of the matrix obtained by taking bases. Thus if (u_i) is a left K-basis of $[V, M]$ and (v_j) a right K-basis of $[M, V]$, then the rank of b is given by the rank of the matrix $(b(u_i, v_j))$. Clearly this is independent of the choice of bases; we shall denote it by $\mu_V(M)$, thus

$$\mu_V(M) = \operatorname{rk}(b(u_i, v_j)). \tag{4.1.3}$$

Suppose now that M is expressed as a direct sum: $M = \sum_1^s \oplus M_i$. It is clear that this gives rise to a direct sum decomposition for both $[V, M]$ and $[M, V]$:

$$[V, M] = \oplus[V, M_i], \quad [M, V] = \oplus[M_i, V].$$

Here a homomorphism $f : V \to M_i$ corresponds to a homomorphism $V \to M$ which is obtained by combining f with the canonical injection $M_i \to M$. Similarly a homomorphism $g : M_j \to V$ corresponds to a homomorphism $M \to V$ obtained by following the canonical projection $M \to M_j$ by g. It follows that the elements of $[V, M].[M, V]$ are $s \times s$ matrices with the members of $[V, M_i][M_i, V]$ along the main diagonal and zeros elsewhere. By Lemma 4.1.3, $[V, M_i][M_i, V]$ can be non-zero only if M_i has V as a direct summand; in particular it will be zero whenever

M_i is indecomposable and not isomorphic to V. All these arguments still apply when M_i is an infinite direct sum, and they allow us to draw the following conclusion:

Proposition 4.1.4. *Let R be a ring and V an R-module with local endomorphism ring. Given an R-module M which is expressed as a direct sum of indecomposable modules, the multiplicity of V in this direct sum is independent of the decomposition chosen for M and is equal to $\mu_V(M)$.*

Proof. Let the given decomposition of M be

$$M = \oplus M_\lambda, \tag{4.1.4}$$

and write $\mu_{M_\lambda}(M) = \mu_\lambda(M)$ for short. Then $[V, M][M, V]$ is a direct sum of terms $[V, M_\lambda][M_\lambda, V]$, by the above remarks, and since M_λ is indecomposable, we have

$$[V, M_\lambda][M_\lambda, V] = \begin{cases} K & \text{if } M_\lambda \cong V, \\ 0 & \text{otherwise.} \end{cases}$$

It follows that the rank $\mu_\lambda(M)$ is equal to the number of terms in (4.1.4) that are isomorphic to V, i.e. the multiplicity of V in (4.1.4). ■

Corollary 4.1.5. *Given any ring R and R-module M, let*

$$M = \oplus_I M_\lambda = \oplus_J N_\mu$$

be two direct decompositions of M into indecomposable modules. If the components in at least one of these decompositions have local endomorphism rings, then there is a bijection $\lambda \mapsto \lambda'$ from I to J such that $M_\lambda \cong N_{\lambda'}$.

Proof. Suppose that $\mathrm{End}_R(M_\lambda)$ is local for all λ. By Proposition 4.1.4, the multiplicity of each M_λ is the same in both decompositions and this allows us to construct the desired bijection. ■

For Artinian rings this conclusion may be stated as follows:

Theorem 4.1.6 (Krull–Schmidt theorem). *Let R be any Artinian ring. Any finitely generated R-module M has a finite direct decomposition*

$$M = M_1 \oplus \ldots \oplus M_r, \tag{4.1.5}$$

where the terms M_i are indecomposable, and this decomposition is unique up to isomorphism and the order of the terms; thus if also $M = N_1 \oplus \ldots \oplus N_s$, where each N_i is indecomposable, then $s = r$ and there is a permutation $i \mapsto i'$ of $1, \ldots, r$ such that $M_i \cong N_{i'}$.

Proof. Any finitely generated R-module over an Artinian ring R is Artinian (BA, Theorem 4.2.3) and so has finite length (BA, Theorem 5.3.9). Hence we can form a direct decomposition (4.1.5) with a maximum number of terms, and all terms are then indecomposable. By Corollary 4.1.2 each endomorphism ring $\mathrm{End}_R(M)$ is local, so we can apply Corollary 4.1.5 to conclude that the decompositions have isomorphic terms, possibly after reordering. ■

Theorem 4.1.6 was stated for finite groups by Joseph H. M. Wedderburn in 1909, and first completely proved by Robert Remak in 1911. It was later extended to abelian groups with operators by Wolfgang Krull in 1928 and to general groups with operators by Otto Yu. Schmidt 1928. In 1950 Goro Azumaya noted that the finiteness condition could be replaced by the condition that the endomorphism rings of the components be local (Corollary 4.1.5). The above presentation is based on a lecture by Sandy Green.

Exercises

1. Show that a ring R with Jacobson radical J is local iff R/J is a skew field.
2. Show that in any ring an idempotent with 1-sided inverse equals 1. Deduce that a ring R in which for each $a \in R$ either a or $1 - a$ has a 1-sided inverse, must be local. (Here 'a has a 1-sided inverse' is taken to mean: either $ax = 1$ or $xa = 1$ has a solution.)
3. Give an example of a non-Artinian non-local ring in which every element is either nilpotent or a unit. (Hint. Try the commutative case.)
4. Show that a ring R is indecomposable as module over itself iff R contains no idempotent $\neq 0, 1$.
5. Use Exercise 4 to give an example of an indecomposable module whose endomorphism ring is not local.
6. Let V be an indecomposable module which is not injective and let I be its injective hull. Show that $[V, I] \neq 0$ but $[V, I].[I, V] = 0$.
7. Let $M = M_1 \oplus \ldots \oplus M_r$ and suppose that V is an indecomposable module which is a direct summand of M. Show that V is a direct summand of M_i for some $i, 1 \leq i \leq r$.

4.2 The projective cover of a module

We have seen in Section 2.3 that every module has an injective hull. Dually one can define the projective cover of a module, but this does not exist for all modules. Later, in Section 4.3, we shall meet general conditions for its existence, but for the moment we shall show that its existence is assured over Artinian rings. Throughout this section we shall limit ourselves to finitely generated modules; we recall that over an Artinian ring every finitely generated module has finite composition length (BA, Theorem 4.2.3 and Theorem 5.3.9). We begin with a couple of auxiliary remarks which hold for general rings. Here a maximal submodule is understood to be among all the proper submodules.

Lemma 4.2.1. *Let R be a ring with Jacobson radical J and let M be a finitely generated R-module. Then*

$$JM \subseteq \cap M_\lambda, \tag{4.2.1}$$

where M_λ ranges over all maximal submodules of M. Moreover, if R/J is semisimple, then equality holds in (4.2.1) and we have

$$M/JM \cong \oplus S_\mu, \tag{4.2.2}$$

where the S_μ are simple modules, quotients of M.

Proof. Let M_1 be a maximal submodule of M. If $JM \not\subseteq M_1$, then $M_1 + JM = M$, hence by Nakayama's lemma (BA, Corollary 5.3.7), $M_1 = M$, which is a contradiction. Thus $JM \subseteq M_1$ and (4.2.1) follows. Assume now that $\bar{R} = R/J$ is semisimple; then $\bar{M} = M/JM$ may be regarded as an $\bar{R}$-module and hence is semisimple, so we obtain (4.2.2), for a family S_μ of simple modules. Combining the isomorphism (4.2.2) with the projection on S_μ we obtain a homomorphism $f_\mu : \bar{M} \to S_\mu$ whose kernel is a maximal submodule of $\bar{M}$ and so is of the form N_μ/JM, where N_μ is a maximal submodule of M. Hence $\cap N_\mu/JM = 0$, i.e. $\cap N_\mu = JM$, and since the N_μ form a subfamily of the M_λ, we now have equality in (4.2.1). ∎

A module homomorphism $f : M \to N$ will be called *essential* if it is surjective but its restriction to any proper submodule of M fails to be surjective. By a *projective cover* of a module M we shall understand a projective module P with an essential homomorphism $P \to M$. The following lemma is useful for testing for essentiality.

Lemma 4.2.2. *Let R be any ring. Given an R-module M, a finitely generated projective R-module P and a surjective homomorphism $\alpha : P \to M$, if $\ker \alpha \subseteq JP$, then α is essential. If R/J is semisimple, this sufficient condition is also necessary.*

Proof. Let P' be any submodule of P such that $P'\alpha = M$. Then for any $x \in P$ there exists $x' \in P'$ such that $x\alpha = x'\alpha$, i.e. $x \in P' + \ker \alpha$. Thus $P' + \ker \alpha = P$ and by Nakayama's lemma, $P' = P$; this shows α to be essential. Suppose now that R/J is semisimple and α is essential. Let P_1 be any maximal submodule of P; if $\ker \alpha \not\subseteq P_1$, then $P_1 + \ker \alpha = P$, hence $P_1\alpha = P\alpha = M$, contradicting the fact that α is essential. Therefore $\ker \alpha \subseteq P_1$ and now $\ker \alpha \subseteq \cap P_\lambda = JP$, by Lemma 4.2.1. ∎

Our first task is to prove the existence of projective covers in the Artinian case.

Proposition 4.2.3. *Let R be a left Artinian ring. Then any finitely generated left R-module has a projective cover.*

Proof. Let M be a finitely generated projective R-module; we can find a finitely generated projective module P mapping onto M. We choose P of shortest length and claim that in this case the homomorphism $\pi : P \to M$ is essential. For take a minimal submodule N of P such that $\pi|N$ is surjective and let $i : N \to P$ be the inclusion map and put $f = i\pi : N \to M$. Since P is projective, there is a map $g : P \to N$ such that $gf = \pi$; now π maps N onto M, so f maps $\mathrm{im}(g|N)$ onto M and by the minimality of N it follows that $\mathrm{im}(g|N) = N$. Thus $g|N$ is a surjective endomorphism of the module N of finite length, therefore it is an automorphism and so $N \cap \ker g = 0$. Given $x \in P$, we have $xg \in N$ and since $g|N$ is an automorphism, we have $xg = yg^2$ for some $y \in P$. Now $x - yg \in \ker g$ and $x = yg + (x - yg)$; this shows that $P = N \oplus \ker g$. Therefore N is projective, but this contradicts the minimality of P, unless $N = P$. ∎

When a projective cover exists, it must be unique; this can be proved quite generally.

Proposition 4.2.4. *If P, Q are projective covers of a module M (over any ring R), with essential homomorphisms $\alpha : P \to M$, $\beta : Q \to M$, then there is an isomorphism $\theta : P \to Q$ such that $\alpha = \theta\beta$.*

Proof. Since P is a projective module and β is surjective, there exists $\theta : P \to Q$ such that $\alpha = \theta\beta$. This map θ must be surjective, because β is essential, and so P splits over $\ker\theta$, say $P = Q_1 \oplus \ker\theta$, where $Q_1 \cong Q$. But then $Q_1\alpha = Q_1\theta\beta = M$, hence $Q_1 = P$ and so $\ker\theta = 0$. Thus θ is an isomorphism, as claimed. ∎

We shall denote the projective cover of a module M by $\mathbf{P}(M)$, bearing in mind that this operation P is not defined for all modules. There is a second operation closely related to P which is defined for all modules. For any module M we define its *top* as

$$\mathbf{T}(M) = M/JM, \tag{4.2.3}$$

where $J = \mathbf{J}(R)$ is the Jacobson radical. If M is finitely generated and $\mathbf{T}(M) = 0$, then $M = 0$; this is just the content of Nakayama's lemma. When M is finitely generated, then the natural homomorphism $\tau : M \to \mathbf{T}(M)$ is essential, for it is clearly surjective, and if $N \subset M$, then N is contained in a maximal submodule N_0 of M. We have $N_0 \supseteq JM$, hence the natural map $\nu : M \to M/N_0$ can be factored as $M \to \mathbf{T}(M) \to M/N_0$, where $\nu|N_0 = 0$, but this contradicts the fact that $\tau|N_0$ is surjective. This shows τ to be essential. When R is Artinian, or more generally, when R/J is semisimple, then $\mathbf{T}(M)$ is semisimple by Lemma 4.2.1.

We note that the projective cover, when it exists, may be obtained as the projective cover of its top:

Proposition 4.2.5. *For any ring R and any R-module M with a projective cover we have $\mathbf{P}(M) \cong \mathbf{P}(\mathbf{T}(M))$.*

Proof. The composition of two essential maps is clearly essential and so we have the essential map

$$\mathbf{P}(M) \to M \to \mathbf{T}(M), \tag{4.2.4}$$

so the result follows. ∎

Similarly, when forming the top, under the right conditions it does not matter whether we start from a given module or its projective cover.

Proposition 4.2.6. *Let R be a ring such that R/J is semisimple, and let M be a finitely generated R-module with a projective cover. Then*

$$\mathbf{T}(\mathbf{P}(M)) \cong \mathbf{T}(M). \tag{4.2.5}$$

Proof. Write $P = \mathbf{P}(M)$ and let N be the kernel of the essential map $\alpha : P \to M$. By Lemma 4.2.1, $N \subseteq JP$ and since α is surjective, it maps JP onto JM. Therefore $\mathbf{T}(P) = P/JP \cong P/N/JP/N \cong M/JM \cong \mathbf{T}(M)$.

In particular this shows that the projective cover of a simple module has a simple top.

Exercises

1. Verify that $\mathbf{T}$ is a functor. Under what circumstances is $\mathbf{P}$ a functor?
2. Show that the only finitely generated $\mathbf{Z}$-modules with a projective cover are the free modules.
3. Show that $\mathbf{T}(\mathbf{T}(M)) = \mathbf{T}(M)$, $\mathbf{P}(\mathbf{P}(M)) = \mathbf{P}(M)$.
4. Show that if $\mathbf{P}(M)$ is indecomposable, then so is M. Does the converse hold?
5. Show that if $\alpha : P \to M$ is a projective cover and $\beta : Q \to M$ is surjective, where Q is a projective module, then $Q = P_0 \oplus P_1$, where $P_1 \cong P$ and $\beta|P_1$ corresponds to α, while $\beta|P_0 = 0$. Use the result to give another proof of Proposition 4.2.4.

4.3 Semiperfect rings

We have seen that Artinian rings have many properties not shared by general rings, but there are certain classes of rings for which at least some of these properties hold. One such class is formed by the semiperfect rings, introduced by Hyman Bass in 1960. We begin by examining the role of idempotents in Artinian rings.

Any Artinian ring has finite composition length as left module over itself and so can be decomposed into a direct sum of indecomposable left modules. Such decompositions can be described by the corresponding decompositions of 1 as a sum of idempotents. Let us recall that two idempotents e, f of a ring R are called *orthogonal* if $ef = fe = 0$; an idempotent e is *primitive* if $e \neq 0$ *and* e cannot be written as a sum of two non-zero orthogonal idempotents. In general rings idempotents can be used to describe direct decompositions, as our first result shows.

Proposition 4.3.1. *Let R be any ring. Any decomposition of R as a direct sum of a finite number of left ideals*

$$R = \mathfrak{a}_1 \oplus \ldots \oplus \mathfrak{a}_r \tag{4.3.1}$$

corresponds to a decomposition of 1 as a sum of pairwise orthogonal idempotents

$$1 = e_1 + \ldots + e_r,\ e_i^2 = e_i,\ e_ie_j = 0 \quad \text{for } i \neq j, \tag{4.3.2}$$

where $\mathfrak{a}_i = Re_i$, *and for any idempotent e,* Re *is indecomposable if and only if e is primitive.*

Proof. Given (4.3.1), we write 1 as Re_i, where $e_i \in \mathfrak{a}_i$. Then $e_k = \sum_i e_k e_i$; since the sum (4.3.1) is direct, we see that $e_k e_i = 0$ for $i \neq k$ and $e_k^2 = e_k$, so (4.3.2) follows. Conversely, given (4.3.2), put $\mathfrak{a}_i = Re_i$; then any $x \in R$ can be written as $x = \sum xe_i$, where $xe_i \in \mathfrak{a}_i$, hence $R = \sum \mathfrak{a}_i$ and this sum is direct, for if $\sum x_i e_i = 0$, then right multiplication by e_k gives $x_k e_k = 0$.

For any idempotent e, Re is a direct summand of $R : R = Re + R(1-e)$; now the above correspondence shows that any direct decomposition of Re corresponds to writing e as a sum of two orthogonal idempotents. It follows that Re is indecomposable iff e is primitive. ■

In the Artinian case one can construct *complete* decompositions (4.3.1) (i.e. into indecomposable terms) by writing the semisimple ring R/J as a direct sum of simple left ideals and then 'lifting' this decomposition to R. The essential step is to lift an idempotent from R/J to R; below we shall see how to do this for Artinian rings and then go on to describe a somewhat larger class of rings for which this is possible.

Let R be any ring and $\mathfrak{N}$ an ideal of R; an element $u \in R$ such that $u^2 \equiv u \pmod{\mathfrak{N}}$ is called an *idempotent mod* $\mathfrak{N}$, and we say that u can be *lifted* to R if there exists $e \in R$ such that $e^2 = e$ and $e \equiv u \pmod{\mathfrak{N}}$. For such a lifting to be possible one usually has to assume that $N \subseteq \mathrm{J}(R)$, but this by itself is not enough. For example, in the ring R of rational numbers with denominators prime to 6, $\mathrm{J}(R) = 6R$ and 3, 4 are idempotents mod $6R$, which however cannot be lifted to R. The next result gives a sufficient condition. We recall that a nil ideal is an ideal consisting of nilpotent elements.

Lemma 4.3.2. *Let R be a ring and* $\mathfrak{N}$ *a nil ideal in R. Then idempotents* mod $\mathfrak{N}$ *can be lifted to R.*

Proof. Let u be an idempotent mod $\mathfrak{N}$; then $(u-u^2)^m = 0$ for some $m \geq 0$. We have

$$1 = [u + (1-u)]^{2m-1} = \sum \binom{2m-1}{i} u^{2m-i}(1-u)^{i-1}.$$

On the right the first m terms are divisible by u^m, while each term after the first m is divisible by $(1-u)^m$, so on denoting the sum of the first m terms by e, we can write $1 = e + (1-u)^m g$, where g is a polynomial in u. Now $u(1-u) \in \mathfrak{N}$, so

$$e = u^{2m-1} + (2m-1)u^{2m-2}(1-u) + \ldots + \binom{2m-1}{m-1} u^m (1-u)^{m-1} \equiv u (\mathrm{mod}\ \mathfrak{N}),$$

and $e(1-e) = e(1-u)^m g = 0$, hence e is an idempotent. ■

In an Artinian ring R the Jacobson radical $\mathrm{J}(R)$ is nilpotent, so in this case any idempotent mod $\mathrm{J}(R)$ can be lifted to R, by Lemma 4.3.2, but there are other cases where this is possible and it is convenient to make a definition at this point, introduced by Hyman Bass [1960]:

Definition. A ring R is said to be *semiperfect* if idempotents (mod $\mathbf{J}(R)$) can be lifted to R and $R/\mathbf{J}(R)$ is semisimple.

It is clear from the definition that this notion is left–right symmetric. Moreover, since for any ring R, $R/J(R)$ has zero radical, the latter is semisimple iff it is right (or left) Artinian, by BA, Theorem 5.3.5. In particular, this establishes

Theorem 4.3.3. *Every left or right Artinian ring is semiperfect.* ■

Of course the converse does not hold; the class of semiperfect rings is much wider than the class of Artinian rings (see Exercise 11). To describe semiperfect rings we shall need to know when two idempotents generate isomorphic ideals. Let us call two idempotents e, f in a ring R *conjugate* if R contains a unit u such that $f = u^{-1}eu$. If there exist $a \in eRf$, $b \in fRe$ such that $ab = e$, $ba = f$, we shall call e and f *isomorphic*. For example, conjugate idempotents are isomorphic, for if $f = u^{-1}eu$, we can take $a = eu$, $b = u^{-1}e$ and then find that $a = euf$, $b = fu^{-1}e$ and $ab = e$, $ba = f$. The next lemma clarifies the relation between these two concepts.

Lemma 4.3.4. *Let e, f be any idempotents in a ring R. Then*

(i) *e, f are isomorphic if and only if $eR \cong fR$ or equivalently, $Re \cong Rf$,*
(ii) *e, f are conjugate if and only if e is isomorphic to f and $1 - e$ is isomorphic to $1 - f$.*

Proof. (i) Assume that $Re \cong Rf$, say $\theta : Re \to Rf$ is an isomorphism and suppose that θ maps e to a, while θ^{-1} maps f to b. Since $a \in Rf$, we have $af = a$ and $e(e\theta) = e\theta$, i.e. $ea = a$. Hence $eaf = af = a$ and similarly $fbe = b$. Further, $e = e\theta\theta^{-1} = a\theta^{-1} = (af)\theta^{-1} = a.f\theta^{-1} = ab$, and similarly $f = ba$, so e is isomorphic to f. Conversely, if e, f are isomorphic, say $ab = e$, $ba = f$, where $a \in eRf$, $b \in fRe$, then $x \mapsto xa$ is a homomorphism from Re to Rf with inverse $y \mapsto yb$. Hence $Re \cong Rf$ iff e, f are isomorphic, and by symmetry this is equivalent to $eR \cong fR$.

(ii) If e, f are conjugate, say $f = u^{-1}eu$, then as we saw, e and f are isomorphic. Further $1 - f = u^{-1}(1 - e)u$, so $1 - e$, $1 - f$ are also isomorphic. Conversely, if e, f are isomorphic and $1 - e$, $1 - f$ are isomorphic, say $ab = e$, $ba = f$, where $a = eaf$, $b = fbe$, and $a'b' = 1 - e$, $b'a' = 1 - f$, $a' = (1 - e)a'(1 - f)$, $b' = (1 - f)b'(1 - e)$, let us put $A = a + a', B = b + b'$. Then $b'(1 - e) = b'$, so $b'e = 0$ and similarly $ea' = 0$ and hence $BeA = bea + b'ea + bea' + b'ea' = bea = f$. For the same reasons $B(1 - e)A = 1 - f$, hence $BA = 1$. By symmetry $AfB = e$, $A(1 - f)B = 1 - e$ and so $AB = 1$. This shows e, f to be conjugate, as claimed. ■

This lemma together with the Krull–Schmidt theorem (Theorem 4.1.6) shows that in an Artinian ring isomorphic idempotents are conjugate, so in this case isomorphism and conjugacy for idempotents mean the same thing.

We shall usually want to lift orthogonal families of idempotents; this can be accomplished without further hypotheses on the ring.

Proposition 4.3.5. *Let R be any ring and e, f idempotents in R. Write $J = \mathrm{J}(R)$ and denote the natural homomorphism $R \to R/J$ by $x \mapsto [x]$. Then*

(i) *$e = 0$ or 1 if and only if $[e] = 0$ or 1 respectively;*
(ii) *e is isomorphic to f if and only if $[e]$ is isomorphic to $[f]$; in particular, if $[e] = [f]$, then e is conjugate to f;*
(iii) *if $ef \equiv fe \equiv 0 \pmod J$, then there exists an idempotent f_1 such that $f_1 \equiv f \pmod J$ and $ef_1 = f_1 e = 0$.*

Proof. (i) If $e \in J$, then $1 - e$ is a unit and since $e(1-e) = 0$, we have $e = 0$. Similarly if $1 - e \in J$ and the converse is clear.

(ii) Let $a, b \in R$ be such that $a \equiv eaf$, $b \equiv fbe$, $ab \equiv e$, $ba \equiv f \pmod J$ and put $a_1 = eaf$, $b_1 = fbe$. Then $a_1 b_1 = e - z$, where $z \in eJe$. Let z' be the quasi-inverse of z (i.e. $z + z' = zz' = z'z$) and put $z'' = ez'e$; then z'' is a quasi-inverse of z in eRe, and it follows that $a_1 b_1 (e - z'') = e$. Putting $a_2 = a_1, b_2 = b_1(e - z'')$, we have $a_2 \equiv a$, $b_2 \equiv b \pmod J$ and $a_2 b_2 = e$. Next write $b_2 a_2 = f - y$; then $y \in fJf$ and since $(b_2 a_2)^2 = b_2 e a_2 = b_2 a_2$, it follows that $f - y = (f - y)^2 = f^2 - fy - yf + y^2 = f - 2y + y^2$. Thus $y(y-1) = 0$, and since $y \in J$, we find that $y = 0$ and so $b_2 a_2 = f$. Thus e is isomorphic to f; the rest is clear.

(iii) If $ef \equiv fe \equiv 0 \pmod J$, then $1 - fe$ is invertible. Put

$$f_0 = (1 - fe)^{-1} f (1 - fe);$$

then f_0 is an idempotent conjugate to f. Moreover, $f_0 \equiv f \pmod J$ and clearly $f_0 e = 0$. Writing $f_1 = (1 - e) f_0$, we have $f_1 = f_0 - ef_0 \equiv f_0 - ef \equiv f_0 \equiv f \pmod J$ and $f_1 e = 0 = ef_1$; moreover, $f_1^2 = (1-e) f_0 (1-e) f_0 = (1-e) f_0^2 = f_1$, so f_1 is the required idempotent. ∎

We shall use this result to lift decompositions of the form (4.3.2), again without further hypothesis:

Proposition 4.3.6. *In any ring R, let $e_1, \ldots, e_r$ be a set of idempotents such that $e_i e_j \equiv 0 \pmod J$ for $i \neq j$. Then there exist idempotents e_i' such that $e_i' \equiv e_i \pmod J$ and $e_i' e_j' = 0$ for $i \neq j$. If moreover,*

$$1 = e_1 + \ldots + e_r, \quad e_i e_j \equiv 0 \pmod J, \quad i \neq j, \tag{4.3.3}$$

then there exist idempotents e_i' such that $e_i' \equiv e_i \pmod J$, $1 = \sum e_i'$ and $e_i' e_j' = 0$ for $i \neq j$.

Proof. For $n = 1$ there is nothing to prove, so we assume that $n > 1$ and use induction on n. This means that we may assume $e_i e_j = 0$ for $i \neq j, i, j > 1$. Put $e = e_2 + \ldots + e_r$; then e is again idempotent and $e_1 e \equiv ee_1 \equiv 0 \pmod J$. By Proposition 4.3.5 there exists an idempotent e_1' such that $e_1' \equiv e_1 \pmod J$ and $ee_1' = e_1' e = 0$. It follows that $e_1', e_2, \ldots, e_r$ are pairwise orthogonal idempotents.

Assume now that (4.3.3) holds and choose the e_i' as in the first part. The $e = e_1' + \ldots + e_r'$ is an idempotent such that $e \equiv 1 \pmod{J}$, hence $e = 1$ by Proposition 4.3.5(i). ■

The first part remains true (with the same proof) for a countable set of idempotents, but it ceases to hold for uncountable sets (Zelinsky, 1954).

Proposition 4.3.6 shows that every semiperfect ring R can be written as $R = \sum Re_i$, where the e_i are primitive and so the Re_i are indecomposable. To establish the uniqueness of such decompositions we shall want to apply the Krull–Schmidt theorem and we need to check that the endomorphism ring of an indecomposable left ideal Re is local. Here we shall need a couple of elementary lemmas:

Lemma 4.3.7. *A semiperfect ring in which 1 is a primitive idempotent is a local ring.*

Proof. The semisimple ring R/J can be written as a direct sum of a finite number of simple left ideals, and this decomposition can be lifted to R. Since 1 is primitive, it is primitive (mod J), hence there is a single summand and so R/J is a skew field. This means that R is a local ring. ■

Lemma 4.3.8. *Let R be any ring, e an idempotent in R and M a left R-module. Then there is an isomorphism of left eRe-modules:*

$$\mathrm{Hom}_R(Re, M) \cong eM. \tag{4.3.4}$$

In particular, taking $M = Re$, we obtain a ring isomorphism

$$\mathrm{End}_R(Re) \cong eRe. \tag{4.3.5}$$

Proof. Each homomorphism $\alpha : Re \to M$ is completely determined by its effect on e. If $e\alpha = u$, then $x\alpha = (xe)\alpha = x(e\alpha) = xu$; in particular, $u = e\alpha = eu \in eM$. Conversely, for any $u \in eM$ the mapping $\alpha_u : xe \mapsto xu$ is a homomorphism from Re to M, and the correspondence $u \mapsto \alpha_u$ is additive, as is easily checked. It is surjective, as we have seen, and if $\alpha_u = 0$, then $u = eu = 0$, so it is injective and hence an isomorphism of abelian groups, or more generally, left eRe-modules. This establishes (4.3.4). If $M = Re$, both sides acquire a multiplicative structure, which is again compatible with the isomorphism, so we have a ring isomorphism (4.3.5). ■

Over a semiperfect ring R, the top of any finitely generated R-module M can be written

$$\mathrm{T}(M) = \oplus S_\mu,$$

where each S_μ is a simple quotient of M, by Lemma 4.2.1. We shall use this remark to show that projective covers exist over a semiperfect ring.

Theorem 4.3.9. *Any finitely generated module over a semiperfect ring R has a projective cover. More precisely, the projective module P is a projective cover for M if and only if* $\mathrm{T}(P) \cong \mathrm{T}(M)$.

Proof. Let P be a projective cover for M, say $P/N \cong M$, where $N \subseteq JP$, by Lemma 4.2.1. Then $JM \cong JP/N$, hence $\mathbf{T}(M) = M/JM \cong P/N/JP/N \cong P/JP = \mathbf{T}(P)$, so the condition is necessary. Now let M be any finitely generated left R-module and put $\bar{R} = R/J$. As we have just seen, $\mathbf{T}(M)$ is a finite direct sum of simple R-modules, which may also be regarded as simple $\bar{R}$-modules; further, any simple $\bar{R}$-module has the form $\bar{R}\bar{e}$, where $\bar{e}$ is a primitive idempotent in $\bar{R}$. By definition of R, $\bar{e}$ lifts to an idempotent e in R, which is again primitive, by Proposition 4.3.5. Write $P = Re$ for the corresponding indecomposable projective and put $P = \oplus P$. Then $\mathbf{T}(P) \cong \oplus\bar{R}\bar{e} \cong \mathbf{T}(M)$. More generally, given any projective module P and an isomorphism $\theta : \mathbf{T}(P) \to \mathbf{T}(M)$, we have a diagram

$$\begin{array}{ccccc} P & \longrightarrow & P/JP & \longrightarrow & 0 \\ \vdots\, \alpha & & \downarrow\, \theta & & \\ M & \longrightarrow & M/JM & \longrightarrow & 0 \end{array}$$

Since P is projective, there exists a to make the diagram commutative. Given $x \in M$, there exists $a \in P$ such that $\bar{a}\theta = \bar{x}$, hence $P\alpha + JM = M$, and so by Nakayama's lemma, $P\alpha = M$. Since θ is injective, $\ker \alpha \subseteq JP$, therefore by Lemma 4.2.2, P is a projective cover for M. ∎

This result and its proof allows us to view finitely generated modules over a semiperfect ring in a new light. With every such module M we associate on the one hand its top T and on the other its projective cover P. We have essential mappings

$$P \to M, \quad P \to T,$$

and P, T are the largest resp. smallest modules for which such essential mappings exist, for a given M. We conclude with an analogue of the Krull–Schmidt theorem for projective modules over semiperfect rings.

Theorem 4.3.10. *Let R be a semiperfect ring. Every finitely generated projective left R-module P can be written as a direct sum*

$$P = P_1 \oplus \ldots \oplus P_r, \tag{4.3.6}$$

where each P_i is isomorphic to an indecomposable left ideal which is a direct summand of R, and the P_i are unique up to isomorphism and order.

Proof. We have seen that R has a direct decomposition into indecomposable left ideals, e.g. by lifting a complete direct decomposition of R/J. Hence for any $n \geq 1$, R^n can be written as a direct sum of indecomposable modules isomorphic to left ideals; by Lemmas 4.3.8 and 4.3.7 each such left ideal has as endomorphism ring a local ring, therefore Corollary 4.1.5 can be applied to establish the uniqueness of this decomposition, up to isomorphism and order. Now if P is any finitely generated projective left R-module, we have $P \oplus P' \cong R^n$ for some $n \geq 1$, and so we obtain a decomposition $P \oplus P' \cong \oplus Q_i$, where each Q_i is isomorphic to an indecomposable left ideal. If we now take a direct decomposition of P with the maximal number of terms and apply Proposition 4.1.4, we obtain a decomposition of the required form. ■

It can be shown that semiperfect rings form the precise class of rings for which every finitely generated module has a projective cover. A ring over which every left module has a projective cover is said to be *left perfect*. Such a ring R is characterized by the fact that R/J is semisimple and J is *right vanishing* (or also *left T-nilpotent*), i.e. for any infinite sequence $\{a_\nu\}$ in J there exists n such that $a_1 \dots a_n = 0$ (see Bass [1960]).

Exercises

1. Show that an idempotent e in a ring R is primitive iff eRe is non-trivial and has no idempotents $\neq 0, 1$.
2. Let R be a ring and $\mathfrak{a}$ a minimal left ideal. Show that either $\mathfrak{a}^2 = 0$ and $\mathfrak{a}R$ is a nilpotent two-sided ideal in R, or $\mathfrak{a} = Re$ for some idempotent e, and hence $\mathfrak{a}$ is a direct summand in R.
3. Find conditions on idempotents e, f for $Re = Rf$ to hold.
4. Show that if $Re \cong Rf$ for idempotents e, f where e is central, then $f = ef = fe$. If f is also central, deduce that $e = f$.
5. Let R be a local ring and P a finitely generated projective left R-module. Show by lifting a basis of $\mathbf{T}(P)$ that P is free.
6. Let R be a semiperfect ring such that R/J is simple. Show that R is a full matrix ring over a local ring. If further, R is an integral domain, deduce that it must be a local ring.
7. Let e be a central idempotent in a ring R. Show that if e_1 is an idempotent such that $e_1 \equiv e \pmod{J}$, then $e_1 = e$.
8. Show that if $1 = \sum e_i = \sum f_i$ are two decompositions of 1 into orthogonal families of idempotents such that $e_i \equiv f_i \pmod{J}$, then $u = \sum e_i f_i$ is a unit and $f_i = u^{-1} e_i u$.
9. Show that if R is semiperfect, then so is R_n for all $n \geq 1$.
10. Show that a commutative Artinian ring is a direct product of completely primary rings (i.e. every non-unit is nilpotent). Give a counter-example in the non-commutative case.
11. Show that a commutative ring is semiperfect iff it is a direct product of finitely many local rings. Show that this ring is (left and right) perfect iff the maximal ideal of each local factor is vanishing.

4.4 Equivalence of module categories

A natural question asks when two rings A, B have equivalent module categories. We recall from BA, Section 4.4 that two rings A, B are called *Morita equivalent*, $A \sim B$, or simply *equivalent* if there is a category equivalence $\mathrm{Mod}_A \cong \mathrm{Mod}_B$. There we saw too that any ring A is equivalent to A_n for all $n \geq 1$; however there are also other cases and in this section and the next we shall find precise conditions for A and B to be

Morita equivalent. We begin by describing the notion of a generator which plays an important role in what follows.

By a *generator* in an abelian category $\mathscr{A}$ one understands an object P in $\mathscr{A}$ such that $h^P = \mathscr{A}(P, -)$ is faithful. An equivalent condition is that every $\mathscr{A}$-object is a quotient of a copower of P (i.e. a direct sum of copies of P). For if P is a generator and for any $\mathscr{A}$-object X we put

$$S = \coprod P_f,$$

where $P_f \cong P$ and f runs over $\mathscr{A}(P, X)$, with natural injection $i_f : P_f \to S$; then the family of maps $f : P_f \to X$ gives rise to a map $F : S \to X$ such that $f = i_f F$. To establish that X is a quotient of S we show that F is epic. Let $g : X \to \operatorname{coker} F$ be the natural map. Then $Fg = 0$, hence $fg = 0$ for all $f \in \mathscr{A}(P, X)$ and since P is a generator, it follows that $g = 0$. Thus $\operatorname{coker} F = 0$ and F is epic, as claimed.

Conversely, assume that X is a quotient of $S = \coprod P_\lambda$, where $P_\lambda \cong P, F : S \to X$ and write $i_\lambda : P_\lambda \to S$ for the natural injection. Given $f : X \to Y$, $f \neq 0$, we have $Ff \neq 0$ because F is epic, so $i_\lambda F f \neq 0$ for some λ, but $i_\lambda F \in \mathscr{A}(P, X)$. This shows P to be a generator.

It is clear that in the category Mod_R of all right R-modules, R is a generator. The existence of a generator gives rise to a useful criterion for a natural isomorphism of functors.

Theorem 4.4.1. *Let $\mathscr{A}$, $\mathscr{B}$ be abelian categories with coproducts and let F, G be functors from $\mathscr{A}$ to $\mathscr{B}$ which are right exact and preserve coproducts. If there is a natural transformation*

$$t : F \to G, \tag{4.4.1}$$

such that for some generator P of $\mathscr{A}$, the map

$$t_P : P^F \to P^G \tag{4.4.2}$$

is an isomorphism, then t is a natural isomorphism.

Proof. For any $X \in Ob\,\mathscr{A}$ we have a short exact sequence

$$S_2 \to S_1 \to X \to 0,$$

where S_1, S_2 are copowers of P. Applying F and G we have the following commutative diagram with exact rows (by the right exactness of F and G):

$$\begin{array}{ccccccc} S_2^F & \longrightarrow & S_1^F & \longrightarrow & X^F & \to & 0 \\ \downarrow t_2 & & \downarrow t_1 & & \downarrow t_X & & \\ S_2^G & \longrightarrow & S_1^G & \longrightarrow & X^G & \to & 0 \end{array}$$

By hypothesis $S_2^F = (\coprod P_\lambda)^F = \coprod P_\lambda^F$, hence $t_2 = t_{s_2}$ is an isomorphism, and likewise t_1. By the 5-lemma, t_X is an isomorphism, as asserted. ∎

Let us now consider right exact functors $\mathrm{Mod}_A \to \mathrm{Mod}_B$ which preserve direct sums (coproducts). An example of such a functor is $\otimes_A M$, where M is an (A, B)-bimodule. The next result shows that this is essentially the only case:

Lemma 4.4.2 (Eilenberg–Watts). *For any functor* $S : \mathrm{Mod}_A \to \mathrm{Mod}_B$ *between module categories the following conditions are equivalent:*

(a) *S has a right adjoint $T : \mathrm{Mod}_B \to \mathrm{Mod}_A$;*
(b) *S is right exact and preserves direct sums;*
(c) $S = \otimes_A P$ for some (A, B)-bimodule P.

Moreover, when (a)–(c) *hold, then the right adjoint T is given by*

$$Y^T = \mathrm{Hom}_B(P, Y),$$

where $P = A^S$, and T is unique up to natural isomorphism.

Proof. (a) $\Rightarrow$ (b) follows by Theorem 2.2.7, since S is a left adjoint.

(b) $\Rightarrow$ (c). Given a functor S, right exact and preserving direct sums, put $P = A^S$. Each A-endomorphism of A_A induces a B-endomorphism of P, but the A-endomorphisms of A_A are just the left multiplications by elements of A (BA, Theorem 5.1.3 or also Lemma 4.3.8 above); thus P is an (A, B)-bimodule. Now consider the functors S and $\otimes_A P$: both are right exact and preserve direct sums, so to show their isomorphism we need only, by Theorem 4.4.1, find a natural transformation between them which, for the generator A of Mod_A, is an isomorphism. Given $X \in \mathrm{Mod}_A$, we have a map

$$X \xrightarrow{\cong} \mathrm{Hom}_A(A, X) \xrightarrow{S} \mathrm{Hom}_B(A^S, X^S) \cong \mathrm{Hom}_B(P, X^S). \tag{4.4.3}$$

The result is a map $f_X : X \to \mathrm{Hom}_B(P, X^S)$ which is an A-module homomorphism. By adjoint associativity we have

$$f_X \in \mathrm{Hom}_A(X, \mathrm{Hom}_B(P, X^S)) \cong \mathrm{Hom}_B(X \otimes P, X^S).$$

Each step in (4.4.3) is natural in X, so f_X is a natural transformation from $X \otimes P$ to X^S. For $X = A$ it clearly reduces to the identity, hence it is a natural isomorphism, as claimed.

(c) $\Rightarrow$ (a). Given (c), we define functor $T : \mathrm{Mod}_B \to \mathrm{Mod}_A$ by

$$Y \longmapsto \mathrm{Hom}_B(P, Y).$$

Then $\mathrm{Hom}_B(X^S, Y) \cong \mathrm{Hom}_B(X \otimes P, Y) \cong \mathrm{Hom}_A(X, \mathrm{Hom}_B(P, Y)) \cong \mathrm{Hom}_B(X, Y^T)$, where each step is natural in X and Y. Thus T is the required right adjoint of S.

When (a)–(c) hold, we have

$$Y^T \cong \mathrm{Hom}_A(A, Y^T) \cong \mathrm{Hom}_B(A^S, Y) \cong \mathrm{Hom}_B(P, Y),$$

where $P = A^S$, and here P is clearly unique up to isomorphism. ■

We shall introduce a preordering of module categories on the basis of the next result:

Proposition 4.4.3. *For any rings A, B the following are equivalent:*

(a) *there exists a functor* $S : \mathrm{Mod}_A \to \mathrm{Mod}_B$ *with right adjoint* $T : \mathrm{Mod}_B \to \mathrm{Mod}_A$ *such that T has a right adjoint and* $ST \cong 1$;
(b) *there exist functors* $S : \mathrm{Mod}_A \to \mathrm{Mod}_B$, $T : \mathrm{Mod}_B \to \mathrm{Mod}_A$, *both right exact and preserving direct sums, such that* $ST \cong 1$;
(c) *there exist modules* ${}_AP_B$, ${}_BQ_A$ *such that* $P \otimes_B Q \cong A$, *as bimodules.*

If one (and hence all) of (a)–(c) holds, we shall call Mod_A a *quotient category* of Mod_B and write $A \prec B$. The functor S is called the *section functor* and T the *retraction functor.*

Proof. (a) $\Rightarrow$ (b) follows by Lemma 4.4.2 and (b) $\Rightarrow$ (a) will follow if we show T to be right adjoint to S. This follows because we have the natural transformation

$$\mathrm{Hom}_B(X^S, Y) \xrightarrow{T} \mathrm{Hom}_A(X^{ST}Y^T) \cong \mathrm{Hom}_A(X, Y^T),$$

which is an isomorphism for $X = A$.

Moreover, if (b) holds, then by Lemma 4.4.2, $S \cong - \otimes_A P$, $T \cong - \otimes_B Q$, for some ${}_AP_B$, ${}_BQ_A$, and $ST \cong 1$ means that $P \otimes_B Q \cong A$, so (c) holds. Conversely, this condition clearly implies (A). ■

We list some properties of quotient categories:

Proposition 4.4.4. *Let A, B be rings such that* $A \prec B$, *with modules* ${}_AP_B$, ${}_BQ_A$ *satisfying* $P \otimes Q \cong A$. *Then*

(i) $Q \cong \mathrm{Hom}_B(P, B)$, $P \cong \mathrm{Hom}_B(Q, B)$;
(ii) $A \cong \mathrm{End}_B(P) \cong \mathrm{End}_B(Q)$;
(iii) P_B *and* ${}_BQ$ *are projective;*
(iv) ${}_AP$ *and* Q_A *are generators;*
(v) *we have the following lattice homomorphisms with right inverses ('retractions'):*

$\mathrm{Lat}(A_A) \to \mathrm{Lat}(P_B)$ *with 2-sided ideals of A corresponding to* (A, B)*-submodules of P,*

$\mathrm{Lat}({}_AA) \to \mathrm{Lat}({}_BQ)$ *with 2-sided ideals of A corresponding to* (B, A)*-submodules of Q.*

Proof. In each case it is enough to prove the first part; the second then follows by symmetry.

(i) Write $S = \otimes P$, $T = \otimes Q$; we have

$$\mathrm{Hom}_B(P, B) \cong \mathrm{Hom}_B(A^S, B) \cong \mathrm{Hom}_A(A, B^T) \cong Q.$$

(ii) We have the bimodule homomorphisms

$$\mathrm{End}_B(P) \cong \mathrm{Hom}_B(A^S, P) \cong \mathrm{Hom}_A(A, P^T) \cong \mathrm{Hom}_A(A, A) \cong A.$$

Each term has a natural multiplication; all these correspond and give a ring isomorphism.

(iii) We have

$$\mathrm{Hom}_B(P, Y) \cong \mathrm{Hom}_A(A, Y^T) \cong Y^T,$$

and by hypothesis the functor $T : Y \mapsto Y^T$ is right exact. Hence $\mathrm{Hom}_B(P, -)$ is right exact and so P_B is projective.

(iv) We have $P \oplus P' \cong {}^I B$, where ${}^I B$ stands for a direct sum of $|I|$ copies of B. Hence

$${}^I Q \cong {}^I B \otimes Q \cong (P \otimes Q) \oplus (P' \otimes Q) \cong A \oplus (P' \otimes Q).$$

This shows that Q is a generator, because A is one.

(v) It is clear that the functor S induces a map from $\mathrm{Lat}({}_A A)$ to $\mathrm{Lat}({}_B P)$ which is order-preserving and has a right inverse, induced by T; further, A-bimodules (= ideals) of A correspond to (A, B)-bimodules in P. ■

Later, in Theorem 4.5.4, we shall find that the modules P and Q are actually finitely generated. For the moment we note that Proposition 4.4.3 leads to a criterion for Morita equivalence:

Theorem 4.4.5. *For any rings A, B the following conditions are equivalent:*

(a) $\mathrm{Mod}_A \cong \mathrm{Mod}_B$,

(a^0) ${}_A\mathrm{Mod} \cong {}_B\mathrm{Mod}$,

(b) *there exist bimodules* ${}_A P_B$, ${}_B Q_A$ *with bimodule isomorphisms*

$$P \otimes_B Q \cong A, \quad Q \otimes_A P \cong B.$$

Moreover, when (b) *holds and* $f : P \otimes Q \to A$, $g : Q \otimes P \to B$ *are bimodule isomorphisms, these may be chosen so as to make the following diagrams commutative:*

$$\begin{array}{ccccccc} P \otimes Q \otimes P & \xrightarrow{f \otimes 1} & A \otimes P & & Q \otimes P \otimes Q & \xrightarrow{g \otimes 1} & B \otimes Q \\ \downarrow 1 \otimes g & & \downarrow \cong & & \downarrow 1 \otimes f & & \downarrow \cong \\ P \otimes B & \xrightarrow{\cong} & P & & Q \otimes A & \xrightarrow{\cong} & Q \end{array}$$

Proof. The equivalence of (a), (b) is clear by the proof of Proposition 4.4.3, and now the equivalence of (a^0), (b) follows by the evident symmetry of (b). Let us pick isomorphisms $f : P \otimes Q \to A$, $g : Q \otimes P \to B$; then all the arrows in the diagrams are isomorphisms. Consider the first diagram. If we take $p \in P$ and move it (anticlockwise) round the square we obtain $\theta p \to P$, where θ is an (A, B)-automorphism of P. Now $\mathrm{End}_B(P) \cong \mathrm{End}_A(A) \cong A^0$, so θ is left multiplication by a unit u in A; since θ is also an A-automorphism, u lies in the centre of A. If we replace f by uf, then the first square becomes commutative. We complete the proof by showing that with this choice of f, g the second square also commutes. For brevity write $f(p \otimes q) = (p, q), g(q \otimes p) = [q, p]$; we have adjusted f so that

$$(p, q)p' = p[q, p'], \tag{4.4.4}$$

and we must show that

$$[q, p]q' = q(p, q'). \tag{4.4.5}$$

Given $p, p' \in P$, $q, q' \in Q$, we have by the left and right B-linearity of g,

$$[[q, p]q', p'] = [q, p][q', p'] = [q, p[q', p']].$$

By (4.4.4) and the fact that g is A-balanced, this is

$$[q, p[q', p']] = [q, (p, q')p'] = [q(p, q'), p'].$$

Thus if $c = [q, p]q' - q(p, q')$, then $c \in Q$ and $[c, p'] = 0$ for all $p' \in P$. Let us define $h : A \to Q$ by $h(a) = ca$; then $(h \otimes 1)g : A \otimes P \to Q \otimes P \to B$ and this map is zero because $[c, p'] = 0$. But g is an isomorphism, so $h \otimes 1 = h^T = 0$ and T is an equivalence, hence $h = 0$. Thus $c = h(1) = 0$, as we had to show. ∎

To illustrate the result, we take $B = A_n$ for some $n > 1$. Then we may choose $P = A^n$, $Q = {}^nA$ and it is clear that $P \otimes Q \cong A$, $Q \otimes P \cong A_n$.

As a first consequence we see how to sharpen Proposition 4.4.4 for Morita equivalent rings.

Corollary 4.4.6. *If A, B are Morita equivalent rings, with bimodules P, Q satisfying $P \otimes Q \cong A, Q \otimes P \cong B$, then* $\mathrm{Lat}(A_A) \cong \mathrm{Lat}(P_B)$, $\mathrm{Lat}({}_AA) \cong \mathrm{Lat}({}_BQ)$, $\mathrm{Lat}({}_BB) \cong \mathrm{Lat}({}_AP)$, $\mathrm{Lat}(B_B) \cong \mathrm{Lat}(Q_A)$.*Moreover,*

$$\mathrm{Lat}({}_AA_A) \cong \mathrm{Lat}({}_AP_B) \cong \mathrm{Lat}({}_BB_B);$$

in other words, A and B have isomorphic ideal lattices. ∎

By a *Morita invariant* we understand a property of rings which is preserved by Morita equivalence. For example, being simple is a Morita invariant, by Corollary 4.4.6.

As another example of a Morita invariant we have the centre. It is convenient to define this notion in the wider context of abelian categories. In any abelian category $\mathscr{A}$ let $\mathrm{Nat}(I)$ be the set of all natural transformations from the identity functor to itself. Since functors and natural transformations themselves form a category $\mathscr{A}^{\mathscr{A}}$, it follows that the set $\mathrm{Nat}(I)$, as the set of all 'endomorphisms' of I is a ring; it is called the *centre* of the category $\mathscr{A}$. Explicitly, $\alpha \in \mathrm{Nat}(I)$ means that for each $X \in \mathrm{Ob}\mathscr{A}$ there is a map $\alpha_X : X \to X$ such that any map $f : X \to Y$ in $\mathscr{A}$ gives rise to a commutative square

$$\begin{array}{ccc} X & \xrightarrow{f} & Y \\ \downarrow{\alpha_X} & & \downarrow{\alpha_Y} \\ X & \xrightarrow{f} & Y \end{array}$$

In the particular case where $\mathscr{A} = \mathrm{Mod}_A$, $\mathrm{Nat}(I)$ consists of all A-endomorphisms which commute with all A-homomorphisms. Writing C for the centre of the ring A, we have for each $c \in C$ an element μ_c of $\mathrm{Nat}(I)$, defined as

$$x\mu_c = xc, \quad \text{for } x \in X,\ X \in \mathrm{Mod}_A.$$

It is clear that the map $\mu : c \mapsto \mu_c$ defines a ring homomorphism

$$C \to \mathrm{Nat}(I).$$

We assert that this is an isomorphism: if $\mu_c = 0$, take $X = A$; then $0 = 1\mu_c = c$, so $c = 0$ and μ is injective. To prove surjectivity, let $f \in \mathrm{Nat}(I)$, say $1f_A = c$. By naturality, $ac = ca$, hence $c \in C$ and so $g = f - \mu_c \in \mathrm{Nat}(I)$. Now on A, $g = 0$ by definition, and given $x \in X$, where $X \in \mathrm{Mod}_A$, we define $\varphi : A \to X$ by $1 \mapsto x$. Then commutativity shows that $g(x) = x.g(1) = 0$, so $g = 0$ and $f = \mu_c$. The result may be stated as follows:

Theorem 4.4.7. *The centre of a ring A is isomorphic to the centre of the category* Mod_A. *Hence Morita equivalent rings have isomorphic centres.* ■

This result shows for example that two commutative rings are equivalent iff they are isomorphic; more generally, a ring R is equivalent to a commutative ring iff it is equivalent to its centre.

The property of being finitely generated can be expressed categorically: M is finitely generated iff M cannot be expressed as the union of a chain of proper submodules. This means that any module corresponding to a finitely generated module under a category equivalence is again finitely generated. However, the cardinal of a minimal generating set may well be different for the two modules, and this fact can be utilized to turn any problem on finitely generated modules into a problem on cyclic modules. In order to show this clearly we examine the equivalence $A \sim A_n$ in greater detail.

Fix $n \geq 1$ and write $P = A^n, Q = {}^nA$. We have the functors

$$M \mapsto M^n = M \otimes_A P \quad (M \in \mathrm{Mod}_A), \tag{4.4.6}$$

and

$$N \mapsto N^\sharp = N \otimes Q \quad (N \in \mathrm{Mod}_{A_n}). \tag{4.4.7}$$

Here $N^\sharp$ may also be defined as $\mathrm{Hom}_{A_n}(P, N)$. It is easily checked that

$$(M^n)^\sharp \cong M, \quad (N^\sharp)^n \cong N,$$

and this provides an explicit form for the equivalence $A \sim A_n$. Given any finitely generated right A-module M, with generating set $u_1, \ldots, u_n$ say, we can apply (4.4.6) and pass to the right A_n-module M^n, which is generated by the single element $(u_1, \ldots, u_n)$. We state the result as

Theorem 4.4.8. *For any ring A, any finitely generated A-module M corresponds to a cyclic A_n-module under the category-equivalence (4.4.6), for suitable n. In fact it is enough to take n equal to the cardinal of a generating set of M.* ■

For example, if A is a principal ideal ring, any submodule of an n-generator A-module can be generated by n elements (as follows from Proposition 8.2.3 below). Applying Theorem 4.4.8, we see that any submodule of a cyclic A_n-module is cyclic, so A_n is again a principal ideal ring. In the opposite direction, if A_n is a principal ideal ring, then any submodule of a cyclic module is cyclic. It follows that any n-generator A-module can be generated by n elements. This can happen for some $n > 1$ for certain rings which are not principal (see Webber [1970]).

Exercises

1. Show that a skew field K is Morita equivalent only to K_n, $n = 1, 2, \ldots$.
2. Verify directly that the centre of A is $\mathrm{Hom}_{A-A}(A, A)$.
3. Verify that for a ring to be Noetherian or Artinian is a Morita invariant.
4. Show that $A \prec B \Leftrightarrow A^0 \prec B^0$.
5. Show that any non-trivial ring without IBN has a simple homomorphic image without IBN. Verify that a simple ring without IBN is Morita equivalent only to a finite number of rings (up to isomorphism).

4.5 The Morita context

We have seen in Theorem 4.4.5 that a Morita equivalence between two rings A and B is determined by two bimodules P, Q, but in practice one wants to know for which pairs of rings A, B it is true that $A \sim B$. The first step is to find conditions on P and Q for Theorem 4.4.5 to apply. Here we need a general property of modules.

For any right A-module M we define its dual as $M^* = \mathrm{Hom}(M, A)$; this is a left A-module in a natural way. Let us write the image of $x \in M$ under $f \in M^*$ as (f, x) and put

$$\tau(M) = \tau_A(M) = \left\{ \sum (f_i, x_i) | f_i \in M^*, \ x_i \in M \right\}.$$

This is a two-sided ideal in A, called the *trace ideal* of M. For example, if F is a non-zero free A-module, then $\tau(F) = A$. The modules for which $\tau = A$ are of particular interest; as we see from the next result, they are just the generators of the category Mod_A.

Lemma 4.5.1. *Let A be any ring. For any right A-module M the following are equivalent:*

(a) *M is a generator,*
(b) $\tau_A(M) = A$,
(c) *$M^n \cong A \oplus N$ for some integer n and some N_A.*

Proof. (a) $\Rightarrow$ (b). Assume that $\tau(M) = \mathfrak{a} \neq A$; then the natural homomorphism $\pi : A \to A/\mathfrak{a}$ is non-zero, hence by (a) the induced map

$$M^* = \mathrm{Hom}(M, A) \to \mathrm{Hom}(M, A/\mathfrak{a}) \tag{4.5.1}$$

is non-zero. But every $f \in M^*$ maps M into $\mathfrak{a}$ by assumption, and this means that (4.5.1) is zero, a contradiction. Hence $\tau(M) = A$ and (b) holds.

(b) $\Rightarrow$ (c). By hypothesis $\tau(M) = A$, hence there exist $f_1, \ldots, f_n \in M^*$, $u_1, \ldots, u_n \in M$ such that $\sum (f_i, u_i) = 1$. We define a homomorphism $\varphi : M^n \to A$ by the rule $(x_1, \ldots, x_n) \mapsto \sum (f_i, x_i)$. Its image in A is a right ideal containing $1 = \sum (f_i, u_i)$, hence φ is surjective and if $\ker \varphi = N$, we have the exact sequence

$$0 \to N \to M^n \to A \to 0.$$

Since A is projective, this sequence splits and (c) follows.

(c) $\Rightarrow$ (a). Given a map $f : X \to Y$ of A-modules, if the induced map $\mathrm{Hom}(M, X) \to \mathrm{Hom}(M, Y)$ is zero, then this also holds for the map $\mathrm{Hom}(M^n, X) \to \mathrm{Hom}(M^n, Y)$, i.e. $\mathrm{Hom}(A \oplus N, X) \to \mathrm{Hom}(A \oplus N, Y)$. But the restriction to the first summand is just the original map $f : X \to Y$ (because $\mathrm{Hom}(A, X) \cong X$), so $f = 0$. This shows h^M to be faithful, so (a) holds. ■

We now consider the following situation. Given two rings A, B and bimodules P, Q, assume that we have two bimodule homomorphisms:

$$\tau : P \otimes Q \to A, \quad f, x \mapsto (f, x), \tag{4.5.2}$$

$$\mu : Q \otimes P \to B, \quad x, f \mapsto [x, f], \tag{4.5.3}$$

such that

$$[x, f]y = x(f, y), \tag{4.5.4}$$

$$f, g \in P, \quad x, y \in Q,$$

$$g[x, f] = (g, x)f. \tag{4.5.5}$$

These rules may be expressed symbolically by saying that $\begin{pmatrix} A & P \\ Q & B \end{pmatrix}$ is a ring under the usual matrix multiplication. This sums up the module laws and (4.5.2), (4.5.3), while (4.5.4), (4.5.5) are instances of the associative law. The 6-tuple (A, B, P, Q, τ, μ) is called a *Morita context*. We remark that $\mathrm{im}\, \tau$ is an ideal in A and $\mathrm{im}\, \mu$ is an ideal in B.

Starting from any module E_A we obtain a Morita context as follows. We put

$$E^* = \mathrm{Hom}_A(E, A), \quad B = \mathrm{End}_A(E),$$

and regard E as a (B, A)-bimodule and E^* as an (A, B)-bimodule in the natural way. Further, we have a natural map $\tau : E^* \otimes E \to A$ given by evaluation, as in (4.5.2). To find μ, we use (4.5.4): for any $x \in E$, $f \in E^*$ we define $[x, f]$ by its effect on E:

$$[x, f]y = x(f, y) \quad (y \in E).$$

This is an A-balanced biadditive map of $E \times E^*$ into B and so defines a homomorphism $\mu : E \otimes E^* \to B$ which is easily verified to be a B-bimodule homomorphism. Further, (4.5.4) holds by the definition of μ and the definition of E^* as B-module shows that for all $y \in E$, $(g[x, f], y) = (g, [x, f]y) = (g, x(f, y)) = (g, x)(f, y) = ((g, x)f, y)$; therefore $g[x, f] = (g, x)f$ and so (4.5.5) is proved.

We thus have a Morita context $(A, B, E^*, E, \tau, \mu)$ starting from E_A; this is called the Morita context derived from E_A. To give an example, if A is a simple Artinian ring, say $A \cong K_n$, where K is a skew field (by Wedderburn's theorem, BA, Theorem 5.2.2) and $E = K^n$ is a simple right A-module, then $E^* = {}^nK$ and the derived Morita context has the form $(K_n, K, {}^nK, K^n, \tau, \mu)$.

For general Morita contexts we shall be interested in the case where τ, μ are isomorphisms. In that case we have a Morita equivalence between A and B, by Theorem 4.4.5, with functors

$$S : \mathrm{Mod}_A \to \mathrm{Mod}_B, \quad M \mapsto M \otimes_A P,$$

$$T : \mathrm{Mod}_B \to \mathrm{Mod}_A, \quad N \mapsto N \otimes_B Q,$$

which are mutually inverse, because $P \otimes Q \cong B$, $Q \otimes P \cong A$. Under S, A corresponds to P and right ideals of A correspond to B-submodules of P, with two-sided ideals corresponding to (A, B)-submodules. Similarly, Q corresponds to B and under T, A corresponds to Q and P to B (see Proposition 4.4.4). Moreover, we have the isomorphisms

$$Q \cong \mathrm{Hom}_A(P, A) \cong \mathrm{Hom}_B(P, B), P \cong \mathrm{Hom}_A(Q, A) \cong \mathrm{Hom}_B(Q, B),$$

$$A \cong \mathrm{End}_B(Q) \cong \mathrm{End}_B(P)^0, \quad B \cong \mathrm{End}_A(P) \cong \mathrm{End}_B(Q)^0.$$

The next lemma shows that to verify that τ or μ is an isomorphism it is enough to check surjectivity:

Lemma 4.5.2. *In any Morita context (A, B, P, Q, τ, μ), if μ is surjective, it is an isomorphism; similarly for τ.*

Proof. If μ is surjective, we have

$$\sum [x_i, f_i] = 1 \quad \text{for some } x_i \in Q,\ f_i \in P.$$

Now assume that $\sum [y_\lambda, g_\lambda] = 0$; then

$$\sum y_\lambda \otimes g_\lambda = \sum y_\lambda \otimes g_\lambda [x_i, f_i] = \sum y_\lambda \otimes (g_\lambda, x_i) f_i = \sum y_\lambda (g_\lambda, x_i) \otimes f_i$$
$$= \sum [y_\lambda, g_\lambda] x_i \otimes f_i = 0.$$

This shows μ to be injective, and hence an isomorphism. The same argument applies to τ. ∎

In the special case of a derived Morita context we can give an explicit criterion for μ to be surjective. We recall the dual basis lemma from BA, Lemma 4.7.5. In the finitely generated case (BA, Corollary 4.7.6) this states that ${}_AP$ is a direct summand of A^n iff there exist $u_1, \ldots, u_n \in P$, $f_1, \ldots, f_n \in P^*$ (the 'projective coordinate system') such that

$$x = \sum (f_i, x) u_i \quad \text{for all } x \in P. \tag{4.5.6}$$

Similarly for a right module this equation takes the form $x = \sum u_i(f_i, x)$.

Lemma 4.5.3. *Given any module Q_A, let (A, B, P, Q, τ, μ) be the derived Morita context. Then $\mu : Q \otimes P \to B$ is an isomorphism if and only if Q_A is finitely generated projective.*

Proof. By the dual basis lemma just quoted, Q_A is finitely generated projective iff there is a finite projective coordinate system

$$x = \sum u_i(f_i, x) \quad \text{for all } x \in Q.$$

Bearing in mind that $Q^* = P$, we can by (4.5.4) write this as $x = \sum [u_i, f_i]x$ for all $x \in Q$, i.e. $\sum [u_i, f_i] = 1$. But this is just the condition for μ to be surjective, and by Lemma 4.5.2, for μ to be an isomorphism. ∎

We can now state a condition on any module Q_A for its derived Morita context to define an equivalence.

Theorem 4.5.4. *Let A be a ring, Q a right A-module and (A, B, P, Q, τ, μ) the Morita context derived from Q. Then this context defines a Morita equivalence between A and B if and only if Q is a finitely generated projective generator.*

A finitely generated projective generator is also called a *progenerator.*

Proof. If we have a Morita equivalence, then Q_A corresponds to B_B in the equivalence, so Q is a progenerator because B is. Conversely, assume this condition for Q. By Lemma 4.5.3 the map $\mu : Q \otimes P \to B$ is an isomorphism. Since Q is a generator, the trace ideal im τ is A (Lemma 4.5.1), so τ is surjective and hence an isomorphism by Lemma 4.5.2. Thus $P \otimes Q \cong A$, and now Theorem 4.4.5 shows that we have indeed a Morita equivalence. ∎

This result shows that every Morita equivalence can be obtained from a particular Morita context. Given $A \sim B$, to find Q we need a finitely generated projective A-module; this is a direct summand of A^n for some $n \geq 1$, and it may be specified by an idempotent e in $A_n \cong \mathrm{End}_A(A^n)$. Suppose first that $n = 1$; this means that $Q_A = eA$, where e is an idempotent in A. By Lemma 4.3.8 (or rather, its left–right dual) we have $P = \mathrm{Hom}_A(eA, A) = Ae$, $B = \mathrm{End}_A(eA) = eAe$. Now the condition for Q to be a generator reads: the natural map $P \otimes Q \to A$ is surjective, i.e. $AeA = A$. The translation to A_n is now clear. We choose $n \geq 1$ and take an idempotent e in A_n such that $A_neA_n = A_n$. Then we have $B = eA_ne$, and all rings Morita equivalent to A are obtained in this way, with the appropriate Morita context $(A, eA_ne, A^ne, e^nA, \tau, \mu)$.

An important particular case is obtained by starting from a commutative ring K say. If Q is any finitely generated projective K-module, then $A = \mathrm{End}_K(Q)$ is a K-algebra and for any K-algebra R we have

$$R \sim \mathrm{End}_R(R \otimes Q) \cong R \otimes_K A,$$

as is easily checked. Thus $A \sim K$ and tensoring with A (over K) converts any K-algebra into one Morita equivalent to it. Such algebras are called *Brauer equivalent*; in the special case when K is a field, we have $A = K_n$ and $R \otimes K_n \cong R_n$, while the general case leads to the study of Azumaya algebras (Azumaya [1950], Auslander and Goldman [1960]; see also Section 8.6 below).

We conclude this section by describing another important Morita invariant, the trace group. Let K be any commutative ring and consider a K-algebra A. With A we associate a K-module, its *trace group*, defined as

$$\mathbf{T}(A) = A/C, \quad \text{where } C = \left\{\sum (xy - yx) | x, y \in A\right\}. \tag{4.5.7}$$

The natural map $A \to \mathbf{T}(A)$ is called the *trace function* and is written $\mathrm{tr}(x)$. Clearly it has the following properties:

T.1 $\mathrm{tr} : A \to \mathbf{T}(A)$ *is K-linear,*
T.2 $\mathrm{tr}(xy) = \mathrm{tr}(yx)$ *for all $x, y \in A$.*

Moreover, any linear map $\alpha : A \to M$ into a K-module such that $\alpha(xy) = \alpha(yx)$ can be written as $\alpha(x) = \alpha'(\mathrm{tr}(x))$ for a unique $\alpha' : \mathbf{T}(A) \to M$. Thus tr is the universal mapping satisfying **T.**1–**T.**2. Let us show that T is a Morita invariant:

Proposition 4.5.5. *For any rings A, B, if $A \sim B$, then* $\mathbf{T}(A) \cong \mathbf{T}(B)$.

Proof. Let $(A, B, P, Q, (), [])$ be the Morita context providing the equivalence and consider the map $A \to \mathbf{T}(B)$ given by

$$\alpha : \sum (f_i, x_i) \mapsto \mathrm{tr}\left(\sum [x_i, f_i]\right), \quad \text{where } f_i \in P,\ x_i \in Q.$$

To show that this is well-defined we must check that the map $f, x \mapsto \mathrm{tr}([x, f])$ is bilinear and B-balanced. The bilinearity is clear, and we have for any $b \in B$,

$$\mathrm{tr}([x, fb]) = \mathrm{tr}([x, f]b) = \mathrm{tr}(b[x, f]) = \mathrm{tr}([bx, f]),$$

by **T.**2, hence the result. Moreover, $\alpha(aa') = \alpha(a'a)$, because $\alpha(\sum (f, x)a) = \mathrm{tr}(\sum [xa, f]) = \mathrm{tr}(\sum [x, af]) = \alpha(\sum a(f, x))$; hence α induces a map $\alpha' : \mathbf{T}(A) \to \mathbf{T}(B)$. By symmetry there is a map $\beta' : \mathbf{T}(B) \to \mathbf{T}(A)$ and these two maps are easily seen to be mutually inverse. ∎

As an illustration, if A is an algebra with centre K, then of the two Morita invariants $\mathbf{C}(A)$, $\mathbf{T}(A)$ the centre is just K, whereas $\mathbf{T}(A)$ is a K-module which in general is larger than K and so will tell us more about A. We remark that in (4.5.7) C is merely a K-module and not an ideal; this means that $\mathbf{T}(A)$ will often be non-zero even if A is simple. However, for some rings $\mathbf{T}(A) = 0$ (see Section 9.3, Exercise 10). The trace function introduced here was first defined by Akira Hattori and independently by John Stallings in 1965.

The notion of Morita equivalence was introduced by Kiiti Morita in 1958. The above account follows the author's notes (Cohn (1966)), which in turn were based on the notes of Hyman Bass (1962).

Exercises

1. Given a Morita context (A, B, P, Q, τ, μ), if $P \otimes Q \cong A$ and $A \cong \mathrm{End}(Q)$, show that Q is finitely generated projective.
2. Let K be a commutative ring and P a finitely generated faithful projective module (P is faithful if $Pa = 0 \Rightarrow a = 0$). Show by using the dual basis lemma that $\tau(P) = K$ and so P is a generator.
3. If (A, B, P, Q, τ, μ) is a Morita context defining a Morita equivalence and $R = \begin{pmatrix} A & P \\ Q & B \end{pmatrix}$ show that $A \prec R$, $B \prec R$.
4. Let K be any ring, E, F any K-modules and define $P = \mathrm{Hom}(E, F)$, $Q = \mathrm{Hom}(F, E)$, $A = \mathrm{End}(E)$, $B = \mathrm{End}(F)$. Show that together with the natural maps $P \otimes Q \to A$, $Q \otimes P \to B$ this defines a Morita context.
5. (Kazuhiko Hirata) Show that the Morita context of Exercise 4 defines a Morita equivalence between A and B iff there exist integers $m, n \geq 1$ and K-modules E', F' such that $E \oplus E' \cong F^n$, $F \oplus F' \cong E^m$.

4.6 Projective, injective and flat modules

We now take a closer look at flat modules and their relation to projective and injective modules.

Let R be any ring. Any left R-module M has a presentation

$$G \xrightarrow{\alpha} F \xrightarrow{\beta} M \to 0, \tag{4.6.1}$$

where F, G are free R-modules. Explicitly, let F have the basis (f_λ) and G the basis (g_i); then α is described by the matrix $A = (a_{i\lambda})$, which is said to *present* M and is called the *presentation matrix* of M, where

$$g_i\alpha = \sum a_{i\lambda} f_\lambda. \tag{4.6.2}$$

In general F, G will not be finitely generated and A may have infinitely many rows and columns, but each row has only finitely many non-zero entries; we say that A is *row-finite.* We note that M is determined up to isomorphism by A as the cokernel of the corresponding map α, by (4.6.1). Moreover, every row-finite matrix A defines a module in this way. If F, G have finite ranks m, n respectively, then the presentation matrix is $m \times n$.

It is clear that F can be taken to be of finite rank iff M is finitely generated. If G can be taken to be of finite rank, M is said to be *finitely related* and the module M is called *finitely presented* if F, G can both be taken to be of finite rank. Given two presentations

$$0 \to K_i \to F_i \to M \to 0 \quad (i = 1, 2) \tag{4.6.3}$$

where F_1, F_2 are free (but K_1, K_2 need not be free), we have by Schanuel's lemma (Lemma 2.4.2), $F_1 \oplus K_2 \cong F_2 \oplus K_1$, hence if M has a presentation with F_1 of finite rank and another with K_2 finitely generated, then F_2, K_1 are also finitely generated and M is finitely presented.

Let us examine the presentation matrix of a projective module.

Proposition 4.6.1. *Let M be an R-module with the presentation (4.6.1). Then M is projective if and only if there exists a mapping $\alpha' : F \to G$ such that $\alpha\alpha'\alpha = \alpha$. Thus a matrix A represents a projective module if and only if there is a matrix A' such that $AA'A = A$.*

Proof. Assume that M is projective; then (4.6.1) splits and so $F \cong M \oplus \ker\beta \cong M \oplus \operatorname{im}\alpha$, so $\operatorname{im}\alpha$ is also projective. Hence the projection $F \to \operatorname{im}\alpha$ can be lifted to a map $\alpha' : F \to G$, whose composition with a is the projection onto $\operatorname{im}\alpha$, i.e. $\alpha\alpha'\alpha = \alpha$.

Conversely, if α' satisfies $\alpha\alpha'\alpha = \alpha$, then $\alpha\alpha'$ is an idempotent endomorphism of G, hence $G = G_1 \oplus G_2$, where $G_1 = \operatorname{im}\alpha\alpha'$, $G_2 = \ker\alpha\alpha'$, and writing $\alpha_1 = \alpha|G$, we have the exact sequence (4.6.1) with G, α replaced by G_1, a_1. Since $\alpha\alpha' = 1$, the sequence splits and so M is projective. The final assertion follows by rewriting the result in terms of matrices. ■

Next we give some conditions for a module to be flat:

Theorem 4.6.2. *For any right R-module U (over any ring R) the following conditions are equivalent:*

(a) $\operatorname{Tor}_1^R(U, -) = 0$,
(b) *U is flat,*
(c) *for any free left R-module F and any submodule G of F, the map $U \otimes G \to U \otimes F$ induced by the inclusion $G \subseteq F$ is injective,*
(d) *given $uc = 0$, where $u \in U^n$, $c \in {}^nR$, there exists $A \in {}^mR^n$ and $v \in U^m$ such that $u = vA$ and $Ac = 0$,*
(e) *for any finitely generated left ideal $\mathfrak{a}$ of R the map $U \otimes \mathfrak{a} \to U$ induced by the inclusion $\mathfrak{a} \subseteq R$ is injective, so that $U \otimes \mathfrak{a} \cong U\mathfrak{a}$,*
(f) *as* (e), *but for any left ideal.*

Condition (d) may be expressed loosely by saying that any relation in U is a consequence of relations in R; this explains the name 'flat', if one thinks of relations in a module as a kind of torsion.

Proof. It is clear from the definition that (a) and (b) are equivalent. When U is flat, the induced sequence

$$0 \to U \otimes G \to U \otimes F$$

is injective, so (b) $\Rightarrow$ (c). To show that (c) $\Rightarrow$ (d), consider the exact sequence

$$0 \to K \xrightarrow{\alpha} R^n \xrightarrow{\beta} R,$$

where $\beta : (x_i) \mapsto \sum x_i c_i$ and $K = \ker\beta$. By (c) the induced sequence

$$0 \to U \otimes K \to U^n \to U$$

is exact; since $\sum u_i c_i = 0$, we have $u_i = \sum v_h a_{hi}$ for some $v_h \in U$, $a_{hi} \in K$, and so $\sum a_{hi} c_i = 0$, which proves (d).

(d) $\Rightarrow$ (e). Every element of $U \in \mathfrak{a}$ has the form $\sum u_i \otimes c_i$ for some $u_i \in U$. If $\sum u_i c_i = 0$, then in vector notation, $u = vA$, where $Ac = 0$, hence in $U \otimes \mathfrak{a}$ we have $u \otimes c = vA \otimes c = v \otimes Ac = 0$, and this shows the mapping $U \otimes \mathfrak{a} \to U$ to be injective.

Now (e) $\Rightarrow$ (f) is clear since any relation involves only finitely many elements of $\mathfrak{a}$, and to prove (f) $\Rightarrow$ (a), we have, for any left ideal $\mathfrak{a}$, an exact sequence

$$0 \to U \otimes \mathfrak{a} \to U \to U \otimes (R/\mathfrak{a});$$

hence $\mathrm{Tor}_1^R(U, C) = 0$ for any cyclic left R-module C. Hence $\mathrm{Tor}_1^R(U, A) = 0$ for any left R-module A, by induction on the number of generators of A, using the exact homology sequence, and then taking the limit over finitely generated left R-modules A. ∎

There is another characterization of flat modules, which clarifies their relation to projective modules.

Proposition 4.6.3. *Let U be an R-module (over any ring R) with a presentation*

$$0 \to K \xrightarrow{\alpha} F \xrightarrow{\beta} U \to 0, \qquad (4.6.4)$$

where F is free. Then the following conditions are equivalent:

(a) *U is flat,*
(b) *given $x \in K$, there exists $\gamma : F \to K$ such that $x\alpha\gamma = x$,*
(c) *given $x_1, \ldots, x_n \in K$, there exists $\gamma : F \to K$ such that $x_i\alpha\gamma = x_i$ for $i = 1, \ldots, n$.*

Proof. (a) $\Rightarrow$ (b). Let (f_λ) be a basis of F; by suitably renumbering the f's we may write $x = f_1c_1 + \ldots + f_rc_r$. Hence $x\beta = \sum (f_i\beta)c_i = 0$, so by Theorem 4.6.2(d) there exist $g_1, \ldots, g_m \in F$ and $a_{hi} \in R$ such that $f_i\beta = \sum (g_h\beta)a_{hi}$, $\sum a_{hi}c_i = 0$. It follows that $f'_i = f_i - \sum g_h a_{hi} \in K$ and $x = \sum f_i c_i = \sum (f'_i + \sum g_h a_{hi})c_i = \sum f'_i c_i$. Now define $\gamma : F \to K$ by

$$f_\lambda \gamma = \begin{cases} f'_\lambda & \text{for } \lambda = 1, \ldots, r, \\ 0 & \text{otherwise.} \end{cases}$$

Then $x\alpha\gamma = (\sum f_i c_i)\alpha\gamma = \sum f'_i c_i = x$, as required.

(b) $\Rightarrow$ (c). By applying the same argument to the exact sequence of R_n-modules

$$0 \to K^n \to F^n \to U^n \to 0,$$

and observing that U^n is flat whenever U is, we obtain (c).

Clearly (c) $\Rightarrow$ (b) holds trivially, and to prove (b) $\Rightarrow$ (a) we shall verify (d) of Theorem 4.6.2. Let (f_λ) be a basis of F and consider a relation in U, which by suitable numbering of the f's may be written $\sum (f_i\beta)c_i = 0$. Then $x = \sum f_i c_i \in K$ and by hypothesis there exists $\gamma : F \to K$ such that $x\alpha\gamma = x$. Put $f_i\gamma = f_i + \sum f_h a_{hi}$; then since $x\alpha\gamma = x$, we have $\sum f_i c_i = \sum (f_i\gamma)c_i = \sum f_i c_i + \sum f_h a_{hi} c_i$, and so $\sum a_{hi} c_i = 0$, because the f_λ are free. Moreover, since $f_\lambda\gamma \in K$, we have $f_i\beta = \sum (f_h\beta)a_{hi}$, so Theorem 4.6.2(d) is satisfied, and U is flat. ■

If U is finitely related, we can in (4.6.4) take a finite generating set of K and so obtain a map $\gamma : F \to K$ which splits (4.6.4). Hence U is then projective, and since any projective module is flat, this proves

Corollary 4.6.4. *A finitely related module is flat if and only if it is projective.* ■

This shows that for example, over a Noetherian ring, every finitely generated flat module is projective. We can now also characterize rings over which every (left or right) module is flat (M. Auslander, 1957):

Theorem 4.6.5. *For any ring R the following conditions are equivalent:*

(a) *every right R-module is flat,*
(a^0) *every left R-module is flat,*
(b) *given $a \in R$, there exists $a' \in R$ such that $aa'a = a$.*

Proof. We shall prove the implications (a^0) $\Rightarrow$ (b) $\Rightarrow$ (a); the theorem then follows by symmetry.

Let us apply Proposition 4.6.3 to the exact sequence of left R-modules

$$0 \to Ra \to R \to R/Ra \to 0. \tag{4.6.5}$$

By hypothesis R/Ra is flat, so there exists a mapping $\gamma : R \to Ra$ such that $a\gamma = a$. Let $1\gamma = a'$; then $a = a\gamma = (a.1)\gamma = a.a'a$ and (b) follows. Conversely, when (b) holds, the argument just given shows that (4.6.5) satisfies Proposition 4.6.3(c); more generally, this holds for any left ideal $\mathfrak{a}$ in place of Ra; thus if $u \in \mathfrak{a}$ and $uu'u = u$, we can define $\gamma : R \to \mathfrak{a}$ by $x \mapsto xu'u$. Hence $R/\mathfrak{a}$ is flat and so $\mathrm{Tor}_1^R(U, R/\mathfrak{a}) = 0$ for any right R-module U, so the mapping $U \otimes \mathfrak{a} \to U$ induced by the inclusion $\mathfrak{a} \subseteq R$ is injective. By Theorem 4.6.2(c) it follows that U is flat and so (a) holds. ■

A ring satisfying condition (b) of this theorem is called (*von Neumann*) *regular* or sometimes *absolutely flat*. Clearly every semisimple ring is regular, but of course the converse does not hold. For example, if K is any field and I a set, then the direct power K^I is a regular ring, but it is not semisimple, unless I is finite.

A link between flat and injective modules is provided by

Proposition 4.6.6. *A left R-module V (over any ring R) is flat if and only if* $\hat{V} = \mathrm{Hom}(V, \mathbf{K})$ *is injective, as right R-module, where* $\mathbf{K} = \mathbf{Q}/\mathbf{Z}$.

Proof. For any right R-module U we have, by adjoint associativity, the natural isomorphism

$$\mathrm{Hom}_R(U, \mathrm{Hom}_\mathbf{Z}(V, \mathbf{K})) \cong \mathrm{Hom}_\mathbf{Z}(U \otimes V, \mathbf{K}). \tag{4.6.6}$$

Now assume that V is flat; then $- \otimes V$ is an exact functor, hence the right-hand side of (4.6.6) is exact as a functor of U, hence so is the left and this means that $\hat{V}$ is injective (by definition). Conversely, when $\hat{V}$ is injective, the two sides of (4.6.6) are exact as functors of U. But the functor $\mathrm{Hom}_{\mathbf{Z}}(-, \mathbf{K})$ is faithful, and so preserves inexact sequences, by Proposition 2.2.6, hence $U \otimes V$ must be exact in U, so V is flat, as claimed. ∎

For example, $\hat{R}$ is always injective; this module has another property that is sometimes useful. We recall that a *cogenerator* U is defined dually to the term 'generator' by the condition that h_U is faithful; explicitly, for $U \in {}_R\mathrm{Mod}$, this means that for any ${}_RM$ and $0 \neq x \in M$ there is a homomorphism $\varphi : M \to U$ such that $x\varphi \neq 0$. For example, $\mathbf{K} = \mathbf{Q}/\mathbf{Z}$ is a cogenerator for $\mathbf{Z}$. For, given any abelian group A and $0 \neq x \in A$, let p be a prime factor of the order of x (or any prime if x has infinite order). Then $x\mathbf{Z}/px\mathbf{Z}$ is of order p and can be embedded in $\mathbf{K}$; since K is injective (i.e. divisible in this case, see Section 2.3), this embedding can be extended to a homomorphism of $A/px\mathbf{Z}$ into $\mathbf{K}$; combined with the natural mapping $A \to A/px\mathbf{Z}$ this gives a homomorphism $A \to \mathbf{K}$ which does not kill x. In the general case we obtain a cogenerator by forming a coinduced extension, as follows.

Theorem 4.6.7. *Let R be a ring and $f : \mathbf{Z} \to R$ the canonical map. Then $\hat{R} = \mathrm{Hom}(R, \mathbf{K})$ is an injective cogenerator.*

Proof. For any right R-module M we have, by adjoint associativity,

$$\mathrm{Hom}_R(M, \hat{R}) \cong \mathrm{Hom}_{\mathbf{Z}}(M \otimes R, \mathbf{K}) = \mathrm{Hom}_{\mathbf{Z}}(M, \mathbf{K}),$$

and this is non-zero whenever $M \neq 0$, because $\mathbf{K}$ is a cogenerator for $\mathbf{Z}$. It follows that there is a non-zero homomorphism from M to $\hat{R}$. Hence for any $x \in M$, $x \neq 0$, we have a non-zero homomorphism from xR to $\hat{R}$; since $\hat{R}$ is injective, by Proposition 4.6.6, this can be extended to a homomorphism from M to $\hat{R}$. This shows $\hat{R}$ to be a cogenerator for R, and by Proposition 4.6.6 it is injective. ∎

From the definition it is clear that the class of projective modules admits direct sums, while that of injective modules admits direct products. In the Noetherian case one can say a little more (E. Matlis, Z. Papp, 1958–59). A module will be called *uniform* if it is non-zero and any two non-zero submodules have a non-zero intersection.

Theorem 4.6.8. *Let R be a ring. Then the direct sum of any family of injective left R-modules is injective if and only if R is left Noetherian. When this is so, every finitely generated injective left R-module is a direct sum of uniform injectives.*

Proof. Suppose that R is left Noetherian, let $\{E_\lambda\}$ be any family of injective left R-modules and put $E = \oplus_I E_\lambda$. We have to show that any homomorphism $f : \mathfrak{a} \to E$ from a left ideal $\mathfrak{a}$ of R into E extends to a homomorphism of R into E (by Theorem 2.3.4). Since R is left Noetherian, $\mathfrak{a}$ is finitely generated, by $u_1, \ldots, u_r$ say and the images $u_1f, \ldots, u_rf$ have only finitely many non-zero components and so lie in a submodule $E' = \oplus_{I'} E_\lambda$, where I' is a finite subset of the index set I. Thus f maps $\mathfrak{a}$ into E'; as a finite direct sum of injective modules, E' is again injective, so f extends to a homomorphism of R into E' and combining this with the inclusion $E' \subseteq E$ we obtain the desired extension of f.

Conversely, assume that any direct sum of injective modules is injective and consider an ascending chain of left ideals in R:

$$\mathfrak{a}_1 \subseteq \mathfrak{a}_2 \subseteq \dots. \tag{4.6.7}$$

Write $\mathfrak{a} = \cup \mathfrak{a}_\lambda$, let I_n be the injective hull of $R/\mathfrak{a}_n$, put $I = \oplus I_n$ and define f: $\mathfrak{a} \to I$ as $xf = \sum xf_n$, where $f_n : \mathfrak{a} \to I_n$ is the homomorphism induced by the natural mapping $\mathfrak{a} \to \mathfrak{a}/\mathfrak{a}_n$. If $x \in \mathfrak{a}$, we have $x \in \mathfrak{a}_r$ for some $r = r(x)$ and so $xf_n = 0$ for all $n \geq r$; hence only finitely many of the xf_n are non-zero and f is well-defined. By hypothesis I is injective, so there is a homomorphism $f' : R \to I$ extending f. Let $1f' = c \in I$; then $c \in I_1 + \dots + I_s$ for some s and so f' followed by the projection on I_n is zero for $n > s$. Hence the same is true of f, so any x in $\mathfrak{a}$ must lie in $\mathfrak{a}_s$, i.e. $\mathfrak{a} = \mathfrak{a}_s$. Thus (4.6.7) breaks off and this shows R to be left Noetherian.

Clearly every indecomposable injective is uniform, so the last part follows in the Noetherian case. ∎

The corresponding problems for projective and flat modules have been solved by Stephen Chase [1960]. A ring R is said to be left *coherent* if every finitely generated left ideal in R is finitely related. Now Chase proves (i) the direct product of any family of flat left R-modules is flat iff R is left coherent, and (ii) the direct product of any family of projective left R-modules is projective iff R is right perfect and left coherent.

For commutative Noetherian rings we can describe the indecomposable injective modules in terms of prime ideals of the ring. We recall from BA, Section 10.8 that a prime ideal $\mathfrak{p}$ of a commutative ring R is meet-irreducible (this is easily verified directly), hence $E(R/\mathfrak{p})$, the injective hull of $R/\mathfrak{p}$, is indecomposable. Further we recall that a maximal annihilator of a module M is a prime ideal; the set of all prime ideals which form annihilators of elements of M is just $\mathrm{Ass}(M)$, and this consists of a single element $\mathfrak{p}$ precisely when 0 is primary in M.

Theorem 4.6.9. *Let R be a commutative Noetherian ring. Then*

(i) *for any prime ideal $\mathfrak{p}$ of R, $E(R/\mathfrak{p})$ is an indecomposable injective module,*
(ii) *any indecomposable injective module E is isomorphic to $E(R/\mathfrak{p})$, for a unique prime ideal $\mathfrak{p}$.*

Proof. Let $\mathfrak{p}$ be a prime ideal in R. Then $\mathfrak{p}$ is meet-irreducible, hence $E(R/\mathfrak{p})$ is indecomposable injective. If E is an indecomposable injective R-module, then 0 is meet-irreducible in any submodule of E, hence 0 is primary and so $\mathrm{Ass}(E)$ consists of a single element $\mathfrak{p}$. Clearly $R/\mathfrak{p}$ is embedded in E, hence $E(R/\mathfrak{p}) \cong E$, and $\mathfrak{p}$ is clearly unique. ∎

Although this result has been extended for certain non-commutative rings, there is no corresponding description in the general case. However, there is a connexion with divisible modules which is quite general and which we shall now explain.

We recall that a left R-module M over an integral domain R is said to be *divisible* if for any $u \in M$, $a \in R^{\times}$ the equation $u = ax$ has a solution for x in M. It is easily verified that over an integral domain every injective module is divisible (see BA, Proposition 4.7.8). Let us examine more closely how the hypothesis that R is a domain was used. Given $u \in M$, $a \in R^{\times}$, we need to be sure that the mapping $ar \mapsto ur$ from aR to M is well defined, and this will be the case if

$$ax = 0 \Rightarrow ux = 0 \quad \text{for all } x \in R. \tag{4.6.8}$$

Let R be any ring and M a right R-module such that for any $u \in M$ and any $a \in R$, the equation $u = va$ has a solution v in M whenever (4.6.8) holds. In that case M is said to be *1-divisible*. Over an integral domain this reduces to the notion of a divisible module and the proof of Proposition 4.7.8 of BA can be adapted to obtain

Proposition 4.6.10. *Over any ring R, an injective module is 1-divisible; over a principal ideal domain the converse holds too.*

Proof. This will follow from the more general result in Theorem 4.6.11 below. ■

In general, being 1-divisible is of course not sufficient for injectivity, but a necessary and sufficient condition is now easily obtained. We recall that for any index set I, M^I denotes the direct product of copies of M indexed by I, while ${}^IM$ denotes the direct sum of I copies. We shall visualize the elements of M^I and ${}^IM$ as rows and columns respectively; thus for any $u \in M^I$, $x \in {}^IR$ we can form $ux = \sum u_i x_i$, because almost all components of x are 0.

A right R-module M is called *fully divisible* if it satisfies the following generalization of (4.6.8):

Given any set I, if $u \in M^I$, $a \in R^I$ are such that

$$ax = 0 \Rightarrow ux = 0 \quad \text{for all } x \in {}^IR, \tag{4.6.9}$$

then there exists $v \in M$ such that $u = va$.

Now the characterization of injective modules can be stated as follows:

Theorem 4.6.11. *A module M over a ring R is injective if and only if it is fully divisible.*

Proof. The necessity follows easily: given $u \in M^I$, $a \in R^I$ satisfying (4.6.9), let $\mathfrak{a}$ be the right ideal generated by the components of $a = (a_i)$ and define a homomorphism $\mathfrak{a} \to M$ by

$$\sum a_i x_i \mapsto \sum u_i x_i.$$

By (4.6.9) this is well defined, clearly it is a homomorphism, so it can be extended to a homomorphism $R \to M$, because M is injective. If $1 \mapsto v$ in this mapping, then $a_i \mapsto va_i = u_i$, and this shows M to be fully divisible.

Conversely, if M is fully divisible, let $\mathfrak{a}$ be any right ideal of R and (a_i) $(i \in I)$ a generating set of $\mathfrak{a}$. Suppose we have a homomorphism $f : \mathfrak{a} \to M$ and let $a_i f = u_i$. If the $x_i \to R$ are such that $\sum a_i x_i = 0$, then $\sum u_i x_i = \sum (a_i f) x_i = (\sum a_i x_i) f = 0$, hence (4.6.9) holds and by full divisibility there exists $v \in M$ such that $u_i = v a_i$. Now the map $x \mapsto vx$ of R into M extends f, because $a_i \mapsto v a_i = u_i$, hence $\sum a_i x_i \mapsto \sum u_i x_i$ for any family $(x_i) \in {}^I R$. By Baer's criterion it follows that M is injective. ■

Let us call M *finitely divisible* if for any integer $n \geq 1$ and any $u \in M^n$, $A \in R_n$ satisfying the condition

$$Ax = 0 \Rightarrow ux = 0 \quad \text{for all } x \in {}^n R. \tag{4.6.10}$$

there exists $v \in M^n$ such that $u = vA$. Essentially this states that M^n is 1-divisible, as R_n-module, for all n. Then we have

Corollary 4.6.12. *A right R-module M over a right Noetherian ring R is injective if and only if it is finitely divisible.*

Proof. The necessity is clear; to prove the sufficiency, we observe that by Morita equivalence we have for all $n \geq 1$,

$$M \cong \mathrm{Hom}_R(R, M) \cong \mathrm{Hom}_{R_n}({}^n R, {}^n M).$$

Let $\mathfrak{a}$ be a right ideal of R, generated by n elements, say. Then $\mathfrak{a}$ corresponds to a submodule T of ${}^n R$ generated by a single element (namely the column with the generating set of $\mathfrak{a}$ as components). As in the proof of the theorem, any homomorphism $T \to M$ extends to a homomorphism ${}^n R \to {}^n M$. Thus in the commutative diagram

$$\begin{array}{ccc} \mathrm{Hom}_R(R, M) & \longrightarrow & \mathrm{Hom}_R(\mathfrak{a}, M) \to 0 \\ \downarrow \cong & & \downarrow \cong \\ \mathrm{Hom}_{R_n}({}^n R, {}^n M) & \longrightarrow & \mathrm{Hom}_{R_n}(T, {}^n M) \to 0 \end{array}$$

the bottom row is exact; hence so is the top row. Now the result follows again by Baer's criterion. ■

Exercises

1. Show that a module is flat whenever every finitely generated submodule is contained in a flat submodule.
2. Show that a flat module over an integral domain is torsion-free, and that the converse holds over a principal ideal domain.
3. Show that a direct sum of modules is flat iff each term is flat.
4. Use Proposition 4.6.6 to show that M_R is flat iff the canonical map $M \otimes \mathfrak{a} \to M$ is injective for every finitely generated left ideal $\mathfrak{a}$ of R. Hence obtain another proof that (b) $\Leftrightarrow$ (d) in Theorem 4.6.2.
5. Show that every finitely related module is a direct sum of a free module and a finitely presented module.

6. Show that every finitely generated projective module is finitely presented.
7. Show that every simple module over a commutative regular ring is injective.
8. Let R be a commutative integral domain with field of fractions K. Show that K is flat as R-module, but not free, unless $K = R$. When is K projective?
9. Show that every right Noetherian regular ring is semisimple.
10. Let E be an injective cogenerator and for any module M define $I^0(M) = E^{\mathrm{Hom}(M,E)}$. Show that $I^0(M)$ is injective and that M may be embedded in it.
11. Let M be a flat module. Show that if M has a resolution of finite length by finitely generated projective modules, then M is projective.
12. Show that every finitely generated projective module over $\mathbf{Z}/p^n$ is free.
13. Show that every non-zero projective module has a maximal (proper) submodule.
14. Let M be an R-module with a presentation (4.6.3) in which F_1 is free and K_1 is finitely generated, and another where F_2 is finitely generated (but not necessarily free). Show that M is finitely presented. Give an example to show that if F_1 is finitely generated free and K_2 is finitely generated (but F_2 is not free), M need not be finitely presented.
15. Show that a presentation matrix A represents a projective right R-module iff there exists a matrix C such that $AC = I$. Show that A represents a flat module iff for any finite set of rows of A there exists C such that the corresponding rows of AC have the form I.

4.7 Hochschild cohomology and separable algebras

Let K be a commutative ring and A a K-algebra; when we speak of an A-bimodule M, it will be understood that the two K-actions on M agree, i.e. $\alpha m = m\alpha$ for all $m \in M$, $\alpha \in K$. It is then clear that an A-bimodule is the same as a right $(A^{\circ} \otimes A)$-module, where A° is the opposite ring of A. We shall write A^{e} for $A^{\circ} \otimes A$ and call it the *enveloping algebra* of A. We shall see in Chapter 5 that for a finite-dimensional simple algebra the enveloping algebra is a full matrix algebra over the centre.

The free right A^{e}-module on one free generator is $A^{\circ} \otimes A$, with multiplication rule

$$(x \otimes y)(a \otimes b) = ax \otimes yb.$$

Similarly A itself is a right A^{e}-module with multiplication rule $x(a \otimes b) = axb$, and the multiplication mapping $\mu : A \otimes A \to A$ defined by $(x \otimes y)\mu = xy$ is an A^{e}-module homomorphism. Its kernel is again denoted by Ω, as in Section 2.7, so that we have an exact sequence

$$0 \to \Omega \to A \otimes A \xrightarrow{\mu} A \to 0. \tag{4.7.1}$$

If this sequence of A^{e}-modules splits, A is said to be *separable* over K. We shall soon see that this is consistent with the use of this term in BA, Section 11.6. We begin by giving some equivalent conditions for separability:

Proposition 4.7.1. *Let K be a commutative ring and A a K-algebra, with enveloping algebra $A^e = A^o \otimes A$. Then the following conditions are equivalent:*

(a) *A is separable,*
(b) *A is projective as A^e-module,*
(c) *there exists $e \in A^e$ such that $e\mu = 1$ and $ae = ea$ for all $a \in A$.*

Proof. (a) $\Leftrightarrow$ (b) is clear from the definition of projective module. Now assume that A is separable and let $\lambda : A \to A \otimes A$ be a section, i.e. an A^e-module homomorphism such that $\lambda\mu = 1_A$. The image of $1 \in A$ under λ is of the form $e = \sum p_i \otimes q_i$ and since $\lambda\mu = 1$, we have $e\mu = \sum p_i q_i = 1$. Further, for any $a \in A$, $ae = a.1\lambda = (a.1)\lambda = (1.a)\lambda = 1\lambda.1 = ea$, and (c) follows. If (c) holds and $e = \sum p_i \otimes q_i$, then $\lambda : x \mapsto xe = ex$ defines a splitting of (4.7.1) and it follows that A is separable, i.e. (a). ∎

The element e in (c) is called a *separator* of A, or also *separating idempotent*; it is in fact an idempotent (see Exercise 1). It has the following property:

Proposition 4.7.2. *Let $e \in A$ be such that $e\mu = 1$. Then e is a separator for A if and only if* $(\ker \mu)e = 0$.

Proof. Let u_λ be a generating set for A as K-module. We claim that

$$\ker \mu = \sum (u_\lambda \otimes 1 - 1 \otimes u_\lambda)A^e. \tag{4.7.2}$$

For suppose that $w = \sum u_\lambda \otimes a_\lambda$ belongs to the left-hand side of (4.7.2), i.e. $w\mu = 0$; then $\sum u_\lambda a_\lambda = 0$ and so $w = \sum (u_\lambda \otimes 1 - 1 \otimes u_\lambda)(1 \otimes a_\lambda)$, which lies in the right-hand side. The converse follows because $(u_\lambda \otimes 1 - 1 \otimes u_\lambda)\mu = 0$ and μ is an A-module homomorphism. Now the conclusion follows because $(u_\lambda \otimes 1 - 1 \otimes u_\lambda)e = u_\lambda e - eu_\lambda$. ∎

The projective dimension of A as A^e-module is sometimes called the *bidimension*, written bidim A; thus bidim $A = 0$ iff A is separable. Let A be a K-algebra; any A-bimodule M can be regarded as a right A^e-module in the usual way: $m(a \otimes b) = amb$, or also as left A^e-module, by the rule $(a \otimes b)m = bma$. We define the n-th *homology group* of M as the derived functor of $M \mapsto A^e \otimes M$, namely

$$H_n(A, M) = \mathrm{Tor}_n^{A^e}(A, M),$$

where M is regarded as left A^e-module. Secondly we define the n-th *cohomology group* of M as the derived functor of $M \mapsto \mathrm{Hom}_{A^e}(A, M)$:

$$H^n(A, M) = \mathrm{Ext}^n_{A^e}(A, M),$$

where M is regarded as right A^e-module. These groups were first introduced (when K is a field) by Gerhard Hochschild in 1945 and are called the *Hochschild groups* of M and A.

To compute these groups we construct the standard resolution for algebras as follows:

$$\dots \to S_n \to S_{n-1} \xrightarrow{d} \dots \xrightarrow{d} S_0 \xrightarrow{\varepsilon} A \to 0, \tag{4.7.3}$$

where S_n is the $(n+2)$-fold tensor product of A with itself over K, defined as A-bimodule in the natural way; thus the A^e-module structure is given by

$$(x_0 \otimes x_1 \otimes \dots \otimes x_{n+1})(a \otimes b) = ax_0 \otimes x_1 \otimes \dots \otimes x_{n+1}b.$$

The map ε is the multiplication and the differential $d : S_n \to S_{n-1}$ is given by

$$(x_0 \otimes \dots \otimes x_{n+1})d = \sum (-1)^i x_0 \otimes \dots \otimes x_i x_{i+1} \otimes \dots \otimes x_{n+1}.$$

It is clearly an A^e-homomorphism, and $d^2 = 0$ can be checked as in Section 3.1, using the associativity of A. To show that (4.7.3) is acyclic, we define a homotopy $h : S_n \to S_{n+1}$ by $u \mapsto 1 \otimes u$ $(u \in S_n)$, together with $\eta : A \to S_0$ given by $x \mapsto 1 \otimes x$. Then it is easily verified that $dh + hd = 1$, $\eta\varepsilon = 1$, $h_0 d + \varepsilon\eta = 1_{S_0}$.

Now $S_0 = A \otimes A$ is free of rank 1, while $S_n = A \otimes S_{n-2} \otimes A = S_{n-2} \otimes A^e$ is A^e-projective whenever S_{n-2} is K-projective, e.g. when K is a field. Thus when A is K-projective, (4.7.3) is a projective resolution of A and may be used to compute the Hochschild groups. We note that $H^0(A, M) = \mathrm{Hom}_{A^e}(A, M) = \{u \in M | au = ua\}$ for all $a \in A$; this group is also denoted by M^A.

The 1-cocycles are functions from A to M such that

$$f(xy) = xf(y) + f(x)y,$$

i.e. derivations, while coboundaries are inner derivations:

$$f(x) = xa - ax.$$

The 2-cocycles are functions $f : A^2 \to M$ such that

$$f(xy, z) + f(x, y)z = f(x, yz) + xf(y, z),$$

while a 2-coboundary is of the form

$$f(x, y) = xg(y) - g(xy) + g(x)y.$$

To interpret $H^2(A, N)$, let us consider algebra extensions of A, where A is K-projective, i.e. short exact sequences

$$0 \to N \xrightarrow{\alpha} B \xrightarrow{\beta} A \to 0, \tag{4.7.4}$$

where B is a K-algebra with A as quotient and N as kernel. For simplicity we shall take N to be contained in B, so that α is the inclusion mapping. Let $\gamma : A \to B$ be a K-linear mapping left inverse to β (which exists since A is K-projective). If γ is a homomorphism, then B is the direct sum of N and the subalgebra $A\gamma$ and we shall say that the sequence (4.7.4) *splits*. In general the failure of γ to be a homomorphism is measured by

$$f(a, b) = (ab)\gamma - (a\gamma)(b\gamma). \tag{4.7.5}$$

Since β is an algebra homomorphism inverse to γ, we have $f(a,b)\beta = ab - ab = 0$, hence $f(a,b) \in N$. Let us assume that $N^2 = 0$; then N, which is a B-bimodule, may be regarded as a (B/N)-bimodule, i.e. an A-bimodule. In that case B is completely determined by A, N and the function $f : A^2 \to N$, as the K-module $A \otimes N$ with the multiplication

$$(a,u)(b,v) = (ab, ub + av + f(a,b)) \quad (a, b \in A, u, v \in N). \qquad (4.7.6)$$

As one might expect, f satisfies a factor set condition, derived from the associative law. Write (4.7.5) as $(ab)\gamma = (a\gamma)(b\gamma) + f(a,b)$; then $(ab.c)\gamma = (ab)\gamma.c\gamma + f(ab,c) = a\gamma.b\gamma.c\gamma + f(a,b)c + f(ab,c)$. Similarly, $(a.bc)\gamma = a\gamma.b\gamma.c\gamma + af(b,c) + f(a,bc)$, hence

$$f(a,b)c + f(ab,c) = af(b,c) + f(a,bc);$$

this shows f to be a cocycle. The extension splits iff there is a section γ for β which is a homomorphism. This means that there is a mapping $g : A \to N$ such that γ, defined by $a\gamma = (a, -g(a))$ is a homomorphism, i.e. by (4.7.6),

$$a\gamma.b\gamma = (ab, -g(a)b - ag(b) + f(a,b)) = (ab)\gamma = (ab, -g(ab)),$$

whence

$$f(a,b) = ag(b) - g(ab) + g(a)b.$$

This is just the condition for f to be a coboundary, so we get

Proposition 4.7.3. *Let A be a K-algebra which is projective as K-module, and let N be an A-bimodule. Then the set of isomorphism classes of algebra extensions of A by N as ideal with zero multiplication is in natural bijection with $H^2(A, N)$.* ∎

In a similar way the first cohomology group describes the module extensions. Given two right A-modules U, V, we can form $\mathrm{Hom}_K(U, V)$, which inherits the right A-module structure from V and a left A-module structure from U and a verification, as in Section 2.3, shows that we have an A-bimodule structure. We now have the following result:

Theorem 4.7.4. *Let A be a K-algebra and U, V right A-modules, where A and U are projective as K-modules. Then*

$$H^n(A, \mathrm{Hom}_K(U, V)) \cong \mathrm{Ext}^n_A(U, V). \qquad (4.7.7)$$

Proof. We start from the situation $({}_AA_A, {}_KU_A, {}_KV_A)$. As we have just seen, there is a natural A-bimodule structure on $\mathrm{Hom}_K(U, V)$ and as is easily verified, we have a natural isomorphism (essentially by adjoint associativity)

$$\mathrm{Hom}_{A^e}(A, \mathrm{Hom}_K(U, V)) \to \mathrm{Hom}_A(U \otimes_A A, V) \cong \mathrm{Hom}_A(U, V). \qquad (4.7.8)$$

Let X be a free resolution of A as A^e-module; then as free A-bimodule X_n has the form $A \otimes_K F_n \otimes_K A$, where F_n is a free K-module. Now the tensor product (over K) of free K-modules is free, hence the tensor product of projective K-modules is projective, so $X_n = (A \otimes_K F_n) \otimes_K A$ is right A-projective and $U \otimes_A X_n = (U \otimes_K F_n) \otimes_K A$ is right A-projective. By (4.7.8) we obtain an isomorphism of complexes

$$\mathrm{Hom}_{A^e}(X, \mathrm{Hom}_K(U, V)) \cong \mathrm{Hom}_A(U \otimes_A X, V).$$

The complex on the left has cohomology groups $\mathrm{Ext}^n_{A^e}(A, \mathrm{Hom}_K(U, V)) \cong H^n(A, \mathrm{Hom}(U, V))$. On the right, since clearly $\mathrm{Tor}^A_n(U, A) = 0$ for $n \geq 1$, $U \otimes_A X$ is a resolution for U. As we saw, it is A-projective, and so we obtain for its cohomology group $\mathrm{Ext}^n_A(U, V)$ and (4.7.7) follows. ■

We observe that the hypotheses of Theorem 4.7.4 on A and U are satisfied when K is a field. The results of Theorem 4.7.4 and Proposition 4.7.3 can now be combined to establish one of the main theorems on the splitting of algebra extensions:

Theorem 4.7.5 (Wedderburn's principal theorem). *Let B be a finite-dimensional algebra over a field k and let I be a nilpotent ideal in B such that* bidim $B/I \leq 1$. *Then B has a subalgebra A which is a complement of I as k-space:*

$$B = A \oplus I. \tag{4.7.9}$$

Proof. If $I^2 = 0$, this follows from Proposition 4.7.3, because $H^2(B/I, I) = 0$, so we shall assume that $I^2 \neq 0$ and use induction on the dimension of B. The algebra $B' = B/I^2$ has lower dimension than B and $B'/I/I^2 \cong B/I$, hence by the induction hypothesis there exists a subalgebra C of B such that $C \supseteq I^2$ and

$$B = C + I, \quad C \cap I = I^2.$$

Now $C/I^2 = C/(I \cap C) \cong (C + I)/I = B/I$ satisfies the same hypothesis; since I is nilpotent, $I^2 \subset I$, hence $C \subset B$ and applying induction again, we find a subalgebra A of C such that $C = A \oplus I^2$. Now $B = C + I = A + I^2 + I = A + I$ and $A \cap I = A \cap C \cap I = A \cap I^2 = 0$, therefore (4.7.9) holds. ■

In particular, taking I to be the Jacobson radical $\mathbf{J}(B)$ of B, we see that $\mathbf{J}(B)$ is complemented whenever $B/\mathbf{J}(B)$ is separable. In that case the result can be proved more explicitly as follows. Let $e = \sum p_i \otimes q_i$ be a separator for B/I. Given the cocycle $f(a, b)$ arising from a section, we put $g(a) = \sum f(a, p_i)q_i$. Then

$$\begin{aligned} ag(b) - g(ab) + g(a)b &= \sum af(b, p_i)q_i - \sum f(ab, p_i)q_i + \sum f(a, p_i)q_i b \\ &= \sum f(a, b)p_i q_i - \sum f(a, bp_i)q_i + \sum f(a, p_i)q_i b. \end{aligned}$$

The last two terms cancel because $be = eb$ and the first reduces to $f(a, b)$ because $e\mu = 1$; hence the right-hand side is just $f(a, b)$ and this shows f to be a coboundary.

It remains to determine the separable algebras. In the first place we note that every separable algebra over a field is finite-dimensional. This follows from

Proposition 4.7.6 (Villamayor–Zelinsky, 1966). *Let A be a separable K-algebra. If A_K is projective, then it is finitely generated as K-module.*

Proof. Since A is separable, it is projective as A^e-module, hence the same is true of A^o. Suppose that $A^o \oplus B = F$, where F is a free K-module with a basis u_λ $(\lambda \in \Lambda)$ and write $\pi : F \to A^o$ for the projection. Let $\alpha_\lambda \in F^*$ be the dual basis and put $\beta_\lambda = \alpha_\lambda | A^o$; then we have

$$x = \sum (\beta_\lambda, x) u_\lambda = \sum (\beta_\lambda, x)\pi(u_\lambda) \quad \text{for all } x \in A^o. \tag{4.7.10}$$

The mapping $(x, y) \mapsto (\beta_\lambda, x)y$ from $A^o \times A$ to A is bilinear, hence there exists $\varphi_\lambda : A^e \to A$ such that

$$(\beta_\lambda, x)y = (\varphi_l, x \otimes y). \tag{4.7.11}$$

Moreover, for any $z \in A$,

$$(\varphi_\lambda, (x \otimes y)z) = (\varphi_\lambda, x \otimes yz) = (\beta_\lambda, x)yz = (\varphi_\lambda, x \otimes y)z.$$

Hence by linearity we have

$$(\varphi_\lambda, wz) = (\varphi_\lambda, w)z \quad \text{for all } w \in A^e,\ z \in A, \tag{4.7.12}$$

and for any w, z the two sides of (4.7.12) vanish for almost all λ, because this is true of the β_λ. Thus φ_λ is a right A-module homomorphism. We claim that there exists a family $w_\lambda \in A^e (\lambda \in \Lambda)$ such that for any $w \in A^e$, $\Lambda(w) = \{\lambda \in \Lambda \mid (\varphi_\lambda, w) \neq 0\}$ is finite and

$$w = \sum w_\lambda (\varphi_\lambda, w). \tag{4.7.13}$$

This is an analogue of the dual basis lemma (BA, Lemma 4.7.5). We define $w_\lambda = \pi(u_\lambda) \otimes 1_A$; then for any $x \in A^o$, $y \in A$,

$$\begin{aligned} x \otimes y &= \sum (\beta_\lambda, x)\pi(u_\lambda) \otimes y = \sum \pi(u_\lambda) \otimes (\beta_\lambda, x)y \\ &= \sum w_\lambda (\varphi_\lambda, x \otimes y), \end{aligned}$$

by (4.7.11); now (4.7.13) follows by linearity.

Now let $e = \sum p_i \otimes q_i$ be a separator for A. By definition $e\mu = 1$ and $ex = xe$ for $x \in A$. Hence for any $y \in A$,

$$\begin{aligned} y = 1_A . y = e\mu . y = (ey)\mu &= \left[\sum w_\lambda (\varphi_\lambda, ey)\right]\mu \quad \text{by (4.7.13)}, \\ &= \left[\sum w_\lambda (\varphi_\lambda, ye)\right]\mu = \left[\sum w_\lambda \sum (\varphi_\lambda, yp_i) q_i\right]\mu, \end{aligned}$$

because μ is a bimodule homomorphism. Further,

$$\left[\sum w_\lambda \sum (\varphi_\lambda, yp_i)q_i\right]\mu = \left[\sum w_\lambda \sum q_i(\varphi_\lambda, yp_i)\right]\mu$$
$$= \sum (w_\lambda \mu)q_i(\beta_\lambda, yp_i).$$

Hence A is generated by the family $(w_\lambda \mu)q_i$, where λ ranges over the finite set $\Lambda(ey)$, but $\Lambda(ey) \subseteq L(e)$, which is finite and independent of y. Thus we have found a finite generating set for A. ■

For the rest of this section we shall confine ourselves to algebras over a field. Our aim will be to show that an algebra over a field k is separable iff it is semisimple and remains so under all extensions of k. We begin with some generalities.

Proposition 4.7.7. (i) *If A, B are separable algebras over a field k, then so are $A \oplus B$ and $A \otimes B$.*

(ii) *Given a k-algebra A and a field extension F of k, the algebra $A_F = A \otimes_k F$ is separable if and only if A is.*

(iii) *For any field k and any $n \geq 1$, the full matrix ring k_n is separable.*

Proof. (i) The separability may be described by the existence of a section λ for the multiplication μ. If λ_A, λ_B are sections for the multiplication in A and B respectively, then $\lambda_A + \lambda_B : A \otimes B \to (A \otimes A) \oplus (A \otimes B) \oplus (B \otimes A) \oplus (B \otimes B)$ is a section for the multiplication in $A \oplus B$, while $\lambda_A \otimes \lambda_B : A \otimes B \to A \otimes B \otimes A \otimes B \cong A \otimes A \otimes B \otimes B$ is a section for $A \otimes B$.

(ii) Let $e = \sum p_i \otimes q_i$ be a separator for A; then e is still a separator for A_F, for $e\mu = 1$ and $ae = ea$ continue to hold for $a \in A_F$. Conversely, if A_F is separable, choose a basis u_j for F over k such that $u_0 = 1$ and write the separator for A_F as

$$e = \sum p_{ij} \otimes q_{ij} \otimes u_j, \quad \text{where } p_{ij}, q_{ij} \in A.$$

Then $\sum p_{ij}q_{ij}u_j = 1$, hence equating coefficients of u_0 we find that $\sum p_{i0}q_{i0} = 1$. Further, for any $a \in A$ we have $ae = ea$, i.e. $\sum ap_{ij} \otimes q_{ij} \otimes u_j = \sum p_{ij} \otimes q_{ij}a \otimes u_j$ and on equating coefficients of u_0 we find that $\sum ap_{0j} \otimes q_{0j} = \sum p_{0j} \otimes q_{0j}a$, hence $e_0 = \sum p_{0j} \otimes q_{0j}$ is a separator for A, showing A to be separable.

(iii) If $A = k_n$, then $A^e = k_n \otimes k_n = k_{n^2}$, hence any A-module is semisimple, and so A is separable. ■

We can now describe separable algebras over a field.

Theorem 4.7.8. *Let A be an algebra over a field k. Then A is separable if and only if A_F is semisimple for all extension fields F of k. Moreover, when this holds, then A is finite-dimensional over k.*

Proof. A is finite-dimensional whenever it is separable or semisimple, by Proposition 4.7.6 and Wedderburn's theorem (BA, Theorem 5.2.4); so we may assume A to be finite-dimensional in what follows. Assume A separable; then by Theorem 4.7.4 all module extensions split, i.e. every A-module is semisimple, so A is semisimple.

By Proposition 4.7.7 A_F is also separable, so A_F is semisimple for every extension F of k.

Conversely, if A_F is semisimple for all $F \supseteq k$, take F to be an algebraic closure of k. Then A_F is a direct product of full matrix algebras F_n. By Proposition 4.7.7, each F_n is separable and A_F is separable, hence so is A. ■

We note that in this theorem it is enough to require A_F to be semisimple for all finite field extensions of k. The theorem also shows that for finite-dimensional commutative algebras over a field the notion of separability introduced here reduces to that of BA, Section 11.6. For a commutative k-algebra A is separable in the sense of this section iff A_F is semisimple for all field extensions F of k, and by the form of the radical in Artinian rings (BA, Theorem 5.3.5) this just means that A_F is a reduced ring.

Exercises

1. Verify that a separating idempotent for A is indeed idempotent. (Hint. Use Proposition 4.7.2 and the fact that $(1-e)\mu = 0$.)
2. Show that for any commutative ring K and any $n \geq 1$, K_n is separable by verifying that $\sum e_{i1} \otimes e_{1i}$ (where the e_{ij} are the usual matrix units) is a separating idempotent.
3. Let A be a separable algebra over a field k. Show that any right A-module M is a direct summand of $M \otimes_k A$ and hence is projective. Deduce that A is semisimple.
4. Show that an algebra over a field k is separable iff it is semisimple and its centre is a direct product of separable field extensions of k.
5. Show that a K-algebra A (over a commutative ring K) is separable iff the functor $M \mapsto M^A$ for any A-bimodule M is exact.
6. Let k be a field of prime characteristic p and $F = k(\alpha)$ a p-radical extension of degree p, where $\alpha^p = a \in k$. Let A be the k-algebra generated by an element u with the defining relation $(u^p - a)^2 = 0$. Show that $J = \mathbf{J}(A)$ is spanned by $v = u^p - a$ and that $\bar{A} = A/J$ is semisimple, but $\bar{A}$ is not. Verify that A contains no subalgebra $\cong \bar{A}$.
7. Show that if E/k is a separable field extension and A is a commutative k-algebra, then A_E is separable over A.
8. Let K be a commutative ring and G a finite group whose order n is a unit in K. Show that the group algebra KG is separable, by verifying that $(1/n)\sum g^{-1} \otimes g$ is a separating idempotent.
9. Let E be a commutative separable K-algebra. Show that for any K-algebra A, $\text{gl.dim}(A \otimes E) = \text{gl.dim}(A)$. (Hint. Use the separating idempotent of E over K to define an averaging operator as in Exercise 8.)
10. Use Exercise 9 to show that a K-algebra A is K-projective, i.e. an exact sequence of A-modules which is K-split is A-split.

Further exercises on Chapter 4

1. Show that if $M_1 \oplus M_2 \cong N_1 \oplus N_2$ and $M_1 \cong N_1$ then $M_2 \cong N_2$. (Hint. Take the isomorphism in the form $\alpha = (\alpha_{ij})$, where $\alpha_{ij} : M_i \to N_j$ and use the fact that α and α_{11} are invertible.)
2. If R is a Dedekind domain and $\mathfrak{a}$, $\mathfrak{b}$ are any non-zero ideals in R, then $\mathfrak{a} \oplus \mathfrak{b} \cong R \oplus \mathfrak{ab}$ (BA, Theorem 10.6.11). Use this result to show that the Krull–Schmidt theorem fails over any Dedekind domain which is not a principal ideal domain.
3. Show that for any projective R-module P, the intersection of all maximal submodules is equal to JP, where $J = \mathrm{J}(R)$. Give an example of a module for which this fails; can this module be finitely generated?
4. Let R be a ring which can be written as a direct sum of indecomposable left ideals with simple tops. Show that R is semiperfect. Deduce that a ring is semiperfect iff every finitely generated module has a projective cover.
5. Let R be a semiperfect ring. Show that any homomorphic image of an indecomposable projective R-module is indecomposable. Does this remain true for more general indecomposable modules?
6. Show that the endomorphism ring of any indecomposable injective module is a local ring and deduce a Krull–Schmidt theorem for injective modules.
7. Show that finitely generated projective modules (over any ring) are isomorphic iff they have isomorphic tops.
8. Let R be any ring and R_ω the ring of all row-finite matrices (with countably many rows and columns) over R. Show that $R \prec R_\omega$ but the R, R_ω are not Morita equivalent. What is the relation between (i) the ring R_ω of all row-finite matrices, (ii) the ring ${}_\omega R$ of all column-finite matrices and (iii) $R_\omega \cap {}_\omega R$?
9. A ring R is called *basic* if $R/\mathrm{J}(R)$ is a direct product of a finite number of skew fields. Show that the ring of all upper triangular matrices over a local ring is basic.
10. Show that for every semiperfect ring A there exists a basic ring B such that $B \sim A$ and that B is unique up to isomorphism.
11. A functor T is called *faithfully exact* if it is faithful, exact and preserves coproducts. An object P in an abelian category $\mathscr{A}$ such that $h^P = \mathscr{A}(P, -)$ is faithfully exact is called *faithfully projective*. Show that P is faithfully projective iff P is a projective generator such that $\mathscr{A}(P, \coprod X_i) \cong \coprod \mathscr{A}(P, X_i)$, for any family (X_i) of $\mathscr{A}$-objects.
12. Let P be a faithfully projective object in an abelian category $\mathscr{A}$. Write $A = \mathscr{A}(P, P)$, $hX = \mathscr{A}(P, X)$ for $X \in \mathrm{Ob}\mathscr{A}$; verify that A is a ring and hX a left A-module, under composition of maps. Verify also that h is a functor from $\mathscr{A}$ to ${}_A\mathrm{Mod}$ and use Theorem 4.4.1 to show that this is an equivalence of functors.

    Deduce that any abelian category with coproducts and a faithfully projective object is equivalent to a category of modules over a ring (Mitchell–Freyd theorem).
13. A short exact sequence is called *pure* if it stays exact under tensoring. Show that a module M is flat iff every short exact sequence with third term M is pure.

14. Show that a short exact sequence of R-modules with first two terms M', M is pure (M' is pure in M) iff $uA = p$, where $u \in M^m$, $p \in M'^n$, $A \in {}^mR^n$ implies that $u'A = p$ for some $u' \in M'^m$.
15. Show that for any R-module U (over any ring R) the correspondence $U \mapsto \hat{U} = \mathrm{Hom}_{\mathbf{Z}}(U, \mathbf{K})$ is a faithful covariant functor. Show also that there is a natural transformation $U \mapsto \hat{\hat{U}}$ a which is an embedding.
16. Show that a module M is flat iff for every homomorphism $\alpha : P \to M$, where P is finitely generated projective and for every $x \in \ker \alpha$ there is a factorization $\alpha = \beta\gamma$, where $\beta : P \to Q$, $\gamma : Q \to M$, Q finitely generated projective, such that $x\beta = 0$.
17. Let A be a K-algebra and M an A-bimodule. Show that the sequence

$$0 \to M^A \to M \to \mathrm{Der}_K(A, M) \to H^1(A, M) \to 0$$

is exact, where $\mathrm{Der}_K(A, M) = \mathrm{Hom}_{A^e}(\Omega, M)$ is the module of derivations of A into M.
18. (A. I. Malcev) Let B be an algebra over a field, with a nilpotent ideal I such that B/I is separable. If $B = A_1 \oplus I = A_2 \oplus I$ are two splittings (Theorem 4.7.4), show that A_1, A_2 are conjugate by an automorphism of the form $x \mapsto (1-u)^{-1}x(1-u)$, where $u \in I$. (Hint. If $a \mapsto a_i$ is an isomorphism $B/I \to A_i$, examine the cochain $f(a) = a_1 - a_2 \in I$.)
19. Show that the injective hull of a PID R (qua left R-module) is just the field of fractions of R.

5 Central simple algebras

Skew fields are more complicated than fields and much less is known about them. However, in the case of division algebras (the case of finite dimension over the centre) the situation is rather better. It is convenient to include full matrix rings over division algebras, thus our topic in Section 5.1 is essentially the class of simple Artinian rings. Although some of our results are proved in this generality we shall soon specialize to the finite-dimensional case over a field.

There is no space to enter into such interesting questions as the discussion of division algebras over number fields, but we shall in Section 5.2 introduce an important field invariant, the Brauer group, show its significance for division algebras and in Section 5.3 describe some of their invariants. Then, after a look at quaternion algebras (Section 5.4), we introduce crossed products (Section 5.5). In Section 5.6 we study the effect of changing the base field, and in Section 5.7 illustrate them on cyclic algebras.

5.1 Simple Artinian rings

It is clear from Wedderburn's theorem that every simple Artinian ring is an algebra over a field; it also contains a skew field and is finite-dimensional over the latter, but it need not be finite-dimensional as an algebra. We begin by not imposing any finiteness restriction, in fact we shall not even assume our rings to be Artinian, although not very much can be said in that generality.

Let A be any ring and denote its centre by C. We may regard A as an A-bimodule or equivalently as a right A^e-module, where $A^e = A^o \otimes A$ as in Section 4.7. The centralizer of A^e acting on A, i.e. of all left multiplications λ_a and right multiplications ρ_b of A, is the intersection of the centralizers of all the λ_a and ρ_b; this is the set of all left multiplications commuting with all left multiplications, i.e. multiplications by elements of C. Thus the centralizer of the action of A^e on A is just C. If further, A is a simple ring, it is simple as A^e-module and by Schur's lemma, C is then a field.

Let k be a field. Given a k-algebra R and a right R-module M, the action of R on M is said to be *dense* if, given $x_1, \dots, x_r \in M$ and $\theta \in \text{End}_k(M)$, there exists $a \in R$ such that $x_i\theta = x_i a$ $(i = 1, \dots, r)$. In this definition we may clearly omit any x_i linearly dependent on the rest. Hence an equivalent definition is obtained by requiring that for any $x_1, \dots, x_r, y_1, \dots, y_r \in M$, where the x_i are linearly independent over

k, there exists $a \in R$ such that $y_i = x_i a$. In particular, when M is finite-dimensional over k, this means that every k-linear transformation of M can be accomplished by acting with some element of R. The next result on dense action is basic in the study of simple algebras.

Theorem 5.1.1 (Density theorem). *The centre of any simple ring is a field. If A is a simple ring with centre k, then* $A^e = A^o \otimes A$ *is dense in the ring of k-linear transformations of A. In particular, when* $[A : k] = n$ *is finite, then*

$$A^e = A^o \otimes A \cong \mathfrak{M}_n(k). \tag{5.1.1}$$

Proof. (Artin–Whaples, 1943) Since A is simple, k is a field. Let $x_1, \ldots, x_r \in A$ be linearly independent over k and $y_1, \ldots, y_r \in A$; we have to find $\varphi \in A^e$ such that $y_i = x_i\varphi$. For $r = 1$ this follows by the simplicity of A, so let $r > 1$. By induction there exists $\varphi'_i \in A$ such that $x_j\varphi'_i = \delta_{ij}$ for $i, j = 2, \ldots, r$. If $x_1\varphi'_i \notin k$ for some i, then for suitable $b \in A$, $\psi = \varphi'_i(b \otimes 1 - 1 \otimes b)$ satisfies $x_1\psi \neq 0$, $x_j\psi = 0$ for $j > 1$, and for some $\theta \in A^e$, $x_1\psi\theta = 1$, hence $\varphi_1 = \psi\theta$ maps x_1 to 1 and the $x_j (j > 1)$ to 0. Now $\varphi_i = \varphi'_i - \varphi_1(1 \otimes x_1\varphi'_i)$ maps x_i to 1 and the other x's to 0, hence $\varphi = \sum \varphi_i(1 \otimes y_i)$ is the required map.

There remains the case where $\lambda_i = x_1\varphi'_i \in k$ for $i = 2, \ldots, r$. Put $\psi = \sum \varphi'_i(1 \otimes x_i) - 1$; then $x_j\psi = 0$ for $j = 2, \ldots, r$, $x_1\psi = \sum \lambda_i x_i - x_1$. By the linear independence of the x's this is not zero and for a suitable $\theta \in A^e$, $x_1\psi\theta = 1$; now the proof can be completed as before.

In the finite case we have a surjective homomorphism $A^e \to \mathrm{End}_k(A) \cong k_n$ and now the isomorphism (5.1.1) follows by counting dimensions, which are n^2 on both sides. ∎

This result will be proved again in a more general context in Section 8.1 below. That (5.1.1) is an isomorphism also follows from the fact, soon to be proved (in Corollary 5.1.3) that A^e is simple, for any simple ring with centre k.

Let A be a k-algebra, where k is a field. Then for any $\alpha \in k$, $a, b \in A$ we have

$$\alpha(ab) = (\alpha a)b = a(\alpha b).$$

Taking $b = 1$, we see that $\alpha a = a\alpha$; thus the mapping $\alpha \mapsto \alpha.1$ is a homomorphism of k into the centre of A; since k is a field, this is actually an embedding whenever $A \neq 0$. Conversely, any ring whose centre contains k as a subfield may be considered as a k-algebra in this way. Thus a non-trivial k-algebra is essentially a ring whose centre contains k as a subfield. If the centre of A is precisely k, A is called a *central* k-algebra. Throughout this section k will be a field, fixed but arbitrary. Our convention will be that an algebra or k-algebra need not be finite-dimensional, but a *division algebra* is finite-dimensional over the ground field. When the dimension is infinite, we shall speak of a *skew field*.

Consider the functor $A \otimes_k -$; we shall show that it preserves the ideal structure when A is a central simple k-algebra.

Theorem 5.1.2 (Azumaya–Nakayama, 1947). *Let A be a central simple k-algebra. Then for any k-algebra B there is a lattice-isomorphism between the ideals of B and those of $A \otimes B$. In particular, the ideal lattice of $\mathfrak{M}_n(B)$ is isomorphic to that of B, for any $n \geq 1$.*

Proof. Consider the mappings

$$\mathfrak{b} \longmapsto A \otimes \mathfrak{b} \quad (\mathfrak{b} \text{ an ideal of } B) \tag{5.1.2}$$

$$\mathfrak{C} \longmapsto \mathfrak{C} \cap B \quad (\mathfrak{C} \text{ an ideal of } A \otimes B), \tag{5.1.3}$$

where B is embedded in $A \otimes B$ by the natural mapping $x \longmapsto 1 \otimes x$. Since k is a field, all these tensor products are exact, so we do have such an embedding. We claim that the mappings (5.1.2), (5.1.3) are mutually inverse; this will prove the result, for they establish a bijection which is clearly inclusion-preserving and hence a lattice isomorphism. The last part then follows because k_n is central simple and $B_n = k_n \otimes B$; of course it is also a consequence of Corollary 4.4.6.

We recall the intersection formula in a tensor product (BA, Section 4.8): If U, V are k-spaces and $U = U' \oplus U''$, $V = V' \oplus V''$, then

$$U' \otimes V \cap U \otimes V' = U' \otimes V'.$$

Since k is field, every subspace is a direct summand. Now if $\mathfrak{b}$ is an ideal in B, then $A \otimes \mathfrak{b}$ is an ideal in $A \otimes B$ and so we have

$$(A \otimes b) \cap B = \mathfrak{b}. \tag{5.1.4}$$

This holds even for left ideals, hence (5.1.2) is injective for any one-sided ideal $\mathfrak{b}$.

Next let $\mathfrak{C}$ be an ideal in $A \otimes B$ and put $\mathfrak{b} = \mathfrak{C} \cap B$; then $A \otimes \mathfrak{b} \subseteq \mathfrak{C}$ and we have to establish equality here. Any $c \in \mathfrak{C}$ can be written in terms of a basis u_i of A as $c = \sum u_i \otimes z_i$, where $z_i \in B$. Only finitely many of the z_i are non-zero, say $z_i \neq 0$ for $i = 1, \dots, r$. By Theorem 5.1.1, A^e acts densely on A, hence there exist x_j, $y_j \in A$ $(j = 1, \dots, s)$ such that $\sum_j x_j u_i y_j = \delta_{i1}$ $(i = 1, \dots, r)$. It follows that $\sum x_j c y_j = \sum_{ij} x_j u_i y_j \otimes z_i = 1 \otimes z_1 \in \mathfrak{C}$; so $z_1 \in \mathfrak{b}$ and similarly $z_i \in \mathfrak{b}$ for $i = 2, \dots, r$. Therefore $\mathfrak{C} = A \otimes \mathfrak{b}$ and this is the desired equality. Thus (5.1.2), (5.1.3) are mutually inverse and the lattice isomorphism follows. ∎

We observe that no finiteness assumptions are needed here. The theorem has a number of important consequences.

Corollary 5.1.3. *If A is a central simple k-algebra and B is any k-algebra, then $A \otimes B$ is simple if and only if B is simple; further the centre of $A \otimes B$ is isomorphic to that of B. In particular, the tensor product of central simple k-algebras is again central simple.*

Proof. The assertion about the centre follows because the centre of a tensor product is the tensor product of the centres, as is easily verified (see BA, Corollary 5.4.4). The rest is a consequence of Theorem 5.1.2. ∎

Sometimes we shall want to know when a given division algebra over k can be embedded in a matrix ring over a skew field D containing k in its centre. There is a simple answer when the centre of D is a regular extension of k; it is given by the following result, taken from Schofield (1985) (a field extension F/k is *regular* if $E \otimes F$ is an integral domain for any field extension E/k).

Proposition 5.1.4. *Let D be a skew field whose centre F is a regular extension of k, and let A be a simple Artinian k-algebra. Then $A^{\circ} \otimes_k D$ is a simple Artinian ring, with a unique simple module S which is finite-dimensional over D, say $[S : D] = s$, and A can be embedded in $\mathfrak{M}_n(D)$ if and only if $s|n$.*

Proof. Let C denote the centre of A. We have

$$A^{\circ} \otimes_k D \cong A^{\circ} \otimes_C (C \otimes_k D),$$

and A is central simple over C; hence by Theorem 5.1.2, the ideals of $A^{\circ} \otimes D$ correspond to those of $C \otimes_k D$. Next we have

$$C \otimes_k D \cong (C \otimes_k F) \otimes_F D,$$

so the ideals correspond to those of $C \otimes F$. By hypothesis this is an integral domain; since $[C \otimes_k F : F] = [C : k]$, which is finite, $C \otimes F$ is a field, and it follows that $A^{\circ} \otimes D$ is simple. It is Artinian because its dimension over D is finite. This also shows that the unique simple $(A^{\circ} \otimes D)$-module S has finite dimension over D.

Suppose now that A is embedded in D_n; we may regard D_n as the endomorphism ring of D^n, qua right D-module. In this way D^n becomes an (A, D)-bimodule, i.e. a right $(A^{\circ} \otimes D)$-module. As such it is isomorphic to S^r for some $r \geq 1$, and a comparison of dimensions shows that $n = rs$. Conversely, if $n = rs$, then $D^n \cong S^r$ and this is an (A, D)-bimodule, hence A is embedded in D_n. ∎

We note that in this result the regularity assumption can be omitted when A is a central k-algebra.

We shall want to know when a given k-algebra can be written as a tensor product. First we shall look at the general case; in the finite-dimensional case we can then easily obtain a complete answer.

Proposition 5.1.5. *Let P be a k-algebra with a central simple subalgebra A, whose centralizer in P is denoted by A'. Then the subalgebra of P generated by A and A' is their tensor product over k.*

If further $[A : k]$ is finite, then $P = A \otimes A'$ and the centre of A' is the centre of P.

Proof. By hypothesis, A and A' commute elementwise, so the mapping $(x, y) \mapsto xy (x \in A, y \in A')$ gives rise to a homomorphism

$$A \otimes A' \to P. \tag{5.1.5}$$

Its kernel is an ideal in $A \otimes A'$, which by Theorem 5.1.2 is of the form $A \otimes \mathfrak{a}$, where $\mathfrak{a}$ is the kernel of the restriction of (5.1.5) to A'. But this is the inclusion mapping, which is injective, so $\mathfrak{a} = 0$ and (5.1.5) is injective. Clearly its image is the subalgebra generated by A and A'.

Suppose now that $[A : k] = n$ is finite. We can regard P as A^e-module, i.e. by Theorem 5.1.1 as k_n-module. This module is semisimple, hence $P = \oplus P_\lambda$, where P_λ is simple, isomorphic to A. Let $u_\lambda \in P_\lambda$ correspond to 1 in this isomorphism; then $u_\lambda a = a u_\lambda$ for all $a \in A$, hence $u_\lambda \in A'$ and so $P = AA'$. Therefore (5.1.5) is surjective, hence an isomorphism in this case. Now the assertion about the centre follows by Corollary 5.1.3. ■

In particular we find the identity

$$k_{rs} \cong k_r \otimes k_s, \tag{5.1.6}$$

which is also easily verified directly. We remark that in general $A \otimes A'$ will be a proper subalgebra of P; for example, if $P = k\langle x, y, z\rangle$, the free algebra on x, y, z and A is the subalgebra generated by x, y, then P and A both have centre k and $A' = k$, so $AA' \neq P$.

We recall from field theory the theorem of the primitive element: A finite separable extension F/k can be generated by a single element over k (see BA, Theorem 7.9.2). Of course a noncommutative algebra cannot be generated by a single element, but as we shall now see, in many cases two elements suffice:

Proposition 5.1.6. *Let D be a central division algebra over a field k and let F be a maximal separable subfield. Then there is an element u in D such that $D = FuF$; in particular, D can be generated by two elements over k.*

Proof. Let $M(D)$ be the multiplication algebra of D, generated by the left and right multiplications λ_a and ρ_a resp. for $a \in D$. Writing D° for the opposite ring of D, we have a homomorphism $D^\circ \otimes D \to M(D)$ mapping $a \otimes b$ to $\lambda_a \rho_b$; it is surjective by definition and since $D^\circ \otimes D$ is simple, it is an isomorphism. Restricting a and b to F, we obtain a faithful action of $F \otimes F$ on D. Now F is separable, so (by BA, Corollary 5.7.4) we have $F \otimes F \cong E_1 \times \ldots \times E_n$, where $n = [F : k]$ and the E_i are fields (composites of F with itself, hence isomorphic to F). Let e_i be the element of $M(D)$ corresponding to the unit element of E_i and choose $u_i \in D$ such that $u_i e_i \neq 0$. If we now write $u = \sum u_i e_i$, then the map $\sum a_r \otimes b_r \mapsto \sum u \lambda_{a_r} \rho_{b_r}$ is injective, because u is not annihilated by any E_i. A comparison of dimensions shows that it is also surjective, and so $D = FuF$. As a separable extension F/k is generated by a single element, c say, hence D is generated by u and c over k. ■

We next come to a basic result in the theory of central simple algebras, the Skolem–Noether theorem, which asserts that every automorphism of a finite-dimensional central simple algebra is inner. It is useful to have a slightly more general form of this result:

Theorem 5.1.7. *Let A be a simple Artinian ring with centre k and B any finite-dimensional simple k-algebra. Given any homomorphisms f_1, f_2 from B into A, there* *exists a unit u of A such that*

$$bf_2 = u^{-1}(bf_1)u \quad \text{for all } b \in B. \tag{5.1.7}$$

Proof. We regard A as $(A^{\circ} \otimes_k B)$-module, i.e. as (A, B)-bimodule in two ways:

$$(a, x, b)_i = ax(bf_i) \quad \text{where } a, x \in A, b \in B, \ i = 1, 2. \tag{5.1.8}$$

By Corollary 5.1.3, $A^{\circ} \otimes B$ is simple, and it is Artinian, for if $A = D_t$, where D is a skew field and $t \geq 1$, then $A^{\circ} \otimes B \cong (D^{\circ} \otimes B)_t$ and this has finite dimension over D°. If V is the unique simple $(A^{\circ} \otimes B)$-module, then every finitely generated $(A^{\circ} \otimes B)$-module has the form V^r for some $r \geq 0$, hence the $(A^{\circ} \otimes B)$-module structures defined on A by (5.1.8) are isomorphic to V^r, V^s respectively. By comparing dimensions over D° (i.e. regarding A as left D-module), we see that $r = s$, hence the structures are isomorphic. This means that there is a bijective linear transformation $\gamma : A \to A$ such that

$$[ax(bf_1)]\gamma = a(x\gamma)(bf_2) \quad \text{where } a, x \in A, b \in B. \tag{5.1.9}$$

Putting $b = 1 = x$ and writing $1\gamma = u$, we find that $a\gamma = au$ for all $a \in A$. Since γ is surjective, there exists $v \in A$ such that $vu = 1$; but γ is also injective and $(uv - 1)\gamma = uvu - u = 0$, hence $uv = 1$ and so $v = u^{-1}$. If we now put $a = x = 1$ in (5.1.9), we obtain $(bf_1)u = u(bf_2)$ and (5.1.7) follows. ■

Two subalgebras or elements are said to be *conjugate* if there is an inner automorphism mapping one to the other. From Theorem 5.1.7 we immediately find

Corollary 5.1.8. *In any simple Artinian ring with centre k, isomorphic finite-dimensional simple k-subalgebras are conjugate and hence have conjugate centralizers.* ■

In particular this shows that every automorphism of a central simple finite-dimensional algebra is inner (Skolem–Noether theorem). More generally we have

Corollary 5.1.9. *Every automorphism of a finite-dimensional semisimple k-algebra which leaves the centre elementwise fixed, is inner.*

Proof. Let θ be the automorphism, write $A = A_1 \oplus \ldots \oplus A_r$, where each A_i is simple and denote the unit-element of A_i by e_i. Then e_i is in the centre of A_i and so $e_i\theta = e_i$, by hypothesis. It follows that θ maps each A_i into itself: if $a \in A_i$, then $a\theta = (ae_i)\theta = a\theta.e_i$, therefore $a\theta \in A_i$. Thus θ induces an endomorphism of A_i, likewise θ^{-1}, so θ in fact defines an automorphism of A_i, θ_i say. Now the centre of A_i, C_i say, is left fixed by θ_i so we may regard θ_i as an automorphism of the central simple C_i-algebra A_i. By Corollary 5.1.8 there exists a unit u_i in A_i such that $x\theta_i = u_i^{-1}xu_i$ $(x \in A_i)$. Now it is easily checked that $u = \sum u_i$ is a unit of A inducing θ. ■

We remark that Corollary 5.1.8 as it stands does not extend to the case of semisimple subalgebras (see Exercise 5).

We next look at the relation between the dimension of a simple subalgebra and that of its centralizer.

Theorem 5.1.10 (R. Brauer, 1932). *Let A be a simple Artinian ring with centre k and B a finite-dimensional simple subalgebra with centre F. Then the centralizer B' of B in A is again simple with centre F and the centralizer B'' of B' equals B, while the centralizer F' of F is given by*

$$F' = B \otimes B'. \tag{5.1.10}$$

Moreover,

$$[A : B'] = [B : k], \tag{5.1.11}$$

and if $[B : k] = r$, then

$$A \otimes B^{\mathrm{o}} \cong B' \otimes_k k_r \cong B'_r. \tag{5.1.12}$$

Proof. We may regard k_r as acting on $B \cong k^r$ by k-linear transformations. As such it contains the subalgebras ρ_B of right multiplications and λ_B of left multiplications. Clearly $\rho_B \cong B$, $\lambda_B \cong B^{\mathrm{o}}$ and $A \otimes k_r$ is central simple, by Corollary 5.1.3. We have the isomorphic simple subalgebras $B \otimes k$, $k \otimes \rho_B$; they are conjugate by Corollary 5.1.8 and so have isomorphic centralizers:

$$B' \otimes k_r \cong A \otimes \lambda_B \cong A \otimes B^{\mathrm{o}}.$$

Since $B' \otimes k_r \cong B'_r$, this proves (5.1.12), and comparing dimensions over B', we find that $r^2 = [A : B'][B : k]$, from which (5.1.11) follows on dividing by r.

Now $A \otimes B^{\mathrm{o}}$ is simple, hence by (5.1.12), so is B' (using Corollary 5.1.3 twice). Clearly $B'' \supseteq B$, $B''' = B'$, and replacing B by B'' in (5.1.11), we find that $[B'' : k] = r$, hence $B'' = B$.

Finally, if the centre of B' is E, then $E \supseteq F$ and since $B'' = B$, we also have $F \supseteq E$, hence $E = F$. Thus B, B' are central simple F-algebras, both subalgebras of F', and now (5.1.10) follows from Proposition 5.1.5. ∎

We observe that $B \otimes B'$ is in general distinct from A, for the two sides have centres F and k respectively. Only when $F = k$ do we have $F' = A$ and the above result then reduces to part of Proposition 5.1.5.

Corollary 5.1.11. *Let A be a simple Artinian ring with centre k and let F be a subfield of A such that $F \supseteq k$ and $[F : k] = r$ is finite. Then*

$$A \otimes_k F \cong F' \otimes_k k_r.$$

If moreover, A has finite dimension n over k, then $r^2 | n$ and writing $B = F'$, we have $A \otimes_k B^{\mathrm{o}} \cong F_{n/r}$.

Proof. This is just the case $B = F$ of Theorem 5.1.10; here $[F' : F] = n/r^2$. ∎

Corollary 5.1.12. *Let A be a finite-dimensional central simple k-algebra and F a subfield of A containing k. Then the following are equivalent:*

(a) $F' = F$,
(b) *F is a maximal commutative subring of A,*
(c) $[A : k] = [F : k]^2$,
(d) $[A : k] = [A : F]^2$.

In particular, every maximal commutative subfield F of a central division algebra D satisfies $[D : k] = [F : k]^2 = [D : F]^2$.

Proof. Clearly F is a maximal commutative subring of A iff $F' = F$ and F is a subfield by hypothesis. Now for any subfield $F, F' \supseteq F$ and $[A : k] = [A : F][F : k] = [F : k][F' : k]$, hence $[A : F]^2 \geq [A : k] \geq [F : k]^2$, with equality in either place iff $F' = F$. ■

We remark that a central simple algebra A may have no subfield F satisfying the conditions of Corollary 5.1.12, e.g. $A = k_n$, where k is algebraically closed and $n > 1$. Nevertheless, the dimension of a central simple algebra is always a perfect square, for by Wedderburn's theorem, $A \cong D_n$ where D is a skew field, again with centre k. If F is a maximal subfield of D, then by Corollary 5.1.12 applied to D we find that $[D : k] = [F : k]^2$, hence $[A : k] = n^2[F : k]^2$. In the next section we shall meet another proof of this important fact.

As another application of Theorem 5.1.10 we have Wedderburn's theorem on finite fields. This was proved in BA, Theorem 7.8.6; below is another proof. We shall need the following remark about finite groups.

Lemma 5.1.13. *Let G be a finite group and H a proper subgroup. Then G cannot be written as the union of all the conjugates of H.*

Proof. Let $|H| = h$, $(G : H) = n$, so that $|G| = hn$. If $a_1, \ldots, a_n$ is a right transversal for H in G, then each conjugate of H has the form $a_i^{-1}Ha_i$. There are n such conjugates and each contains h elements, but the unit element is common to all of them, so their union contains at most $(h - 1)n + 1$ elements in all. Since $n > 1$, this is less than $hn = |G|$, so not every element of G is included. ■

Theorem 5.1.14 (Wedderburn's theorem on finite fields). *Any finite skew field is commutative.*

Proof. Suppose that D is a finite skew field; let k denote its centre and let F be a maximal subfield. Then F is a finite field; all maximal subfields of D have the same degree r, say, over k (Corollary 5.1.12) and hence are isomorphic, as minimal splitting fields of $x^{q^r} - x$ where $q = |k|$. By Corollary 5.1.8 they are conjugate to F. Now each element of D lies in some maximal commutative subfield of D, so D is the union of conjugates of F. It follows that the multiplicative group $D^\times$ is a finite group, equal to the union of the conjugates of $F^\times$. But this is impossible by Lemma 5.1.13, hence D must be commutative. ■

Exercises

1. Show that every finite-dimensional central simple algebra over a finite field F has the form $\mathfrak{M}_n(F)$, for some $n \geq 1$.
2. Let A be a finite-dimensional k-algebra. Show that if $A_E \cong E_n$ for some extension field E/k and some $n \geq 1$, then A is central simple over k.
3. Let D be a skew field with centre k and let E be a finite-dimensional subalgebra (necessarily a skew field). Show that if E' is the centralizer of E, then the centralizer of E' is E and $[D : E'] = [E : k]$.
4. Let C be a finite-dimensional k-algebra and A a central simple subalgebra. Show that $[A : k]$ divides $[C : k]$.
5. Show that Corollary 5.1.8 no longer holds for semisimple subalgebras. (Hint. Take appropriate diagonal subalgebras isomorphic to k^2 of k_3.)
6. Let R, S be k-algebras, where k is a field, and let ${}_RU$, V_S be modules as indicated, where R acts densely on U with centralizer k. Show that for $0 \neq u_0 \in U$ the mapping $v \mapsto u_0 \otimes v$ embeds V in $U \otimes_k V$. Construct a lattice-isomorphism between S-submodules of V and (R, S)-subbimodules of $U \otimes V$.
7. Let D be a skew field with centre k and let F be a maximal subfield of D. Show that D_F is a dense ring of linear transformations on D as F-space. Show that if either $[D : F]$ or $[F : k]$ is finite, then so is the other and D_F is then a full matrix ring over F. (Hint. Use the regular representation of D to get a homomorphism $D \otimes_k F \to F_n$, where n is the dimension of D as right F-space.)
8. Let δ be a derivation of a central simple algebra A. By representing δ as an isomorphism of (triangular) subalgebras of $\mathfrak{M}_2(A)$ show that δ is an inner derivation.
9. (Wedderburn, 1921) Let D be a skew field with centre k. Show that any two elements of D with the same minimal equation over k are conjugate.
10. (A. Kupferoth) Let D be a skew field with centre C and K a subfield with centre F. Show that $[K : F] \leq [D : C]$ and when both are finite and equal then $F \subseteq C$ and $D \cong K \otimes_F C$.
11. Show that if A is a finite-dimensional central simple k-algebra and B is any k-algebra, then $A \otimes B$ is semisimple iff B is.
12. Let D be a skew field with centre k and E a skew subfield with centre C. Show that E and the subfield generated by C and k are linearly disjoint over C.

5.2 The Brauer group

In this section all algebras will be finite-dimensional over the ground field k.

Let A be a central simple k-algebra. We know that A is a full matrix ring over a skew field:

$$A \cong D_m \cong D \otimes k_m.$$

Here m is unique and D is a central division algebra over k, unique up to k-isomorphism. We shall call D the *skew field component* of A. Two central simple k-algebras A, B are said to be *similar*, $A \sim B$, if their skew field components are isomorphic. Clearly this is an equivalence relation; we shall denote the class of A by (A). We now show that tensor multiplication induces a multiplication of these equivalence classes.

If A, B are central simple k-algebras, then so is $A \otimes B$, by Corollary 5.1.3. Moreover, if $A \sim A'$, $B \sim B'$, say $A \cong C \otimes k_m$, $B \cong D \otimes k_n$, where C, D are skew fields, then $A \otimes B \cong (C \otimes k_m) \otimes (D \otimes k_n) \cong C \otimes D \otimes k_{mn}$, hence $A \otimes B \sim C \otimes D$, and similarly $A' \otimes B' \sim C \otimes D$, whence $A \otimes B \sim A' \otimes B'$. The multiplication of similarity classes is associative and commutative, by the corresponding laws for tensor products. Moreover, for any A, we have $A \otimes k \cong A$ and $A \otimes A^{\circ} \cong k_n$ for some $n \geq 1$, hence

$$(A)(k) = (A), \quad (A)(A^{\circ}) = (k).$$

This shows that the class (k) is the neutral element for multiplication and (A°) is the inverse of (A), and it proves

Theorem 5.2.1. *For any field k, the similarity classes of finite-dimensional central simple k-algebras form an abelian group with respect to the multiplication induced by the tensor product.*

We still need to check that the collection of all classes is actually a set; this is easily seen if we observe that the central simple algebras are finite-dimensional over k. ■

The group so obtained is called the *Brauer group* of k and is written $\mathbf{B}_k$; its elements are the Brauer classes of central simple k-algebras. The Brauer group is an invariant of the field k which provides information about the central division algebras over k. Later, in Section 5.5, we shall meet another description of $\mathbf{B}_k$, as a cohomology group.

As an example take an algebraically closed field F. If D is a division algebra over F, then for any $a \in D$, $F(a)$ is a finite extension field, hence $F(a) = F$ because F is algebraically closed, and so $a \in F$, i.e. $D = F$. Thus there are no division algebras over F apart from F itself, and we conclude that $\mathbf{B}_F = 1$, i.e. the Brauer group of an algebraically closed field is trivial. Of course once we drop the restriction on the dimension, we can find skew fields with centre F, see Section 7.3 and Section 9.1.

For a closer study of $\mathbf{B}_k$ we need to examine the behaviour of algebras under ground field extension. Let A be a (finite-dimensional) k-algebra and F an extension field of k (not necessarily finite-dimensional over k). Then the F-algebra defined by

$$A_F = A \otimes_k F$$

is again finite-dimensional; in fact any k-basis of A will be an F-basis of A_F, so that we have

$$[A_F : F] = [A : k]. \tag{5.2.1}$$

We note that $x \longmapsto x \otimes 1$ defines an embedding of A in A_F. If A has centre k, then A_F has centre F, by Corollary 5.1.3. From this fact we can deduce

Proposition 5.2.2. *If A is a finite-dimensional central simple k-algebra, then $[A:k]=r^2$ for some integer r, and there is a finite extension F of k such that $A_F \cong F_r$; hence A can be embedded in F_r.*

Proof. If E is an algebraic closure of k, then A_E is a full matrix algebra over E, say $A_E \cong E_r$. Comparing dimensions and remembering (5.2.1), we find that $[A:k] = [A_E:E] = [E_r:E] = r^2$. Now there is an embedding

$$A \to A_E = E_r, \tag{5.2.2}$$

by the above remark. It remains to show that we can replace E by a finite extension. Let $u_1, \dots, u_n$ be a basis of A and $U_1, \dots, U_n$ the matrices over E which correspond to the u's under the mapping (5.2.2). Then we can express the matrix units in E_r as $e_{ij} = \sum \alpha_{ij\nu} U_\nu$ for some $\alpha_{ij\nu} \in E$. Denote the finite set of entries of the U_ν and the $\alpha_{ij\nu}$ by X. Since E is algebraic over k, X generates a finite extension F of k and it is clear that $A_F \cong F_r$. Since A is embedded in A_F, it can also be embedded in F_r. ∎

We note that this result could not be deduced from Proposition 5.1.4, because here F/k is never regular. Proposition 5.2.2 provides no explicit bound on $[F:k]$, but we shall soon meet such a bound, in Corollary 5.2.7.

For any central simple k-algebra A the integer $\sqrt{[A:k]}$ is called the *degree* of A, written Deg A. If $A \cong D \otimes k_m$, then $\text{Deg}\, A = m(\text{Deg}\, D)$; clearly the degree of D is also an invariant of A, called the *Schur index*, or simply the *index* of A. It is a measure of how far A deviates from being a full matrix algebra over k. We note that the index is an invariant of the Brauer class of A, while the degree determines A up to isomorphism within its Brauer class.

If F is an extension field of k, then the mapping $A \mapsto A_F$ induces a mapping of Brauer classes, for if $A \cong D \otimes k_m$, then $A_F \cong D \otimes k_m \otimes F \cong D \otimes F_m$, so that $(A_F) = (D_F)$. The mapping is a homomorphism, for $(A \otimes_k B)_F \cong A \otimes_k B \otimes_k F \cong A_F \otimes_F B_F$, so we have a group homomorphism

$$\mathbf{B}_k \to \mathbf{B}_F.$$

Its kernel $\mathbf{B}(F/k)$, the *relative Brauer group*, consists of those classes (A) over k for which $A_F \cong F_m$ for some m. Such a class is said to be *split* by F and F is called a *splitting field* for this class, or for the algebra A. If $A_F = A \otimes F \cong F_m$, then on taking anti-isomorphisms, we find that $A^\circ \otimes F \cong F_m$, hence F splits A iff it splits A°. Any central simple k-algebra has a splitting field, which may be taken of finite degree over k, by Proposition 5.2.2.

Next we examine more closely which fields split a given Brauer class, but first we establish a relation between the indices of the extensions.

Proposition 5.2.3 (Index reduction lemma, A. A. Albert). *Let F/k be a finite field extension, say $[F:k]=r$, and A a central simple k-algebra. If A, A_F have indices m, μ, then $\mu|m$ and $m|\mu r$.*

Proof. We may take $A = D$ to be a division algebra, without loss of generality. If the skew field component of D_F is denoted by C, then

$$D \otimes F = D_F \cong C \otimes_F F_q; \tag{5.2.3}$$

comparing dimensions over k, we find $m^2 r = \mu^2 q^2 r$, hence $m = \mu q$ and so $\mu | m$.

Secondly the regular representation of F (by right multiplication) defines an embedding of F in k_r, because $F \cong k^r$ as k-space. Thus F is embedded in k_r and hence $D \otimes F$ is embedded in $D \otimes k_r$. By (5.2.3) we obtain an embedding of F_q in $D \otimes k_r$, but $F \supseteq k$, so $D \otimes k_r$ contains k_q as central simple subalgebra. If the centralizer of k_q in $D \otimes k_r$ is denoted by B, then by Proposition 5.1.5,

$$D \otimes k_r \cong k_q \otimes B \cong k_q \otimes G \otimes k_s,$$

where G is the skew field component of B (in fact $G \cong D$ by uniqueness). Comparing dimensions, we find that $r = qs$, thus $m/\mu = q | r$, i.e. $m | \mu r$. ■

The factor q in $m = \mu q$ is called the *index reduction factor*; we note that $q = r$ whenever F is isomorphic to a subfield of D.

The next corollaries are immediate consequences of Proposition 5.2.3.

Corollary 5.2.4. *Let A be a central simple k-algebra and F/k a field extension. If $[F : k]$ is prime to the index of A, then A and A_F have the same index. In particular, if A is a division algebra of degree prime to $[F : k]$, then A_F is again a division algebra.* ■

Corollary 5.2.5. *The degree of a splitting field (over k) of a central simple k-algebra A is divisible by the index of A.* ■

We now obtain a criterion for a given finite extension field of k to split a given Brauer class over k.

Theorem 5.2.6. *Let $w \in \mathbf{B}_k$ and let F be an extension field of k, of degree r. Then F splits w if and only if some algebra in w contains F as a maximal subfield; this algebra necessarily has degree r.*

Proof. Let D be the division algebra in the class w and let n be the least integer such that F can be embedded in D_n. Then the centralizer F' of F in D_n is a skew field, by the minimality of n, and F is the centre of F'. By Corollary 5.1.11,

$$D \otimes_k F \sim D_n \otimes F \cong F' \otimes_k k_r \sim F'. \tag{5.2.4}$$

Since F' has centre F, this shows that F splits D iff $F' = F$, i.e. iff F is a maximal subfield of D_n. ■

Corollary 5.2.7. *Let D be a central division algebra over k. Then any maximal subfield of D is a splitting field for D.*

Proof. By Corollary 5.1.12 any maximal subfield F satisfies $[D : k] = [F : k]^2$, so the theorem may be applied. ■

The next step is to show that the splitting field of a central simple algebra can always be taken to be a separable extension of the ground field. We prove more than this, namely that we can actually find a separable splitting field in the skew field component. More precisely, the proof below shows that every algebraic skew field extension contains a separable element.

Theorem 5.2.8 (Köthe, 1932). *Every central division algebra D over k contains a maximal commutative subfield (hence splitting D) which is separable over k.*

Proof. (Herstein) Clearly we may assume that char $k = p \neq 0$. Our first task will be to find a separable extension F of k in D. Since $[D : k]$ is finite, each element of D is algebraic over k; if some $a \notin k$ is separable over k, then $k(a)$ is the required extension. Otherwise there are no separable extensions of k, so each element of D is p-radical over k, say $x^{p^r} \in k$ for some $r = r(x)$. Hence we can find $a \notin k$ such that $a^p \in k$. Denote by δ the inner derivation induced by $a, \delta : x \mapsto xa - ax$. Then $x\delta^p = xa^p - a^p x = 0$, but of course $\delta \neq 0$, because $a \notin k$. Choose $b \in D$ such that $c = b\delta \neq 0$, $b\delta^2 = 0$. Then $c\delta = 0 = a\delta$, so if $u = bc^{-1}a$, then $u\delta = cc^{-1}a = a$. Writing this out, we have $ua - au = a$, i.e. $u = 1 + aua^{-1}$, but $u^q \in k$ for some $q = p^s$, hence $u^q = 1 + (aua^{-1})^q = 1 + u^q$ (because $u^q \in k$). We obtain $1 = 0$, a contradiction. It follows that D contains a proper separable extension.

Taking a separable extension of maximal degree in D, we obtain a maximal separable extension F. By Theorem 5.1.10, its centralizer F' is simple with centre F, but as centralizer in a division algebra F' is itself a division algebra, so if $F' \neq F$, then F has a proper separable extension E, by the first part. But then E is separable over k, which contradicts the maximality of F. Hence $F' = F$, i.e. F is a maximal subfield and by Corollary 5.2.7, a splitting field of D, separable by construction. ■

We remark that the result holds for any skew fields that are algebraic over k but not necessarily finite-dimensional. For we can use Zorn's lemma instead of a dimension argument to obtain a maximal separable extension.

Corollary 5.2.9. *Every Brauer class of k has a splitting field which is a finite Galois extension of k.*

Proof. Take a division algebra D in $w \in \mathbf{B}_k$ and let F be a maximal separable extension in D. Then F is contained in a finite Galois extension of k, and this will also split D. ■

Of course the Galois splitting field need not be contained as a subfield in D. Later, in Section 5.5, we shall find that splitting subfields that are Galois lead to the crossed product construction.

A basic question in the theory of central simple algebras is this: when is the tensor product of two division algebras again a division algebra? In essence this is a question about the index of a tensor product, and little explicit information is available. We first take a simple case, where there is a complete answer.

Proposition 5.2.10. *If C, D are two central division k-algebras of coprime degrees, then $C \otimes D$ is again a division algebra.*

Proof. By Corollary 5.1.3, $A = C \otimes D$ is central simple, say $A \cong K_n$ for a skew field K. If the simple A-module is denoted by V, then $A \cong V^n$; here V is isomorphic to a minimal right ideal, hence a C-space, and so $n|[A : C] = [D : k]$; similarly, $n|[C : k]$, so $n = 1$. ■

Our next result provides a description of the index of a tensor product, even when only one of the factors has finite degree. We remark that if R is a simple Artinian ring and $R_r \cong D_n$, where D is a skew field, then $r|n$, say $n = rs$ and $R \cong D_s$. For by Wedderburn's theorem, R has the form $R \cong K_s$ where K is a skew field. It follows that $K_{rs} \cong D_n$ and by uniqueness, $n = rs$ and $K \cong D$, and so $R \cong D_s$.

Theorem 5.2.11. *Let F/k be a field extension of degree d, and let C, D be skew fields with centres k, F respectively. If either C or D is of finite degree, then*

$$C \otimes_k D \cong G_m, \tag{5.2.5}$$

where G is a skew field with centre F. Moreover,

(i) *if* Deg $C = r$, *then*

$$D_q \cong C^{\circ} \otimes_k G, \quad \text{where } mq = r^2, \tag{5.2.6}$$

and q is the least integer such that C° can be embedded in D_q;

(ii) *if* Deg $D = s$, *then D° can be embedded in C_n for $n = s^2d/m$ but no smaller n, the centralizer of D° in C_n is isomorphic to G and if F′ denotes the centralizer of F in C_n, then*

$$F' \cong D^{\circ} \otimes_F G, \quad \text{and } mn = s^2 d. \tag{5.2.7}$$

In particular, in case (i) *$q|r^2$ and $C \otimes D$ is a skew field if and only if $q = r^2$, while in case* (ii) *$n|s^2d$ and $C \otimes D$ is a skew field if and only if $n = s^2d$.*

(iii) *When both C, D have finite degrees r, s respectively, and n, q are as in* (i), (ii), *then n, q are related by the equation*

$$nr^2 = qs^2 d, \tag{5.2.8}$$

and Deg $G = t$, *where*

$$t = sq/r = rn/sd, \quad m = r^2/q = s^2 d/n. \tag{5.2.9}$$

From (i), (ii) it is clear that r, s, d are independent, while n, q are related as in (5.2.8), and m, t are determined by (5.2.9) in terms of r, s, d, n, q.

Proof. The algebra $C \otimes D$ is simple with centre F, by Theorem 5.1.2. If C has finite degree, then $C \otimes D$ is finite-dimensional over D; if D has finite degree, $C \otimes D$ is finite-dimensional over C. In either case it is Artinian and by Wedderburn's theorem it has the form G_m for some m, where G is a skew field with centre F.

(i) Suppose now that Deg $C = r$; *then* C and hence C° can be embedded in k_{r^2} and hence in D_{r^2}. Let q be the least integer such that C can be embedded in D_q as k-algebra. Then by Proposition 5.1.5, $D_q \cong C^{\circ} \otimes_k E$, where E is a simple algebra with centre F, by Theorem 5.1.2. Moreover, E is Artinian, for if $\mathfrak{a}$ is a left ideal, then $C^{\circ} \otimes \mathfrak{a}$ is a left D-space, hence the length of chains of left ideals is bounded by q. Thus E is a matrix ring over a skew field. Taking Brauer classes we have $(E) = (C)(D) = (G)$, therefore $E \cong G_h$ for some h, but if $h > 1$, we can replace q by q/h, which contradicts the minimality of q. Hence $h = 1$ and $D_q \cong C^{\circ} \otimes G$. It follows that $D_{qm} \cong C^{\circ} \otimes G_m \cong C^{\circ} \otimes C \otimes D \cong k_{r^2} \otimes D \cong D_{r^2}$ and so $qm = r^2$; thus (5.2.6) is established.

(ii) Next assume that Deg $D = s$; then D and with it D° can be embedded in F_{s^2} and hence in k_{s^2d}. Let n be the least integer for which D° can be embedded in C_n and denote by H the centralizer of D° in C_n. Then by Theorem 5.1.10, $F' \cong D^{\circ} \otimes_F H$, where F' is the centralizer of F, and H is simple with centre F, while $C_n \otimes_k D \cong H_{s^2d}$, by (5.1.12). Comparing this relation with (5.2.5), we see that $s^2d = mn$ and $H \cong G$. Finally when both C and D have finite degrees, then by combining (5.2.6) and (5.2.7) we see that $m = r^2/q = s^2d/n$, therefore $nr^2 = qs^2d$, and if Deg $G = t$, then a comparison of degrees in (5.2.5), (5.2.6) yields $t = sq/r = rn/sd$. ∎

Consider the special case when $D = F$. Then $s = 1$ and $mn = d$, so we obtain a result essentially contained in Proposition 5.2.3:

Corollary 5.2.12. *Let C be a skew field with centre k and F a finite extension of k. Then $C_F \cong G_m$ for some skew field G, where m divides $[F : k]$, with equality if and only if F can be embedded in C.* ∎

We remark that here C need not be finite-dimensional over k.

Examples of Brauer groups

1. We have already seen that the Brauer group of an algebraically closed field is trivial, e.g. $\mathbf{B}_{\mathbf{C}} = 0$. More generally, this holds for any field which is separably closed, by Theorem 5.2.8.
2. The Brauer group of any finite field is trivial. For any division algebra over a finite field F is itself a finite skew field, hence commutative by Theorem 5.1.14, so it reduces to F. This case will be generalized in the next section.
3. The Brauer group of the real numbers has order 2. For the only algebraic extensions of $\mathbf{R}$ are $\mathbf{R}$, $\mathbf{C}$, so any division algebra has degree 1 or 2, and as we shall see in Section 5.4, the only algebra of degree 2 is the algebra of quaternions.
4. If F is a complete field for a discrete valuation with finite residue class field (e.g. the p-adic field $\mathbf{Q}_p$), then $\mathbf{B}_F = \mathbf{Q}/\mathbf{Z}$.
5. If F is an algebraic number or function field, then $\mathbf{B}_F$ is a subgroup of the direct sum of the Brauer groups of the corresponding local fields, described in examples

3 and 4 above (see Weil (1967)or Reiner (1976)). More precisely, we have an exact sequence (Hasse reciprocity)

$$0 \to \mathbf{B}_F \to \oplus \mathbf{B}_{F_p} \to \mathbf{Q}/\mathbf{Z} \to 0,$$

where F_p are the completions of F.

Exercises

1. Define Brauer classes for any central simple algebra, not necessarily finite-dimensional, and show that these classes form a monoid whose group of units is $\mathbf{B}_k$.
2. In Proposition 5.2.3 show that q is the degree of the largest subfield common to F and D.
3. Show that any skew field p-radical over its centre is commutative.
4. Prove Theorem 5.2.8 in detail for skew fields algebraic over k.
5. Let D be a central division k-algebra. For any automorphism α of D as a skew field define the *inner order* as the least r such that α^r is inner. Show that the inner order of any such α divides the order of the restriction $\alpha|k$.
6. Show that a central simple algebra of degree n is split iff it has a left ideal of dimension n.

5.3 The reduced norm and trace

In BA, Section 5.5, we met the notions of norm and trace of an element in a finite-dimensional k-algebra. We recall that for an m-dimensional algebra A with basis $u_1, \ldots, u_m$ the right multiplication is represented by a matrix $\rho(a) = (\rho_{ij}(a))$, where

$$u_i a = \sum \rho_{ij}(a) u_j \quad \text{for all } a \in A. \tag{5.3.1}$$

Here $\rho : A \to \mathfrak{M}_m(k)$ is the regular representation and the norm and trace are defined in terms of it by

$$\mathrm{Nm}(a) = \det(\rho(a)), \quad \mathrm{Tr}(a) = \sum \rho_{ii}(a). \tag{5.3.2}$$

When A is central over k, we have $m = n^2$ and for a splitting field F of A we have

$$A_F \cong F_n, \quad \text{where } n^2 = [A : k]. \tag{5.3.3}$$

Let e_{ij} be the standard basis of matrix units for F_n and write $a \in A$ as $a = \sum a_{ij} e_{ij}$. Then the equation (5.3.1) takes the form

$$e_{rs} a = \sum_j a_{sj} e_{rj};$$

hence the matrix $\rho(a)$ has as (rs, uv)-entry $(a_{sv}\delta_{ur})$ and the equations (5.3.2) become in this case

$$\mathrm{Nm}(a) = \det(a_{sv}\delta_{ur}) = (\det(a_{rs}))^n, \quad \mathrm{Tr}(a) = \sum_{rs} a_{ss} = n\sum a_{ss}. \tag{5.3.4}$$

This is most easily seen by writing A as V^n, where V is a minimal right ideal, corresponding to a single row of F_n. The right action is right multiplication by a, so that the matrix $\rho(a)$ is the diagonal sum of n terms a, and we obtain (5.3.4). In particular this shows that $\mathrm{Tr} = 0$ whenever n is a multiple of char k, and it suggests that we can get more information by taking the determinant and trace of a itself as invariants.

This is accomplished as follows, for any central simple k-algebra. Let F/k be a Galois extension which splits A:

$$A \xrightarrow{\lambda} A \otimes_k F \xrightarrow{\mu} F_n,$$

where λ is the natural embedding and μ is an isomorphism. Then $\lambda\mu$ embeds A in F_n and there we have the usual norm and trace. We define the *reduced norm* $N(a)$ and the *reduced trace* $T(a)$ of any $a \in A$ as

$$N_{A/k}(a) = N(a) = \det(a\lambda\mu), \quad T_{A/k}(a) = T(a) = \mathrm{tr}(a\lambda\mu) \quad \text{for } a \in A. \tag{5.3.5}$$

We note that whereas λ is the canonical mapping $a \mapsto a \otimes 1$, μ is not uniquely determined, but the definition (5.3.5) is independent of the choice of μ, for two isomorphisms of A_F with F_n differ by an automorphism of F_n which must be inner, by the Skolem–Noether theorem, and so leave N, T unaffected. From the definition $N(a)$, $T(a)$ lie in F, but if σ is a k-automorphism of F, then σ induces a k-automorphism of F_n which gives another representation $a \mapsto (a\lambda\mu)$. Since A is a k-algebra, σ leaves $a \in A$ fixed and so $N(a)^\sigma = N(a)$, $T(a)^\sigma = T(a)$. This holds for all $\sigma \in \mathrm{Gal}(F/k)$, hence $N(a)$, $T(a) \in k$. Further, if F' is another separable splitting field of A, we can find a Galois extension E to contain both F and F', and it follows that F and F' give rise to the same N and T. The following familiar properties of norm and trace are easily verified; here $[A : k] = n^2$.

R.1 $N(ab) = N(a)N(b)$, $N(\alpha a) = \alpha N(a)$, $N(1) = 1$, where $\alpha \in k$,
R.2 $T(a + b) = T(a) + T(b)$, $T(\alpha a) = \alpha T(a)$, $T(1) = n$,
R.3 $T(ab) = T(ba)$,
R.4 $\mathrm{Nm}(a) = N(a)^n$, $\mathrm{Tr}(a) = n.T(a)$.

We also have a product formula for the reduced norm and trace. For a field extension F/k we shall write $N_{F/k}$ and $T_{F/k}$ for the usual norm and trace.

Proposition 5.3.1. *Let A be a central simple K-algebra and B a simple subalgebra of A, with centre F. Suppose that* $\mathrm{Deg}\, A = n$, $\mathrm{Deg}\, B = r$, $[F : k] = t$; *then $rt|n$, say $n = rst$, and for any $b \in B$,*

$$N_{A/k}(b) = N_{F/k}(N_{B/F}(b))^s, \quad T_{A/k}(b) = s.T_{F/k}(T_{B/F}(b)). \tag{5.3.6}$$

Proof. Let E be a Galois splitting field of B which also splits A. Then $A_E \cong E_n$ and under the mapping $A \to A_E$, B becomes

$$B \otimes_k E = (B \otimes_k F) \otimes_F E \cong (B \otimes_F E) \otimes_k F \cong (E \otimes_k F)_r. \tag{5.3.7}$$

We thus have an embedding of $(E \otimes_k F)_r$ in E_n, so E_n is an $r \times r$ matrix ring, $E_n \cong C_r$, where C is simple Artinian, hence $C \cong E_{n/r}$ by uniqueness.

Since $A_E \cong E_n$, there is a unique simple right A_E-module $V \cong E^n$. By (5.3.7), $B \otimes E$ is faithfully represented by endomorphisms of $U = (E \otimes F)^r$. Now $[U : E] = r.[E \otimes F : E] = rt$ and $B \otimes E$ acts on V, hence $V \cong U^s$ for some s, and a comparison of dimensions shows that $n = rst$. For any $b \in B$ we have $b \otimes 1 \in B \otimes_F E \cong E_r$, so $b \otimes 1$ is represented by an $r \times r$ matrix and $N_{B/F}(b) = \det(b \otimes 1)$, $T_{B/F}(b) = \mathrm{Tr}(b \otimes 1)$. If we now tensor with F and consider $b \otimes 1 \otimes 1$ in $(B \otimes_F E) \otimes_k F = (E \otimes_k F)_r$, we have

$$\det(b \otimes 1 \otimes 1) = N_{F/k}(N_{B/F}(b)), \quad \mathrm{Tr}(b \otimes 1 \otimes 1) = T_{F/k}(T_{B/F}(b)),$$

and since $V \cong U^s$, we obtain (5.3.6). ■

We note the special case $B = F$:

Corollary 5.3.2. *Let A be a central simple k-algebra of degree n and F a subfield of A, of degree t over k. Then $t|n$, say $n = st$, and for any $a \in F$,*

$$N_{A/k}(a) = N_{F/k}(a)^s, \quad T_{A/k}(a) = s.T_{F/k}(a). \quad ■$$

In particular, when A is a division algebra and F a maximal subfield, then $s = 1$ and the norm and trace in F coincide with the reduced norm and trace in A. Further, if in Corollary 5.3.2 we take $F = k(a)$, then $N_{F/k}(a)$ is (up to sign) the constant term of the minimal polynomial of a over k, and a is invertible precisely when the constant term is non-zero. We deduce

Proposition 5.3.3. *Let A be a central simple algebra. Then an element a of A is invertible if and only if $N(a) \neq 0$.* ■

Let us denote the group of units of A by $\mathbf{U}(A)$. The reduced norm defines a mapping $\mathbf{U}(A) \to k^\times$ which by **R**.1 above is a homomorphism. Since k is commutative, the commutator subgroup $\mathbf{U}(A)'$ is mapped to 1 in this homomorphism. Let us define the *Whitehead group* of A as

$$\mathbf{K}_1(A) = \mathbf{U}(A)^{ab} = \mathbf{U}(A)/\mathbf{U}(A)'.$$

By what has been said, the reduced norm induces a homomorphism $\nu : \mathbf{K}_1(A) \to k^\times$. The kernel of this homomorphism is called the *reduced Whitehead group* and is denoted by $\mathbf{SK}_1(A)$. We thus have the exact sequence

$$1 \to \mathbf{SK}_1(A) \to \mathbf{K}_1(A) \xrightarrow{\nu} k^\times \to \operatorname{coker} \nu \to 1. \tag{5.3.8}$$

For a matrix ring over a field, $\mathbf{K}_1$ is easily determined:

Lemma 5.3.4. *For any field k and any $n \geq 2$, except $n = 2$ and $k = \mathbf{F}_2, \mathbf{F}_3$,*

$$\mathbf{SK}_1(k_n) = 1 \quad \text{and} \quad \mathbf{K}_1(k_n) \cong k^\times. \tag{5.3.9}$$

Proof. We have a surjective homomorphism $\mathbf{GL}_n(k) \to k^\times$, given by the determinant, with kernel $\mathbf{SL}_n(k)$; hence $\mathbf{SK}_1(k_n) \cong \mathbf{SL}_n(k)/\mathbf{GL}_n(k)'$. Now by Proposition 3.5.2 we have $\mathbf{SL}_n(k) = \mathbf{GL}_n(k)'$ except when $n = 2$ and $k = \mathbf{F}_2, \mathbf{F}_3$, so (5.3.9) follows with these exceptions. ■

The exceptions in Lemma 5.3.4 are treated in Exercises 1 and 2. We next describe the diagram resulting from an algebra homomorphism.

Theorem 5.3.5. *Let F/k be a field extension and let A, B be simple algebras with centres k, F respectively. If there is a k-algebra homomorphism $\theta : A \to B$, then* Deg $B = d.$Deg A *for some $d \geq 1$, and there are homomorphisms such that the diagram*

$$\begin{array}{ccccccccc} 1 \to & \mathbf{SK}_1(A) & \to & \mathbf{K}_1(A) & \to & k^\times & \to & \text{coker}\ \nu_{A/k} & \to 1 \\ & \downarrow & & \downarrow{\scriptstyle \mathbf{K}_1(\theta)} & & \downarrow & & \downarrow & \\ 1 \to & \mathbf{SK}_1(B) & \to & \mathbf{K}_1(B) & \to & F^\times & \to & \text{coker}\ \nu_{B/F} & \to 1 \end{array} \tag{5.3.10}$$

commutes, where the mapping $k^\times \to F^\times$ is the power mapping $x \mapsto x^d$.

Proof. It is clear that θ maps $\mathbf{U}(A)$ to $\mathbf{U}(B)$ and so induces a homomorphism $\mathbf{K}_1(\theta) : \mathbf{K}_1(A) \to \mathbf{K}_1(B)$. The F-subalgebra of B generated by $A\theta$ is a homomorphic image of A_F and hence, by the simplicity of the latter, isomorphic to A_F. If $a \in A$, then $N_{A/k}(a) = N_{A_F/F}(a) = N_{F(a)/F}(a)^r$, where $r.[F(a) : F] = \text{Deg}\, A_F$, and $N_{B/F}(a\theta) = N_{F(a)/F}(a)^s$, where $s.[F(a) : F] = \text{Deg}\, B$. Now $\text{Deg}\, A_F = \text{Deg}\, A$ is a divisor of Deg B, because A_F is embedded in B. Thus $\text{Deg}\, B = d.\text{Deg}\, A$ and so $s.[F(a) : F] = d.\text{Deg}\, A = dr.[F(a) : F]$; therefore $s = rd$ and $N_{B/F}(a) = N_{A/k}(a)^d$. This shows that the central square in (5.3.1) commutes and this determines the outer vertical arrows. ■

We note the special case when A is any central simple k-algebra and $B = A_F$. Then we obtain an exact commutative diagram of the form (5.3.10) with $B = A_F$, where the mapping $k^\times \to F^\times$ is the inclusion mapping (because now $d = 1$). In particular, taking F to be a splitting field of A, we have an isomorphism $\mathbf{K}_1(A_F) \cong F^\times$, by Lemma 5.3.4 (with the exceptions listed), hence $\mathbf{SK}_1(A_F)$ and coker $\nu_{A_F/F}$ are then trivial. We remark that any $a \in K$ satisfies $N(a) = a^n$; it follows that coker ν is a group of exponent dividing n, the index of A. It can be shown that $\mathbf{SK}_1(A)$ has finite exponent dividing $\prod p_i^{\alpha_i - 1}$ where $\prod p_i^{\alpha_i}$ is the index of A (see Draxl (1983)). In fact for many ground fields, e.g. all algebraic number fields, it can be shown that $\mathbf{SK}_1(A) = 1$ and it was an open problem for many years whether (apart from the trivial exceptions of Exercises 1 and 2) algebras with non-trivial reduced Whitehead group exist (Tannaka–Artin problem). In 1975 Vladimir Platonov gave examples of algebras with non-trivial reduced Whitehead group; we shall meet some simple examples due to Peter Draxl later, in Section 7.3.

The reduced norm can be used to show that $\mathbf{B}_k$ is trivial for certain fields k. In any central division algebra A of degree r we have $N(x) \neq 0$ for $x \neq 0$, and taking a basis $u_1, \ldots, u_n$ $(n = r^2)$ of A, we can write the general element of A as $x = \sum \xi_i u_i$. Now $N(x)$ become a *form*, i.e. a homogeneous polynomial of degree r in the r^2 variables ξ_i. A field k is said to be *quasi-algebraically closed* or a *C_1-field* if every form of degree d in $n > d$ variables has a non-trivial zero. With this definition we have

Theorem 5.3.6. *Every C_1-field has a trivial Brauer group.*

Proof. Let k be a C_1-field; we have to show that there are no central division algebras other than k. Let D be a central division algebra of degree r over k. The reduced norm $N(x)$ is a form of degree r in r^2 variables, and $N(x) = 0$ has no non-trivial solutions, hence $r^2 \leq r$, so $r = 1$ and $D = k$, as claimed. ∎

An obvious example of C_1-fields are the algebraically closed fields; we shall soon meet other examples. For the moment we note a reduction that is sometimes useful:

Proposition 5.3.7. *Any finite extension of a C_1-field is a C_1-field.*

Proof. Let F/k be an extension of degree r and take a basis $v_1, \ldots, v_r$ of F over k. If $f(x_1, \ldots, x_n)$ is a form of degree $d < n$ with coefficients in F, let us write $x_\lambda = \sum \xi_{\lambda i} v_i$ and consider

$$g(\xi_{11}, \ldots, \xi_{nr}) = N_{F/k}\Big(f\Big(\sum \xi_{1i} v_i, \ldots, \sum \xi_{ni} v_i\Big)\Big).$$

We claim that g is homogeneous of degree dr in the ξ's:

$$\begin{aligned} g(\lambda\xi_{11}, \ldots, \lambda\xi_{nr}) &= N_{F/k}\Big(f\Big(\sum \lambda\xi_{1i} v_i, \ldots, \sum \lambda\xi_{ni} v_i\Big)\Big) \\ &= N_{F/k}\Big(\lambda^d f\Big(\sum \xi_{1i} v_i, \ldots, \sum \xi_{ni} v_i\Big)\Big) = \lambda^{dr} N_{F/k}(f). \end{aligned}$$

Clearly g has coefficients in k, and it is of degree dr in the nr variables $\xi_{\lambda i}$. Since $dr < nr$ and k is C_1, it follows that $g(\xi') = 0$ for some $\xi'_{\lambda i} \in k$, not all 0. This gives $x'_\lambda = \sum \xi'_{\lambda i} v_i \in F$ not all 0, such that $N_{F/k}(f(x')) = 0$, hence $f(x') = 0$, so f has a non-trivial zero in F, as claimed. ∎

Let us show that every finite field is C_1. By Proposition 5.3.7 we can limit ourselves to $\mathbf{F}_p$, but that is no easier. We shall need a formula for power sums in $\mathbf{F}_q$:

Lemma 5.3.8. *Let $k = \mathbf{F}_q$ be the field of $q = p^r$ elements. Then*

$$S_m = \sum_{x \in k} x^m = \begin{cases} -1 & \text{if } q - 1 | m, \\ 0 & \text{otherwise.} \end{cases}$$

Proof. We have $x^{q-1} = 1$ for $x \neq 0$ and $x^{q-1} = 0$ for $x = 0$, hence $S_{q-1} = \sum x^{q-1} = q - 1 = -1$; similarly, $S_{d(q-1)} = -1$. When m is not divisible by $q - 1$, then $a^m \neq 1$ for some $a \in k^\times$, hence $S_m = \sum x^m = \sum (ax)^m = a^m S_m$, so $(1 - a^m)S_m = 0$ and since $a^m \neq 1$, we conclude that $S^m = 0$. ■

With the help of this formula we can show that the number of points on the hypersurface over $\mathbf{F}_q$ defined by a polynomial in more variables than its degree is divisible by p:

Theorem 5.3.9 (Chevalley–Warning, 1934). *Let f be a polynomial in n variables over $k = \mathbf{F}_q$, where $q = p^r$. Write $V(f)$ for the set of zeros of f. If $n > d$, where d is the degree of f, then*

$$|V(f)| \equiv 0 \pmod{p}. \tag{5.3.11}$$

In particular, if f has zero constant term, then it has a non-trivial zero.

Proof. For each $x \in k^n$ we have

$$1 - f(x)^{q-1} \equiv \begin{cases} 1 & \text{if } x \in V(f), \\ 0 & \text{if } x \notin V(f). \end{cases}$$

Thus $1 - f(x)^{q-1}$ is the characteristic function of $V(f)$, and summing over all points x of $V(f)$, we find

$$|V(f)| \equiv \sum (1 - f(x)^{q-1}) = -\sum f(x)^{q-1},$$

because the total number of points in k^n is $q^n \equiv 0$. Now $f(x)$ is a linear combination of terms $x_1^{\nu_1} \dots x_n^{\nu_n}$. We have

$$\sum_x x_1^{\nu_1} \dots x_n^{\nu_n} = \prod_{i=1}^{n} \sum_{x_i \in k} x_i^{\nu_i} = \prod_{i=1}^{n} S_{\nu_i}. \tag{5.3.12}$$

If $\nu_i = 0$ for some i, then $S_{\nu_i} \equiv 0 \pmod{q}$ and we get zero, so we may assume that $\nu_i > 0$ for $i = 1, \dots, n$. But by Lemma 5.3.8, $S_m = 0$ unless $q - 1|m$, and since $\sum \nu_i \leq d(q-1) < n(q-1)$, it follows that some ν_i is not divisible by $q - 1$. So in any case the sum in (5.3.12) is zero and (5.3.11) follows.

Moreover, if $f(0) = 0$, then the number of non-zero roots of $f = 0$ is $\equiv -1 \pmod{p}$, hence it is non-zero. ■

If f is homogeneous of positive degree, its constant term is 0 and we obtain

Corollary 5.3.10. *Every finite field is C_1.* ■

This then shows that $\mathbf{B}_{\mathbf{F}_q} = 0$ in Exercise 4 we shall meet another proof of this fact. As a third example of C_1-fields we consider function fields of degree 1:

Theorem 5.3.11 (Tsen's theorem). *Let k be an algebraically closed field and F a field of functions in one variable over k. Then $\mathbf{B}_F = 0$.*

Proof. F is a finite algebraic extension of the rational function field $k(t)$. By Proposition 5.3.7 it will be enough to show that $k(t)$ is a C_1-field. Let $f(x_1, \ldots, x_n)$ be a polynomial over $k(t)$, homogeneous of degree $d < n$. We shall show that $f(x) = 0$ has a solution when the x are polynomials in t. Write

$$x_i = \xi_{i0} + \xi_{i1}t + \ldots + \xi_{ir}t^r.$$

The coefficients of f are rational functions of t and on multiplying f by an element of $k(t)$ we may take them to be polynomials in t, of degree $\leq k$, say. Then

$$f(x_1, \ldots, x_n) = p_0 + p_1 t + \ldots + p_{rd+k}t^{rd+k},$$

where p is a form, i.e. a homogeneous polynomial in the ξ's with coefficients in k. We thus have $rd + k + 1$ forms p_λ in the $(r+1)n$ variables ξ_{ij}. We eliminate ξ_{11} by taking a form in which it occurs and forming resultants with the remaining p_λ; this diminishes the forms and variables by one. By continuing this process we eventually obtain a form in at least two variables, provided that the number of variables is greater than the number of forms, i.e. $(r+1)n > rd + k + 1$, and this holds for suitable r because $d < n$. This means that all the p have a common zero, as we had to show. ■

Exercises

1. Show that for $A = \mathfrak{M}_2(\mathbf{F}_2)$, $\mathbf{SK}_1(A) = \mathbf{K}_1(A) = \mathbf{C}_2$, the cyclic group of order 2.
2. Show that for $A = \mathfrak{M}_2(\mathbf{F}_3)$, $\mathbf{SK}_1(A) = \mathbf{C}_3$, $\mathbf{K}_1(A) = \mathbf{C}_6$. (Hint. Verify that the mapping $\begin{pmatrix} x & 1 \\ -1 & 0 \end{pmatrix} \mapsto x$ extends to a homomorphism $\mathbf{SL}_2(\mathbf{F}_3) \to \mathbf{F}_3^+$ (Cohn [1966].)
3. Let A be a central simple k-algebra and let p be a prime dividing the index of A. Show that there is a finite extension F of k such that A_F has index p.
4. Use Theorem 5.1.14 to show that every central simple algebra over a finite field F splits and deduce that F has trivial Brauer group.
5. Show that a field is quasi-algebraically closed iff, for all $n > 1$, every form of degree $n - 1$ in n variables has a non-trivial solution.
6. Show that a field is algebraically closed iff, for all $n > 1$, every form of degree n in n variables has a non-trivial zero. (Hint. if k is not algebraically closed and f is an irreducible polynomial of degree > 1, take the field $F = k[t]/(f)$ and consider the norm of F/k.)

5.4 Quaternion algebras

The first skew field was discovered by William Rowan Hamilton in 1843. His aim had been to find a generalization of the complex numbers, to represent three-dimensional vectors; it took some 12 years to realize that four dimensions rather than three were needed and that the commutative law had to be given up. The algebra he found, which he called the *quaternions*, was a four-dimensional $\mathbf{R}$-algebra H with basis $1, i, j, k$, and multiplication table:

	1	i	j	k
1	1	i	j	k
i	i	-1	k	$-j$
j	j	$-k$	-1	i
k	k	j	$-i$	-1

(5.4.1)

The group of order 8 generated by i, j, k is called the *quaternion group*. Hamilton and his followers developed an elaborate geometrical calculus on the basis of the quaternions, but this will not concern us here. For us the quaternions form the simplest division algebra and an important tool in the general theory.

Let k be any field of characteristic not 2 and let $a, b \in k^\times$. We define the *quaternion algebra* $(a, b; k)$ *as the* k-algebra with basis $1, u, v, uv$ and multiplication rules

$$u^2 = a, \; v^2 = b, \; vu = -uv. \tag{5.4.2}$$

In this notation Hamilton's algebra becomes $(-1, -1; \mathbf{R})$. When char $k = 2$, we define the *quaternion algebra* $(a, b; k|$ as the algebra with basis $1, u, v, uv$ and multiplication rules

$$u^2 = a, \; v^2 + v = b, \; vu = uv + u. \tag{5.4.3}$$

It is easily checked that in each case the quaternion algebra is central simple; hence by Wedderburn's theorem, it is either a division algebra or it is split, i.e. a full 2×2 matrix ring over k.

Let A be a quaternion algebra. Then any element α not in k is quadratic over k; its equation may be written

$$\alpha^2 t(\alpha)\alpha + n(\alpha) = 0, \tag{5.4.4}$$

where $t(\alpha)$ and $n(\alpha)$ are the trace and norm respectively. Explicitly, if $\alpha = t + xu + yv + zuv$, then for char $k \neq 2$,

$$t(\alpha) = 2t, \; n(\alpha) = t^2 x^2 a - y^2 b - z^2 ab,$$

while for char $k = 2$ we have

$$t(a) = y, \; n(\alpha) = t^2 + x^2 a + y^2 b + z^2 ab + ty + xza.$$

This is most easily seen by observing that A has an involution, i.e. an anti-automorphism whose square is 1, $\alpha \mapsto \bar{\alpha}$, such that $t(\alpha) = \alpha + \bar{\alpha}$, $n(\alpha) = \alpha\bar{\alpha}$.

Our first result shows that the quaternion algebras effectively include all four-dimensional division algebras.

Theorem 5.4.1. *Let k be any field. Then any central simple k-algebra A possessing a two-dimensional splitting field is either split or a quaternion algebra.*

Proof. Since A has a two-dimensional splitting field, it is four-dimensional by Theorem 5.2.6, and so is either k_2 or a division algebra. Leaving the first case aside, we see by Theorem 5.2.8 that the splitting field may be taken to be a subfield of A and separable over k. Suppose first that char $k \neq 2$; then we may take the splitting field to be $F = k(u)$, where $u^2 = a \in k$. Clearly $u \mapsto -u$ defines an automorphism of F, which by Corollary 5.1.8 is induced by an inner automorphism of A, say $x \mapsto v^{-1}xv$. Hence $v^{-1}uv = -u$, and v^2 centralizes A and so lies in k, say $v^2 = b \in k$. Now it is easily verified that 1, u, v, uv are linearly independent and clearly span A, and they satisfy (5.4.2) by construction.

When char $k = 2$, we can take as our splitting field $F = k(v)$, where $v^2 + v = b \in k$. Now $v \mapsto v + 1$ is an automorphism, so for some $u \in A$ we have $u^{-1}vu = v + 1$. Again u centralizes A, so $u^2 = a \in k$, and as before 1, u, v, uv form a basis for A, and (5.4.3) holds. ■

As a consequence we have

Corollary 5.4.2 Frobenius' theorem, 1886). *The only division algebras over the real field are* **R**, **C** *and* **H**, *the Hamilton quaternions.*

Proof. Let D be a division algebra over **R**. Since **C** is the only proper algebraic field extension of **R**, if $D \neq$ **R**, **C**, then it must be non-commutative. Let F be a maximal subfield of D; F is a proper extension of **R**, hence $F \cong$ **C** and by Corollary 5.2.7, F is a splitting field of D. Thus D is a quaternion algebra, $(a, b; \mathbf{R})$ say, by Theorem 5.4.1. If a or b is positive, it is easily seen to split, hence $a, b < 0$ and on dividing the basis elements u by $\sqrt{-a}$ and v by $\sqrt{-b}$, we reach the form $(-1, -1; \mathbf{R})$. ■

In general $(a, b; k)$ may be split; conditions for this to happen are given by

Proposition 5.4.3. *Let $H = (a, b; k)$ be a quaternion algebra over a field k of characteristic not 2. Then the following conditions are equivalent:*

(a) $H = (1, -1; k)$,
(b) *H splits, i.e. $H \cong k_2$,*
(c) *H is not a skew field,*
(d) *$n(x) = 0$ for some $x \neq 0$ in H,*
(e) *$a = N(\alpha)$ for some $\alpha \in k(\sqrt{b})$, where N is the norm from $k(\sqrt{b})$ to k,*
(f) *$ax^2 + by^2 = z^2$ has a non-zero solution in k.*

Proof. (a) ⇒ (b). Consider the map $H \to k_2$ defined by

$$u \mapsto \begin{pmatrix} 1 & 0 \\ 0 & -1 \end{pmatrix}, \quad v \mapsto \begin{pmatrix} 0 & 1 \\ -1 & 0 \end{pmatrix}, \quad uv \mapsto \begin{pmatrix} 0 & 1 \\ 1 & 0 \end{pmatrix}.$$

where $u^2 = 1$, $v^2 = -1$. It is easily checked that this map preserves the defining relations of H and so defines a homomorphism from H to k_2. It is clearly surjective, and so it is an isomorphism by a comparison of dimensions.

(b) ⇒ (c) is clear, as is (c) ⇒ (d), for if $n(x) \neq 0$, then x has an inverse, as we see from (5.4.4).

(d) $\Rightarrow$ (e). If b is a square in k, then $k(\sqrt{b}) = k$ and the conclusion follows. Otherwise take $q \neq 0$ with $n(q) = 0$; on writing $q = t + xu + yv + zuv$, we have $0 = n(q) = t^2 - ax^2 - by^2 - abz^2$, hence

$$a = (t^2 - by^2)/(x^2 + bz^2),$$

and this shows a to be a norm in $k(\sqrt{b})$.

(e) $\Rightarrow$ (f). If $a = N(\alpha) = x^2 - by^2$, then $a + by^2 = x^2$, so (f) holds for $(1, y, x)$.

Finally (f) $\Rightarrow$ (a): Let $ax^2 + by^2 = z^2$, where x, y are not both zero, say $x \neq 0$. Then

$$\frac{1}{a} + b(\frac{y}{ax})^2 = (\frac{z}{ax})^2;$$

changing variables, we have $z^2 - by^2 = a^{-1}$. Taking the basis of H to be $1, i, j, k$, we put $u = zi + yk$; then $u^2 = az^2 - aby^2 = a(z^2 - by^2) = 1$. Thus $u^2 = 1$; further, we have $ju = -uj$, so if $v = [(1-b) + (1+b)u]j/2b$, then $uv = -vu$ and $v^2 = -1$. This shows that $H = (1, -1; k)$. ∎

Exercises

1. Verify directly that $(1, 1; k)$ is split.
2. Show that in the Hamilton quaternions the equation $x^2 + 1 = 0$ has infinitely many solutions, all conjugate.
3. Show that two elements of a quaternion algebra satisfying the same irreducible equation either commute or are conjugate. Deduce that every quaternion of norm 1 is a commutator, i.e. $\mathbf{SK}_1(H) = 1$.
4. Show that in characteristic 2, $(a, b; k|$ splits iff $a = N(\alpha)$ for some $\alpha \in k(\wp^{-1}(b))$, where $\wp(x) = x^p - x$.
5. Show that $(a, b; k)$ is multiplicative in each factor, i.e. $(a, b; k) \otimes (a', b; k) \cong (aa', b; k)$ and similarly for the other factor. Likewise for $(a, b; k|$ when char $k = 2$.
6. Show that if $H \cong (a, b; k)$ is not split but is split by $k(\sqrt{a'})$, then $H \cong (a', b'; k)$ for suitable $b' \in k$.
7. (A. A. Albert, P. K. Draxl) Show that if $(a', b'; k) \otimes (a'', b''; k)$ is similar to a quaternion algebra $(a, b; k)$, then there exist $c', c'', d \in k$ such that $(a', b'; k) \cong (c', d; k)$ and $(a'', b''; k) \cong (c'', d; k)$. Deduce that a tensor product of two quaternion algebras H, K is split iff H, K have a common splitting field which is separable quadratic over k.

5.5 Crossed products

For a closer study of the Brauer group there is a concrete representation of central simple algebras which is often useful, namely as a crossed product. Such a representation does not exist for each algebra, but there is one in every Brauer class.

Definition. A central simple k-algebra is called a *crossed product* if it contains a maximal subfield F such that F/k is a Galois extension. As maximal subfield F will then be a splitting field (Theorem 5.2.6).

It is easy to see that every Brauer class contains a crossed product: if D is a central division algebra, then D has a separable splitting field F, by Theorem 5.2.8, and the normal closure E of F/k is a Galois extension of k. Let $[F:k]=r$, $[E:F]=n$; then $E\subseteq D_n$ and $[E:k]=nr$, $[D_n:k]=n^2r^2$, hence E is a maximal subfield of D_n, by Corollary 5.1.12. Thus D_n is a crossed product, though D itself need not be (because the maximal subfield may not be Galois over k). The situation was first studied in the 1930s by Helmut Hasse, Adrian Albert and others, and it was found that every central division algebra over Q is a crossed product, but it was only much later that Shimshon Amitsur [1972] gave examples of central division algebras that are not crossed products.

Crossed products have an explicit description which is of importance (and which accounts for the name). Let A be a crossed product, with Galois splitting field F over k as subfield. Denote by U the group of units of A and by N the normalizer of $F^\times$ in U:

$$N=\{u\in U|u^{-1}Fu\subseteq F\}.$$

Then $F^\times$ is a normal subgroup of N and we have the exact sequence

$$1\to F^\times\to N\to\Gamma\to 1,$$

where $\Gamma=N/F$. Thus A determines a group extension N of $F^\times$ by Γ. We shall show that (i) $\Gamma\cong\mathrm{Gal}(F/k)$, (ii) every extension of $F^\times$ by $\mathrm{Gal}(F/k)$ determines a crossed product (up to isomorphism). This will also provide an explicit form for A.

Any $u\in N$ defines an automorphism $\alpha(u)$ of $F^\times$ by the rule $x^{\alpha(u)}=u^{-1}xu$; this automorphism leaves k elementwise fixed, so we have a mapping

$$\alpha: N\to\mathrm{Gal}(F/k), \tag{5.5.1}$$

which is clearly a homomorphism. Its kernel is the centralizer of $F^\times$ in N, which is $F^\times$, because F is a maximal subfield. To show that α is surjective, let $\sigma\in\mathrm{Gal}(F/k)$; then by Skolem–Noether (Corollary 5.1.8) σ is induced by an inner automorphism of A, say $\sigma=\alpha(u)$, where $u\in U$. By definition, $u\in N$, so (5.5.1) is surjective and this shows that $\Gamma\cong\mathrm{Gal}(F/k)$.

Returning to our crossed product A, let us take a transversal $\{u_\sigma\}$ of $\Gamma=\mathrm{Gal}(F/k)$ in N, so that the elements of N have the form $u_\sigma a$ ($a\in F^\times,\sigma\in\Gamma$). We shall indicate the action of Γ on F by putting exponents, thus

$$au_\sigma=u_\sigma a^\sigma\quad(a\in F,\sigma\in\Gamma). \tag{5.5.2}$$

Further, we have

$$u_\sigma u_\tau=u_{\sigma\tau}c_{\sigma,\tau}\quad\text{for some } c\in F^\times, \tag{5.5.3}$$

where the $c_{\sigma,\tau}$ satisfy the factor set condition (by the associativity of N):

$$c_{\rho,\sigma\tau}c_{\sigma\omega,\tau} = c_{\rho\sigma,\tau}c^{\tau}_{\rho,\sigma}. \tag{5.5.4}$$

We assert that A is determined completely as right F-space with basis u_σ ($\sigma \in \Gamma$) and the multiplication rules (5.5.2), (5.5.3). We know that $[A : k] = n^2 = [A : F][F : k]$ and $[F : k] = n$, hence $[A : F] = n$, so the dimension is correct, since there are $n = |\Gamma|$ basis elements. It only remains to show that the u_σ are right linearly independent over F. If there is a non-trivial relation

$$\sum u_\sigma a_\sigma = 0, \quad \text{where } a_\sigma \in F, \tag{5.5.5}$$

let us take such a relation with the fewest non-zero coefficients. Pick $\rho \in \Gamma$ such that $a_\rho \neq 0$ and multiply on the left by u_ρ^{-1} so as to obtain a relation (5.5.5) with $a_1 \neq 0$. The left-hand side of (5.5.5) cannot consist of a single term, hence $a_\tau \neq 0$ for some $\tau \neq 1$. Let $b \in F$ be such that $b^\tau \neq b$ and take the commutator of (5.5.5) with b:

$$0 = \sum u_\sigma b^\sigma a_\sigma - \sum u_\sigma a_\sigma b = \sum u_\sigma a_\sigma (b^\sigma - b).$$

The coefficient of u_τ is $a_\tau(b^\tau - b) \neq 0$, so this relation is non-trivial, but it has fewer terms than (5.5.5), because the coefficient of u_1 is $a_1(b - b) = 0$. This contradicts the minimality of (5.5.5) and it shows that the u_σ are right F-linearly independent. We note that this is essentially the argument of Dedekind's lemma (BA, Lemma 7.5.1).

Suppose now that we are given a (finite) Galois extension F/k with group Γ, and a group extension N of $F^\times$ by Γ, where Γ acts on F by automorphisms. Let us take a transversal $\{u_\sigma\}$ of Γ in N; this determines a factor set for which (5.5.3) holds. We define an algebra A by taking the right F-space on the u_σ as basis, with multiplication defined by (5.5.2), (5.5.3). Then we claim that A is a crossed product.

In the first place, A is simple, for if $\bar{A}$ is a non-zero quotient, it is spanned by the $\bar{u}_\sigma$ ($\sigma \in \Gamma$) over F and $\bar{u}_\sigma \neq 0$ because u_σ is a unit in A and so cannot map to 0. Now the same argument as before shows that the $\bar{u}_\sigma$ are linearly independent over F, hence the mapping $\sum u_\sigma a_\sigma \mapsto \sum \bar{u}_\sigma a_\sigma$ is injective and A is simple.

Next we note that A has centre k. For if $x = \sum u_\sigma a_\sigma$ lies in the centre, then $xb = bx$ for all $b \in F$, so $\sum u_\sigma a_\sigma(b^\sigma - b) = 0$. Hence $a_\sigma(b^\sigma - b) = 0$ for all $b \in F$ and $\sigma \in \Gamma$, therefore $a_\sigma = 0$ for $\sigma \neq 1$, and $x = u_1 a_1 \in F$. Now $u_\sigma x = x u_\sigma = u_\sigma x^\sigma$, hence $x^\sigma = x$ for all $\sigma \in \Gamma$, and so $x \in k$. Thus k is the centre of A. Finally $[A : F] = [F : k]$ by construction, so F is a splitting field of A. This proves

Theorem 5.5.1. *Any crossed product A over k with Galois splitting field F contained in A is defined up to isomorphism by an extension N of $F^\times$ by* Gal(F/k) *and conversely, any such extension N defines a crossed product.* ■

We now examine when two factor sets define isomorphic crossed products. Identifying our two isomorphic algebras, we have to compare two transversals $\{u_\sigma\}$, $\{u'_\sigma\}$ in our crossed product A; two such transversals are related by equations

$$u'_\sigma = u_\sigma a_\sigma \quad \text{for some } a_\sigma \in F^\times,$$

and if the factor set for $\{u'_\sigma\}$ is $\{c'\} = \{c'_{\sigma,\tau}\}$, then

$$c'_{\sigma,\tau} = c_{\sigma,\tau} a_{\sigma\tau}^{-1} a_\sigma^\tau a_\tau;$$

hence the factor sets c, c' are associated (see (3.1.13)). The factor sets form a group C under multiplication, the group of 2-cocycles, in which the bounding cocycles form a subgroup B. These are the cocycles associated to 1:

$$c_{\sigma,\tau} = a_{\sigma\tau}^{-1} a_\sigma^\tau a_\tau.$$

The quotient C/B is just $H^2(\Gamma, F^\times)$, the second cohomology group of Γ with coefficients in $F^\times$, and by Theorem 5.5.1 we have a mapping

$$H^2(\Gamma, F^\times) \to \mathbf{B}_k. \tag{5.5.6}$$

The above remarks show this mapping to be injective and its image is the relative Brauer group $\mathbf{B}(F/k)$, the subgroup of Brauer classes split by F, already encountered in Section 5.2. Since each central simple k-algebra has a separable splitting field, contained in a Galois extension of k, it follows that $\mathbf{B}_k$ is a union of the $\mathbf{B}(F/k)$, as F ranges over the finite Galois extensions of k.

It remains to show that (5.5.6) is a homomorphism. To establish this fact we need to verify that the tensor product of algebras corresponds to the Baer product of the extensions. Take $w \in \mathbf{B}_k$, with Galois splitting field F and let $B \in w$. Put $[F : k] = n$, $[B : k] = r^2$ and let V be an F-space of dimension r; then

$$B^0 \otimes F \cong B_F^0 \cong F_r \cong \mathrm{End}_F(V).$$

Let A be the centralizer of B^0 in $\mathrm{End}_k(V)$; then $A \sim B$ and $[A : k] = n^2$. Since A contains F, we have $[A : F] = n$, so F is a maximal subfield and A is a crossed product. We can realize this situation by taking V to be a simple left ideal in B_F; then V is a (B, F)-bimodule, i.e. a right $(B^0 \otimes F)$-module, and $[V : F] = r$. Moreover, A has a right F-basis $\{u_\sigma\}$ and each u_σ defines an F-semilinear transformation with automorphism σ:

$$(\alpha x + \beta y)u_\sigma = \alpha^\sigma x u_\sigma + \beta^\sigma y u_\sigma \quad \text{for any } x, y \in V, \alpha, \beta \in F.$$

Thus A is spanned over F by semilinear transformations.

Given two Brauer classes w, w', let us take $B \in w$, $B' \in w'$ and let V, V' be simple left ideals in B_F, B'_F respectively, where F is a Galois splitting field for w and w'. Then $B_F^0 = \mathrm{End}_F(V)$, $B'^0_F = \mathrm{End}_F(V')$, hence

$$\mathrm{End}_F(V \otimes_F V') \cong B_F^0 \otimes_F B'^0_F \cong (B^0 \otimes_k B'^0)_F \cong (B \otimes_k B')_F^0.$$

Denote the respective centralizers of $B^0, B'^0, (B \otimes B')^0$ by A, A', A''; then $A \sim B$, $A' \sim B'$, $A'' \sim B \otimes B'$ and so $(A)(A') = (A'')$. Let N, N', N'' be the normalizers of $F^\times$ in A, A', A'' respectively; to show that (5.5.6) is a homomorphism we must prove that N'' is just the Baer product of N and N'. To find this Baer product, let $N \circ N'$ be the pullback of the mappings $N \to \Gamma$, $N' \to \Gamma$, i.e. the subgroup of $N \times N'$ of elements of the form $(u_\sigma \alpha, u'_\sigma \beta)$, where $\alpha, \beta \in F$, u_σ, u'_σ are transversals of Γ in N, N' respectively and $u_1 = u'_1 = 1$ for simplicity. The set

$L = \{(\lambda, \lambda^{-1}) | \lambda \in F^\times\}$ is a normal subgroup of $N \circ N'$, the elements of $N \circ N'/L$ can uniquely be written as $(u_\sigma, u'_\sigma \alpha)$ and it is easily verified that this is an extension of $F^\times$ by Γ with factor set equal to the product of those of N and N', thus it is the Baer product.

Now each element of $N \circ N'$ defines a semilinear transformation on $V \otimes V'$:

$$(u_\sigma \alpha, u'_\sigma \alpha') : v \otimes v' \longmapsto v(u_\sigma \alpha) \otimes v'(u'_\sigma \alpha').$$

Hence we have a homomorphism $f : N \circ N' \to N''$. Clearly it is surjective and the kernel consists of all $(u_\sigma \alpha, u'_\sigma \alpha')$ inducing the identity, i.e. $\sigma = 1$, $\alpha\alpha' = 1$. Hence $\ker f = L$ and so N'' is isomorphic to $N \circ N'/L$, the Baer product; this shows (5.5.6) to be a homomorphism. We sum up the result as

Theorem 5.5.2. *Let F/k be a finite Galois extension with group Γ. Then*

$$\mathbf{B}(F/k) \cong H(\Gamma, F^\times). \qquad \blacksquare \qquad (5.5.7)$$

Once we have the homomorphism property, the injectivity also follows from the explicit criterion for splitting:

Proposition 5.5.3. *Let A be a central simple k-algebra of degree n which is a crossed product with maximal subfield F. Then $A \cong \mathfrak{M}_n(k)$ if and only if the extension of $F^\times$ by* Gal(F/k) *in A splits.*

Proof. Write $\Gamma = \mathrm{Gal}(F/k)$. If the extension of $F^\times$ by Γ splits, then we can realize it with 1 as factor set. Put $A = \mathrm{End}_k(F) \cong k_n$. Each $\sigma \in \Gamma$ acts as k-linear transformation of $F \cong k^n$, while F itself acts by right multiplication. By Dedekind's lemma (BA, Lemma 7.5.1) the σ are linearly independent over F, so we obtain an n-dimensional F-space, i.e. an n^2-dimensional k-space, which must be all of A, by a comparison of dimensions. Thus A is realized as a crossed product. Conversely, any factor set for the algebra k is associated to the trivial factor set (which we saw defines k) and hence itself corresponds to a split extension. $\blacksquare$

Let us examine the Brauer group in a little more detail.

Theorem 5.5.4. *For any field k the Brauer group $\mathbf{B}_k$ is a torsion group. More precisely, if $w \in \mathbf{B}_k$ has index r, then $w^r = 1$. Hence for an extension F/k of degree n we have $n.\mathbf{B}(F/k) = 0$.*

Proof. Let $w \in \mathbf{B}_k$ have index r and take a Galois splitting field F of w, where $[F : k] = n$, say. By Theorem 5.5.2, w corresponds to an element c of $H^2(\Gamma, F^\times)$ and it follows from Proposition 3.1.6 that the order of c and hence of w divides n, but we want to get the sharper bound r.

Let $A \in w$ be a crossed product with F as maximal subfield and let V be a minimal right ideal of A; then $[V : F] = r$ and we can represent A by F-linear transformations of V. With a right F-basis $v_1, \ldots, v_r$ of V we have $v_j a = \sum v_i \alpha_{ij}$ for any $a \in A$, or in matrix form

$$(v)a = (v)\alpha,$$

where $(\nu) = (\nu_1, \ldots, \nu_r)$ and $\alpha = (\alpha_{ij})$. In particular, if $u_\sigma \mapsto U_\sigma$, we have

$$(\nu)u_\sigma u_\tau = (\nu)U_\sigma u_\tau = (\nu)u_\tau U_\sigma^\tau = (\nu)U_\tau U_\sigma^\tau,$$

$$(\nu)u_\sigma u_\tau = (\nu)u_{\sigma\tau}c_{\sigma,\tau} = (\nu)U_{\sigma\tau}c_{\sigma,\tau},$$

hence

$$U_{\sigma\tau}c_{\sigma,\tau} = U_\tau U_\sigma^\tau, \tag{5.5.8}$$

where the U_σ are $r \times r$ matrices over F. Now write $d_\sigma = \det U_\sigma$ and take determinants in (5.5.8):

$$d_{\sigma\tau}c_{\sigma,\tau}^r = d_\tau d_\sigma^\tau,$$

hence $\{c_{\sigma,\tau}^r\}$ is a splitting factor set; therefore $(A)^r = 1$. ■

If k is a perfect field of characteristic p, then it can be shown that the p-component of $\mathbf{B}_k$ is trivial (see Exercise 2).

The order of w as an element of $\mathbf{B}_k$ is called its *exponent*. For any central simple algebra A, its *exponent* is defined as the exponent of its Brauer class, and Theorem 5.5.4 may be expressed by saying that for any Brauer class, the exponent divides the index. The question naturally arises whether the exponent is always equal to the index. For rational algebras this is true, but not in general; however we do have the following connexion:

Proposition 5.5.5. *For any $w \in \mathbf{B}_k$ the index and the exponent have the same prime factors.*

Proof. Let w have index m and exponent t, so that $t|m$, by Theorem 5.5.4. We have to show that any prime factor p of m also divides t. Take a Galois splitting field F of w, with group Γ and let S be a Sylow p-subgroup of Γ with fixed field E. Then $[E:k] = (\Gamma : S) = \nu$ is prime to p, while $|S| = p^\alpha$. For any $A \in w$, the index reduction factor from A to A_E divides ν (Proposition 5.2.3) and so is prime to p, hence the index of A_E is still divisible by p and it is enough to show that p also divides the exponent. Now A_E is a central simple E-algebra which does not split but which is split by F. Since $[F:E] = p^\alpha$, its exponent is a positive power of p, as we had to show.

This result leads to a remarkable decomposition formula:

Theorem 5.5.6. *Any central division algebra D of degree $m = q_1 \ldots q_r$, where the q_i are powers of distinct primes, has the decomposition*

$$D \cong D^{(1)} \otimes \ldots \otimes D^{(r)}, \tag{5.5.9}$$

where $D^{(i)}$ is a central division algebra of degree q_i.

We note that the assertion is that (5.5.9) is an isomorphism. The corresponding assertion, with 'isomorphism' replaced by 'similarity' is a trivial consequence of the basis theorem for abelian groups, applied to the cyclic subgroup generated by the Brauer class of D.

Proof. The class (D) has exponent $n = q'_1 \dots q'_r$, where $q'_i | q_i$ and by Proposition 5.5.5, $q'_i > 1$. By the basis theorem for abelian groups, (D) can be written as a product of classes which are powers of (D) with prime power exponent. Let $D^{(i)}$ be a division algebra similar to a power of D with exponent q'_i; then

$$D^{(1)} \otimes \dots \otimes D^{(r)} \sim D.$$

By Proposition 5.5.5 the $D^{(i)}$ have coprime degrees and by Proposition 5.2.10 we have a division algebra on the left, hence the two sides are isomorphic. ■

A central simple k-algebra is called *primary* if it is not equal to k and it contains no proper central simple subalgebra. Thus if A is not primary, it is either k or it has a central simple subalgebra $B \neq k, A$. By Proposition 5.1.5 we have $A = B \otimes B'$, where B' is the centralizer of B in A. Bearing in mind Theorem 5.5.6 and the relation $k_{rs} \cong k_r \otimes k_s$, we see that any primary algebra is a division algebra of prime power degree or of the form k_p. Thus we obtain

Proposition 5.5.7. *Every central simple algebra is a tensor product of primary algebras. The primary k-algebras are $\mathfrak{M}_p(k)$, where p is a prime, and certain division algebras of prime power degree.* ■

A division algebra of prime power degree is not necessarily primary, though this does hold over an algebraic number field.

Exercises

1. Let F/k be a finite Galois extension with group Γ of order n. Show that $F \otimes_k k\Gamma \cong k_n$. (Hint. Use a normal basis for F/k.)
2. Let k be a perfect field of prime characteristic p and D a central division algebra. Show that Deg D is prime to p. (Hint. Use Theorem 5.2.8 to show that D contains no proper extension of degree p over k.) Deduce that $\mathbf{B}_k$ has trivial p-component.
3. Show that every division k-algebra has a splitting field which is a tensor product of extensions of k with prime power degrees.
4. Let G be a group whose centre Z is free abelian and of finite index n in G. By constructing a suitable crossed product with group G/Z, show that G can be embedded in a division algebra of degree n.
5. Let A, B be central division k-algebras that are crossed products with groups G, H. Show that if $A \otimes B$ is a division algebra, then it is a crossed product with group $G \times H$.

5.6 Change of base field

Let us consider the effect of changes in the base field on a crossed product. We begin by recalling a result from Galois theory (BA, Theorem 7.10.3):

Proposition 5.6.1. *Let F/k be a Galois extension and E any field extension of k, where E, F are both contained in the same field. The EF/E is Galois, with group isomorphic to the subgroup of* $\mathrm{Gal}(F/k)$ *corresponding to* $E \cap F$.

Proof. This is essentially a translation of the parallelogram rule applied to the Galois groups. The isomorphism is obtained by taking an automorphism of *EF/E* and restricting it to *F*; this provides an isomorphism with $\mathrm{Gal}(F/E \cap F)$. ■

In the above situation let us write $G = \mathrm{Gal}(F/k)$ and denote by H the subgroup leaving $E \cap F$ fixed, so that $H \cong \mathrm{Gal}(EF/E)$. Any factor set $\{c\} : G \times G \to F^\times$ when restricted to H yields a factor set $\{c'\} : H \times H \to (EF)^\times$. It is clear that a split factor set has a split restriction, hence the inclusion $H \subseteq G$ gives rise to a homomorphism, the *restriction*

$$\mathrm{res} : H^2(G, F^\times) \to H^2(H, (EF)^\times). \tag{5.6.1}$$

In what follows we shall write $(F/k, c)$ for the crossed product over *F/k* with factor set $\{c\}$.

Theorem 5.6.2 (Restriction theorem). *Let F/k be a finite Galois extension with group* $G = \mathrm{Gal}(F/k)$, *let E/k be any extension (within a field containing F) and let H be the subgroup corresponding to* $K = E \cap F$, *so that* $H \cong \mathrm{Gal}(EF/E)$. *Then*

$$(F/k, c)_E \sim (EF/E, c'), \tag{5.6.2}$$

where $\{c'\}$ *is the factor set* $\{c\}$ *restricted to H:*

$$\begin{array}{ccc} H^2(G,F) & \xrightarrow{\mathrm{res}} & H^2(H,F) \\ \downarrow & & \downarrow \\ \mathbf{B}(F/k) & \xrightarrow{\otimes K} & \mathbf{B}(F/K) \end{array}$$

Proof. We shall show

$$(F/k, c) \otimes_k K \cong (F/K, c') \otimes k_r, \quad \text{where } r = [K : k], \tag{5.6.3}$$

$$(F/K, c') \otimes_K E \cong (EF/E, c'). \tag{5.6.4}$$

It is clear that (5.6.2) is a consequence of (5.6.3) and (5.6.4). Let us write $A = (F/k, c)$; this algebra contains F and hence K as a subfield. If K' denotes the centralizer of K in A, then by Brauer's theorem (Theorem 5.1.10),

$$K' \otimes k_r \cong A \otimes_k K,$$

and this will establish (5.6.3) if we can show that $K' \cong (F/K, c')$. In A take $x = \sum u_\sigma a_\sigma$ $(a_\sigma \in F)$; we have $x \in K'$ iff $xy = yx$ for all $y \in K$, i.e. $\sum u_\sigma a_\sigma y = \sum y u_\sigma a_\sigma = \sum u_\sigma y^\sigma a_\sigma$. Thus we must have $a_\sigma = 0$ whenever $y^\sigma \neq y$ for some y, i.e. when $\sigma \notin \mathrm{Gal}(F/K) = H$. This shows that $K' = \{\sum u_\sigma a_\sigma | a_\sigma \in F,\ \sigma \in H\}$ and (5.6.3) follows.

To prove (5.6.4) we note that the K-algebra homomorphism

$$E \otimes_K F \to EF$$

is surjective and both sides have dimension $[EF : E] = |H| = [F : k]$ over E, so it is an isomorphism (thus the composite EF is independent of the choice of embedding; this depends on F being normal over K). Hence

$$(F/K, c') \otimes_K E = \sum_{\sigma \in H} (u_\sigma \otimes 1)EF \cong (EF/E, c'),$$

and (5.6.4) follows. ∎

There is a second operation called *inflation*, corresponding to the natural homomorphism $G \to G/N$, for $N \triangleleft G$. Given a factor set on $\bar{G} = G/N$, we define a factor set $\{\bar{c}\}$ on G by the inflation rule (where $\sigma \mapsto \bar{\sigma}$ is the natural homomorphism from G to G/N):

$$\bar{c}_{\sigma,\tau} = c_{\bar{\sigma},\bar{\tau}}. \tag{5.6.5}$$

Theorem 5.6.3 (Inflation theorem). *Let $k \subseteq K \subseteq F$, where F/k, K/k are Galois extensions, $G = \mathrm{Gal}(F/k)$ and N is the (normal) subgroup of G corresponding to K.*

Given any factor set $\{c\}$ on G/N and the corresponding factor set $\{\bar{c}\}$ on G derived by the inflation rule (5.6.5), then

$$(F/k, \bar{c}) \cong (K/k, c) \otimes k_r, \quad \text{where } r = [F : K]. \tag{5.6.6}$$

Proof. Let $[F : K] = r$, $[K : k] = s$, $\bar{G} = G/N$ and define $B = (K/k, c) \otimes k_r$. The field F can be embedded in K and hence in B; now B is a central simple k-algebra split by K, hence also by F and $\mathrm{Deg}\, B = rs = [F : k]$, so F is a maximal subfield of B, therefore B is a crossed product. We shall prove that $B \cong (F/k, \bar{c})$ by constructing an explicit embedding of F in B; this will establish (5.6.6).

Take a K-basis $e_1, \dots, e_r$ for F and define $T(x) = (t_{ij}(x))$ by

$$e_i x = \sum t_{ij}(x) e_j \quad \text{for } x \in F.$$

On writing $e = (e_1, \dots, e_r)^T$, we can express this equation in matrix form as

$$ex = T(x)e, \quad \text{where } T(x) \in K_r. \tag{5.6.7}$$

Since $e_i^\sigma \in F$ for all $\sigma \in G$, we have

$$e^\sigma = P_\sigma e, \quad \text{where } P_\sigma \in K_r. \tag{5.6.8}$$

It follows that $P_{\sigma\tau} e = e^{\sigma\tau} = (P_\sigma e)^\tau = P_\sigma^\tau e^\tau = P_\sigma^\tau P_\tau e$, hence

$$P_{\sigma\tau} = P_\sigma^{\bar{\tau}} P_\tau, \tag{5.6.9}$$

where we have replaced τ by $\bar{\tau}$ in the action on P because the latter has entries in K. Applying σ to (5.6.7), we find $e^\sigma x^\sigma = T(x)^\sigma e^\sigma$, i.e. $P_\sigma T(x^\sigma) = T(x)^\sigma P_\sigma$ and again $T(x)$ has entries in K, so that

$$P_\sigma T(x^\sigma) = T(x)^{\bar{\sigma}} P_\sigma. \tag{5.6.10}$$

We claim that for any right K-basis $u_{\bar{\sigma}}$ of $(K/k, c)$,

$$v_\sigma = u_{\bar{\sigma}} P_\sigma \tag{5.6.11}$$

is a right F-basis for $B \cong (K/k, c) \otimes k_r$, with the isomorphism

$$\sum v_\sigma a_\sigma \leftrightarrow \sum u_{\bar{\sigma}} P_\sigma T(a_\sigma).$$

For the proof we need only verify the conditions on v_σ; using (5.6.10), we have

$$T(x)v_\sigma = T(x)u_{\bar{\sigma}}P_\sigma = u_{\bar{\sigma}}T(x)^{\bar{\sigma}}P_\sigma = u_{\bar{\sigma}}P_\sigma T(x^\sigma) = v_\sigma T(x^\sigma)$$

and

$$v_\sigma v_\tau = u_{\bar{\sigma}}P_\sigma u_{\bar{\tau}}P_\tau = u_{\bar{\sigma}}u_{\bar{\tau}}P_\sigma^{\bar{\tau}}P_\tau = u_{\bar{\sigma}\bar{\tau}}c_{\bar{\sigma},\bar{\tau}}P_{\sigma\tau} = v_{\sigma\tau}c_{\bar{\sigma},\bar{\tau}}.$$

This shows that $B \cong (F/k, \bar{c})$, as claimed, and it proves (5.6.6). ■

Since $r > 1$ except in the trivial case $F = K$, we obtain from Theorem 5.6.3,

Corollary 5.6.4. *A central simple algebra obtained by inflation is never a division algebra.* ■

We note that the natural homomorphism $G \to G/N$ induces the inflation homomorphism $H^2(G/N, K^\times) \to H^2(G, F^\times)$ and Theorem 5.6.3 can be expressed as a commutative square, which together with the previous ones gives the commutative diagram

$$\begin{array}{ccccc}
0 \to H^2(G/N, K^\times) & \xrightarrow{\text{inf}} & H^2(G, F^\times) & \xrightarrow{\text{res}} & H^2(N, F^\times) \\
\downarrow & & \downarrow & & \downarrow \\
0 \to \mathbf{B}(K/k) & \xrightarrow{\text{inc}} & \mathbf{B}(F/k) & \longrightarrow & \mathbf{B}(F/K)
\end{array}$$

The bottom row is easily seen to be exact: a central simple k-algebra split by F will split as K-algebra iff it is split by K. Hence the top row is also exact.

As a third operation we have the *corestriction* (or *transfer*), which for $k \subseteq F$ provides a homomorphism $\mathbf{B}_F \to \mathbf{B}_k$.

Let B be an F-algebra with a finite group G of automorphisms such that each element of G other than 1 restricts to a non-trivial automorphism of F. As usual we write B^G and F^G for the subset fixed by G. Given a k-algebra A, if $B = A_F$, where $k = F^G$, then $B^G = A$, as is easily verified.

We begin by showing that B^G can be expressed in terms of the trace, where for $b \in B$ we define $\text{tr}\, b = \sum_{\sigma \in G} b^\sigma$ and $\text{tr}\, B = \{\text{tr}\, b \mid b \in B\}$.

Lemma 5.6.5. *Let B be an F-algebra with a finite group of automorphisms which induce distinct automorphisms on F. Then* tr B *coincides with the fixed algebra B^G and if $F^G = k$, then*

$$B \cong B^G \otimes_k F. \tag{5.6.12}$$

Proof. Clearly tr $B \subseteq B^G$ *and both tr* B, B^G *are vector spaces over* k. Let C be the F-space spanned by tr B; if $C \subset B$, then there is an F-linear functional $\varphi : B \to F$ such that $\varphi(C) = 0$ but $\varphi \neq 0$, say $\varphi(u) \neq 0$. For any $\alpha \in F$ we have

$$0 = \varphi(\mathrm{tr}(\alpha u)) = \varphi\left(\sum \alpha^\sigma u^\sigma\right) = \sum \alpha^\sigma \varphi(u^\sigma).$$

By hypothesis the automorphisms σ of F are distinct and hence by Dedekind's lemma they are linearly independent, contradicting the assumption that $\varphi(u) \neq 0$. Hence φ must vanish and $C = B$.

We thus have a canonical map $B^G \otimes F \to B$, which we claim is injective. Any element of $B^G \otimes F$ can be written as $x = \sum b_i \otimes \alpha_i$, where $b_i \in B^G$, $\alpha_i \in F$ and we may take the b_i to be linearly independent over k. Suppose that $x \neq 0$, but that $\sum b_i\alpha_i = 0$; then for any $\beta \in F$,

$$0 = \mathrm{tr}\left(\beta \sum b_i\alpha_i\right) = \mathrm{tr}\left(\sum \beta\alpha_i b_i\right) = \sum \mathrm{tr}(\beta\alpha_i)b_i.$$

By the linear independence of the b_i we have $\mathrm{tr}(\beta\alpha_i) = 0$ for all $\beta \in F$, hence

$$\sum \beta^\sigma \alpha_i^\sigma = 0 \quad \text{for all } \beta \in F,$$

and this again contradicts Dedekind's lemma, unless $\alpha_i = 0$ for all i, so $x = \sum b_i \otimes \alpha_i = 0$ and our mapping is injective. By comparing dimensions in (5.6.12), we see that this is an isomorphism. ■

If B is simple with centre F, then (5.6.12) shows that B^G is a central simple k-algebra such that $[B : F] = [B^G : k]$.

Let E/k be a Galois extension with group G and let B be an E-algebra. We define B^σ, for $\sigma \in G$, as an E-algebra on B with the same ring structure as B but with scalar multiplication

$$\alpha.x = \alpha^\sigma x \quad \text{for } \alpha \in E,\ x \in B.$$

Given any separable extension F/k, let E/k be a Galois extension containing F/k, with group G, and denote by H the subgroup corresponding to F. Put $n = [F : k] = (G : H)$ and let $\sigma_1, \ldots, \sigma_n$ be a transversal of H in G: $G = \cup H\sigma_i$. Then for any F-algebra B the corestriction from F to k is defined as follows. Put

$$B^{(G:H)} = B_E^{\sigma_1} \otimes \ldots \otimes B_E^{\sigma_n}, \tag{5.6.13}$$

and define a G-action on $B^{(G:H)}$ by writing, for any $\sigma \in G$, $\sigma_i\sigma = t_i(\sigma)\sigma_{i'}$, where $t_i(\sigma) \in H$ and $i \mapsto i'$ is a permutation of $1, \ldots, n$ determined by σ. Now put

$$(b_i \otimes \alpha_i)^\sigma = b_{i'} \otimes \alpha_{i'}^{t_i(\sigma)}. \tag{5.6.14}$$

To check that this indeed defines a G-action, let $\sigma_i\sigma = t_i(\sigma)\sigma_{i'}$, $\sigma_{i'}\tau = t_{i'}(\tau)\sigma_{i^*}$. Then $\sigma_i\sigma\tau = t_i(\sigma)t_{i'}(\tau)\sigma_{i^*}$ and

$$\bigotimes_{i=1}^n ((b_i \otimes \alpha_i)^\sigma)^\tau = \bigotimes_{i=1}^n (b_{i'} \otimes \alpha_{i'}^{t_i(\sigma)})^\tau = \bigotimes_{i=1}^n (b_{i^*} \otimes \alpha_{i^*}^{t_i(\sigma)t_{i'}(\tau)}) = \bigotimes_{i=1}^n (b_i \otimes \alpha_i)^{\sigma\tau}.$$

Now the corestriction of B is the subalgebra of $B^{(G:H)}$ fixed by G:

$$\mathrm{cor}_{F/k}B = (B^{(G:H)})^G. \tag{5.6.15}$$

For example, taking $B = F$, we have $F^{(G:H)} = \otimes F^{\alpha_i}$ and the fixed ring is $k^{\otimes n} = k$, hence we have

$$\mathrm{cor}_{F/k}(F) = k. \tag{5.6.16}$$

Proposition 5.6.6. *Let F/k be a separable extension of degree n and B a central simple F-algebra. Then* $\mathrm{cor}_{F/k}B$ *is a central simple k-algebra which depends only on F/k, not on the Galois extension or the choice of transversal. Moreover, the correspondence* $\mathrm{cor}_{F/k}$ *is a homomorphism from* $\mathbf{B}_F$ *to* $\mathbf{B}_k$.

Proof. If B is any central simple F-algebra, then $C = B^{(G:H)}$ is simple with centre E, by Corollary 5.1.3. Hence C^G has centre k; now any ideal of C^G gives rise to an ideal of C, and the simplicity of the latter shows C^G to be simple.

Now let N be a normal subgroup of G contained in H and let L be the corresponding subfield; thus $F \subseteq L \subseteq E$ and L/k is Galois, with group G/N. The transversal $\sigma_1, \ldots, \sigma_n$ of H in G is still a transversal of $\bar{H} = H/N$ in $\bar{G} = G/N$. Suppose that $A = \mathrm{cor}_{F/k}(B)$ is formed as above, going via E; going via L we obtain $C = B^{(\bar{G}:\bar{H})}$, and it is clear that $C = (B^{(G:H)})^N$. Therefore

$$C^{\bar{G}} = ((B^{(G:H)})^N)^{\bar{G}} = (B^{(G:H)})^G,$$

so we reach the same algebra going via L. Given any two extensions L_1, L_2 of F, we can find a Galois extension E/k containing both, and the above argument shows that using L_1 or L_2 we obtain the same algebra $\mathrm{cor}_{F/k}(B)$ as if we had used E, hence all three cases give the same result. Further, suppose that $B = B' \otimes B''$, write $C = B^{(G:H)}$ and define C', C'' similarly in terms of B', B''. Then by the associativity and commutativity of the tensor product, $C = C' \otimes C''$, hence $C^G = C'^G \otimes C''^G$ and this shows the corestriction to be a homomorphism. ∎

For our next result we note that if K is any commutative ring and E is a commutative K-algebra, then for any K-modules U, V, writing again $U_E = U \otimes_K E$ etc., we have the K-module isomorphism

$$U_E \otimes_E V_E \cong (U \otimes_K V)_E \tag{5.6.17}$$

given by the mapping $(u \otimes \alpha) \otimes (v \otimes \beta) \mapsto (u \otimes v) \otimes \alpha\beta$.

Proposition 5.6.7. *Let F/k be a separable extension of degree n and let A be a central simple k-algebra. Then*

$$\mathrm{cor}_{F/k}(A) \cong A^{\otimes n}. \tag{5.6.18}$$

Proof. Take a Galois extension E/k containing F, with group G and subgroup H corresponding to F. We have $A_F \otimes_F E \cong A \otimes_k E = A_E$ and $A_E^\sigma = A \otimes E^\sigma$, for any $\sigma \in G$, hence

$$(A_F)^{(G:H)} = \otimes_E (A_E)^{\sigma_i} \cong \otimes_E (A \otimes_k E^{\sigma_i}) \cong A^{\otimes n} \otimes (\otimes E^{\sigma_i}),$$

in terms of a transversal of H in G, as before. Taking fixed subrings and using (5.6.16), we obtain (5.6.18). ■

We remark that in the isomorphism with cohomology groups, (5.6.18) corresponds to the formula cor $\circ$ res $= n$.

Exercises

1. Let F/k be a finite Galois extension. Show that for any k-algebra A, tr $A_F = A$.
2. Given a field F of characteristic $p \neq 0$, an F-algebra B and a group G of automorphisms of B inducing distinct automorphisms of F, show that the order of G is prime to p.
3. Show that for a central simple F-algebra B, if $[F : k] = n$, then $[\text{cor}\, B : k] = [B : F]^n$.
4. Let A be a central simple k-algebra of degree p^r, where p is prime. Show that A is similar to a crossed product of degree p^s, for some s. (Hint. Take a Galois splitting field and a subfield corresponding to a Sylow p-subgroup.) What can be said about the relations of r and s?

5.7 Cyclic algebras

The simplest crossed products are those with cyclic groups; they are called *cyclic* algebras and can be brought to the following simple form. Let F/k be a cyclic Galois extension of degree n, with group G generated by σ. We shall choose an F-basis for our algebra A as follows: $u_0 = 1$, $u_i = u^i$ $(i = 1, \ldots, n-1)$, where u is an element of A inducing σ. Since u^i induces σ^i, we have indeed an F-basis; moreover, $u^n = \alpha \in F$ and since $u\alpha^\sigma = \alpha u = u^{n+1} = u\alpha$, we have $\alpha^\sigma = \alpha$, so $\alpha \in k$. Thus the multiplication in A is given by

$$u^i u^j = \begin{cases} u^{i+j} & \text{if } i+j < n, \\ \alpha u^{i+j-n} & \text{if } i+j \geq n. \end{cases} \tag{5.7.1}$$

The element u is called the *canonical generator* of A. We see that A is determined up to isomorphism by F/k, α and σ, and one also writes $A = (F/k, \sigma, \alpha)$.

Our first concern is to find when two presentations give isomorphic algebras:

Proposition 5.7.1. *Let F/k be a cyclic Galois extension of degree n. Two cyclic algebras $(F/k, \sigma, \alpha)$ and $(F/k, \sigma, \beta)$ are isomorphic if and only if $\beta/\alpha = N_{F/k}(c)$, where $c \in F$. In particular, $(F/k, \sigma, \alpha)$ splits precisely when $\alpha = N_{F/k}(c)$ $(c \in F)$.*

This condition is more briefly expressed by saying that α (resp. β/α) is a norm from F to k.

Proof. Assume that $(F/k, \sigma, \alpha) \cong (F/k, \sigma, \beta)$; then the canonical generators u, v are related by an equation $v = uc$, where $c \in F$. Hence $v^i = (uc)^i = u^i cc^\sigma \dots c^{\sigma^{i-1}}$; for $i = n$ we find $v^n = (uc)^n = u^n N(c)$, i.e. $N(c) = \beta/\alpha$. Conversely, if $\beta/\alpha = N(c)$, then the same calculation shows that $(uc)^n = \beta$, so that uc is a canonical generator for the isomorphic algebra $(F/k, \sigma, \beta)$. ■

For example, $(F/k, \sigma, 1) \cong k_n$ (by Proposition 5.5.3); this algebra can be realized as endomorphism ring of $F = k^n$, e.g. by taking a normal basis in F. Then σ acts by cyclic permutation of the coordinates.

The above presentation of a cyclic k-algebra A only provides a basis over F, but frequently one needs to have a basis over k. Such a basis takes a simple form if k contains a primitive n-th root of 1, say ω. Then a k-basis for A can be formed as follows. The splitting field F is of the form $F = k(v)$, where $v^n = \beta \in k$. Taking $u \in A$ such that $u^{-1}vu = \omega v$, we have $u^n = \alpha \in k$ and so A has the k-basis $u^i v^j$ $(i, j = 1, \dots, n-1)$ with the defining relations

$$u^n = \alpha, \ v^n = \beta, \ vu = \omega uv. \tag{5.7.2}$$

By a *symbol* $(\alpha, \beta; k)_n$ or $(\alpha, \beta)_n$ one understands a cyclic algebra over k with the presentation (5.7.2). We note the following consequence of (5.7.2):

$$(u + v)^n = \alpha + \beta. \tag{5.7.3}$$

For on expansion the left-hand side of (5.7.3) is a sum of products of degree n. Consider a product P involving i factors u and $n - i$ factors v; by moving the last factor to the first place we obtain $\omega^i P$, whether this factor was u or v. Now P occurs with all its cyclic conjugates and by what has been said, their sum is $(1 + \omega^i + \dots + \omega^{i(n-1)})P$ which is 0 except when $i = 0$ or n, so the sum reduces to $u^n + v^n$, which is $\alpha + \beta$, as claimed.

By applying Proposition 5.7.1 to a symbol, we obtain

Corollary 5.7.2. *Let k be a field containing a primitive n-th root of 1. Then a symbol $(\alpha, \beta; k)_n$ splits if and only if α is a norm from $k(\beta^{1/n})$ to k.* ■

There remains the case when char k divides the degree of the algebra. We shall only consider the case of a cyclic algebra A of prime degree p over a field k of characteristic p. A Galois splitting field F of degree p over k contains an element v such that $v^p - v = \beta \in k$ and u such that $u^{-1}vu = v + 1$. Hence $u^p = \alpha \in k$ and so A has the basis $u^i v^j$ $(i, j = 0, 1, \dots, p-1)$ with the defining relations

$$u^p = \alpha, \ v^p - v = \beta, \ vu = u(v + 1). \tag{5.7.4}$$

This algebra will be denoted by $(\alpha{\cdot}\beta; k|_p$.

We obtain a more symmetric form by putting $t = u^{-1}v$. Then $tu - ut = u^{-1}vu - v = v + 1 - v = 1$. To find t we note that in k,

$$\prod (x - i) = x^p - x;$$

further, $vu^{-1} = u^{-1}(v - 1)$, therefore $t^p = u^{-1}v \ldots u^{-1}v = u^{-p}(v - (p-1)) \ldots (v-1)v = u^{-p}(v^p - v) = \alpha\beta$. Thus A may be defined by t, u with the defining relations

$$t^p = \gamma, \quad u^p = \alpha, \quad tu - ut = 1,$$

where we have put $\gamma = \beta/\alpha$.

In Section 5.2 we saw that every central simple algebra is similar to a crossed product. It can actually be shown that when k has a primitive m-th root of 1, then every central division k-algebra of exponent m is similar to a tensor product of cyclic algebras of degree m. This is the content of the Merkuryev–Suslin [1986] theorem, whose proof is beyond the level of this book (see Rowen (1988) for an illuminating discussion and a proof of various special cases).

Exercises

1. Show that $(\alpha, \beta; k)_n$ splits if it contains an element x such that $1, x, \ldots, x^{n-1}$ are linearly independent and x^n is an n-th power in k.
2. Show that if $(F/k, \sigma, \alpha)$ is cyclic of degree n, then $x^n - \alpha$ is irreducible over k and D contains a maximal subfield generated by a root of $x^n = \alpha$.
3. Prove that $(F/k, \sigma, \alpha) \otimes (F/k, \sigma, \beta) \sim (F/k, \sigma, \alpha\beta)$.
4. Show that if $(F/k, \sigma, \alpha)$ has degree n, then for any r prime to n, $(F/k, \sigma^r, \alpha^r) \cong F/k, \sigma, \alpha)$.
5. Show that $(F/k, \sigma, \alpha)$ has exponent e, where e is the least number for which α^e is a norm from F.
6. (Wedderburn) Show that a cyclic algebra $(F/k, \sigma, \alpha)$ of degree n is a division algebra if α^n is the least power of α which is a norm.
7. Let $k \subseteq F \subseteq E$, where E/k is cyclic of degree n and $[E : F] = d$. Show that if Gal(E/k) is generated by σ and $\sigma|F = \bar{\sigma}$, then $(F/k, \bar{\sigma}, \alpha) \sim (F/k, \sigma, \alpha^d)$.
8. With the notation of Theorem 5.6.2, show that if E/k is cyclic, then $(E/k, \sigma, \alpha)_F \sim (EF/F, \sigma^r, \alpha)$, where $r = [E \cap F : k]$.
9. Let k be a field with a primitive n-th root of 1. Show that if $A = (F/k, \sigma, \alpha)$ is a cyclic division algebra of degree n^2, then A can also be represented as a crossed product with group $\mathbf{C}_n \times \mathbf{C}_n$.

Further exercises on Chapter 5

1. Let D be a skew field, α an endomorphism and δ a $(1, \alpha)$-derivation of D. Show that if D is finite-dimensional over its centre, then either α or δ must be inner.
2. Let D be a skew field with centre k but not algebraic over k. Show that $D \otimes k(t)$ is a simple Noetherian ring, not a skew field (and not a full matrix ring over a skew field).

3. (A. A. Albert) Let D be a skew field which is totally ordered (BA, Section 8.8), and suppose that D is algebraic over its centre k. Show that the conjugates of any positive element are again positive. Deduce that the sum of the conjugates of any non-zero element cannot be zero, and hence prove that D must be commutative. (Hint. Use Exercise 9 of Section 5.1.)
4. (Kharchenko) Let A be a central simple k-algebra of finite degree. By regarding A as a right A^e-module show that every k-linear mapping of A into itself has the form $f: x \mapsto \sum a_i x b_i$ $(a_i, b_i \in A)$. Deduce the existence of a non-constant central polynomial, i.e. a polynomial with values in k (see Section 7.7 below).
5. Let F/k be a field extension. Show that there is an exact sequence
$$0 \to \mathbf{B}(F/k) \to \mathbf{B}_k \xrightarrow{f} \mathbf{B}_F.$$
Identify coker f in case F/k is a Galois extension.
6. (J.-P. Serre) Suppose that in a central division k-algebra D of degree n every extension is p-radical. Show that $x^n \in k$ for all $x \in D$. By extension to a splitting field obtain a contradiction, and hence give another proof of Theorem 5.2.8.
7. Let A, B be crossed products with factor sets (a), (b) respectively. Given a Galois splitting field F for both A and B, write $F = k(\theta)$, put $P = A \otimes B$ and
$$e = \prod (\theta \otimes 1 - 1 \otimes \theta^\sigma)/(\theta - \theta^\sigma) \otimes 1,$$
where the product is taken over all $\sigma \neq 1$ in Gal(F/k). Verify that for the minimum polynomial f of θ over k, $f(\theta \otimes 1) = f(\theta) \otimes 1 = 0$ and $(c \otimes 1)e = (1 \otimes c)e$ for all $c \in F$. Hence show that e is idempotent and that ePe is a crossed product with factor set $(a)(b)$.
8. Let $F = k(\sqrt{c})$ (char $k \neq 2$). Show that there is a central division k-algebra with F as maximal subfield iff the form $x^2 + cy^2$ is not universal over k (i.e. it does not represent every $a \in k^\times$). Similarly in characteristic 2, if F is generated over k by a root of $x^2 + x + c = 0$, the same holds iff $x^2 + xy + cy^2$ is not universal.
9. Show that $\mathbf{SL}_2(\mathbf{F}_3)$ has order 24 but is not isomorphic to Sym$_4$. (Hint. Show that its derived group is the quaternion group. It is known as the *binary tetrahedral* group.)
10. The Hilbert norm residue symbol $(a, b)_p$ is defined to be 1 or -1 according as $(a, b; \mathbf{Q}_p)$ does or does not split, where $\mathbf{Q}_p$ is the p-adic field when p is prime and $\mathbf{R}$ when $p = \infty$. Using Proposition 5.4.3(e), show that for fixed b and p the a with $(a, b)_p = 1$ form a group under multiplication. (Note that the law of quadratic reciprocity, BA, Chapter 7, Further Exercise 23, may be expressed as $\prod (a, b)_p = 1$, where the product is taken over all primes and over $p = \infty$.)
11. Let k be a field with a primitive n-th root of 1 and $F \supseteq k$. Show that for any $\alpha \in k$, $\beta \in F$, $\text{cor}_{F/k}(\alpha, \beta; F)_n \sim (\alpha, N_{F/k}(\beta); k)_n$.
12. Let A be a central simple k-algebra of index $p^\alpha m$, where p is a prime not dividing m and $\alpha \geq 1$. Show that there is a separable extension F of degree prime to p over k such that A_F has index p^α.
13. Show that the Brauer classes split by a cyclic extension F/k of degree n form a group $H^2(\mathbf{C}_n, F^\times) \cong k^\times/N_{F/k}(F^\times)$.

14. Show that if x, y are regular elements over a field of prime characteristic p, satisfying $xy - yx = 1$, then $(xy)^p = x^p y^p + xy$; deduce further that $(xy)^{p-1} = y^{p-1}x^{p-1} + 1$.
15. Let D be a central division k-algebra of prime degree p. If D has a maximal subfield E not its own normalizer in $D^\times$, show that E/k is Galois and deduce that D is a crossed product.
16. (L. E. Dickson) Let D be a central division k-algebra of degree 3 and suppose $u_1 \in D \backslash k$ has the minimal polynomial $(x - u_1)(x - u_2)(x - u_3)$. Find $v \in D$ such that $u_i v = v u_{i+1}$ (i mod 3) and show that either $k(v)$ or $k(u_1 v)$ is not its own normalizer. Deduce that D is a crossed product. (Hint. Try a quadratic polynomial in the u's for v.)
17. Show that every central division algebra of degree 6 is cyclic.

6 Representation theory of finite groups

Although much of the theory of finite-dimensional algebras had its origins in the theory of group representations, it seems simpler nowadays to develop the theory of algebras first and then use it to give an account of group representations. This theory has been a powerful tool in the study of groups, especially the modular theory (representations over a field of finite characteristic), which has played a key role in the classification of finite simple groups. The theory also has important applications to physics: quantum mechanics describes physical systems by means of states which are represented by vectors in Hilbert space (infinite-dimensional complete unitary space). Any group which may act on the system, such as the rotation group or a permutation group of the constituent particles, acts by unitary transformations on this Hilbert space and any finite-dimensional subspace admitting the group leads to a representation of the group. If we know the irreducible representations of our group, this will often allow us to classify these spaces

Of course an introductory chapter like the present one is not the place to develop modular representations, nor the applications to physics. The plan of the chapter is as follows. The first four sections give a concise account of the theory based on the Wedderburn theorems (Chapter 5 of BA), including the basic results on orthogonality and completeness (Section 6.3) and in Section 6.4 we explain the role of characters. Some simplifications can be made over the complex numbers and they are described in Section 6.5. The rest of the chapter deals with representations and characters of the symmetric group in Section 6.6, and in Section 6.7 describes induced representations, an important technique which is illustrated in Section 6.8 by the theorems of Burnside and Frobenius.

6.1 Basic definitions

Let G be any group (not necessarily finite). By a *representation* of G over a field k one understands a homomorphism

$$\rho : G \to \mathbf{GL}_d(k), \tag{6.1.1}$$

where $\mathbf{GL}_d(k)$ is the *general linear group* of *degree d* over k, i.e. the group of all invertible $d \times d$ matrices over k. Thus we have a mapping $x \mapsto \rho(x)$ such that

$$\rho(xy) = \rho(x)\rho(y) \quad \text{for all } x, y \in G. \tag{6.1.2}$$

Since each matrix $\rho(x)$ is invertible, we have $\rho(1) = I$, where 1 is the neutral element of G, and $\rho(x^{-1}) = \rho(x)^{-1}$. The integer d is called the *degree* of the representation.

For example, to find a representation of the cyclic group $\mathbf{C}_3 = \{1, t, t^2\}$ over $\mathbf{R}$ of degree 2, we need to find $A \in \mathbf{GL}_3(\mathbf{R})$ such that $A^3 = I$. We may take $A = \begin{pmatrix} 0 & -1 \\ 1 & -1 \end{pmatrix}$ and then ρ is defined by

$$\rho(t) = \begin{pmatrix} 0 & -1 \\ 1 & -1 \end{pmatrix}, \quad \rho(t^2) = \begin{pmatrix} -1 & 1 \\ -1 & 0 \end{pmatrix}, \quad \rho(1) = \begin{pmatrix} 1 & 0 \\ 0 & 1 \end{pmatrix}. \tag{6.1.3}$$

Every group has the *trivial representation*, obtained by mapping each element of G to I.

At the other extreme we have the *faithful* representations, defined as homomorphisms with trivial kernel; e.g. (6.1.3) is a faithful representation of $\mathbf{C}_3$.

Two representations ρ, σ of a group G are said to be *equivalent*, if they have the same degree, d say, and there exists $P \in \mathbf{GL}_d(k)$ such that

$$\sigma(x) = P^{-1}\rho(x)P \quad \text{for all } x \in G. \tag{6.1.4}$$

It is clear that this is indeed an equivalence relation on the set of all representations of G. For example, if ω is a primitive cube root of 1, then

$$\begin{pmatrix} 0 & -1 \\ 1 & -1 \end{pmatrix}\begin{pmatrix} 1 & 1 \\ -\omega & -\omega^2 \end{pmatrix} = \begin{pmatrix} 1 & 1 \\ -\omega & -\omega^2 \end{pmatrix}\begin{pmatrix} \omega & 0 \\ 0 & \omega^2 \end{pmatrix},$$

therefore the representation ρ of $\mathbf{C}_3$ given by (6.1.3) is equivalent to σ, where σ is given by

$$\sigma(t) = \begin{pmatrix} \omega & 0 \\ 0 & \omega^2 \end{pmatrix}, \quad \sigma(t^2) = \begin{pmatrix} \omega^2 & 0 \\ 0 & \omega \end{pmatrix}, \quad \sigma(1) = \begin{pmatrix} 1 & 0 \\ 0 & 1 \end{pmatrix}.$$

If we interpret the matrices of a representation of G as linear transformations of a vector space, we reach the notion of a G-module. A *G-module* is a vector space V over k such that each $x \in G$ defines a linear mapping $v \mapsto vx$ on V satisfying

$$v(xy) = (vx)y, \quad v1 = v, \quad \text{for all } v \in V, x, y \in G. \tag{6.1.5}$$

Given two G-modules U and V, a *homomorphism* or *G-homomorphism* from U to V is a k-linear mapping $f : U \to V$ such that

$$f(ux) = (fu)x \quad \text{for all } u \in U, x \in G. \tag{6.1.6}$$

If a homomorphism from U to V is bijective, its inverse is easily seen to be a homomorphism from V to U; this is called an *isomorphism* or a *G-isomorphism*. We then say that U and V are isomorphic and write $U \cong V$

To establish the link between representations of G and G-modules, let us take a finite-dimensional G-module V, with basis $v_1, \ldots, v_d$ over k. The action of G on V is completely described by the equations

$$v_i x = \sum_j \rho_{ij}(x) v_j \quad (x \in G), \tag{6.1.7}$$

where $\rho_{ij}(x) \in k$, and it is easily checked that the matrices $\rho(x) = (\rho_{ij}(x))$ form a representation of G; we shall say that the G-module V with the basis $v_1, \ldots, v_d$ *affords* the representation ρ. Conversely, given a representation $\rho = (\rho_{ij})$ of G of degree d and a d-dimensional vector space V over k, we can turn V into a G-module by defining the action of $x \in G$ on a basis $v_1, \ldots, v_d$ by (6.1.7) and generally putting

$$\left(\sum \alpha_i v_i\right) x = \sum \alpha_i \rho_{ij}(x) v_j.$$

The verification that this provides a G-module is straightforward and may be left to the reader.

To see that the operations of passing between representations and modules are mutually inverse we need to examine the effect of a change of basis on the representation. Let V be a G-module affording the representation ρ relative to a basis $v_1, \ldots, v_d$. Thus the equations (6.1.7) hold, which may be written concisely as

$$vx = \rho(x) v, \tag{6.1.8}$$

where $v = (v_1, \ldots, v_d)^T$ stands for the column of basis vectors $v_1, \ldots, v_d$. Suppose that $u = (u_1, \ldots, u_d)^T$ is a second basis of V, affording the representation σ, so that

$$ux = \sigma(x) u. \tag{6.1.9}$$

If the matrix of transformation from u to v is denoted by P, we have

$$v = Pu, \quad u = P^{-1} v. \tag{6.1.10}$$

Hence $\sigma(x)u = ux = (P^{-1}v)x = P^{-1}(vx) = P^{-1}\rho(x)v = P^{-1}\rho(x)Pu$. It follows that

$$\sigma(x) = P^{-1}\rho(x)P; \tag{6.1.11}$$

thus ρ and σ are equivalent, and what we have shown is that different bases of a G-module afford equivalent representations. Moreover, since P may be any invertible matrix, we see that representations of G that are equivalent are afforded by the same G-module, for suitable bases. Further, if two G-modules afford the same representation, they must be isomorphic. For take the modules to be V, W with bases $v = (v_1, \ldots, v_d)^T$, $w = (w_1, \ldots, w_d)^T$ and let

$$vx = \rho(x) v, \quad wx = \rho(x) w.$$

Then the mapping $\sum \alpha_i v_i \mapsto \sum \alpha_i w_i$ is easily verified to be a G-isomorphism between V and W. It follows that isomorphic modules afford equivalent representations, for by changing the basis in one of the modules we can make the representations equal. Thus we have proved

Proposition 6.1.1. *For any group G there is a natural bijection between the sets of equivalence classes of representations and isomorphism classes of G-modules.* ■

By this result we can use G-modules and representations of G interchangeably; we shall examine various concepts from both points of view, but first we must clarify the connexion with modules over a ring, discussed in Chapter 4 of BA. In order to do so we shall recall the notion of a group algebra. For any group G and any field k we can form the vector space over k on G as basis, denoted by kG. Its elements have the form $\sum \alpha_x x$, where the summation is over all $x \in G$, and if G is infinite, almost all the α_x are zero. Using the multiplication in G and distributivity, we thus obtain an algebra kG which is known as the *group algebra* of G.

We can form modules over kG, as for any ring, and it is clear that a kG-module is also a G-module. Conversely, a G-module V becomes a kG-module by the rule

$$v\left(\sum \alpha_x x\right) = \sum \alpha_x vx, \quad (v \in V),$$

where the summation is over all $x \in G$. In particular, kG itself may be regarded as a right kG-module; this is the *regular representation*, which provides a faithful representation of G, since $xg = x$ for all $x \in kG$ implies that $1 = 1g = g$.

Given a G-module V, we can define a submodule as for modules over a ring, as a subspace V' of V admitting the G-action, i.e. such that $vx \in V'$ for $v \in V'$, $x \in G$. Alternatively we may regard V as kG-module and look for its kG-submodules; clearly they are just the G-submodules of V. Likewise the homomorphisms between G-modules are nothing other than the module homomorphisms between kG-modules.

Let us now examine the form taken by the representations corresponding to submodules. We consider a G-module V with a submodule V'. If $v_1, \ldots, v_d$ is a basis of V, adapted to V', say $v_1, \ldots, v_t$ $(t \leq d)$ is a basis of V' and the action is given by (6.1.8), then for $i \leq t$ we have $v_i x \in V'$ and so $\rho_{ij}(x) = 0$ for $i \leq t < j$. Thus ρ has the form

$$\rho(x) = \begin{pmatrix} \rho'(x) & 0 \\ \theta(x) & \rho''(x) \end{pmatrix}. \tag{6.1.12}$$

We note that $\rho'(x)$ is a representation afforded by V', while ρ'' is afforded by the quotient module V/V' relative to the basis $\bar{v}_{t+1}, \ldots \bar{v}_d$, where $\bar{u}$ denotes the residue class of u. Both ρ' and ρ'' are sometimes called *subrepresentations* of ρ.

We note that if the basis of V is chosen so that the *last* t members form a basis of V' (instead of the first t), then ρ takes the form

$$\rho(x) = \begin{pmatrix} \rho'(x) & \theta(x) \\ 0 & \rho''(x) \end{pmatrix}.$$

A representation is said to be reducible if it is equivalent to a representation of the form (6.1.12), where $0 < t < d$. Thus ρ is reducible iff the corresponding G-module V has a non-zero proper submodule, i.e. it is not simple. In the contrary case, when V is simple, the representation is called *irreducible*.

If ρ can be written in the form of a diagonal sum:

$$\rho(x) = \begin{pmatrix} \rho'(x) & 0 \\ 0 & \rho''(x) \end{pmatrix},$$

it is said to be *completely reduced.* Clearly this corresponds to V being directly decomposable. We observe that any finite-dimensional G-module V has a composition series:

$$V = V_0 \supset V_1 \supset V_2 \supset \ldots \supset V_r = 0,$$

such that V_{i-1}/V_i is simple. The corresponding representation can then be taken in the form

$$\rho(x) = \begin{pmatrix} \rho_1(x) & & & 0 \\ & \rho_2(x) & & \\ & & \cdots & \\ * & & & \rho_r(x) \end{pmatrix}. \tag{6.1.13}$$

If ρ is *completely reducible,* i.e. we can find an equivalent representation of the form (6.1.13) in which $* = 0$, this means that the corresponding G-module is a direct sum of simple modules, i.e. *semisimple.*

Just as for modules over a ring we can define left G-modules; they are vector spaces V with a G-action $v \mapsto xv$ such that

$$x(yv) = (xy)v, \quad 1v = v.$$

However, any such left G-module may be regarded as a right G-module by defining $v.x = x^{-1}v\ (x \in G)$. For we have $v.(xy) = (xy)^{-1}v = (y^{-1}x^{-1})v = y^{-1}(x^{-1}v) = y^{-1}(v.x) = (v.x).y$.

In terms of the group algebra this may be expressed as follows: the group algebra kG has an anti-automorphism $\blacklozenge$, i.e. a linear mapping satisfying $(ab)^{\blacklozenge} = b^{\blacklozenge}a^{\blacklozenge}$, given by

$$\left(\sum \alpha_x x\right)^{\blacklozenge} = \sum \alpha_x x^{-1}. \tag{6.1.14}$$

Now any left kG-module becomes a right kG-module on putting $v.a = a^{\blacklozenge}v$.

We remark that the mapping defined by (6.1.14) has the property $a^{\blacklozenge\blacklozenge} = a$. An antiautomorphism of order two is called an *involution,* thus kG is an algebra with an involution.

Exercises

1. Let G be a group and consider the regular representation of kG (defined by right multiplication). Show that this representation always has the trivial representation $x \mapsto 1$ as a subrepresentation.
2. Let F be a field containing a primitive n-th root of 1, say ω (hence of characteristic 0 or prime to n) and let $\mathbf{C}_n$ be the cyclic group of order n, with generator t. Show that $\rho_k : t^r \mapsto \omega^{kr}$ is a representation of degree 1 of $\mathbf{C}_n$ for $k = 0, 1, \ldots, n-1$.

Show that if ρ is any representation of $\mathbf{C}_n$ over F, then $\rho(t)$ can be transformed to diagonal form. Deduce that ρ can be written as a diagonal sum of the ρ_k.

3. Show that if ρ is any representation of a group G, then $\rho^\blacklozenge$ defined as $\rho^\blacklozenge(x) = \rho(x^{-1})^T$ is again a representation of G (it is called the *contragredient* of G). What are the conditions for $\rho^\blacklozenge$ to coincide with ρ?
4. Let G be a group and ρ, δ representations of G, where δ is of degree 1. Show that $\delta\rho$ is again a representation of G, and this is irreducible iff ρ is.
5. Let ρ be an irreducible representation of a (finite) p-group G, acting on a vector space V over a field k of characteristic p. Show that V contains a vector $v \neq 0$ which is fixed under the action of G. Show that the matrices $\rho(x)$ $(x \in G)$ have a common eigenvector and deduce that ρ must be the trivial representation. (Hint. Use the fact that $\rho(x) - 1$ is nilpotent; a matrix $\rho(x)$ with this property is called *unipotent*.)
6. Let G be a p-group and k a field of characteristic p. Suppose that G acts transitively on a finite set S and let V be a k-space on S as basis, with the G-action defined by the permutations of S. Show that V has a unique maximal submodule; deduce that V is indecomposable, but not simple, unless it is one-dimensional.
7. Let ρ be an irreducible representation of degree d of a finite group G over a field of characteristic p. Show that any normal subgroup of index p^d lies in the kernel of ρ. Deduce that if ρ is faithful, then G has no non-trivial normal p-subgroup.

6.2 The averaging lemma and Maschke's theorem

For a closer study of representations we need to assume that our group is finite and we shall make this assumption from now on. The first important fact to note is that for a finite group every representation over a field of characteristic 0 is completely reducible.

Theorem 6.2.1 (Maschke's theorem, 1899). *Let G be a finite group and k a field of characteristic 0 or prime to the order of G. Then every representation of G over k is completely reducible.*

Proof. Let ρ be a representation of G and suppose that ρ is reduced:

$$\rho(x) = \begin{pmatrix} \rho'(x) & 0 \\ \theta(x) & \rho''(x) \end{pmatrix}, \tag{6.2.1}$$

where ρ', ρ'' are subrepresentations of degrees d', d'' respectively. To establish complete reducibility it will be enough to find a $d'' \times d'$ matrix μ such that

$$\begin{pmatrix} \rho'(x) & 0 \\ \theta(x) & \rho''(x) \end{pmatrix} \begin{pmatrix} I & 0 \\ \mu & I \end{pmatrix} = \begin{pmatrix} I & 0 \\ \mu & I \end{pmatrix} \begin{pmatrix} \rho'(x) & 0 \\ 0 & \rho''(x) \end{pmatrix}.$$

When we multiply out, only the (2, 1)-block gives anything new:

$$\theta(x) = \mu\rho'(x) - \rho''(x)\mu, \tag{6.2.2}$$

and we shall complete the proof by finding a matrix μ to satisfy this equation. By substituting from (6.2.1) in the relation $\rho(xy) = \rho(x)\rho(y)$ we obtain the following equation for $\theta(x)$:

$$\theta(xy) = \theta(x)\rho'(y) + \rho''(x)\theta(y). \tag{6.2.3}$$

Writing $|G| = m$ and noting that $m \neq 0$ in k, by hypothesis, we have

$$\begin{aligned} m\theta(x) &= \sum_y \theta(x)\rho'(y)\rho'(y^{-1}) \\ &= \sum_y [\theta(xy) - \rho''(x)\theta(y)]\rho'(y^{-1}). \end{aligned}$$

Put $z = xy$; then $y^{-1} = z^{-1}x$, and as y runs over G, so does z, for fixed x. Hence we can rewrite this last sum as

$$\sum_z \theta(z)\rho'(z^{-1}x) - \sum_y \rho''(x)\theta(y)\rho'(y^{-1}),$$

so

$$m\theta(x) = \sum_z \theta(z)\rho'(z^{-1})\rho'(x) - \sum_y \rho''(x)\theta(y)\rho'(y^{-1}),$$

and this has the form (6.2.2), if we abbreviate $m^{-1}\sum_y \theta(y)\rho'(y^{-1})$ as μ. ■

In view of its importance we shall give a second proof of this result, or rather, restate the same proof in module terms. The essential step is a lemma which is also used elsewhere, but first we shall need to introduce some notation. If U, V are G-modules over k and α is a mapping from U to V, we shall write $\alpha : U \to {}_kV$, $\alpha : U \to {}_GV$ to indicate that α is k-linear or a G-homomorphism respectively. The space of all k-linear mappings from U to V is denoted by $\mathrm{Hom}_k(U, V)$ and the subspace of G-homomorphisms by $\mathrm{Hom}_G(U, V)$.

In the next lemma we shall (exceptionally) write mappings between right G-modules on the right, so that for $\alpha : U \to {}_kV$ the condition for a G-homomorphism is that

$$(ux)\alpha = (u\alpha)x \quad \text{for all } u \in U,\ x \in G.$$

Lemma 6.2.2 (Averaging lemma). *Let G be a finite group and k a field of characteristic 0 or prime to $|G|$. Given any two G-modules U, V and $\alpha : U \to {}_kV$, the mapping*

$$\alpha^* : u \mapsto |G|^{-1}\sum_x ((ux^{-1})\alpha)x \tag{6.2.4}$$

is a G-homomorphism from U to V. Moreover,

(i) *if α is a G-homomorphism, then $\alpha^* = \alpha$,*
(ii) *if $\alpha : U \to {}_kV$, $\beta : V \to {}_GW$, then $(\alpha\beta)^* = \alpha^*\beta$,*
(iii) *if $\alpha : U \to {}_GV$, $\beta : V \to {}_kW$, then $(\alpha\beta)^* = \alpha\beta^*$.*

Proof. Let us fix $a \in G$ and write $y = xa$, $x = ya^{-1}$. Then as one of x, y runs over G, so does the other. Now for $\alpha : U \to {}_kV$ we have

$$|G|.u\alpha^* a = \sum_x ux^{-1}\alpha xa = \sum_y uay^{-1}\alpha y = |G|.ua\alpha^*. \tag{6.2.5}$$

This shows α^* to be a G-homomorphism. If α is a G-homomorphism, each term in the sum in (6.2.5) is $u\alpha a = ua\alpha$, so $\alpha^* = \alpha$ in this case and (i) follows. Now let $\beta : V \to {}_GW$; then

$$|G|.u(\alpha\beta)^* = \sum_x ux^{-1}\alpha\beta x = \sum_x ux^{-1}\alpha x\beta = |G|.u\alpha^*\beta.$$

Hence (ii) follows; (iii) is proved similarly. ∎

We note that if neither α nor β is a G-homomorphism, there is nothing we can say. We can now prove the module form of Maschke's theorem, which states that every module extension splits, or equivalently, that the group algebra kG is semisimple.

Theorem 6.2.3 (Maschke's theorem, form 2). *Let G be a finite group and k a field of characteristic 0 or prime to |G|. Then kG is semisimple.*

Proof. We shall show that every (finite-dimensional) G-module is semisimple, or equivalently, that every short exact sequence of G-modules

$$0 \to V' \xrightarrow{\alpha} V \xrightarrow{\beta} V'' \to 0. \tag{6.2.6}$$

splits. Such a sequence certainly splits as a sequence of k-spaces, for this just means that V' as k-subspace of V has a vector space complement. Thus we have a k-linear splitting map $\gamma : V \to V'$. We have $\alpha\gamma = 1_{V'}$; therefore $1 = 1^* = (\alpha\gamma)^* = \alpha\gamma^*$, and so γ^* is the desired G-homomorphism splitting the sequence (6.2.6). ∎

Exercises

1. Let G be a finite group and V a finite-dimensional G-module over a field of characteristic prime to $|G|$. Show that if G acts trivially on every simple composition factor of V then the G-action on V is trivial.
2. Show that for any (finite-dimensional) left G-modules U, V, $\mathrm{Hom}_k(U, k) \otimes_k V \cong \mathrm{Hom}_k(U, V)$.
3. For $G = \mathbf{C}_p = \mathrm{gp}\{t|t^p = 1\}$ and $k = \mathbf{F}_p$ define a two-dimensional space V with basis v_1, v_2 as G-module by $v_1t = v_1 + v_2$, $v_2t = v_2$. Verify that V is not semisimple; calculate the corresponding representation.
4. Show that the infinite cyclic group has, over a field of characteristic 0, a faithful two-dimensional representation which is not completely reducible.
5. Let G be a finite group and k a field of characteristic dividing $|G|$. Show that the element $z = \sum_x x$ is central and nilpotent in kG. Deduce that kG is not semisimple.

6. Let k be a field of characteristic p and G a finite group. Show that for any element g of p-power order, $g - 1$ is nilpotent in kG. Deduce that for a finite p-group G the radical of kG is the augmentation ideal. (Hint. Find a basis of nilpotent elements for the radical and use Theorem 5.5.4.) Deduce further that kG is completely primary.

6.3 Orthogonality and completeness

The representation theory of finite groups was developed by Georg Frobenius in the 1880s and 1890s, using the determinant of the matrix of a general group element in the regular representation. Issai Schur in his dissertation in 1901 greatly simplified the theory by using his lemma, in the form given below, and the averaging lemma, Lemma 6.2.2.

Lemma 6.3.1 (Schur's lemma). *Let R be any ring and U, V two simple R-modules. Then*

(i) $\mathrm{Hom}_R(U, V) = 0$ *unless* $U \cong V$,
(ii) $\mathrm{End}_R(U)$ *is a skew field.*

Proof. (i) If $f : U \to V$ is a non-zero homomorphism, then $\ker f$ is a proper submodule of U, and hence is 0, while $\mathrm{im}\, f$ is a non-zero submodule of V and so is equal to V. Thus f is an isomorphism, as claimed.

(ii) When $V = U$, this argument shows that every non-zero endomorphism of U is an automorphism, and (ii) follows. ∎

When the ground field k is algebraically closed, every matrix over k has an eigenvalue, so for each automorphism f of U there exists $\lambda \in k$ such that $f - \lambda.1$ is non-invertible, and hence zero, i.e. $f = \lambda.1$. This proves the following sharper form of Lemma 6.3.1:

Lemma 6.3.2 (Schur's lemma for algebraically closed fields). *Let k be an algebraically closed field, A a k-algebra and U, V any two simple A-modules, finite-dimensional over k. Then*

$$\mathrm{Hom}_A(U, V) = \begin{cases} k & \text{if } U \cong V \\ 0 & \text{otherwise.} \end{cases} \qquad ∎ \qquad (6.3.1)$$

Let G be a finite group; we shall define an inner product on the group algebra kG by the rule

$$\left(\sum a(x)x,\ \sum b(y)y\right) = |G|^{-1} \sum a(x^{-1})b(x). \qquad (6.3.2)$$

It is clear that this product is bilinear; it is not symmetric, but satisfies the equation

$$(g, f) = (f^{\blacklozenge}, g^{\blacklozenge}), \qquad (6.3.3)$$

where ◆ is the involution defined by (6.1.14). The product is regular, i.e. non-singular: if $(f, x) = 0$ for all $x \in G$, then $f(x^{-1}) = 0$ for all $x \in G$ and so $f = 0$. Of course from the point of view of the inner product (6.3.2) the multiplication on kG is immaterial, and kG may be thought of as the space of all k-valued functions on G.

Our next aim is to show that the different representation coefficients, regarded as functions on G, are orthogonal.

Theorem 6.3.3 (Orthogonality relations for representations). *Let G be a finite group and k an algebraically closed field of characteristic 0 or prime to $|G|$. If ρ, σ are irreducible representations of degrees c, d respectively, then we have the relations*

$$\frac{1}{|G|}\sum_{x\in G}\rho_{ij}(x^{-1})\sigma_{pq}(x) = \begin{cases} 0 & \text{if } \rho \text{ and } \sigma \text{ are inequivalent,} \\ \frac{1}{d}\delta_{jp}\delta_{iq} & \text{if } \rho = \sigma. \end{cases} \tag{6.3.4}$$

Thus different representation coefficients are orthogonal. We note that the alternatives on the right of (6.3.4) are not exhaustive: the representations ρ, σ may be equivalent but distinct. In that case (6.3.4) will not apply (but of course we can use (6.3.4) even then, after transforming one of ρ, σ into the other).

Proof. Take spaces U, V affording ρ, σ with bases $u_1, \ldots, u_c, v_1, \ldots, v_d$ and let $\alpha_{jp} : U \to {}_kV$ be the linear mapping defined by

$$u_i\alpha_{jp} = \delta_{ij}v_p.$$

Explicitly the (i, q)-entry of the matrix for α_{jp} is

$$(\alpha_{jp})_{iq} = \delta_{ij}\delta_{pq}. \tag{6.3.5}$$

By Lemma 6.2.2, α^*_{jp} is a G-homomorphism from U to V, and its matrix is given by applying * to (6.3.5):

$$\begin{aligned} |G|.(\alpha^*_{jp})_{iq} &= \sum_{h,r,x}\rho_{ih}(x)\delta_{hj}\delta_{pr}\sigma_{rq}(x) \\ &= \sum_{x}\rho_{ij}(x^{-1})\sigma_{pq}(x). \end{aligned}$$

If ρ, σ are inequivalent, then $\alpha^*_{jp} = 0$ by Lemma 6.3.2, and this proves the first line of (6.3.4).

Next we take $\rho = \sigma$. By Lemma 6.3.2, $\alpha^*_{jp} = \lambda_{jp} \in k$, hence we have

$$\sum_{x}\rho_{ij}(x^{-1})\rho_{pq}(x) = |G|.\lambda_{jp}\delta_{qi}. \tag{6.3.6}$$

To find λ_{jp} we put $q = i$ and sum over i:

$$|G|.d.\lambda_{jp} = \sum_{i,x}\rho_{pi}(x)\rho_{ij}(x^{-1}) = \sum_{x}\rho_{pj}(1) = |G|.\delta_{jp}.$$

By the hypothesis on k we can divide by $|G|$, hence $d \neq 0$ in k and $\lambda_{jp} = d^{-1}\delta_{jp}$. Inserting this value in (6.3.6) we obtain the second line of (6.3.4). ∎

To illustrate this result, let us take the trivial representation for σ. Then $\sigma(x) = 1$ for all $x \in G$ and we find that every non-trivial irreducible representation ρ of G satisfies

$$\sum \rho_{ij}(x) = 0 \quad \text{for all } i, j. \tag{6.3.7}$$

More generally, if ρ is any representation, we can take it to be completely reduced, and we then see that (6.3.7) holds precisely when ρ does not contain the trivial representation.

In terms of the inner product (6.3.2) the relation (6.3.4) may be written

$$(\rho_{ij}, \sigma_{pq}) = (1/d)\delta_{jp}\delta_{iq} \text{ or } 0;$$

this shows the ρ_{ij} to be a linearly independent set of functions; in particular their number cannot exceed $\dim kG = |G|$. Hence we have

Corollary 6.3.4. *The coefficients of inequivalent irreducible representations of a finite group G are linearly independent; hence the number of such representations is finite, and if their degrees are $d_1, \dots, d_t$ then*

$$\sum_1^t d_i^2 \leq |G|. \qquad \blacksquare \tag{6.3.8}$$

Our next task is to show that equality holds in (6.3.8) if we take enough representations. This means that every k-valued function on G can be written as a linear combination of irreducible representation coefficients; this is expressed by saying that these coefficients form a *complete system* of functions on G.

To see this, let us go back to the group algebra kG. We have seen that this is semisimple, hence a direct product of full matrix rings over skew fields. But as we saw, the only skew field finite-dimensional over k is k itself, because k is algebraically closed. Thus kG is a direct product of full matrix rings over k:

$$kG \cong \prod_{i=1}^s \mathfrak{M}_{d_i}(k). \tag{6.3.9}$$

Here each factor provides an irreducible representation of G, of degree d_i, and these representations are inequivalent because the product is direct, so the coefficients corresponding to different factors are linearly independent. On counting dimensions in (6.3.9) we obtain the desired equality (first obtained by Frobenius in 1896):

$$\sum_1^s d_i^2 = |G|. \tag{6.3.10}$$

Moreover, a comparison with (6.3.8) shows that the set of representations provided by (6.3.9) is complete, so $s = t$. Of course we can also take the regular representation of G, i.e. we take kG as G-module under right multiplication by G. Each irreducible representation ρ_i occurs d_i times, representing the d_i rows of the corresponding matrix. Thus we again obtain the equation (6.3.10).

Let G be a group and U, V any G-modules. Then $U \otimes V$ may be defined as a G-module by the equation

$$(u \otimes v)g = ug \otimes vg.$$

Since the right-hand side is bilinear in u and v, this defines an action and $U \otimes V$ is easily verified to be a G-module. If the representations afforded by U, V are ρ, σ relative to bases $u_1, \ldots, u_m, v_1, \ldots, v_n$ respectively, then

$$(u_i \otimes v_p)g = \sum \rho_{ij}(g)\sigma_{pq}(g)u_j \otimes v_q;$$

hence the representation of G afforded by $U \otimes V$ is the tensor product of the matrices: $\rho \otimes \sigma$. When G is finite, then by Maschke's theorem, each representation is a diagonal sum of irreducible ones, and if $\rho_1, \ldots, \rho_r$ are all the inequivalent irreducible representations of G, it is enough to determine the products $\rho_i \otimes \rho_j$. We have

$$\rho_i \otimes \rho_j = \sum_k g_{ijk}\rho_k, \qquad (6.3.11)$$

where the g_{ijk} are non-negative integers, indicating how often a given representation ρ_k occurs in the tensor product.

For each representation ρ of G we define its *kernel* as

$$K = \{x \in G | \rho(x) = 1\}.$$

Thus K is a normal subgroup of G and G has a faithful irreducible representation iff some irreducible representation of G has a trivial kernel. This need not be the case, but at any rate we have

Theorem 6.3.5. *For any finite group G the intersection of the kernels of all the irreducible representations over a field of characteristic zero or prime to $|G|$ is trivial.*

Proof. The regular representation of G is faithful; since it is a direct sum of irreducible representations, the conclusion follows. ■

Exercises

1. Find all irreducible representations of Sym_3 by reducing the regular representation.
2. Show that for any representation ρ of a finite group G the set $N = \{x \in G | \det \rho(x) = 1\}$ is a normal subgroup of G with cyclic quotient.
3. Let G be a finite group, k an algebraically closed field of characteristic 0 and ρ, σ inequivalent irreducible representations of G of degrees c, d. Show that for any $c \times d$ matrix T we have $\sum_x \rho(x^{-1})T\sigma(x) = 0$. Further show that for a $d \times d$ matrix P we have $\sum_x \sigma(x^{-1})P\sigma(x) = d^{-1}.|G|.\mathrm{Tr}(P).I$.
4. Show that if $\rho_1, \ldots, \rho_r$ are irreducible pairwise inequivalent representations of a group and $\rho = \oplus c_i\rho_i$, then the centralizer of ρ has dimension $\sum c_i^2$. Use this fact to obtain another proof of (6.3.10).

5. Let G be a finite group and let d be the degree of an irreducible representation of G over $\mathbf{Z}$, i.e. a homomorphism $G \to \mathbf{GL}_d(\mathbf{Z})$. Show that every prime dividing d must also divide $|G|$.
6. Show that the only matrix of order 2 in $\mathbf{SL}_2(\mathbf{C})$ is $-I$. Show that for any integers $k, m, n > 1$ there is a group $\mathrm{gp}\{a, b|(ab)^k = a^m = b^n = 1\}$. (Hint. Find a faithful representation ρ in $\mathbf{PSL}_2(\mathbf{C})$ with $\mathrm{tr}\ \rho(a) = \lambda + \lambda^{-1}$, $\mathrm{tr}\ \rho(b) = \mu + \mu^{-1}$, $\mathrm{tr}\ \rho(ab) = \nu + \nu^{-1}$, where λ, μ, ν are primitive $2m$-th, $2n$-th, $2k$-th roots of 1 respectively, and take $\rho(a)$ upper and $\rho(b)$ lower triangular.)
7. Show that $\varphi(g) = |\{(x, y) \in G \times G | g = x^{-1}y^{-1}xy\}$ is a character, where (x, y) is the form defined by (6.3.2), and that $\varphi = \sum |G|\chi_i/d_i$, where the χ_i, d_i are as in Corollary 6.3.4.
8. Prove the converse of Schur's lemma: If every endomorphism of a G-module V (over $\mathbf{C}$) is scalar, then V is simple.
9. Show that each irreducible representation is contained in the regular representation. Deduce that the number of isomorphism types of irreducible representations of a finite group is finite.

6.4 Characters

A one-dimensional representation is also called a *linear character* or simply a *character* if we are dealing with an abelian group. Such characters have already been discussed in section Section 4.9 of BA. We recall that an irreducible representation of an abelian group over $\mathbf{C}$ (an algebraically closed field) is necessarily one-dimensional, by Schur's lemma. For a non-abelian group there will always be irreducible representations of degrees greater than 1 (see Proposition 6.4.2 below) and the definition then runs as follows. Given any representation ρ of a group G over $\mathbf{C}$, its *character* is defined as

$$\chi(x) = \mathrm{tr}\ \rho(x), \quad x \in G, \tag{6.4.1}$$

where tr denotes the trace of the matrix $\rho(x)$; thus if $\rho(x) = (\rho_{ij}(x))$, then $\mathrm{tr}\ \rho(x) = \sum_i \rho_{ii}(x)$. When χ and ρ are related as in (6.4.1), ρ is said to afford the character χ. For example, any representation of degree 1 is its own character; in particular, the function $\chi_1(x) = 1$ for all $x \in G$ is the character afforded by the trivial representation,, and is called the *trivial* or also the *principal* character.

Some obvious properties of characters are collected in

Proposition 6.4.1. *The character of a representation is independent of the choice of basis, i.e. equivalent representations have the same character. Moreover, each character is a class function on G, i.e. it is constant on conjugacy classes:*

$$\chi(y^{-1}xy) = \chi(x), \quad x, y \in G. \tag{6.4.2}$$

The degree of χ is $\chi(1)$ and for any $x \in G$ of order n, $\chi(x)$ is a sum of n-th roots of 1.

If ρ_1, ρ_2 are any representations with the characters χ_1, χ_2 then the characters afforded by $\rho_1 \oplus \rho_2$ and $r_1 \otimes r_2$ are $\chi_1 + \chi_2$ and $\chi_1\chi_2$ respectively.

Proof. Let χ be the character of ρ; any equivalent representation has the form $T^{-1}\rho(x)T$ and since $\mathrm{tr}(BA) = \mathrm{tr}(AB)$ for any square matrices of the same size, we have

$$\mathrm{tr}(T^{-1}\rho(x)T) = \mathrm{tr}\ \rho(x),$$

so both ρ and $T^{-1}\rho T$ afford the same character. For the same reason, $\mathrm{tr}\ \rho(y^{-1}xy) = \mathrm{tr}(\rho(y)^{-1}\rho(x)\rho(y)) = \mathrm{tr}\ \rho(x)$, and (6.4.2) follows. $\chi(1)$ equals the degree because char $\mathbf{C} = 0$, while $A = \rho(x)$ satisfies $A^n = I$ if $x^n = 1$. Thus A satisfies an equation with distinct roots and so can be transformed to diagonal form over $\mathbf{C}$; its diagonal elements λ again satisfy $\lambda^n = 1$, so they are n-th roots of 1, and $\chi(x)$ is the sum of these diagonal elements.
The final assertion follows because $\mathrm{tr}(A \oplus B) = \mathrm{tr}\ A + \mathrm{tr}\ B$ and $\mathrm{tr}(A \otimes B) = \mathrm{tr}\ A . \mathrm{tr}\ B$. ■

The next result may be regarded as a generalization of the fact that a finite abelian group is isomorphic to its dual.

Proposition 6.4.2. *For any finite group G the number of linear characters is* $(G : G')$, *where* G' *is the derived group. Hence every non-abelian group has irreducible representations of degree greater than 1.*

Proof. Every homomorphism $\alpha : G \to \mathbf{C}^{\times}$ corresponds to a homomorphism from G/G' to $\mathbf{C}^{\times}$ and conversely. But we know from Theorem 4.9.1 of BA that the number of such homomorphisms is $|G/G'| = (G : G')$, so the result follows by (6.3.10). ■

In Section 6.3 we defined an inner product on kG; we shall now see how in the case of $k = \mathbf{C}$ we can define a hermitian inner product on $\mathbf{C}G$. Let us put

$$(f, g) = |G|^{-1} \sum_x \bar{f}(x) g(x). \tag{6.4.3}$$

Since every character α is a sum of roots of 1, we have $\bar{\alpha}(x) = \alpha(x^{-1})$. Hence for characters the formula (6.4.3) can also be written

$$(\alpha, \beta) = |G|^{-1} \sum_x \alpha(x^{-1})\beta(x); \tag{6.4.4}$$

so in this case it agrees with the inner product introduced in Section 6.3. From the orthogonality relations in Theorem 6.3.3 we obtain the following orthogonality relations for irreducible characters, by putting $j = i$, $q = p$ in (6.3.4) and summing over i and p:

$$(\chi, \psi) = \begin{cases} 1 & \text{if } \chi = \psi, \\ 0 & \text{otherwise.} \end{cases} \tag{6.4.5}$$

Thus under the metric (6.4.4) the irreducible characters form an orthonormal system. Suppose that all the irreducible representations of G are $\rho_1, \ldots, \rho_r$ with characters $\chi_1, \ldots, \chi_r$. Any representation of G is equivalent to a direct sum of irreducible ones, by complete reducibility (Theorem 6.2.1):

$$\rho = \nu_1\rho_1 \oplus \ldots \oplus \nu_r\rho_r.$$

Hence its character is $\chi = \nu_1\chi_1 + \ldots + \nu_r\chi_r$, and here ν_i is given by

$$\nu_i = (\chi, \chi_i),$$

using (6.4.5). This shows that any representation of G (over **C**) is determined up to equivalence by its character. For example, the regular representation $\mu(x) : a \mapsto ax$ has the form $\mu = \oplus d_i\rho_i$ and so we again find

$$|G| = \mu(1) = \sum d_i^2.$$

The inner product (6.4.4) can also be expressed directly in terms of the modules affording the representation:

Proposition 6.4.3. *Let G be a finite group and U, V any G-modules over an algebraically closed field k of characteristic 0, affording representations with characters α, β respectively. Then*

$$(\alpha, \beta) = \dim_k(\mathrm{Hom}_G(U, V)). \tag{6.4.6}$$

Proof. Suppose first that U, V are simple. Then by Lemma 6.3.2 the right-hand side of (6.4.6) is 1 or 0 according as U, V are or are not isomorphic, and this is just the value of the left, by (6.4.5). Now the general case follows because every G-module is a direct sum of simple modules. ■

The number on the right of (6.4.6) is also called the *intertwining number* of U and V.

Above we have found the character of the regular representation, and it is not difficult to obtain an explicit expression for it. Sometimes we shall want an expression for the character of the representation afforded by a given right ideal of the group algebra. Since the latter is semisimple, each right ideal is generated by an idempotent, and the next result expresses the character in terms of this idempotent.

Proposition 6.4.4. *Let G be a finite group, $A = kG$ its group algebra and $I = eA$ a right ideal in A, with idempotent generator $e = \sum e(x)x$. Then the character afforded by I is*

$$\chi(g) = \sum_x e(xg^{-1}x^{-1});$$

more generally, we have for any $b \in A$,

$$\chi(b) = \sum_{x,y} e(xy^{-1}x^{-1})b(y).$$

Proof. Consider the operation $\rho(b) : a \mapsto eab$, representing the projection on I followed by the right regular representation. We have

$$a.\rho(b) = eab = \sum e(uv^{-1}w^{-1})a(w)b(v).u. \tag{6.4.7}$$

Now $x.\rho(b) = \sum_y \rho_{xy}(b)y$, by expressing ρ in terms of the natural basis, hence on putting $a = x$, $u = y$ in (6.4.7), we find

$$\rho_{xy}(b) = \sum e(yv^{-1}x^{-1})b(v).$$

Therefore the character afforded by I is

$$\chi(b) = \text{tr}(\rho(b)) = \sum_x \rho_{xx}(b) = \sum_{x,v} e(xv^{-1}x^{-1})b(v). \qquad \blacksquare$$

So far we have regarded the characters as functions on G, but as we saw in Proposition 6.4.1, they are really class functions and we may regard them as functions on the set of all conjugacy classes of G. We shall now interpret the orthogonality relations in this way and in the process find that the number of irreducible characters of G is equal to the number of conjugacy classes.

Our first remark is that $a(x)\,(x \in G)$ is a class function iff the element $a = \sum a(x)x$ lies in the centre of the group algebra. For we have

$$y^{-1}ay = \sum_y a(x)y^{-1}xy = \sum_z a(yzy^{-1})z \quad \text{for all } y \in G;$$

hence

$$y^{-1}ay = a \quad \text{for all } y \in G \Leftrightarrow a(yzy^{-1}) = a(z) \quad \text{for all } y, z \in G.$$

Thus an element $a = \sum a(x)x$ lies in the centre of kG iff $a(x)$ is constant on conjugacy classes. This just means that we can write $a = \sum a_\lambda c_\lambda$, where c_λ is the sum of all elements in a given conjugacy class C_λ. It follows that these class sums form a basis for the centre of kG. This proves the first part of our next result:

Theorem 6.4.5 *Let G be a finite group and k an algebraically closed field of characteristic 0 or prime to $|G|$. An element $a = \sum a(x)x$ of kG lies in the centre if and only if $a(x)$ is a class function. Moreover, the class sums c_λ form a basis of the centre of kG and the irreducible characters over k form a basis for the class functions; thus if $\chi_1, \ldots, \chi_r$ are the different irreducible characters, then any class function α on G may be written in the form*

$$\alpha = \sum_i (\chi_i, \alpha)\chi_i. \tag{6.4.8}$$

Hence the number of irreducible characters equals the number of conjugacy classes of G.

To complete the proof we denote the number of irreducible characters by r and the number of conjugacy classes by s; as we have seen, s is the dimension of the centre of kG. Now kG is the direct product of r full matrix rings over k. Clearly each matrix ring has a one-dimensional centre and the centre of the direct product is easily seen to be the direct product of the centres. Hence the centre of kG is r-dimensional over k, and it follows that $r = s$. The characters are independent, by the orthogonality relation (6.4.5), hence they form a basis, which is orthonormal by (6.4.5), and we therefore have (6.4.8). ■

Let us consider the multiplication table for the basis $c_1, \ldots, c_r$ of the centre of kG. Any product of classes $C_\lambda C_\mu$ is a union of a number of classes, hence we have

$$c_\lambda c_\mu = \sum_\nu \gamma_{\lambda\mu\nu} c_\nu, \tag{6.4.9}$$

where the $\gamma_{\lambda\mu\nu}$ are non-negative integers. If ρ is any irreducible representation of G, then $\rho(c_\lambda) = \eta_\lambda I$ by Schur's lemma, where $\eta_\lambda \in k$. Let χ denote the character of ρ, d its degree and write $h_\lambda = |C_\lambda|$. Taking traces in the last equation and writing $\chi^{(\lambda)}$ for the value of χ on C_λ, we find

$$h_\lambda \chi^{(\lambda)} = \eta_\lambda d. \tag{6.4.10}$$

If we apply ρ to (6.4.9) we obtain $\eta_\lambda \eta_\mu = \sum \gamma_{\lambda\mu\nu}\eta_\nu$ and it follows that η_μ is a root of the equation

$$\det(xI - \Gamma_\mu) = 0, \quad \text{where } \Gamma_\mu = (\gamma_{\lambda\mu\nu}). \tag{6.4.11}$$

This shows η_μ to be an algebraic integer. Further, (6.4.10) shows that for each irreducible character χ, $h_\mu \chi^{(\mu)}/d$ is a root of (6.4.11); since (6.4.11) is of degree r, its roots are the values $h_\mu \chi_i^{(\mu)}/d_i$ for the different irreducible characters of G.

As a consequence of this development we can show that the degrees of the irreducible representations divide the group order. We recall from Section 9.4 of BA that the sum and product of algebraic integers are again integral.

Proposition 6.4.6 (Frobenius). *For any finite group G the degree of each irreducible representation over* **C** divides $|G|$.

Proof. Let χ be an irreducible character and d its degree. As a sum of roots of 1, χ is an algebraic integer, and so is $h_\lambda \chi^{(\lambda)}/d$, as we saw above. By the orthogonality relations (6.4.5) we have

$$\sum_\lambda \frac{h_\lambda \chi^{(\lambda)}}{d} \bar{\chi}^{(\lambda)} = \frac{|G|}{d},$$

and since sums and products of algebraic integers are integral, it follows that $|G|/d$ is integral, as we had to show. ■

We note that the relations (6.4.5) can be written $|G|^{-1}\sum_\lambda \overline{\chi}_i^{(\lambda)}\chi_j^{(\lambda)}h_\lambda = \delta_{ij}$, where $h_\lambda = |C_\lambda|$. This tells us that the $r \times r$ matrix $([h_\lambda/|G|]^{1/2}\chi_i^{(\lambda)})$ is unitary; hence so is its conjugate transpose and we have

$$|G|^1 \sum_i h_\lambda^{1/2} h_\mu^{1/2} \bar{\chi}_i^{(\lambda)} \chi_i^{(\mu)} = \delta_{\lambda\mu}.$$

When $\lambda \neq \mu$, we can omit h_λ, and so we obtain the second orthogonality relation for characters:

Proposition 6.4.7. *For any finite group G, if $\chi_i^{(\lambda)}$ is the value of the i-th irreducible character on the conjugacy class C_λ and $|C_\lambda| = h_\lambda$, then*

$$\sum_i \bar{\chi}_i^{(\lambda)} \chi_i^{(\mu)} = \begin{cases} |G|/h_\lambda & \text{if } \lambda = \mu, \\ 0 & \text{if } \lambda \neq \mu. \end{cases}$$ ■

The character of a representation may also be used to describe its kernel.

Proposition 6.4.8. *Let G be a finite group and ρ a representation of G over* **C** *with character χ. Then the kernel of r is determined by its character and is given by*

$$K_\rho = \{x \in G | \chi(x) = \chi(1)\}.$$

Proof. Denote the degree of ρ by d, so that $\chi(1) = d$. If $x \in G$ has order n, then x is represented by a matrix $\rho(x)$, whose eigenvalues are n-th roots of 1. Thus we have

$$\chi(x) = \omega_1 + \ldots + \omega_d, \quad \text{where } \omega_i^n = 1.$$

If $\chi(x) = \chi(1) = d$, it follows that

$$d = |\omega_1 + \ldots + \omega_d| \leq |\omega_1| + \ldots + |\omega_d| = d.$$

Hence equality must hold, which is possible only if $\omega_1 = \ldots = \omega_d$ and since $\sum \omega_i = d$, each ω_i must be 1. Since $\rho(x)$ satisfies an equation with distinct roots, it can be transformed to diagonal form. Hence $\rho(x) = I$ and so x is in the kernel of ρ. The converse is clear. ■

Combining this result with Theorem 6.3.5, we obtain

Corollary 6.4.9. *Let G be a finite group. Given $x \in G$, if $\chi(x) = \chi(1)$ for all irreducible characters χ of G over* **C**, *then $x = 1$.* ■

As a final result on characters we give a formula for characters of permutation modules. By a *permutation module* for a group G we understand a G-module V with a basis B, the *natural basis*, on which G acts by permutations. The character afforded by V, $\chi(x)$, is just the number of points of B fixed by $x \in G$. Suppose that G is transitive on B and let H be the stabilizer of a point $p \in B$. Then each point of B is fixed by $|H|$ elements, for its stabilizer is conjugate to H. Recalling the orbit formula from BA, Section 2.1: $|B| = (G : H)$, we therefore have

$$\sum_x \chi(x) = |B|.|H| = |G|.$$

For general permutation modules we can apply the result to each orbit and obtain

Proposition 6.4.10. *Let G be a finite permutation group acting on a set B and denote the character afforded by the corresponding permutation module by χ. Then*

$$\sum_x \chi(x) = n.|G|,$$

where n is the number of orbits of B under the action of G. ■

An elaboration of this result allows us to evaluate (χ, χ):

Theorem 6.4.11. *Let G be a transitive permutation group, denote by H the stabilizer of a point and let r be the number of double cosets in the decomposition $G = \cup Hs_iH$. If χ is the character afforded by the corresponding permutation module, then*

$$(\chi, \chi) = |G|^{-1} \sum \chi(x)^2 = r. \tag{6.4.12}$$

Proof. Let G act on B and consider the action of H on B. We may replace B by the coset decomposition with respect to $H : G = \cup Hx_\lambda$. Here each orbit of H corresponds to a double coset Hs_iH. Now take $a \in G$, say $a = hs_ih'$, where $h, h' \in H$. If $\chi(a) = t$, then a leaves t cosets Hx_λ fixed, i.e. $a \in x_\lambda^{-1}Hx_\lambda$ for t values of λ. Now the action of $x_\lambda^{-1}Hx_\lambda$ on B yields $r.|H|$ for the sum of its characters, by Proposition 6.4.10. There are $(G : H)$ such conjugates, so taking all these characters, we obtain $r.|H|.(G : H) = r.|G|$. The character sum includes each value $\chi(a)$, and if $\chi(a) = t$, this term occurs t times, so in all we obtain $\sum \chi(a)^2$, i.e. (6.4.12). ■

Examples

We end this section with some examples of representations and characters; throughout, k is algebraically closed of characteristic 0.

1. Let A be a finite abelian group. Then kA is a commutative semisimple algebra, hence a direct product of copies of k, and all its irreducible representations are of degree 1. We take a basis $a_1, \ldots, a_m$ of A (in multiplicative notation), where a_i has order n_i, and denote by ε_i any primitive n_i-th root of 1. Then for any integers $\nu_1, \ldots, \nu_m$, the mapping

$$a_1^{\alpha_1} \ldots a_m^{\alpha_m} \longmapsto \varepsilon_1^{\alpha_1\nu_1} \ldots \varepsilon_m^{\alpha_m\nu_m}$$

is a representation. We get distinct characters for n_i different values of ν_i, and the $n_1 \ldots n_m$ characters so obtained are all different and constitute all the irreducible characters of A. This corresponds to the fact that the dual of A, i.e. its group of characters, is isomorphic to A itself (see Theorem 4.9.1 of BA).

2. Consider $\mathbf{D}_m$, the dihedral group of order $2m$, with generators a, b and defining relations $a^m = 1$, $b^2 = 1$, $b^{-1}ab = a^{-1}$. Every element can be uniquely expressed in the form $a^\alpha b^\beta$, where $0 \le \alpha < m$, $0 \le \beta < 2$. It is easily verified that the conjugacy classes are for odd $m : \{a^r, a^{-r}\}$ $(r = 1, \ldots, (m-1)/2)$, $\{1\}$, $\{a^\alpha b\}$; and for even $m : \{a^r, a^{-r}\}$ $(r = 1, \ldots, m/2 - 1)$, $\{1\}$, $\{a^{m/2}\}$, $\{a^{2\alpha b}\}$, $\{a^{2a+1}b\}$. Further it may be checked that the index of the derived group in $\mathbf{D}_m$ is 2 when m is odd and 4 when m is even.

   To find the representations of $\mathbf{D}_m$ we have a homomorphism $\mathbf{D}_m \to \mathbf{C}_2$ obtained by mapping $a \mapsto 1$, which gives rise to two representations, the trivial representation and $a \mapsto 1$, $b \mapsto -1$. Further representations are obtained by taking a primitive m-th root of 1, say ω, and writing

$$a \mapsto \begin{pmatrix} \omega^i & 0 \\ 0 & \omega^{-i} \end{pmatrix}, b \mapsto \begin{pmatrix} 0 & 1 \\ 1 & 0 \end{pmatrix} \quad (i = 1, \ldots, [m/2]). \tag{6.4.13}$$

When m is odd, we thus obtain $(m-1)/2$ irreducible representations of degree 2 and two of degree 1, and this is a complete set, because the total number is $(m-1)/2 + 2$, which is the number of conjugacy classes. We also note the degree equation

$$\frac{1}{2}(m-1).2^2 + 2.1^2 = 2m.$$

When m is even, (6.4.13) becomes reducible for $i = m/2$ and we obtain two more representations of degree 1, $a \mapsto -1$, $b \mapsto \pm 1$. We now have $m/2 + 3$ classes and the degree equation becomes

$$\left(\frac{1}{2}m - 1\right).2^2 + 4.1^2 = 2m.$$

3. Character tables for the symmetric groups Sym_3, Sym_4. In the tables below the rows indicate the different characters, while the columns indicate the conjugacy classes, headed by a typical element and the order of each class. As is well known and easily checked, the conjugacy class of each permutation is determined by its cycle structure, hence the number of conjugacy classes of Sym_n is the number of partitions of n into positive integers. Moreover, the derived group is the alternating group Alt_n and its index in Sym_n is 2; hence there are just two linear characters, the trivial character and the sign character.

	1 1	3 (12)	2 (123)
χ_1	1	1	1
χ_2	1	-1	1
χ_3	2	0	-1

	1 1	6 (12)	8 (123)	6 (1234)	3 (12)(34)
χ_1	1	1	1	1	1
χ_2	1	-1	1	-1	1
χ_3	3	1	0	-1	-1
χ_4	3	-1	0	1	-1
χ_5	2	0	-1	0	2

In each table the first row is the trivial character and the second row the sign character. The degrees in the first column are found by solving the degree equation $\sum d_i^2 = n!$. Each character χ gives rise to a 'conjugate' character $\chi\chi_2$, corresponding to the tensor product $\rho \otimes \rho_2$ of the representations. Thus χ_3 and χ_4 in the second table are conjugate and χ_3 in the first table and χ_5 in the second table are self-conjugate, hence they vanish on odd classes. Now the remaining values are found by orthogonality, using Proposition 6.4.7.

We note that the characters for Sym_3, Sym_4 are rational. This is a general feature of symmetric groups; in fact, as we shall see in Section 6.6, all the irreducible representations of Sym_n can be expressed over **Q**.

Exercises

1. Verify the calculations in the above examples, and make a character table for the quaternion group of order 8.
2. Show that any two elements x, y of a group G are conjugate iff $\chi(x) = \chi(y)$ for all irreducible characters χ of G.
3. Show that any character which is zero on all elements $\neq 1$ of G is a multiple of the regular representation.
4. Show that if $S = \{x_1, \ldots, x_n\}$ is a transitive G-set, then the vector space spanned by the $x_i - x_j$ is a G-module affording a representation which does not contain the trivial representation.
5. Let χ be a character of a finite group over a field of characteristic 0. Show that (χ, χ) is a positive integer.
6. Let ρ be an irreducible representation of degree d of G, with character χ. Show that the simple factor of the group algebra corresponding to ρ has the unit element e given by $|G|.e = d.\sum \chi(x^{-1})x$.
7. If g_{ijk} is defined as in (6.3.11), show that $g_{ijk} = (\chi_i\chi_j, \chi_k)$. Use Proposition 6.4.7 to evaluate the sum of all the g_{ijk}^2 and deduce the formula $\sum_{ijk} g_{ijk}^2 = |G| \sum_\lambda h_\lambda^{-1}$. Show that the value of this sum for Sym_3 is 11 and for Sym_4 is 43, and verify the formula in these cases.
8. (Theodor E. Molien, 1897) Let G be a finite group with irreducible characters $\chi_1, \ldots, \chi_r$ and U a G-module with basis $u_1, \ldots, u_n$. Show that the character afforded by the G-module $U \otimes U \otimes \ldots \otimes U$ (n factors) is $\sum n_i\chi_i$, where n_i is the coefficient of t^n in $|G|^{-1} \sum_x \bar{\chi}_i(x) \det(I - t\rho(x))^{-1}$, ρ being the representation afforded by U. Deduce that the number of invariants of degree n in the u's is the coefficient of t^n in $|G|^{-1} \sum_x \det(I - t\rho(x))^{-1}$.

6.5 Complex representations

When we restrict our representations to be over the complex numbers, several simplifications can be made. In the first place we can then confine attention to unitary representations; secondly, by examining the different types of irreducible representations we obtain an estimate for the sum of their degrees in terms of the number of elements of order 2 in the group.

Throughout this section we only consider complex representations, thus we take **C** as the ground field. We recall that a square matrix P over C is said to be *unitary* if $PP^H = I$, where $P^H = \bar{P}^T$ is the transpose of the complex conjugate of P. The $d \times d$ unitary matrices over **C** form a group, the *unitary group*, written $\mathbf{U}_d(\mathbf{C})$. By a *unitary representation* of G, of degree d, we understand a homomorphism $G \to \mathbf{U}_d(\mathbf{C})$. In the special case where the representation is real, we have an *orthogonal* representation, because $\mathbf{U}_d(\mathbf{C}) \cap \mathbf{GL}_d(\mathbf{R}) = \mathbf{O}_d(\mathbf{R})$, the orthogonal group.

Any G-module U affording a unitary representation of G has a hermitian metric defined on it which is invariant under G. Thus there is a positive definite hermitian form (u, v) on U:

$$(\lambda u + \lambda' u', v) = \lambda(u, v) + \lambda'(u', v), \ (v, u) = \overline{(u, v)}, \ (u, u) > 0 \text{ for } u \neq 0, \quad (6.5.1)$$

which in addition satisfies

$$(ux, vx) = (u, v) \quad \text{for all } u, v \in U, \ x \in G. \quad (6.5.2)$$

For, given a unitary representation ρ, relative to the basis $v_1, \dots, v_d$ of U, let us define a hermitian form by writing $(v_i, v_j) = \delta_{ij}$. If the vectors u, v have coordinate rows α, β, then $(u, v) = \alpha\beta^H$ and $(ux, vx) = \alpha\rho(x)(\beta\rho(x))^H = \alpha\rho(x)\rho(x)^H\beta^H = \alpha\beta^H$, because $\rho(x)\rho(x)^H = I$ by unitarity. Conversely, if we have a positive definite hermitian form satisfying (6.5.2), then transformation by x preserves the metric and so must be unitary.

Any unitary representation is completely reducible. To verify this assertion we take the corresponding module U. If W is any submodule, then its orthogonal complement $W^\perp = \{u \in U | (u, w) = 0 \text{ for all } w \in W\}$ is again a G-submodule and is complementary to W. In terms of representations we can also verify this fact by noting that a reduced matrix $\rho(x)$ must be fully reduced, because its transpose is $\overline{\rho(x^{-1})}$.

The importance of unitary representations is underlined by the following result, which incidentally provides another proof of Maschke's theorem.

Proposition 6.5.1. *Every complex representation of a finite group G is equivalent to a unitary representation; every real representation is equivalent to an orthogonal representation. In particular every real or complex representation is completely reducible.*

Proof. Let ρ be the representation of G and U a G-module affording ρ. We have to find a positive definite hermitian form on U which is invariant under G. Take any positive definite hermitian form $h(u, v)$ on U and define

$$(u, v) = \sum_{x \in G} h(ux, vx).$$

For any $a \in G$ we have

$$(ua, va) = \sum_x h(uax, vax) = \sum_y h(uy, vy) = (u, v).$$

Thus (u, v) is invariant under G, and it is clearly positive definite hermitian, as a sum of such forms. On choosing an orthonormal basis, we obtain the desired unitary representation; starting from a real representation, we obtain an orthogonal representation of G. Now the last part follows by the earlier remarks. ∎

In dealing with unitary representations, we usually restrict equivalence to be by unitary matrices. Thus two unitary representations ρ, σ are unitarily equivalent if $\sigma(x) = P\rho(x)P^{-1}$ for a unitary matrix P. For irreducible representations this is automatic, for if ρ, σ are equivalent, we have $\sigma(x)S = S\rho(x)$ for an invertible matrix S. Taking hermitian conjugates, we have $S^H\sigma(x)^H = \rho(x)^H S^H$, hence

$$\rho(x)S^H S = \rho(x)S^H\sigma(x)^H\sigma(x)S = \rho(x)\rho(x)^H S^H S\rho(x) = S^H S\rho(x).$$

Since ρ is irreducible, we have $S^H S = \lambda I$ by Schur's lemma, and here $\lambda > 0$, because $S^H S$ is positive definite. Writing $\lambda = \mu^2$ $(\mu > 0)$ and $T = \mu^{-1}S$, we obtain a unitary matrix T such that $\sigma(x) = T\rho(x)T^{-1}$.

The irreducible complex representations may be classified as follows. Let ρ be an irreducible complex representation (not necessarily unitary). If ρ is equivalent to a real representation, it is said to be of the *first kind*. If ρ is not of the first kind, but is equivalent to its conjugate $\bar{\rho}$, it is said to be of the *second kind*; in the remaining case ρ is of the *third kind*. Our next result shows how to distinguish the first two of these cases. In the proof we shall need the elementary fact that a symmetric unitary matrix can always be written as the square of a symmetric unitary matrix. We recall the proof.

Let P be symmetric and unitary; as a unitary matrix P is similar to a diagonal matrix, say $S^{-1}PS = D$ for a unitary matrix S, and

$$SDS^{-1} = P = P^T = (S^T)^{-1}DS^T,$$

i.e. $S^T SD = DS^T S$. Now D is diagonal and again unitary, so its diagonal elements have absolute value 1 and we can find a diagonal matrix E such that $E^2 = D$ and

$$S^T SE = ES^T S. \tag{6.5.3}$$

Put $Q = SES^{-1}$; then Q is unitary, because E is, and $Q^2 = SE^2S^{-1} = P$. Moreover, $Q^T = (S^T)^{-1}ES^T = SES^{-1}$ by (6.5.3), hence $Q^T = Q$, so Q is also symmetric.

Proposition 6.5.2. *Let ρ be a complex irreducible representation of a group G such that*

$$\overline{\rho(x)} = P^{-1}\rho(x)P \quad \text{for some } P \in \mathbf{U}_d(\mathbf{C}). \tag{6.5.4}$$

Then either $P^T = P$ and ρ is of the first kind, or $P^T = -P$ and ρ is of the second kind.

Proof. Taking complex conjugates in (6.5.4) we have $\rho(x) = P^T\overline{\rho(x)}(P^T)^{-1}$, because $P^{-1} = P^H = \bar{P}^T$; therefore

$$P^TP^{-1}\rho(x) = P^T\overline{\rho(x)}P^{-1} = \rho(x)P^TP^{-1},$$

hence $P^TP^{-1} = \lambda I$, so $P^T = \lambda P$. Transposing, we find $P = \lambda P^T = \lambda^2 P$, hence $\lambda^2 = 1$ and so $\lambda = \pm 1$. This shows that either $P^T = P$ or $P^T = -P$.

Now it is clear from (6.5.4) that ρ cannot be of the third kind, so it is enough to show that ρ is of the first kind iff $P^T = P$. If ρ is of the first kind, then there is an invertible matrix L such that $L^{-1}\rho(x)L$ is real for all $x \in G$, hence by (6.5.4),

$$L^{-1}\rho(x)L + \bar{L}^{-1}\overline{\rho(x)L} = \bar{L}^{-1}P^{-1}\rho(x)P\bar{L}.$$

It follows that $P\bar{L}L^{-1}$ commutes with $\rho(x)$ and so $P\bar{L}L^{-1} = \alpha I$, i.e. $P = \alpha L\bar{L}^{-1}$. Now if $P^T = -P$, then $P^{-1} = P^H = -\bar{P}$ and so $I = PP^{-1} = -\alpha L\bar{L}^{-1}.\bar{\alpha}\bar{L}L^{-1} = -\alpha\bar{\alpha}I$, which is a contradiction. Therefore $P^T = P$ in this case.

Conversely, assume that $P^T = P$. Then P is symmetric unitary and by the above remark, $P = Q^2$, for a symmetric unitary matrix Q. Hence $\bar{Q} = Q^H = Q^{-1}$, and $\overline{Q^{-1}\rho(x)Q} = QP^{-1}\rho(x)PQ^{-1} = Q^{-1}\rho(x)Q$. Therefore ρ is equivalent to a real representation, and so is of the first kind. ■

If in Proposition 6.5.2, ρ is of the second kind and its degree is denoted by d, we have $P^T = -P$, hence $\det P = (-1)^d \det P$, and it follows that $(-1)^d = 1$. Hence d must be even and we have

Corollary 6.5.3. *Any complex irreducible representation of the second kind has even degree.* ■

For any character χ of a finite group G let us define its *indicator* as

$$\nu(\chi) = |G|^{-1}\sum_{x \in G}\chi(x^2). \tag{6.5.5}$$

The three kinds of irreducible representation may be described in terms of the indicator:

Theorem 6.5.4. *Let χ be a complex irreducible character of a finite group G and $\nu(\chi)$ its indicator as in (6.5.5). Then χ is of the first kind if $\nu(\chi) = 1$, of the second kind if $\nu(\chi) = -1$ and of the third kind if $\nu(\chi) = 0$.*

Proof. Let ρ be a representation affording χ and denote its degree by d. We may take ρ to be unitary, and then have

$$|G|.\nu(\chi) = \sum_{i,x}\rho_{ii}(x^2) = \sum_{ij}\sum_{x}\rho_{ij}(x)\rho_{ji}(x)$$
$$= \sum_{ij}\sum_{x}\rho_{ij}(x)\overline{\rho_{ij}(x^{-1})}.$$

If ρ is of the third kind, ρ and $\bar{\rho}$ are inequivalent and then $\nu(\chi) = 0$ by the orthogonality relations (Theorem 6.3.3). If ρ is of the first kind, we may assume it to be real; in that case

$$|G|.\nu(\chi) = \sum\sum\rho_{ij}(x)\rho_{ij}(x^{-1}) = \sum\delta_{ji}\delta_{ij}.|G|/d = |G|,$$

and hence $\nu(\chi) = 1$. Finally, if ρ is of the second kind, then by Proposition 6.5.2 there is a unitary matrix P such that $P^T = -P$ and $P^{-1}\rho(x)P = \overline{\rho(x)}$; hence on writing $P = (p_{ij})$, $P^{-1} = (q_{ij})$ we have, again by the orthogonality relations,

$$\begin{aligned}\nu(\chi) = |G|^{-1}.\sum_{irsj}\sum_{x}\rho_{ij}(x)q_{ir}\rho_{rs}(x^{-1})p_{sj} &= d^{-1}\sum q_{ir}p_{sj}\delta_{jr}\delta_{is}\\ &= d^{-1}\sum q_{ir}p_{ir}\\ &= d^{-1}\,\mathrm{tr}(P^{-1}P^T) = \frac{\mathrm{tr}(-I)}{d} = -1.\end{aligned}$$

Thus $\nu(\chi) = -1$, as we wished to show. ■

Finally we show how the indicator is related to the solution of the equation $x^2 = a$ in G.

Proposition 6.5.5. *Let G be a finite group. Given $a \in G$, let $t(a)$ be the number of solutions of the equation $x^2 = a$ in G. If $\chi_1, \ldots, \chi_r$ are all the inequivalent complex irreducible characters of G, then*

$$t(a) = \sum_{i=1}^{r}\nu(\chi_i)\chi_i(a). \tag{6.5.6}$$

Proof. It is clear that $t(a)$ is a class function on G, so it can be written in the form $t(a) = \sum c_i\chi_i(a)$, and it only remains to show that $c_i = \nu(\chi_i)$. The sets $T(a) = \{x \in G | x^2 = a\}$ form a partition of G, and we have by Theorem 6.4.5, using (6.5.8) to determine the coefficient c_i:

$$\begin{aligned}|G|.c_i = \sum t(a)\overline{\chi_i(a)} &= \sum_{a\in G}\sum_{x\in T(a)}\overline{\chi_i(x^2)}\\ &= \sum_{x\in G}\overline{\chi_i(x^2)} = |G|.\overline{\nu(\chi_i)},\end{aligned}$$

by (6.5.5). Since the indicator is always real, the desired relation $c_i = \nu(\chi_i)$ follows. ■

Let us apply the result for $a = 1$. In this case $\chi(1) = d$ is the degree of χ. Bearing in mind that $\nu(\chi) \leq 1$, with equality precisely when χ is of the first kind, by Theorem 6.5.4, we obtain

Corollary 6.5.6. *If t is the number of elements of order 2 in the finite group G, and the degrees of its complex irreducible representations are $d_1, \ldots d_r$, then*

$$t + 1 = \sum\nu(\chi_i)d_i \leq \sum d_i, \tag{6.5.7}$$

with equality if and only if all the irreducible characters are of the first kind. ■

As an example consider $\mathbf{D}_m$, the dihedral group of order $2m$, where m is odd. We saw that there are $(m-1)/2$ characters of degree 2 and two linear characters, and there are m elements of order 2, namely $a^\alpha b$ in the notation of Section 6.4. The inequality (6.5.7) in this case reads

$$m+1 \leq \lfloor (m-1)/2 \rfloor .2+2.1.$$

Since equality holds here, all the representations must be of the first kind, and going through the proof of Proposition 6.5.2, we find that the representation given at the end of Section 6.4 is equivalent to the real representation

$$a \mapsto \frac{1}{2}\begin{pmatrix} \lambda+\lambda^{-1} & \lambda-\lambda^{-1} \\ \lambda^{-1}-\lambda & \lambda+\lambda^{-1} \end{pmatrix}, \ b \mapsto \begin{pmatrix} 0 & 1 \\ 1 & 0 \end{pmatrix}, \quad \lambda = \omega^j, \ j=1, \ldots, (m-1)/2.$$

As a second example consider the degrees of the irreducible representations of Sym_5. We shall see in Section 6.6 that all irreducible representations of the symmetric group can be taken to be rational (even integral). The elements of order 2 in Sym_5 are transpositions or products of two transpositions. There are $5.4/2 = 10$ transpositions $(i\ j)$ and $10.3/2 = 15$ products of the form $(i\ j)(k\ l)$; hence $t = 25$, $t+1 = 26$. To find the number of conjugacy classes we count the partitions of 5: $(5), (4,1), (3,2), (3,1^2), (2,1^3), (2^2,1), (1^5)$. So we have to look for seven positive integers, divisors of $5! = 120$, whose sum is 26 and whose squares have the sum 120, by (6.3.10). Proposition 6.4.2 tells us that just two of the numbers are 1, because $(\mathrm{Sym}_5 : \mathrm{Alt}_5) = 2$, and this leaves $d_1, \ldots, d_5$ such that

$$\sum_1^5 d_i = 24, \quad \sum_1^5 d_i^2 = 118.$$

Further, we can rule out the low dimensions 2, 3 because an integral representation can be reduced mod 2, and we cannot have a homomorphism of Sym_5 into $\mathbf{GL}_2(\mathbf{F}_2)$ (order $(2^2-1)(2^2-2) = 6$) or into $\mathbf{GL}_3(\mathbf{F}_2)$) order $(2^3-1)(2^3-2)(2^3-2^2) = 192$) which maps Alt_5 non-trivially. This leaves as the only possibility for the degrees 1, 1, 4, 4, 5, 5, 6.

Exercises

1. Use the information on Sym_5 to make a character table for it.
2. Find the matrix of transformation which reduces the representation (6.4.14) for $\mathbf{D}_m$ to the form given above.
3. Let V be a simple G-module with real character. Show that there is just one invariant bilinear form b on V, up to scalar multiples. Show further that b is either symmetric or antisymmetric, and that the latter can happen only when $\dim V$ is even.
4. Show that in a group G of odd order no element $\neq 1$ is conjugate to its inverse. Deduce that for any $y \neq 1$, $\sum_i \chi_i(y)^2 = \sum_i \chi_i(y)\chi_i(y^{-1}) = 0$. Hence show that G has an irreducible character of the third kind.

5. Show that every non-trivial irreducible character of a group of odd order is of the third kind. (Hint. Use the methods of Exercise 4 and the fact that $\chi(1)$ is an odd integer.)
6. Find a quadratic polynomial f such that a complex irreducible representation of the n-th kind has the indicator $f(n)$. For what numbering of the different kinds of representations can f be chosen as a linear polynomial?

6.6 Representations of the symmetric group

In principle all the irreducible representations of a finite group can be obtained by taking the regular representation and reducing it completely. This is a lengthy undertaking, but for certain types of groups there are more direct methods. We shall describe such a method for symmetric groups; it is due mainly to Georg Frobenius and Alfred Young, with some simplifications by John von Neumann.

We recall that the symmetric group of degree n, Sym_n or simply S_n is the group of all permutations of $1, 2, \ldots, n$. It has order $n!$ and each permutation can be written as a product of disjoint cycles, e.g.

$$\begin{pmatrix} 1 & 2 & 3 & 4 & 5 & 6 & 7 & 8 & 9 \\ 7 & 5 & 6 & 8 & 1 & 9 & 2 & 4 & 3 \end{pmatrix} = (1\ 7\ 2\ 5)(3\ 6\ 9)(4\ 8).$$

The cycles have no digits in common, so they commute and we can arrange them by decreasing length. If the lengths are $\alpha_1, \ldots, \alpha_h$, we may thus suppose that

$$\alpha_1 + \alpha_2 + \ldots + \alpha_h = n, \tag{6.6.1}$$

and

$$\alpha_1 \geq \alpha_2 \geq \ldots \geq \alpha_h. \tag{6.6.2}$$

If λ_1 of the α_i are 1, λ_2 are 2, etc., we can also write $1^{\lambda_1}2^{\lambda_2}\ldots r^{\lambda_r}$ (where r is the largest α_i) for the set of α's. This is called the *cycle structure* of the permutation. Two permutations have the same cycle structure iff they are conjugate in S_n: If

$$f = (a_1 \ldots a_{\alpha_1})(a_{\alpha_1+1} \ldots a_{\alpha_1+\alpha_2}) \ldots (\ldots a_n),$$
$$g = (b_1 \ldots b_{\alpha_1})(b_{\alpha_1+1} \ldots b_{\alpha_1+\alpha_2}) \ldots (\ldots b_n),$$

then $g = p^{-1}fp$, where $p : a_i \mapsto b_i$. Hence two permutations with the same cycle structure are conjugate; the converse is clear.

It follows that the number of conjugacy classes of S_n equals the number of sequences $(a_1, \ldots, a_h)$ of positive integers satisfying (6.6.1) and (6.6.2); by Theorem 6.4.5 this is also the number of inequivalent irreducible representations of S_n. To get a complete system of irreducible representations of S_n we need only construct for each sequence $(\alpha_1, \ldots, \alpha_h)$ an irreducible representation, such that representations corresponding to different sequences are inequivalent. This will be our aim in what follows.

We shall write $\alpha \vdash n$ to indicate that $\alpha = (\alpha_1, \ldots, \alpha_h)$ is a descending partition on n, as in (6.6.1), (6.6.2). To each such partition there corresponds a diagram of n squares arranged in h rows of α_i squares in the i-th row. For example $(4, 3^2, 1)$ corresponds to

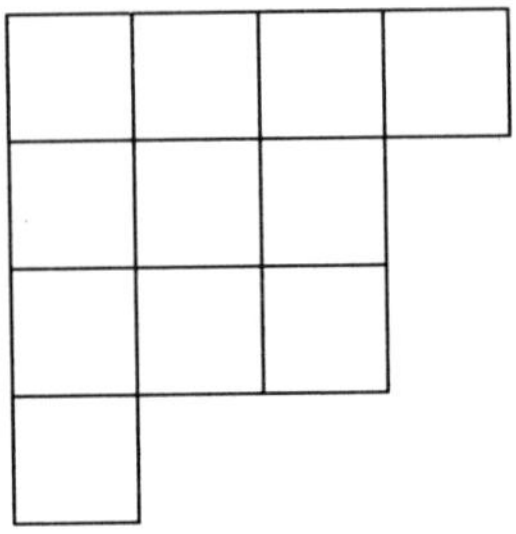

This is called a *Young diagram* and is again denoted by α. Since $\alpha \vdash n$, we can write the numbers 1 to n in these squares (in some order). The result is called a *Young tableau*. For example

(3, 2, 1)

2	6	5
1	3	
4		

We see that for each diagram there are $n!$ distinct tableaux. If T_α is a Young tableau and $g \in S_n$, then $T_\alpha g$ denotes the tableau obtained from T_α by applying g.

We can let each Young tableau represent a permutation by regarding the rows as cycles. In this way the tableau just illustrated represents the permutation (2 6 5)(1 3). If T_α represents c in this way, then $T_\alpha g$ represents $g^{-1}cg$, as is easily verified.

We now fix a tableau T_α and define two subgroups of S_n as follows: P_α is the set of all permutations leaving each symbol in its row, briefly the set of *row permutations* of T_α, and Q_α is the set of all permutations leaving each symbol in its column, the set of *column permutations* of T_α. Here P_α and Q_α depend of course on T_α and not merely on the diagram α. For example, in the above tableau P_α is generated by (2 6), (2 6 5), (1 3), while Q_α is generated by (2 1), (2 1 4), (3 6). If we apply g to T_α and use the remark made earlier, we obtain

Lemma 6.6.1. *Let T_α be a Young tableau with groups P_α, Q_α and let $g \in S_n$. Then the groups for $T_\alpha g$ are $g^{-1}P_\alpha g$ and $g^{-1}Q_\alpha g$.* ■

Let A be the group algebra of S_n (over the rational numbers, say) and write $\varepsilon(g)$ or ε_g for the sign of the permutation g. Writing p, q for the typical elements of P_α, Q_α

respectively, we define two elements of A, the sum of the row permutations and the alternating sum of the column permutations:

$$f_\alpha = \sum_p p, \quad g_\alpha = \sum_q \varepsilon_q q. \tag{6.6.3}$$

Lemma 6.6.2. *Let T_α be a Young tableau and f_α, g_α as in (6.6.3), and write $|P_\alpha| = r_\alpha$, $|Q_\alpha| = s_\alpha$. Then*

$$pf_\alpha = f_\alpha p = f_\alpha, \quad \text{for } p \in P_\alpha, \tag{6.6.4}$$

$$qg_\alpha = g_\alpha q = \varepsilon_q g_\alpha \quad \text{for } q \in Q_\alpha, \tag{6.6.5}$$

$$f_\alpha^2 = r_\alpha f_\alpha, \; g_\alpha^2 = s_\alpha g_\alpha. \tag{6.6.6}$$

Proof. We have $pf_\alpha = \sum_{p'} pp' = \sum p' = f_\alpha$, $\varepsilon_q q g_\alpha = \sum \varepsilon_q \varepsilon_{q'} qq' = g_\alpha$, which establishes one half of (6.6.4), (6.6.5); the other half follows similarly. Now $f_\alpha^2 = \sum_p pf_\alpha = \sum f_\alpha = r_\alpha f_\alpha$, $g_\alpha^2 = \sum \varepsilon_q q g_\alpha = \sum g_\alpha = s_\alpha g_\alpha$. ∎

Next comes a basic combinatorial lemma on which everything else depends. We order the partitions lexicographically, by writing $\alpha > \beta$ if the first non-zero difference $\alpha_i - \beta_i$ is positive (where α or β is completed by 0's if necessary). This provides a total ordering of the set of all partitions of n.

Lemma 6.6.3. *Let $\alpha, \beta \vdash n$ and take any tableaux T_α, T'_β. If $\alpha \geq \beta$ and no two numbers in the same row of T_α occur in the same column of T'_β, then* (i) $\alpha = \beta$ *and* (ii) $T'_\beta = T_\alpha qp$ *for some $p \in P_\alpha$, $q \in Q_\alpha$.*

Some care is needed here, for we have abused notation by writing P_α instead of the more accurate $P(T_\alpha)$. In fact we shall take P_α, Q_α to be the groups associated with T_α and write P'_β, Q'_β for the groups of T'_β.

Proof. Since $\alpha \geq \beta$, we have $\alpha_1 \geq \beta_1$. The first row of T_α has α_1 numbers, which must be in different columns of T'_β, so $\beta_1 \geq \alpha_1$, and hence $\beta_1 = \alpha_1$. Now for a certain column permutation $q'_1 \in Q'_\beta$ we can bring these numbers into the top row, though possibly in a different order from that in T_α.

Leaving out the top row in $T'_\beta q'_1$ and T_α we can repeat the argument, showing that $\beta_2 = \alpha_2$ and finding q'_2 such that $T'_\beta q'_1 q'_2$ has the same numbers as T_α in the second row as well as the top row. After h steps (if α has h parts) we get $q' = q'_1 q'_2 \ldots q'_h$ such that $T'_\beta q'$ differs from T_α only by a row permutation: $T'_\beta q' = T_\alpha p$, where $p \in P_\alpha$, $q' \in Q'_\beta$. Now $T_\alpha = T'_\beta q' p^{-1}$, hence

$$Q_\alpha = (q'p^{-1})^{-1} Q'_\beta q' p^{-1} = pQ'_\beta p^{-1};$$

it follows that $q = pq'^{-1}p^{-1} \in Q_\alpha$. Therefore $qp = pq'^{-1}$ and we have $T'_\beta = T_\alpha qp$, as claimed, with $p \in P_\alpha$, $q \in Q_\alpha$. ∎

Corollary 6.6.4. *Given $\alpha, \beta \vdash n$ and any tableaux T_α, T'_β, if $\alpha > \beta$, then*

$$f_\alpha x^{-1} g_\beta x = 0 \quad \text{for all } x \in S_n, \tag{6.6.7}$$

$$f_\alpha A g_\beta = 0. \tag{6.6.8}$$

Proof. We begin by showing that

$$f_\alpha g_\beta = 0 \quad \text{for } \alpha > \beta. \tag{6.6.9}$$

By Lemma 6.6.3 there must be two numbers i, k which lie in the same row of T_α and in the same column of T'_β. Write $t = (i\ k)$; then $f_\alpha t = f_\alpha$, $t g_\beta = -g_\beta$, hence $f_\alpha g_\beta = f_\alpha t^2 g_\beta = -f_\alpha g_\beta$ and (6.6.9) follows. Now $x^{-1} g_\beta x$ corresponds to the tableau $T'_\beta x$ and (6.6.7) follows by applying (6.6.9) with $g'_\beta = x^{-1} g_\beta x$. Replacing x^{-1} by y and multiplying by y on the right, we have $f_\alpha y g_\beta = 0$, and now (6.6.8) follows by summing over y. ■

Given a Young tableau T_α, let us put

$$h_\alpha = f_\alpha g_\alpha = \sum_{p,q} \varepsilon_q pq. \tag{6.6.10}$$

Clearly $h_\alpha \neq 0$, because the coefficient of 1 is 1:

$$h_\alpha(1) = 1. \tag{6.6.11}$$

We may consider h_α as an operator symmetrizing the rows and antisymmetrizing the columns; it is called the *Young symmetrizer* associated with the tableau T_α.

Proposition 6.6.5. *Let h_α be the Young symmetrizer associated with a tableau T_α. Then the relation*

$$pa\varepsilon_q q = a \quad \text{for all } p \in P_\alpha,\ q \in Q_\alpha \tag{6.6.12}$$

holds for $a = h_\alpha$, and any a satisfying (6.6.12) must be of the form $a = \lambda h_\alpha$, where λ is a scalar. Moreover,

$$h_\alpha b h_\beta = 0 \quad \text{for } \alpha > \beta \text{ and any } b \in A, \tag{6.6.13}$$

$$h_\alpha b h_\alpha = \mu h_\alpha \quad \text{for any } b \in A. \tag{6.6.14}$$

Proof. By (6.6.4), $p f_\alpha g_\alpha = f_\alpha g_\alpha$, while (6.6.5) yields $f_\alpha g_\alpha q = \varepsilon_q f_\alpha g_\alpha$; hence (6.6.12) holds for $a = h_\alpha$. Now let $a = \sum a(x)x$ satisfy (6.6.12). Then

$$\sum_x \varepsilon_q a(x) pxq = \sum a(x)x \quad \text{for all } p \in P_\alpha,\ q \in Q_\alpha. \tag{6.6.15}$$

Comparing coefficients for $x = pq$, we find

$$\varepsilon_q a = a(pq); \tag{6.6.16}$$

we claim that $a(x) = 0$ when x is not of the form pq. Consider T_α and $T'_\alpha = T_\alpha x^{-1}$; by Lemma 6.6.3 there are two numbers j, k in the same row in T_α and in the same column in T'_α (because T'_α is not of the form $T_\alpha pq$). Put $t = (j\ k)$; then we have $t \in P_\alpha$, $t \in Q'_\alpha$, where Q'_α corresponds to T'_α; therefore $t \in xQ_\alpha x^{-1}$ or also $x^{-1}tx \in Q_\alpha$. In (6.6.15) let us take $p = t$, $q = x^{-1}tx$; comparing coefficients of x we find $(-1)a(x) = a(x)$, hence $a(x) = 0$. Together with (6.6.16) this shows that $a = a(1). \sum pq\varepsilon_q = a(1).h_\alpha$ as claimed.

Now when $\alpha > \beta$, then by Corollary 6.6.4, $h_\alpha b h_\beta = f_\alpha g_\alpha b f_\beta g_\beta \in f_\alpha A g_\beta = 0$. This proves (6.6.13), and (6.6.14) follows because $h_\alpha b h_\alpha$ satisfies (6.6.12). ∎

We can now obtain the irreducible representations of S_n by expressing the group algebra as a direct sum of minimal right ideals.

Theorem 6.6.6. *With each Young diagram α let us associate one definite Young tableau T_α and construct the corresponding Young symmetrizer h_a as an element of the group algebra $A = \mathbf{Q}S_\alpha$:*

$$h_\alpha = \sum \varepsilon_q pq, \quad (p \in P_\alpha, q \in Q_\alpha).$$

Then the $I^\alpha = h_\alpha A$ are simple submodules for the right regular representation of S_n; they are pairwise non-isomorphic and afford a complete system of irreducible representations for S_n.

Proof. We first show that $I^\alpha = h_\alpha A$ is a minimal right ideal of A. Given a right ideal $\mathfrak{m} \subseteq I^\alpha$, we have $\mathfrak{m}h_\alpha \subseteq I^\alpha h_\alpha \subseteq \mathbf{Q}h_\alpha$. We distinguish two cases: (i) $\mathfrak{m}h_\alpha = \mathbf{Q}h_\alpha$, then $I^\alpha = h_\alpha A = \mathfrak{m}h_\alpha A \subseteq \mathfrak{m}$, hence $I^\alpha = \mathfrak{m}$; (ii) $\mathfrak{m}h_\alpha = 0$, then $\mathfrak{m}^2 = \mathfrak{m}I^\alpha = \mathfrak{m}h_\alpha A = 0$, so $\mathfrak{m}^2 = 0$ and therefore $\mathfrak{m} = 0$. It follows that I^α is minimal and the representation induced by the regular representation is irreducible.

We next show that I^α, I^β are not isomorphic for $\alpha \neq \beta$. If $\alpha > \beta$, say, then by Corollary 6.6.4, $h_\alpha A h_\beta = 0$, hence $I^\alpha h_\beta = 0$, but $I^\alpha h_\alpha \neq 0$, because $h_\alpha \neq 0$, and the conclusion follows. Now the number of distinct diagrams is the number of partitions of n, which is just the number of conjugacy classes of S_n, as we have seen; hence by Theorem 6.4.5, we have a complete set of irreducible representations. ∎

It still remains to calculate the irreducible characters. By Proposition 6.6.5 we have

$$h_\alpha^2 = \mu_\alpha h_\alpha, \tag{6.6.17}$$

therefore $e_\alpha = \mu_\alpha^{-1} h_\alpha$ is an idempotent and the right ideal of A generated by e_α is minimal. To find μ_α consider the operation of left multiplication by h_α on I^α. By (6.6.17) this is

$$\lambda(h_\alpha) = \mu_\alpha.1. \tag{6.6.18}$$

On the other hand, for $a \in I^\alpha$, $a\lambda(h_a) = h_\alpha a = \sum h_\alpha(uv^{-1})a(v).u$, hence the matrix of $\lambda(h_\alpha)$ (in the natural basis) is given by

$$\lambda_{xy}(h_\alpha) = h_\alpha(yx^{-1}),$$

and so $\mathrm{tr}(\lambda(h_{\alpha})) = \sum h_\alpha(xx^{-1}) = \sum h_\alpha(1) = \sum 1 = n!$, by (6.6.11). A comparison with (6.6.18) shows that $n! = \mathrm{tr}(\lambda(h_\alpha)) = d_\alpha \mu_\alpha$, where d_α is the degree of the representation. Thus we obtain

$$\mu_\alpha = n!/d_\alpha.$$

Recalling Proposition 6.4.4, we find that the corresponding character is given by (6.6.19)

$$\chi_\alpha(x) = \frac{d_\alpha}{n!}\sum_y h_\alpha(yx^{-1}y^{-1}).$$

From this formula it is possible to calculate χ_α explicitly in terms of the partitions α of n; we shall give the result here without proof and refer to Weyl (1939), Chapter VII or James and Kerber (1981), Chapter 2 for details.

Let $x \in S_n$ have the cycle structure $\beta = (1^{\beta_1}2^{\beta_2}\dots t^{\beta_t})$; since the value of a character χ at x depends only on β, we shall write it as $\chi(\beta)$. Let $x_1, \dots, x_h$ be any variables and denote the Vandermonde determinant formed from them by writing down the terms on the main diagonal:

$$|x_1^{h-1}x_2^{h-2}\dots x_h^0|.$$

This is an alternating function of the x's, briefly denoted by Δ in what follows. More generally, for any positive integers $\alpha_1, \dots, \alpha_h$ the function

$$|x_1^{\alpha_1+h-1}x_2^{\alpha_2+h-2}\dots x_h^{\alpha_h}|$$

formed in the same way, is an alternating function of the x's, zero unless all the α's are in descending order, and it may be written as $S(\alpha_1, \dots, \alpha_h)\Delta$, where S as a quotient of two alternating functions is symmetric in the x's. Such a function S is called a *bialternant* or *S-function.*

Given $\alpha = (\alpha_1, \dots, \alpha_h) \vdash n$, let us write $r_1 = \alpha_1 + (h-1)$, $r_2 = a_2 + (h-2), \dots, r_h = \alpha_h$. We define the power sums in the x's as $s_i = \sum x_\nu^i$ and for a given $\beta = (\beta_1, \dots, \beta_t)$ satisfying $\sum i\beta_i = n$ put $\sigma(\beta) = s_1^{\beta_1}s_2^{\beta_2}\dots s_t^{\beta_t}$. With these notations we have the following relation for the characters of S_n corresponding to the partitions into at most h parts:

$$\sigma(\beta).|x_1^{h-1}x_2^{h-2}\dots x_h^0| = \sum_\alpha \chi_\alpha(\beta)|x_1^{r_1}x_2^{r_2}\dots x_h^{r_h}|. \tag{6.6.20}$$

Thus $\chi_\alpha(\beta)$ is the coefficient of $x_1^{r_1}x_2^{r_2}\dots x_h^{r_h}$ in the expansion of the left-hand side of (6.6.20). On dividing by Δ we may write (6.6.20) as

$$\sigma(\beta) = \sum_\alpha \chi_\alpha(\beta)S(\alpha_1, \dots, \alpha_h); \tag{6.6.21}$$

this shows that $\chi_\alpha(\beta)$ is the coefficient of $x_1^{\alpha_1}x_2^{\alpha_2}\dots x_h^{\alpha_h}$ in the expansion of the left-hand side of (6.6.21). Further, the degree of the irreducible character χ_α is given by

$$d_\alpha = n!\frac{\prod_{j<i}(r_j - r_i)}{r_1!r_2!\dots r_h!}.$$

Exercises

1. Compute the characters of S_4, S_5 by the methods of this section.
2. For each partition α of n define the *conjugate* partition as α', where α'_i is the number of parts of α that are $\geq i$. Describe the relation between the Young diagrams of α and α', and show that $\alpha'' = \alpha$.
3. Show that the character corresponding to the conjugate partition is given by $\chi'_\alpha = \varepsilon\chi_\alpha$, where ε is the sign character.
4. Show that for $n \geq 4$, S_n has at least two irreducible representations of degree $n-1$, and show that for $n = 6$ it has four.
5. A group is called *ambivalent* if each element is conjugate to its inverse. Such a group, if non-trivial, must have even order. Show that an ambivalent group has a real character table.
6. Show that the symmetric group S_n of any degree n is ambivalent (it can be shown that the alternating group Alt_n is ambivalent precisely when $n = 1, 2, 5, 6, 10, 14$; see James and Kerber (1981), Chapter 1).
7. Show that any irreducible representation of S_n of degree greater than 1 is faithful except the representation corresponding to 2^2 for $n = 4$.

6.7 Induced representations

There are various relations between the representations of a group and those of its subgroups which are often needed. In the first place, if G is a group and H a subgroup, then each G-module V is an H-module by restriction of the action to H. This H-module is denoted by $\text{res}_H^G V$ or res V or also V_H; if the representation afforded by V is ρ, then the corresponding representation of V is written $\text{res}_H^G \rho$ or ρ_H.

We next describe a process of passing from a representation of a subgroup of G to one of G itself. Let H be a subgroup of finite index r in G and consider a right H-module U. From it we can form the right G-module

$$\text{ind}_H^G U = U^G = U \otimes_H kG, \tag{6.7.1}$$

called the *induced* G-module. To find the representation afforded by U^G we note that the subspace $U \otimes H = \{u \otimes 1 | u \in U\}$ of U^G is an H-module in a natural way: for any $h \in H$ we have $(u \otimes 1)h = uh \otimes 1$. More generally we can define the subspace $U \otimes Ha = \{u \otimes a | u \in U\}$ of (6.7.1) as an H^a-module, where $H^a = a^{-1}Ha$, by the rule

$$(u \otimes a)x = uaxa^{-1} \otimes a \quad \text{for } x \in H^a.$$

This is well-defined because $axa^{-1} \in H$ precisely when $x \in H^a$. When $U \otimes Ha$ is considered as a right H^a-module in this way we shall denote it by U^a.

Now take a coset representation of G:

$$G = Ht_1 \cup \ldots \cup Ht_r. \tag{6.7.2}$$

With its help we can write (6.7.1) as

$$U^G = (U \otimes Ht_1) \oplus \ldots \oplus (U \otimes Ht_r) = U^{t_1} \oplus \ldots \oplus U^{t_r}. \tag{6.7.3}$$

Each U^{t_λ} is a H^{t_λ}-module and under the action of G these terms are permuted among themselves.

Given $x \in G$, we can for each $\lambda = 1, \ldots, r$ find a unique $h \in H$ and μ, $1 \leq \mu \leq r$, such that $t_\lambda x = ht_\mu$. Then we have

$$(u \otimes t_\lambda)x = uh \otimes t_\mu;$$

this shows how the action of x permutes the terms in (6.7.3), as well as acting on each. To find the representation afforded by U^G, let us take a basis $u_1, \ldots, u_n$ of U and let $\rho(h) = (\rho_{ij}(h))$ $(h \in H)$ be the corresponding representation of H. Then the elements $u_i \otimes t_\lambda$ form a basis of U^G and we have, for any $x \in G$,

$$(u_i \otimes t_\lambda)x = u_i \otimes t_\lambda x = u_i \otimes ht_\mu,$$

where μ is chosen so that $t_\lambda x t_\mu^{-1} = h \in H$. Hence we find

$$(u_i \otimes t_\lambda)x = \sum_j \rho_{ij}(h)u_j \otimes t_\mu.$$

Thus from a representation $(\rho(h))$ of degree d we obtain the induced representation $(\rho(t_\lambda x t_\mu^{-1}))$ of degree rd, where $r = (G : H)$. We note that this representation consists of r $d \times d$ blocks, one in each row and one in each column:

$$\rho^G = \begin{pmatrix} P_{11} & \ldots & P_{1r} \\ \ldots & \ldots & \ldots \\ P_{r1} & \ldots & P_{rr} \end{pmatrix},$$

where

$$P_{\lambda\mu} = \begin{cases} \rho(t_\lambda x t_\mu^{-1}) & \text{if } t_\lambda x t_\mu^{-1} \in H, \\ 0 & \text{otherwise.} \end{cases}$$

This is called the *induced* representation and is denoted by $\mathrm{ind}_H^G\, \rho$ or ρ^G. Hence if the character of ρ is α, then ρ^G has the character $\mathrm{ind}_H^G\, \alpha = \alpha^G$, given by

$$\begin{aligned} \alpha^G(x) &= \mathrm{tr}(\rho^G(x)) \\ &= \sum_\lambda \mathrm{tr}\ P_{\lambda\lambda} \\ &= \sum \mathrm{tr}(\rho(t_\lambda x t_\lambda^{-1})), \end{aligned}$$

where the sum is over all λ such that $t_\lambda x t_\lambda^{-1} \in H$. Let us define $\alpha^\bullet$ on G by

$$\alpha^\bullet(x) = \begin{cases} \alpha(x) & \text{if } x \in H, \\ 0 & \text{otherwise.} \end{cases}$$

Bearing in mind that $\alpha(h) = \mathrm{tr}(\rho(h))$, we can rewrite the equation for $\alpha^G(x)$ as

$$\alpha^G(x) = \sum_\lambda \alpha^\bullet(t_\lambda x t_\lambda^{-1}). \tag{6.7.4}$$

Since α is a class function on H, we have $\alpha(ht_\lambda x t_\lambda^{-1}h^{-1}) = \alpha(t_\lambda x t_\lambda^{-1})$ for any $h \in H$. It follows that

$$\begin{aligned} |H|.\alpha^G(x) &= \sum_{h\in H}\sum_\lambda \alpha^\bullet(ht_\lambda x t_\lambda^{-1}h^{-1}) \\ &= \sum_{g\in G} \alpha^\bullet(gxg^{-1}). \end{aligned}$$

If we define α^g by $\alpha^g(x) = \alpha^\bullet(gxg^{-1})$, we can express this as a formula

$$\alpha^G = |H|^{-1}\sum_{g\in G}\alpha^g. \tag{6.7.5}$$

We remark that for any class function a on H this formula defines a class function α^G on G. To illustrate (6.7.5), consider the case $H = 1$. There is only one character, namely the trivial character 1. We see from (6.7.1) that the induced representation is just the regular representation of G. More generally, let us take an arbitrary subgroup H of G but take the trivial character 1_H on H. Then we obtain an induced character

$$\mathrm{ind}_H^G(1_H(x)) = n_x,$$

where for any $x \in G$, n_x is the number of elements $t_\lambda x t_\lambda^{-1}$ $(\lambda = 1, \dots, r)$ that lie in H.

As another example take the dihedral group of order $2m$, $\mathbf{D}_m = \mathrm{gp}\{a, b \mid a^m = b^2 = (ab)^2 = 1\}$, with subgroup $H = \mathrm{gp}\{a\}$ of order m. H has the representation $a \mapsto \omega$, where $\omega^m = 1$. With the transversal $t_1 = 1$, $t_2 = b$ of H in $\mathbf{D}_m$ we have

$$at_1 = t_1 a,\ \ at_2 = t_2 a^{-1},\ \ bt_1 = t_2 1,\ \ bt_2 = t_1 1.$$

Hence we obtain the induced representation

$$\rho(a) = \begin{pmatrix} \omega & 0 \\ 0 & \omega^{-1} \end{pmatrix},\ \ \rho(b) = \begin{pmatrix} 0 & 1 \\ 1 & 0 \end{pmatrix},$$

and the character $\chi(a) = \omega + \omega^{-1}$, $\chi(b) = 0$.

It is clear from (6.7.1) that ind_H^G is a covariant functor; thus

$$\mathrm{ind}_H^G(U \oplus V) \cong \mathrm{ind}_H^G U \oplus \mathrm{ind}_H^G V. \tag{6.7.6}$$

Further, if $K \subseteq H \subseteq G$ and U is a K-module, then $U \otimes_K kG \cong U \otimes_K kH \otimes_H kG$; hence

$$\mathrm{ind}_H^G(\mathrm{ind}_K^H U) \cong \mathrm{ind}_K^G U. \tag{6.7.7}$$

It follows that for any character α of K (indeed for any class function),

$$(\alpha^H)^G = \alpha^G. \tag{6.7.8}$$

The following important relation between induction and restriction was proved by Georg Frobenius in 1898:

Theorem 6.7.1 (Frobenius reciprocity theorem). *Let G be a finite group and H a subgroup.*

(i) *Given a G-module V and an H-module U, there is a natural isomorphism*

$$\mathrm{Hom}_H(U, \mathrm{res}_H^G V) \cong \mathrm{Hom}_G(\mathrm{ind}_H^G U, V). \tag{6.7.9}$$

(ii) *If α, β are class functions on H, G respectively and $\alpha^G = \mathrm{ind}_H^G \alpha$ is defined as in (6.7.5), then*

$$(\alpha, \mathrm{res}_H^G \beta)_H = (\mathrm{ind}_H^G \alpha, \beta)_G, \tag{6.7.10}$$

where the subscripts denote the groups on which the scalar products are taken.

For irreducible characters ρ on G and σ on H this tells us that the multiplicity of σ in ρ_H is equal to that of ρ in σ^G.

Proof. By adjoint associativity (see Section 2.3) we have

$$\begin{aligned}\mathrm{Hom}_G(\mathrm{ind}\, U, V) &\cong \mathrm{Hom}_G(U \otimes_H kG, V)\\ &\cong \mathrm{Hom}_G(kG, \mathrm{Hom}_H(U, \mathrm{res}\, V))\\ &\cong \mathrm{Hom}_H(U, \mathrm{res}\, V).\end{aligned}$$

This establishes (6.7.9). For (6.7.10) we have

$$\begin{aligned}(\alpha^G, \beta)_G = |G|^{-1}\sum_{x\in G}\alpha^G(x)\overline{\beta(x)} &= |G|^{-1}|H|^{-1}\sum_{x,g}\alpha^\bullet(x^g)\overline{\beta(x)}\\ &= |G|^{-1}|H|^{-1}\sum \alpha^\bullet(x^g)\overline{\beta(x^g)}\\ &= |H|^{-1}\sum_{x\in H}\alpha(x)\overline{\beta(x)}\\ &= (\alpha, \mathrm{res}_H\, \beta)_H.\end{aligned}$$

When α, β are characters, (6.7.10) is of course an immediate consequence of (6.7.9). Since every class function is a linear combination of characters, this provides another proof of (6.7.10) ∎

We next derive a formula described by George Mackey in 1951, on the effect of inducing and then restricting a representation. Before stating it we recall that for any group G with subgroups H, K we have a double coset decomposition

$$G = \cup Ha_iK \tag{6.7.11}$$

into disjoint sets. It is obtained by letting the direct product $H \times K$ act on G by the rule: $x \mapsto h^{-1}xk$ for $x \in G$, $(h, k) \in H \times K$. Each orbit has the form HaK, for some

$a \in G$, and (6.7.11) is the partition of G into orbits. The element $(h, k) \in H \times K$ leaves $a \in G$ fixed iff $h^{-1}ak = a$, i.e. $k = a^{-1}ha$; hence the stabilizer of a is $H^a \cap K$, and so we obtain the following formula for the size of a double coset:

$$|HaK| = |H|.|K|/|H^a \cap K|.$$

Theorem 6.7.2 (Mackey). *Let G be a finite group, H, K subgroups of G and (6.7.11) a double coset decomposition of G for H, K. If V is any H-module and $H_i = H^{a_i} \cap K$; then*

$$\mathrm{res}_K \,\mathrm{ind}_H^G(V) = \oplus_i \mathrm{ind}_{H_i}^K \,\mathrm{res}_{H_i}(V^{a_i}). \tag{6.7.12}$$

Hence if α is the character of the representation afforded by V and α^g is again defined by $\alpha^g(x) = \alpha^\bullet(x^{g^{-1}})$, for $x, x^g \in H$, then

$$\mathrm{res}_K \,\mathrm{ind}_H^G(\alpha) = \sum_i \mathrm{ind}_{H_i}^K \,\mathrm{res}_{H_i}(\alpha^{a_i}). \tag{6.7.13}$$

Proof. Take a coset decomposition $G = \cup Ht$ of G for H; then $\mathrm{ind}_H^G V = HV^t$, and for fixed i, the set $\{Ht | t \in Ha_iK\}$ is an orbit for the action of K on the coset space G/H. Hence we find

$$\mathrm{ind}_H^G V = HW_i,$$

where $W_i = \oplus\{V^t | t \in Ha_iK\}$ is a right K-module. Fix $a = a_i$ and put $W = W_i$. For any $y, z \in K$ we have

$$Hay = Haz \Leftrightarrow yz^{-1} \in H^a \cap K = D,$$

say. Hence if $K = \cup Dy$ is a coset decomposition of K over D, then $HaK = \cup Hay$, and so $H^aK = \cup H^a y$. Now

$$W \oplus (V \otimes ay) = \oplus V^a y;$$

here $V \oplus a = V^a$ is an H^a-module, hence also a D-module and we have

$$W = ((V^a)_D)^K = \mathrm{ind}_D^K(V^a)_D,$$

by the definition of induced module. Now (6.7.12) follows by summing over i and restricting to K, and (6.7.13) follows by taking characters in (6.7.12). ∎

Let us apply this result in the special case of $K = H$:

Proposition 6.7.3. *Let G be a finite group, H a subgroup with double coset decomposition $G = \cup Ha_iH$ and α a character of H. Then the induced character $\mathrm{ind}_H^G\, \alpha$ is irreducible if and only if α is irreducible and $(\alpha^{a_i}, \mathrm{res}_{H_i}(\alpha))_{H_i} = 0$ for all $a_i \notin H$, where $H_i = H^{a_i} \cap H$.*

Proof. Write $\beta = \mathrm{ind}_H^G\alpha$; by Frobenius reciprocity (6.7.10) we have

$$(\beta, \beta)_G = (\alpha, \mathrm{res}_H\, \beta)_H, \tag{6.7.14}$$

and by Mackey's formula (6.7.13),

$$\mathrm{res}_H \beta = \sum_i \mathrm{ind}_{H_i}^H \mathrm{res}_{H_i}(\alpha^{a_i}).$$

Applying Frobenius reciprocity once again, we have

$$(\alpha, \mathrm{ind}_{H_i}^H \mathrm{res}_{H_i}(\alpha^{a_i}))_H = (\mathrm{res}_{H_i}(\alpha), \alpha^{a_i})_{H_i} = d_i,$$

say. Substituting into (6.7.14), we obtain

$$(\beta, \beta)_G = \sum_i d_i.$$

If the double coset $H = HH$ is represented by a_1, then $d_1 = (\alpha, \alpha) \geq 1$. For β to be irreducible, we must have $(\beta, \beta)_G = 1$, i.e. $d_1 = 1$ and $d_i = 0$ for $i > 1$, and these are just the conditions stated. ■

In particular, when H is normal in G, then $H_i = H$, $\mathrm{res}_{H_i} \alpha = \alpha$ and we find

Corollary 6.7.4. *Let G be a finite group and H a normal subgroup. For any character α of H, $\mathrm{ind}_H^G \alpha$ is irreducible if and only if α is irreducible and different from α^a for all $a \notin H$.* ■

Exercises

1. Show that if $N \lhd G$ and α is a class function on N, then $\mathrm{ind}_H^G \alpha$ vanishes outside N.
2. Let α be a character of a subgroup H of G and define α^G as in (6.7.5). Show directly that $(\alpha^G, \chi)_G$ is a non-negative integer for every irreducible character χ of G, and deduce that α^G is a character of G.
3. If $\alpha^{\blacklozenge}$ denotes the contragredient of ρ (see Exercise 3 of 6.1), show that $(\rho^G)^{\blacklozenge} = (\rho^{\blacklozenge})^G$.
4. Given a group G with subgroup H and characters α, β on H, G respectively, show that $\mathrm{ind}_H^G(\alpha, \mathrm{res}_H \beta) = (\mathrm{ind}_H^G \alpha)\beta$.
5. Given a group G and subgroups H, K, show that if α, β are characters of H, K respectively, then

$$(\mathrm{ind}_H^G \alpha, \mathrm{ind}_K^G \beta) = \sum (\alpha^x, \beta^y)_{H^x \cap K^y},$$

where $x^{-1}y$ runs over a set of double coset representatives.
6. (A. H. Clifford) Let G be a finite group and H a normal subgroup. Show that if V is a simple G-module and U a simple H-submodule of V, qua H-module, then Ux is a simple H-module for each $x \in G$. Deduce that V is semisimple as H-module. Show also that for each simple H-type α (i.e. isomorphism class of simple submodules) the α-socle has the same length.

6.8 Applications: The theorems of Burnside and Frobenius

In this section we shall apply the earlier work to prove some results of pure group theory. In the first place we have the $p^\alpha q^\beta$-theorem of Burnside, which states that any finite group whose order is divisible by only two distinct primes is soluble. This was proved by William Burnside in 1904 using character theory; for odd primes a purely group theoretic proof was found by John Thompson about 1960, but this is far more complicated.

Let G be a finite group; we denote its conjugacy classes again by $\{C_\lambda\}$ and denote the sum of the elements in C_λ (in the group algebra kG) by c_λ and their number by h. We recall from (6.4.10) that for any irreducible character χ of degree d,

$$h_\lambda \chi^{(\lambda)} = \eta_\lambda d, \tag{6.8.1}$$

where η_λ is given by $\rho(c_\lambda) = h_\lambda I$, ρ being the corresponding representation. Further we saw in Section 6.4 that both $\chi^{(\lambda)}$ and h_λ are algebraic integers.

Lemma 6.8.1. *Let G be a finite group, ρ a complex irreducible representation of G of degree d with character χ and let C_λ be a conjugacy class of G with h_λ elements. If h_λ is prime to d, then either χ vanishes on C_λ or $\rho(x) = c.I$ for all $x \in C_\lambda$, for some $c \in \mathbf{C}$.*

Proof. We have $\chi^{(\lambda)} = \omega_1 + \ldots + \omega_d$, where each ω_i is a root of 1; hence $|\chi^{(\lambda)}| \leq d$, with strict inequality unless all the ω_i are equal. In the latter case the elements of C_λ are represented by scalars; otherwise we have $|\chi^{(\lambda)}/d| < 1$, and the same holds for all conjugates of $\chi^{(\lambda)}$ over $\mathbf{Q}$. Since d is prime to h_λ, there exist integers a, b such that $ah_\lambda + bd = 1$, and so

$$\begin{aligned}\chi^{(\lambda)} = (ah_\lambda + bd)\chi^{(\lambda)} &= ad\eta_\lambda + bd\chi^{(\lambda)} \quad \text{by} \quad (6.8.1)\\ &= d(a\eta_\lambda + b\chi^{(\lambda)}).\end{aligned}$$

Now both η_λ and $\chi^{(\lambda)}$ are algebraic integers, hence so is $\chi^{(\lambda)}/d$, and the same holds for its conjugates. Therefore the norm $N(\chi^{(\lambda)}/d)$ is an integer; as we have seen, it is less than 1 in absolute value, so $N(\chi^{(\lambda)}/d) = 0$, and it follows that $\chi^{(\lambda)} = 0$, as claimed. ∎

This lemma has the following important consequence. We recall that for any element x of a finite group G the number of conjugates of x in G is the index $(G : C_x)$, where C_x is the centralizer of x in G (see BA, Section 2.1). In what follows we understand by a *simple* group a simple non-abelian group.

Theorem 6.8.2 (Burnside). *If a finite group G has a conjugacy class C_λ consisting of p^m elements, where $m \geq 1$ and p is prime, then G cannot be simple.*

Proof. Let $d_1, \ldots, d_r$ be the degrees of the irreducible representations of G; we have seen in (6.3.10) that

$$\sum d_i^2 = |G|, \tag{6.8.2}$$

and here the right-hand side is divisible by p, because $p^m = |C_\lambda|$ is the index of a centralizer. On the left of (6.8.2) we have $d_1 = 1$, so two cases can arise. Either there are more linear representations of G, in which case by Proposition 6.4.2 their number is $(G : G')$, where G' is the derived group, hence $G' \neq G$ and so G is not simple; or all the non-trivial representations have degree greater than 1; then by (6.8.2) there must be a representation ρ_i of degree d_i prime to p. By Lemma 6.8.1, either $\chi_i^{(\lambda)} = 0$ or $\rho_i(x)$ for $x \in C_\lambda$ is a scalar. In the latter case either ρ_i is not faithful or x lies in the centre of G; each time it follows that G is not simple. This only leaves the alternative $\chi_i^{(\lambda)} = 0$. Thus for any character χ_j we have either $\chi_j^{(\lambda)} = 0$ or $\chi_j^{(1)} \equiv 0 \pmod p$, except for $j \neq 1$, when $\chi_1^{(1)} = \chi_1^{(\lambda)} = 1$. By orthogonality we have $\sum_j \chi_j^{(1)} \chi_j^{(\lambda)} = 0$, hence $1 \equiv 0 \pmod p$, which is a contradiction. ■

Now it is an easy matter to deduce the solubility criterion:

Theorem 6.8.3 (Burnside). *Every group of order $p^\alpha q^\beta$, where $\alpha, \beta \geq 1$ and p, q are primes, is soluble.*

Proof. Take a simple factor H of order $p^a q^b$; by Theorem 6.8.2 no conjugacy class can have prime power order, hence $a, b \geq 1$ and moreover, any conjugacy class has order divisible by p and q. The class equation for H reads

$$p^a q^b = 1 + \sum p^{a_i} q^{b_i},$$

where a, b and all the a_i, b_i are positive. This is a contradiction and the result follows. ■

Secondly we give a result proved by Georg Frobenius in 1901 on the existence of complements in a group. If G is any group and H, K are subgroups such that $H \cap K = 1$, $HK = G$, then each of H, K is called a *complement* of the other; this term is used particularly when one of H, K is normal in G. The main result is preceded by a lemma which is needed for the proof.

Lemma 6.8.4. *Let G be a finite group and H a non-trivial subgroup such that $H^x \cap H = 1$ for all $x \notin H$. If α is a class function on H such that $\alpha(1) = 0$ and $\alpha^G = \text{ind}_H^G \alpha$, then $(\alpha^G)_H = \alpha$ and for any class function β on H, $(\alpha^G, \beta^G)_G = (\alpha, \beta)_H$.*

Proof. We have $\alpha^G(1) = (G : H)\alpha(1) = 0$, by hypothesis. Let us fix $h \in H$, where $h \neq 1$. Then by definition, we have

$$\alpha^G(h) = |H|^{-1} \sum_{x \in G} \alpha(h^x). \tag{6.8.3}$$

Consider a term in this sum; if $\alpha(h^x) \neq 0$, then $h^x \in H^x \cap H$, so $x \in H$ and $\alpha(h^x) = \alpha(h)$. Thus all the non-zero terms in the sum on the right of (6.8.3) are $\alpha(h)$, and there is a term for each $x \in H$. Hence

$$\alpha^G(h) = |H|^{-1} \sum_{x \in H} \alpha(h^x) = \alpha(h),$$

and this proves the first assertion. For any other class function β on H, Frobenius reciprocity gives $(\alpha^G, \beta^G)_G = ((\alpha^G)_H, \beta)_H = (\alpha, \beta)_H$, by what has been proved. ■

The actual result of Frobenius can be stated in two equivalent forms, in terms of permutation groups or abstractly.

Theorem 6.8.5 (Frobenius). *Let G be a finite transitive permutation group such that no element $\neq 1$ has more than one fixed point. Then the elements without fixed point, together with 1, form a normal subgroup N of G.*

The normal subgroup N is called the *Frobenius kernel.* We remark that if H is the stabilizer of a point, then

$$NH = G$$

by transitivity (see the proof below), while $N \cap H = 1$ holds by hypothesis. Thus N is a complement of H, and this theorem may also be stated in the following form:

Theorem 6.8.6 (Frobenius). *Let G be a finite group with a subgroup H such that*

$$H^x \cap H = 1 \quad \text{for all } x \notin H, \tag{6.8.4}$$

i.e. H meets each of its conjugates in 1 and is its own normalizer. Then H has a normal complement in G.

A group G with these properties is called a *Frobenius group* and the subgroup H with the property (6.8.4) is called a *Frobenius subgroup* of G. So Theorem 6.8.6 states that every Frobenius subgroup has a normal complement.

Proof. Let us first show the equivalence of Theorems 8.5 and 8.6. Under the hypothesis of Theorem 6.8.5 let H be the stabilizer of a point. If $H \lhd G$, then all points have the same stabilizer. In that case $H = G$ or $H = 1$ and the conclusion follows with $N = 1$ or G respectively, so we may exclude this case. Further, the conditions of Theorem 6.8.5 mean that any $x \in G\backslash H$ moves the point fixed by H to another point, not fixed by any element in H^+ (where $H^+ = H\backslash\{1\}$), so that $H^x \cap H = 1$, while the elements moving all points make up precisely the set $G\backslash \cup H^x$. Conversely, given the hypotheses of Theorem 6.8.6, we see that on taking the coset representation on G/H that no element $\neq 1$ fixes more than one point and the elements moving all points comprise the complement of the union of all the conjugates of H in G. If they, together with 1, form a subgroup (and this is what we have to prove), then it must be a complement of H in G.

Let us denote this set, namely $(G\backslash \cup H^x) \cup \{1\}$, by N. It is clear that N is a normal set (i.e. a union of conjugacy classes), and writing $|G| = n$, $|H| = m$, $n = mr$, we have $|N| = n - r(m-1) = r$, so if N is a subgroup, it will be a complement of H because $r = (G : H)$ and clearly $H \cap N = 1$. So all we need to show is that N is a subgroup. We shall obtain it as the kernel of a certain representation.

Let α be any class function on H and put $\bar{\alpha} = \alpha - \alpha(1).1_H$, where 1_H is the trivial character on H. Then $\bar{\alpha}(1) = 0$, and $\bar{\alpha}$ is a class function on G. Put

$$\alpha^* = \bar{\alpha}^G + \alpha(1).1_G. \tag{6.8.5}$$

Then

$$\begin{aligned}(\alpha^*, \alpha^*)_G &= (\bar{\alpha}^G, \bar{\alpha}^G)_G + 2\alpha(1)(\bar{\alpha}^G, 1_G)_G + \alpha(1)^2 \\ &= (\bar{\alpha}, \bar{\alpha})_H + 2\alpha(1)(\bar{\alpha}, 1_H)_H + \alpha(1)^2,\end{aligned}$$

by Lemma 6.8.4. Hence

$$(\alpha^*, \alpha^*)_G = (\bar{\alpha} + \alpha(1)1_H, \bar{\alpha} + \alpha(1)1_H)_H = (\alpha, \alpha)_H.$$

If α is an irreducible character, then α^* is either a character or a difference of characters of G, i.e. a *virtual* character, and moreover $(\alpha^*, \alpha^*)_G = (\alpha, \alpha)_H = 1$, so either α^* or $-\alpha^*$ is an irreducible character of G; in fact it must be α^*, because $\alpha^*(1) = \bar{\alpha}^G(1) + \alpha(1) > 0$. Further we have by Lemma 6.8.4,

$$\text{res}_H\, \alpha^* = \text{res}_H\, \bar{\alpha} + \alpha(1).1_H = \bar{\alpha} + \alpha(1)1_H = \alpha.$$

Now consider $K = \bigcap \ker \alpha^*$, where the intersection is taken over all irreducible characters α on H and ker indicates the kernel of the corresponding representation. Then $K \lhd G$ and for any irreducible character α on H and $x \in H \cap K$ we have

$$\alpha(x) = \alpha^*(x) = \alpha^*(1);$$

hence $H \cap K = 1$ by Corollary 6.4.9. Since K is normal in G, we also have $H^x \cap K = 1$ for all $x \in G$, and so $K \subseteq N$. Here equality holds, for if $x \in N^+$, then $x \notin H^y$ for all $y \in G$, hence $\bar{\alpha}^G(x) = 0$ for all irreducible characters α of H. So in that case $\alpha^*(x) = \alpha^*(1)$ by (6.8.5), and so $x \in K$ by Proposition 6.4.8. This shows that $N = K$ and it proves N to be a subgroup. ∎

So far no proof of this result without representation theory is known, although special cases (e.g. for soluble groups) have been proved in this way. Burnside has shown that the Sylow p-subgroups of the Frobenius subgroup are either cyclic, or in case $p = 2$, generalized quaternion groups, and Thompson has shown that the Frobenius kernel is nilpotent (see Huppert (1967)).

Exercises

1. Let the finite group G be a split extension of N by a subgroup H. Show that H is a Frobenius subgroup iff H acts freely on $H^+ = H\backslash\{1\}$, i.e. for any $x \in N^+$, $h \in H^+$, then $x^h \neq x$.
2. Verify that in a dihedral group of order $2m$, where m is odd, the cyclic subgroup of order m is a Frobenius kernel.
3. Let G be a finite group with Frobenius subgroup H. Show that under the action of H on the coset space G/H there is one orbit of length 1, while all the others have

length $|H|$. Deduce that $|H|$ divides $(G : H) - 1$ and that the Frobenius kernel N has order prime to its index.

4. Show that if a conjugacy class C of a non-trivial group G has p^m elements, where $m \geq 1$ and p is prime, then the set $C^{-1}C$ generates a proper normal subgroup of G.
5. Give a direct proof of Theorem 6.8.5 in the case where the set acted on by G has p^m elements, where $m \geq 1$ and p is prime.
6. Show that every normal Hall subgroup (i.e. of order prime to its index, see Section 3.2) is characteristic in G (i.e. admitting all automorphisms of G).

Further exercises on Chapter 6

1. Let G be a soluble group of order divisible by a prime p. If $N = G_{i-1}/G_i$ is a chief factor of order p^r of G (see Section 3.1), show that the action of G on N induced by inner automorphisms is a representation of degree r over the field $\mathbf{F}_p$ of p elements.
2. Show that an irreducible representation of a p-group over a field of characteristic p is necessarily trivial. Find a (reducible) representation of degree 2 of $\mathbf{C}_p$ over $\mathbf{F}_p$ which is faithful.
3. Show that any representation of a finite group G over $\mathbf{Q}$ is equivalent to a representation over $\mathbf{Z}$. (Hint. Take a basis (u_i) of the representation module, let N be a common denominator for all the representation coefficients and consider the subgroup A of $\sum \mathbf{Z}u_i$ generated by all the expressions Nu_ix $(x \in G)$; show that A is again a G-module and find a $\mathbf{Z}$-basis for it.)
4. Show that in an irreducible representation of a finite group G over an algebraically closed field k, each element of the centre of G is represented by a scalar matrix. Show that this holds even if k is not algebraically closed but contains a primitive $|G|$-th root of 1.
5. Let G be a group with a faithful irreducible representation over a field of characteristic 0. Show that the centre of G is cyclic. (Hint. Use Exercise 4 to prove first the special case where the ground field contains a primitive $|G|$-th root of 1.)
6. Show that for a permutation representation of a group G with character χ the number of orbits is $(\chi, 1_G)$, where 1_G is the trivial character.
7. Show that if a character χ of a faithful representation assumes r distinct values on G, then the Vandermonde determinant $(\chi^{(\lambda)i})$ is non-zero. Deduce that every irreducible character is a constituent of at least one of $1_G, \chi, \chi^2, \ldots, \chi^{r-1}$.
8. Show that a doubly transitive permutation representation (i.e. transitive on the pairs) is a sum of the trivial representation and one other irreducible representation. (Hint. The stabilizer must be transitive; now use Theorem 6.4.11 and the orthogonality relations.)
9. For each representation $\rho(x) = (\rho_{ij}(x))$ of degree d define d^2 elements of the group algebra as $\rho_{ij} = \sum \rho_{ij}(x)x$. Show that the orthogonality relations take the form: for irreducible representations ρ, σ that are inequivalent, $\rho_{ij}\sigma_{pq} = 0$, while $\rho_{ij}\rho_{pq} = (|G|/d)\delta_{jp}\rho_{iq}$.

10. Let G be a finite group. Show that if every irreducible representation of G over $\mathbf{C}$ is 1-dimensional, then G is abelian. (Hint. Diagonalize a matrix of the regular representation.)
11. (J. A. Green) Show that if V is a simple G-module over a finite field F, then $E = \mathrm{End}_G(V)$ is a finite field. Further, if $|F| = q$, then $V \otimes V \otimes \ldots \otimes V$ (n factors) has exactly $(q^n - 1)/(q - 1)$ simple submodules.
12. Show that the affine group $A = \mathbf{Aff}_1(\mathbf{F}_q)$ of all transformations $x \mapsto ax + b$ ($a, b \in \mathbf{F}_q$, $a \neq 0$) is a Frobenius group with the translation group as kernel and the stabilizer of 0 as complement.
13. Let G be a finite group and $X = (\chi_i^{(\lambda)})$ its character table. Show that any automorphism of G induces a permutation of the rows of X, and a permutation of the columns, where the effect of α on X is defined by $\chi_i^\alpha(x) = \chi_i(x^\alpha)$. If $P(\alpha)$, $Q(\alpha)$ are the permutation matrices describing the effects of α on the rows and columns of X respectively, then $X^\alpha = P(\alpha)X = XQ(\alpha)$. Deduce that $P(\alpha)$, $Q(\alpha)$ are conjugate and hence have the same trace. Hence show that for any group A of automorphisms acting on G the number of orbits of the set of irreducible characters is the same as the number of orbits of the conjugacy classes of G under the action of A.
14. Let G be a Frobenius group with kernel K and complement H. Show that for any non-trivial irreducible character α of K, $\mathrm{ind}_K^G\, \alpha$ is an irreducible character of G. (Hint. Consider the group of automorphisms of K induced by H; show that $c \in H^+$ fixes no conjugacy class $\neq \{1\}$ of K, and use Exercise 13 to show that for any non-trivial character χ of K, $\chi^c \neq \chi$; now apply reciprocity to show that $\mathrm{ind}_H^G\, \alpha$ is irreducible.)
15. Let G, H, K be as in Exercise 14. Show that any irreducible character χ of G is either trivial on K or induced up from an irreducible character ψ of K. Deduce that for such χ, ψ and any $c \in H$, $\mathrm{res}_H^G\, \chi(c) = \delta_{c1}\psi(1).|H|$.
16. Let H be a subgroup of S_n and for $h \in H$ denote by F_h the set of numbers left fixed by h. Show that $\varphi(h) = |F_h| - 1$ is a character of H. (Hint. Let S_n act by permutations on a basis $u_1, \ldots, u_n$ of an n-dimensional vector space V and take a decomposition of V including the subset spanned by $\sum u_i$.)

7 Noetherian rings and polynomial identities

The Artinian condition on rings leads to a very satisfactory theory, at least in the semisimple case, yet it excludes such familiar examples as the ring of integers. This ring is included in the wider class of Noetherian rings, which has been much studied in recent years. We shall present some of the highlights, such as localization (Section 7.1), non-commutative principal ideal domains (Section 7.2) and Goldie's theorem (Section 7.4), and illustrate the theory by examples from skew polynomial rings and power series rings in Section 7.3.

Another condition which helps to make rings amenable is the existence of a polynomial identity. The topics treated include generic matrix rings and central polynomials (Section 7.7) and the theorems of Regev (Section 7.6) and Amitsur (Section 7.8), as well as some generalities in Section 7.5, while Kaplansky's basic theorem on PI-rings is reserved for Chapter 8.

7.1 Rings of fractions

In BA, Section 10.3 we constructed a ring of fractions R_S for a commutative ring R with respect to the multiplicative subset S. This construction can be carried out quite generally, as we shall see in Theorem 7.1.1 below, which should be compared with Theorem 10.3.1 of BA. It is not even necessary to assume R commutative. But for the result to be of use, we must have a means of comparing expressions in R_S and this will require further hypotheses to be imposed. This will be done in Theorem 7.1.3 below. For a general treatment it is necessary to invert matrices rather than just elements, but this will not be needed here (see Cohn (1985)).

Given any ring R and a subset X of R, we say that a homomorphism $f : R \to R'$ is *X-inverting* if it maps each element of X to an invertible element of R'. For fixed R and X, the pairs (R, f), where f is X-inverting, can be regarded as objects in a category whose morphisms are commutative triangles, and an initial object in this category is called a *universal X-inverting homomorphism*. An element c in a ring R is called *left regular* if $xc = 0$ implies $x = 0$, i.e. c is not a right zero-divisor; *right regular* elements are defined similarly, and a *regular* element is one which is left and right regular.

Theorem 7.1.1. *Let R be any ring and S a subset of R. Then there exists a ring R_S and a homomorphism $\lambda : R \to R_S$ which is universal S-inverting.*

Proof. In detail the assertion means that λ is S-inverting and every S-inverting homomorphism can be factored uniquely by λ. To construct R_S we take for each $a \in R$ a symbol p_a and for each $s \in S$ an additional symbol q_s and form the ring R_S on all these symbols as generators, with the defining relations

$$p_1 = 1, p_a + p_b = p_{a+b}, p_a p_b = p_{ab}, p_s q_s = q_s p_s = 1, \text{ for all } a, b \in R, s \in S. \quad (7.1.1)$$

The first three sets of equations ensure that the mapping $\lambda : a \mapsto p_a$ is a homomorphism of R into R_S, while the fourth shows that λ is S-inverting. Now given any S-inverting homomorphism $f : R \to R'$, we can define a homomorphism g of the free ring F on the p's and q's into R' by the rules $p_a \mapsto af$, $q_s \mapsto (sf)^{-1}$. It is clear that g preserves the relations (7.1.1); hence it induces a homomorphism $g_1 : R \to R'$, and it is easily checked that $\lambda g_1 = f$. Moreover, g_1 is uniquely determined by this equation, since its value on p_a is given, while its value on q_s is determined by (7.1.1): if $q_s g_1 = c$, then $sf.c = (p_s q_s)g_1 = 1 = (q_s p_s)g_1 = c.sf$, so $c = (sf)^{-1}$. Thus g_1 is determined on a generating set of R_S and hence is unique. ∎

Corollary 7.1.2. *Let R, R' be rings with subsets S, S' respectively. Then any homomorphism $f : R \to R'$ which maps S into S' can be extended in a unique way to a homomorphism of R_S into $R'_{S'}$.*

Proof. The composition $R \xrightarrow{f} R' \xrightarrow{\lambda'} R'_{S'}$ is S-inverting and so can be factored uniquely by $\lambda : R \to R_S$ to give the required homomorphism $f_1 : R_S \to R'_{S'}$. ∎

The ring R_S constructed in Theorem 7.1.1, together with the homomorphism $\lambda : R \to R_S$ is called the *universal S-inverting ring* for R or also the *localization* of R at the set S. To study it in more detail we need to make some simplifying assumptions. A very simple but most fruitful idea, due to Oystein Ore [1931] (and independently, to Emmy Noether, unpublished) is to look at the case where all the elements of R_S can be written as simple fractions $(a\lambda)(s\lambda)^{-1}$. If this is to be possible, we must in particular be able to express $(s\lambda)^{-1}(a\lambda)$ in this form: $(s\lambda)^{-1}(a\lambda) = (a_1\lambda)(s_1\lambda)^{-1}$, hence on multiplying up, we obtain

$$(as_1)\lambda = (sa_1)\lambda, \quad (7.1.2)$$

and this provides a clue to the condition needed. Of course, if every element is to be expressed as a fraction with denominator in S, we must also assume S to be *multiplicative*, i.e. to contain 1 and be closed under products.

Theorem 7.1.3. *Let R be a ring and S a multiplicative subset of R such that*

O.1 *$aS \cap sR \neq \emptyset$ for all $a \in R, s \in S$,*
O.2 *for each $a \in R, s \in S$, if $sa = 0$, then $at = 0$ for some $t \in S$.*

Then the elements of the universal S-inverting ring R_S can be constructed as fractions a/s, where

$$a/s = a'/s' \Leftrightarrow au = a'u' \text{ and } su = s'u' \in S \text{ for some } u, u' \in R. \qquad (7.1.3)$$

Moreover, the kernel of the natural homomorphism $\lambda : R \to R_S$ is

$$\ker \lambda = \{a \in R | at = 0 \text{ for some } t \in S\}. \qquad (7.1.4)$$

Condition **O**.1 is called the *right Ore condition.* A multiplicative subset S of R satisfying **O**.1 is called a *right Ore set*; if S also satisfies **O**.2, it is called (right) *reversible* or also a *right denominator set.* Such a set allows the construction of *right* fractions $a/s = (a\lambda)(s\lambda)^{-1}$, by Theorem 7.1.3. They must be carefully distinguished from *left* fractions $(s\lambda)^{-1}(a\lambda)$. By symmetry we have the notion of a (reversible) *left Ore set*, which allows us to construct all the elements of R as left fractions, and the set S in Theorem 7.1.3 may well be a right but not left Ore set.

Proof. The proof of Theorem 7.1.3 is similar to the commutative case (BA, Theorem 10.3.1), though more care is needed, owing to the lack of commutativity. Guided by (7.1.3), we define a relation on $R \times S$ by writing

$$(a, s) \sim (a', s') \Leftrightarrow \text{ there exist } u, u' \in R \text{ such that } au = a'u',\ su = s'u' \in S.$$

We claim that this is an equivalence. Clearly it is reflexive and symmetric. To prove transitivity, let $(a, s) \sim (a', s')$, $(a', s') \sim (a'', s'')$; say $au = a'u'$, $su = s'u' \in S$, $a'v = a''v'$, $s'v = s''v' \in S$. By **O**.1 there exist $z \in S$, $z' \in R$ such that $s'u'z = s'vz'$, hence $s'u'z \in S$ (by multiplicativity) and moreover, $s'(u'z - vz') = 0$, hence by **O**.2 there exists $t \in S$ such that $u'zt = vz't$. Now we have $auzt = a'u'zt = a'vz't = a''v'z't$, $suzt = s'u'zt = s'vz't = s''v'z't$, and this lies in S because $s'u'z \in S$ and $t \in S$. Thus $(a, s) \sim (a'', s'')$.

We thus have an equivalence on $R \times S$; let us write a/s for the equivalence class containing (a, s) and call a the *numerator* and s the *denominator* of this expression. We note that (7.1.3) now holds by definition, and it may be interpreted as saying that two fractions are equal iff when they are brought to a common denominator, their numerators agree. Of course it follows from **O**.1 that any two expressions can be brought to a common denominator. For this reason we can define the addition of fractions by the rule

$$a/s + b/s = (a + b)/s. \qquad (7.1.5)$$

Here it is necessary to check that the expression on the right depends only on a/s, b/s and not on a, b, s, a task which may be left to the reader. To define the product of a/s and b/t we determine $b_1 \in R$ and $s_1 \in S$ such that $bs_1 = sb_1$ and then put

$$(a/s)(b/t) = ab_1/ts_1.$$

Again the proof that this is well-defined is left to the reader. Now it is easy to check that with these operations the classes a/s form a ring T say, and the mapping $a \mapsto a/1$ is an S-inverting homomorphism from R to T. Moreover, if $f : R \to R'$ is any S-inverting homomorphism, then the mapping $f_1 : R \times S \to R'$ given by $(a, s) \mapsto (af)(sf)^{-1}$ is constant on each equivalence class and so can be factored via T

to provide a homomorphism $f' : T \to R'$ such that $f = \lambda f'$. Here f' is unique, because it is determined on $a/1$ and $1/s$, so by uniqueness T is indeed the universal S-inverting ring. Finally, $\ker \lambda$ consists of all $a/1 = 0/1$, i.e. by (7.1.3), all a such that $at = 0$ for some $t \in S$. ■

An important case is that where S lies in the centre of R. Then **O**.1–**O**.2 are automatic and we have

Corollary 7.1.4. *Let R be a ring and S any multiplicative subset of the centre of R. Then S is a reversible Ore set and the universal S-inverting ring R consists of all fractions a/s ($a \in R, s \in S$), where*

$$a/s = a'/s' \Leftrightarrow (as' - sa')t = 0 \text{ for some } t \in S. \quad ■$$

The conditions of Theorem 7.1.3 simplify slightly when S consists entirely of regular elements. Then **O**.2 is superfluous and $\ker \lambda = 0$. We state this as

Corollary 7.1.5. *Let R be a ring and S a right Ore subset of regular elements. Then the natural homomorphism $\lambda : R \to R_S$ is injective.* ■

The subset T of all regular elements in R is always multiplicative and satisfies **O**.2. When it satisfies **O**.1, we can form R_T; this is called the *total* (or *classical*) quotient ring. Generally one understands by a *quotient ring* a ring in which every regular element is a unit.

Finally we note the special case of integral domains.

Corollary 7.1.6. *Let R be an integral domain such that*

$$aR \cap bR \neq 0 \text{ for all } a, b \in R^\times. \tag{7.1.6}$$

Then R is a regular Ore set, $K = R_{R^\times}$ is a skew field and the natural mapping $\lambda : R \to K$ is an embedding. Conversely, if R is an integral domain with an embedding in a skew field whose elements all have the form ab^{-1} ($a, b \in R, b \neq 0$), then (7.1.6) holds. ■

The skew field K occurring here is called the *field of fractions* of R. The special case (7.1.6) of **O**.1 was used by Ore [1931] in his proof of Corollary 7.1.6; since then there have been many papers generalizing Ore's construction to the case of Theorem 7.1.3 or a special case. An integral domain satisfying (7.1.6) is called a *right Ore domain*; *left Ore domains* are defined similarly and an *Ore domain* is a domain which is left and right Ore.

The following property of fractions is often useful.

Proposition 7.1.7. *Let R be a ring with a reversible right Ore set S and universal S-inverting ring R_S. Then any finite set in R_S can be brought to a common denominator.*

Proof. We shall use induction on the number of elements. Let $a_i/s_i \in R$ ($i - 1, \dots, n$) be given. For $n = 1$ there is nothing to prove, so by induction we may assume the fractions to be in the form $a_1/s_1, a_2/s, \dots, a_n/s$. By **O**.1 there exist $t \in S$, $c \in R$ such that $sc = s_1 t = u \in S$, hence the fractions can be written $a_1 t/u$, $a_2 c/u, \dots, a_n c/u$. ■

In any integral domain R (commutative or not) the additive subgroup generated by 1 is a commutative subring, necessarily a homomorphic image of $\mathbf{Z}$, hence it is either $\mathbf{Z}$ or $\mathbf{Z}/p$ for some prime p. Accordingly we say that R has *characteristic* 0 or p; this generalizes the customary usage for fields. If K is any commutative ring, a K-algebra A is said to be *faithful* if the natural homomorphism $K \to A$ is injective. Thus we can say that every integral domain can be regarded as a faithful K-algebra, where $K = \mathbf{Z}$ or $\mathbf{F}_p$.

Suppose now that R is an integral domain, hence a faithful K-algebra ($K = \mathbf{Z}$ or $\mathbf{F}_p$), but not right Ore. Then R contains $a, b \neq 0$ such that $aR \cap bR = 0$. We claim that the K-subalgebra of R generated by a and b is free on a, b. For let $f(x, y)$ be a non-zero polynomial in the non-commuting variables x, y with coefficients in K, such that $f(a, b) = 0$, and choose f to have the least possible degree. We can write

$$f(x, y) = \alpha + x f_1(x, y) + y f_2(x, y), \quad \text{where } \alpha \in K,$$

and f_1, f_2 are not both zero, for otherwise $\alpha = 0$ and f would be the zero polynomial. Suppose that $f_1 \neq 0$; then $f_1(a, b) \neq 0$, by the minimality of deg f, but $f(a, b) = 0$, hence $f(a, b)b = 0$ and so

$$a f_1(a, b)b = b(-\alpha - f_2(a, b)b),$$

and this is non-zero because the left-hand side is non-zero. This contradicts the fact that $aR \cap bR = 0$, and it proves

Proposition 7.1.8. *Any integral domain R is a faithful K-algebra, where K is $\mathbf{Z}$ or $\mathbf{F}_p$, according as R is of characteristic 0 or p. Further, R is either a left and right Ore domain, or it contains a free K-algebra on two free generators.* ■

We observe that the last two possibilities are not mutually exclusive; an Ore domain may well contain a free algebra. We also recall from BA, Further Exercise 9 of Chapter 6, that an algebra containing a free algebra of rank 2 contains a free algebra of countable rank.

An interesting observation made by Alfred Goldie [1958] (actually the simplest case of Proposition 7.4.8 below) is that the Ore condition is a consequence of the Noetherian condition.

Proposition 7.1.9. *An integral domain either is a right Ore domain or it contains free right ideals of infinite rank. In particular, any right Noetherian domain is right Ore.*

Proof. Let R be an integral domain and suppose that R is not right Ore. Then there exist $a, b \in R^{\times}$ such that $aR \cap bR = 0$; now the conclusion will follow if we show that the elements $b, ab, a^2 b, \dots$ are right linearly independent over R. Suppose that $\sum a^i bc_i = 0$ and let c_ν be the first non-zero coefficient. We can cancel a^ν and obtain the relation $bc_\nu + abc_{\nu+1} + \ldots + a^{n-\nu}bc_n = 0$, i.e.

$$a(bc_{\nu+1} + \ldots + a^{n-1-\nu}bc_n) = -bc_\nu \neq 0,$$

and this contradicts the assumption on a, b. ■

A ring is said to be *prime* if $R \neq 0$ and $xRy = 0$ implies $x = 0$ or $y = 0$. Such rings will be discussed in Section 8.5; for the moment let us apply the results found to prime rings.

Proposition 7.1.10. *Let R be a prime ring and S a right Ore set consisting of regular elements of R. Then R is embedded in R_S and R_S is again prime.*

Proof. The mapping $R \to R_S$ is an embedding, by Corollary 7.1.5. Suppose that $uRv = 0$, where $u = as^{-1}$, $v = bt^{-1}$ ($a, b \in R$, $s, t \in S$). Then for any $x \in R$, $axb = as^{-1}.sx.bt^{-1}.t = 0$, hence $a = 0$ or $b = 0$ and accordingly $u = 0$ or $v = 0$. ■

Corollary 7.1.11. *Let R be a prime ring with centre C. Then $C^\times$ consists of regular elements in R; in particular, C is an integral domain, and if its field of fractions is denoted by K, then R is embedded in $R_{C^\times} \cong R \otimes_C K$, and the latter ring is again prime, with centre K.*

Proof. We first show that $C^\times$ consists of regular elements. If $c \in C^\times$ and $ca = 0$ ($a \in R$), then for all $x \in R$, $cxa = xca = 0$, hence $a = 0$. Since $ac = ca$ for $a \in R$, this shows c to be regular. Since K is universal $C^\times$-inverting, it follows that $R \otimes K$ is universal $C^\times$-inverting and by Proposition 7.1.10, the mapping $R \to R_{C^\times} \cong R \otimes K$ is an embedding. Clearly K is contained in the centre of $R \otimes K$; conversely, if ac^{-1} is in the centre, then for any $x \in R$, $ax = ac^{-1}.cx = cx.ac^{-1} = xa$, because $c \in C$. Hence $a \in C$ and so $ac^{-1} \in K$. ■

Exercises

1. Show that every right Artinian ring is its own quotient ring.
2. Show that an integral domain in which every finitely generated right ideal is principal is a right Ore domain (such a ring is called a *right Bezout domain*).
3. Show that in a right Noetherian ring every right Ore set is reversible.
4. Show that for any reversible right Ore set S in a ring R, the ring R_S is flat as left R-module.
5. Show that for any ring R and subset S the natural mapping $R \to R_S$ is an epimorphism in the category of rings.
6. Show that the characteristic can be defined for simple rings as for integral domains, and that a simple ring is a $\mathbf{Q}$-algebra or an $\mathbf{F}_p$-algebra according as the characteristic is 0 or p.
7. Let R be a simple ring and S a reversible right Ore subset. Show that the centre of R_S coincides with the centre of R.

8. Let R be a ring, X a subset consisting of regular elements and S the submonoid of $R^\times$ generated by X. Show that S is right Ore provided that the Ore condition holds in the form $xR \cap rS \neq \emptyset$ for all $x \in X$, $r \in R$.
9. Let F be the free group on x and y. Show that the group ring $\mathbf{Z}F$ is an integral domain, but not left or right Ore.
10. Let R be a ring with right total quotient ring. Show that any right factor of a regular element of R is regular.
11. Let R be a right Ore domain and K its field of fractions. Show that the centre of K consists of all ab^{-1} $(b \neq 0)$ such that $axb = bxa$ for all $x \in R$.

7.2 Principal ideal domains

An important example of Noetherian domains are the *principal ideal domains*, i.e. integral domains in which every right ideal and every left ideal is principal. By Proposition 7.1.9 and Corollary 7.1.6 every principal ideal domain (PID for short) can be embedded in a skew field. It follows that for a square matrix over a PID a left (or right) inverse is in fact two-sided, since this holds over a skew field.

In BA, Section 2.4 we saw that every finitely generated abelian group can be written as a direct sum of cyclic groups of prime power orders. A corresponding result can be proved more generally for finitely generated modules over a (commutative) principal ideal domain. It depends on the fact that every square matrix over such a ring is associated to a diagonal matrix; this is known as the Smith normal form, which is well known for $\mathbf{Z}$ and more generally, any Euclidean domain. Surprisingly there is an analogue for non-commutative PIDs, with a rather strong uniqueness condition which is sometimes useful. To state the result, let us define an *invariant* element in a ring R as a regular element c such that $cR = Rc$, thus the left (or right) ideal generated by c is two-sided. If a, b are regular elements of R, then a is called a *total divisor* of b, in symbols $a\|b$, if there exists an invariant element c such that $a|c|b$. We observe that an element is not generally a total divisor of itself; in fact $a\|a$ iff a is invariant. A simple ring has no non-unit invariant elements and so $a\|b$ in a simple ring implies that either a is a unit or $b = 0$. Further, in a principal ideal domain R, every two-sided ideal is of the form $cR = Rc$, where c is 0 or an invariant element.

We shall write $\text{diag}(d_1, \dots, d_r)$ for a matrix with $d_1, \dots, d_r$ along the main diagonal and zeros elsewhere; this notation will be used even for matrices that are not square. The exact size is usually clear from the context, or will be indicated explicitly.

Theorem 7.2.1. *Let R be a principal ideal domain and $A \in {}^mR^n$. Then there exist $P \in \mathbf{GL}_m(R)$, $Q \in \mathbf{GL}_n(R)$ such that for some integer $r \leq \min(m, n)$,*

$$PAQ = \text{diag}(e_1, \dots, e_r, 0, \dots, 0), \quad e_i\|e_{i+1} \neq 0. \tag{7.2.1}$$

Proof. The aim will be to reduce A to the required form by a number of invertible operations on the rows and columns. In the first place there are the elementary operations; for the columns they are

(i) interchange two columns,
(ii) multiply a column on the right by a unit factor,
(iii) add a right multiple of one column to another.

For a Euclidean domain these operations are enough, but in the general case another operation is needed; this is best illustrated by reducing a 1×2 matrix. We have a matrix $(a\ b)$ and need an invertible 2×2 matrix Q such that

$$(a\ \ b)Q = (k\ \ 0). \tag{7.2.2}$$

Clearly k, the highest common left factor (HCLF) of a and b, will be a generator for the right ideal generated by a and b. We may exclude the case $k = 0$, for then $a = b = 0$ and there is nothing to prove. Thus we have $aR + bR = kR$, say $a = ka_1$, $b = kb_1$, and there exist c_1', d_1' such that $ka_1d_1' - kb_1c_1' = k$, hence $a_1d_1' - b_1c_1' = 1$. By hypothesis $Rc_1' \cap Rd_1'$ is principal, with generator $d_1c_1' = c_1d_1'$ and c_1, d_1 have no common left factor, so there exist $a_1', b_1' \in R$ such that $d_1a_1' - c_1b_1' = 1$. Thus we have

$$\begin{pmatrix} a_1 & b_1 \\ c_1 & d_1 \end{pmatrix}\begin{pmatrix} d_1' & -b_1' \\ -c_1' & a_1' \end{pmatrix} = \begin{pmatrix} 1 & m \\ 0 & 1 \end{pmatrix}.$$

This shows that the first matrix on the left, C say, is invertible. It follows that $(k\ 0)C = (a\ b)$ and $(a\ b)C^{-1} = (k\ 0)$, so C^{-1} is the required matrix. Thus we have a fourth operation

(iv) multiply two columns on the right by an invertible 2×2 matrix.

We can now proceed with the reduction. If $A = 0$, there is nothing to prove; otherwise we bring a non-zero element to the $(1, 1)$-position in A, by permuting rows and permuting columns, using (i). Next we use (iv) to replace a_{11} successively by the HCLF of a_{11} and a_{12}, then by the HCLF of the new a_{11} and a_{13}, and so on. After $n - 1$ steps we have transformed A to a form where $a_{12} = a_{13} = \ldots = a_{1n} = 0$. By symmetry the same process can be applied to the first column of A; in the course of the reduction the first row of A may again become non-zero, but this can happen only if the length (i.e. the number of factors) of a_{11} is reduced; therefore by induction on the length of a_{11} we reach the form $a_{11} \oplus A_1$, where A_1 is $m - 1 \times n - 1$. We now apply the same process to A_1 and by induction on $\max(m, n)$ reach the form

$$\operatorname{diag}(a_1, a_2, \ldots, a_r, 0, \ldots, 0).$$

Consider a_1 and a_2; for any $d \in R$ we have

$$\begin{pmatrix} 1 & d \\ 0 & 1 \end{pmatrix}\begin{pmatrix} a_1 & 0 \\ 0 & a_2 \end{pmatrix} = \begin{pmatrix} a_1 & da_2 \\ 0 & a_2 \end{pmatrix};$$

now we can further diminish the length of a_1 unless a_1 is a left factor of da_2 for all $d \in R$, i.e. unless $a_1R \supseteq Ra_2$. But in that case $a_1R \supseteq Ra_2R \supseteq Ra_2$; thus $a_1|c|a_2$, where c is the invariant generator of the ideal Ra_2R. Hence $a_1||a_2$, and by repeating the argument we obtain the expression on the right of (7.2.1). The totality of column operations amount to right multiplication by an invertible matrix Q, while the row operations yield P and we thus have the equation (7.2.1). ■

We remark that most of the PIDs we encounter will be integral domains with a norm function satisfying the Euclidean algorithm, i.e. Euclidean domains (possibly non-commutative). In that case we can instead of (iv) use the Euclidean algorithm, with an induction on the norm instead of the length, to accomplish the reduction in Theorem 7.2.1 (cf. the proof of Theorem 3.5.1). Further we can dispense with (ii), so P, Q can in this case be taken to be products of elementary matrices.

In the case of simple rings Theorem 7.2.1 still simplifies, since then every invariant element is a unit.

Corollary 7.2.2. *If R is a simple principal ideal domain, then every matrix over R is associated to a matrix of the form* $\mathrm{diag}(1, 1, \dots, 1, a, 0, 0, \dots, 0)$ $(a \in R)$.

Proof. If $a||b$, then either $b = 0$ or a is a unit. Now any unit can be transformed to 1 by applying (ii), so there can only be one diagonal element not 1 or 0. ■

We note the consequences for modules over a PID:

Proposition 7.2.3. *Let R be a principal ideal domain. Then every submodule of a free module of finite rank n is free of rank at most n. Further, any finitely generated R-module is a direct sum of cyclic modules*

$$M = M_1 \oplus \dots \oplus M_r, \quad \text{where } M_{i+1} \text{ is a quotient of } M_i.$$

If moreover, R is simple, then every finite generated R-module is the direct sum of a cyclic module and a free module of finite rank.

Proof. Let F be a free right R-module of finite rank n; we shall use induction on n. If α_1 denotes the projection on the first component R in F, then $\alpha_1 M$ is a right ideal, principal by hypothesis, and we have the exact sequence

$$0 \to M' \to M \to \alpha_1 M \to 0.$$

Here M' is a submodule of $\ker \alpha_1$, while $\alpha_1 M$ is free (of rank 1 or 0), hence the sequence splits and $M \cong M' \oplus \alpha_1 M$, and since M' is free of rank $\leq n-1$ by the induction hypothesis, the first assertion follows. Now the rest is clear from Theorem 7.2.1 and Corollary 7.2.2. ■

In a skew field every non-zero element is a unit, so then Corollary 7.2.2 yields the well-known result that every matrix over a skew field is associated to $I_r \oplus 0$, where r is the rank of the matrix. We can also use Theorem 7.2.1 to describe the rank of a matrix over the rational function field $K(t)$, defined as the field of fractions of $K[t]$, where K is any skew field and t an indeterminate. In $K[t]$ we have the Euclidean algorithm relative to the degree function, hence $K[t]$ is a PID. We further remark that if C is the centre of K, then for any polynomial f of degree n over K and any $\lambda \in C$ such that $f(\lambda) = 0$, we can write $f = (t - \lambda)g$, where g has degree $n - 1$. By induction it follows that f cannot have more than n zeros in C.

Lemma 7.2.4. *Let K be a skew field with infinite centre C, and consider the polynomial ring $K[t]$, with field of fractions $K(t)$. If $A = A(t)$ is a matrix over $K[t]$, then the rank of A over $K(t)$ is the supremum of the ranks of $A(\alpha)$, $\alpha \in C$. In fact, this supremum is assumed for all but a finite number of values of α.*

Proof. Since $K[t]$ is a PID we can by Theorem 7.2.1 find invertible matrices P, Q such that

$$PAQ = \mathrm{diag}(f_1, \ldots, f_r, 0, \ldots, 0), \quad \text{where } f_i \in K[t]. \qquad (7.2.3)$$

The product of the non-zero diagonal terms on the right gives us a polynomial f whose zeros in C are the only points of C at which $A = A(t)$ falls short of its maximum rank, and the number of these values cannot exceed the degree of f. ■

Exercises

1. Show that over a PID any submodule of a free module (even of infinite rank) is free.
2. Give the details of the proof that for a Euclidean domain operations (i), (iii) suffice to accomplish the reduction to the form (7.2.1).
3. Let K be a skew field with centre C. Show that a polynomial over C may well have infinitely many zeros in K. How is this to be reconciled with the remark before Lemma 7.2.4?
4. Let R be a PID and M a right R-module. An element x of M is called a *torsion element* if $xc = 0$ for some $c \in R^{\times}$. Verify that the set tM of all torsion elements of M is a submodule of M (called the *torsion submodule*).
5. If $tM = 0$, M is called *torsion free*. Show that any finitely generated torsion free module over a PID R is free and deduce that any finitely generated R-module splits over its torsion submodule.
6. Let R be any ring and M a right R-module. Show that for any invariant element c of R, Mc is a submodule. If R is a PID and M is presented by a matrix $\mathrm{diag}(a_1, \ldots, a_r, 0, \ldots, 0)$, $a_i \| a_{i+1}$, then for any invariant element c such that $a_i | c | a_{i+1}$ we have $M/Mc \cong R/a_1R \oplus \ldots \oplus R/a_iR \oplus R/cR \oplus \ldots \oplus R/cR$. Deduce that the a_i are unique up to similarity, where a, b are *similar* if $R/aR \cong R/bR$.

7.3 Skew polynomial rings and Laurent series

In commutative ring theory the polynomial ring $R[x]$ in an indeterminate x plays a basic role. The corresponding concept for general rings is the ring freely generated by an indeterminate x over R; in the notation of Section 2.7 this is the tensor ring $R_A\langle x\rangle$, where A is the subring of R generated by 1. This looks very different from $R[x]$; its elements are not at all like the usual polynomials, but we can simplify matters by taking the special case of those rings, whose elements can be written in the form of polynomials. Thus for a given ring R we consider a ring P whose elements can be written uniquely in the form

$$f = a_0 + xa_1 + \ldots + x^n a_n, \quad \text{where } a_i \in R. \tag{7.3.1}$$

As usual we define the *degree* of f as the highest power of x which occurs with a non-zero coefficient:

$$d(f) = \max\{i \mid a_i \neq 0\}. \tag{7.3.2}$$

We shall characterize the ring P under the assumption that the degree has the usual properties:

D.1 $d(f) \geq 0$ *for* $f \neq 0$, $d(0) = -\infty$,
D.2 $d(f - g) \leq \max\{d(f), d(g)\}$,
D.3 $d(fg) = d(f) + d(g)$.

An integer-valued function d on a ring satisfying **D**.1–**D**.3 is called a *degree function* (essentially this means that $-d$ is a valuation, see Section 9.4 or BA, Chapter 9). Leaving aside the trivial case $R = 0$, we see from **D**.3 that P is an integral domain and moreover, for any $a \in R^\times$, ax has the degree 1, so there exist $a^\alpha, a^\delta \in R$ such that

$$ax = xa^\alpha + a^\delta. \tag{7.3.3}$$

By the uniqueness of the form (7.3.1), the elements a^α, a^δ are uniquely determined by a, and $a^\alpha = 0$ iff $a = 0$. By (7.3.3) we have $(a + b)x = x(a + b)^\alpha + (a + b)^\delta$, $ax + bx = xa^\alpha + a^\delta + xb^\alpha + b^\delta$, hence on comparing the right-hand sides we find

$$(a + b)^\alpha = a^\alpha + b^\alpha, \quad (a + b)^\delta = a^\delta + b^\delta, \tag{7.3.4}$$

so α and δ are additive mappings of R into itself. Next we have $(ab)x = x(ab)^\alpha + (ab)^\delta$, $a(bx) = a(xb^\alpha + b^\delta) = (xa^\alpha + a^\delta)b^\alpha + ab^\delta$, hence

$$(ab)^\alpha = a^\alpha b^\alpha, \quad (ab)^\delta = a^\delta b^\alpha + ab^\delta. \tag{7.3.5}$$

Finally, $1.x = x.1 = x.1^\alpha + 1^\delta$, so

$$1^\alpha = 1, \quad 1^\delta = 0. \tag{7.3.6}$$

The first equation in (7.3.4)–(7.3.6) shows that α is a ring homomorphism, and by the remark following (7.3.3) it is injective. The remaining equations show δ to be a $(1, \alpha)$-derivation; here we shall refer to δ more briefly as an *α-derivation*.

We next observe that the commutation rule (7.3.3), with the uniqueness of (7.3.1) is enough to determine the multiplication in P. For by the distributive law we need only know $x^m a.x^n b$, and by (7.3.3) the effect of moving a past a factor x is

$$x^m a x^n b = x^{m+1} a^\alpha x^{n-1} b + x^m a^\delta x^{n-1} b.$$

Now an induction on n allows us to write $x^m ax^n b$ as a polynomial in x. Thus P is completely determined when α, δ are given. We shall call P a *skew polynomial ring* in x over R relative to the endomorphism α and α-derivation δ, and write it as $P = R[x; \alpha, \delta]$. Thus we have proved the first part of

Theorem 7.3.1. *Let P be a ring whose elements can be expressed uniquely as polynomials in x with coefficients in a non-trivial ring R, as in (7.3.1), with a degree function defined by (7.3.2), and satisfying the commutation rule (7.3.3). Then R is an integral domain, α is an injective endomorphism, δ is an α-derivation and $P = R[x; \alpha, \delta]$ is the skew polynomial ring in x over R, relative to α, δ. Conversely, given an integral domain R with an injective endomorphism α and an α-derivation δ, there exists a skew polynomial ring $R[x; \alpha, \delta]$.*

Proof. It only remains to prove the converse. Consider the set $R^{\mathbf{N}}$ of all sequences $(a_i) = (a_0, a_1, \ldots)$ $(a_i \in R)$, as right R-module. Besides the right multiplication by R we have the additive group endomorphism

$$x : (a_i) \longmapsto (a_i^\delta + a_{i-1}^\alpha), \quad \text{where } a_{-1} = 0. \tag{7.3.7}$$

Clearly R acts faithfully on $R^{\mathbf{N}}$ by right multiplication, so we may identify R with its image in $E = \mathrm{End}(R^{\mathbf{N}})$. Let P be the subring of E generated by R and x; we claim that P is the required skew polynomial ring. To verify (7.3.3) we have

$$(c_i)ax = (c_i a)x = ((c_i a)^\delta + (c_{i-1}a)^\alpha) = (c_i^\delta a^\alpha + c_i a^\delta + c_{i-1}^\alpha a^\alpha)$$

$$(c_i)(xa^\alpha + a^\delta) = (c_i^\delta a^\alpha + c_{i-1}^\alpha a^\alpha + c_i a^\delta).$$

Hence $ax = xa^\alpha + a^\delta$ in P, and (7.3.3) holds. It follows that every element of P can be written as a polynomial (7.3.1), and this expression is unique, for we have

$$(1, 0, 0, \ldots)(a_0 + xa_1 + \ldots + x^n a_n) = (a_0, a_1, \ldots, a_n, 0, \ldots),$$

so distinct polynomials represent different elements of P. Finally it is clear that $d(f)$ defined as in (7.3.2) is a degree function, because R is an integral domain and α is injective. So P is indeed a skew polynomial ring. ∎

Restricting attention to polynomial rings is analogous to singling out Ore domains among general integral domains, and in fact the skew polynomial rings were first considered by Ore in 1933; the general form of Theorem 7.3.1 was obtained by Jacobson in 1934.

It is important to observe that the construction of the skew polynomial ring is not left–right symmetric, and besides $R[x; \alpha, \delta]$ we can also introduce the left skew polynomial ring, in which the coefficients are written on the left. The commutation rule (7.3.3) then has to be replaced by

$$xa = a^\alpha x + a^\delta. \tag{7.3.8}$$

In general the left and right skew polynomial rings are distinct, but when α is an automorphism of R, with inverse β, say, then on replacing a by a^β we can write (7.3.3) as $a^\beta x = xa + a^{\beta\delta}$, i.e.

$$xa = a^\beta x - a^{\beta\delta},$$

which is of the form (7.3.8). Hence we obtain

Proposition 7.3.2. *The ring $R[x; \alpha, \delta]$ is a left skew polynomial ring whenever α is an automorphism of R.* ■

In the construction of skew polynomial rings it was necessary to start from an integral domain because we insisted on a degree function; this is not essential, but it is the case mostly used in applications. Frequently the coefficient ring will be a field, possibly skew. In that case the skew polynomial ring is a principal right ideal domain; this follows as in the commutative case, using the division algorithm. Thus we are given $f, g \in P = K[x; \alpha, \delta]$ of degrees m, n, where $n \geq m$ say. We have $f = x^m a_0 + \ldots$, $g = x^n b_0 + \ldots$, where $a_0, b_0 \neq 0$. Hence $g - fa_0^{-1}x^{n-m}b_0$ is of degree $< n$. Given any right ideal $\mathfrak{a}$ in P, let $f \in \mathfrak{a}$ be non-zero of least degree; then for any $g \in \mathfrak{a}$ we have $d(g) \geq d(f)$, hence $d(g - fh) < d(g)$ for some $h \in P$; we can repeat this process until we reach a term $g - fh_1$ of degree $< d(f)$. But then $g - fh_1 = 0$, because $g - fh_1 \in \mathfrak{a}$ and $f \in \mathfrak{a}$ was of least degree. It follows that $\mathfrak{a} = fP$, so every right ideal is principal. This proves

Theorem 7.3.3. *Let K be a skew field, a an endomorphism and δ an α-derivation of K. Then the skew polynomial ring $K[x; \alpha, \delta]$ is a Euclidean domain and hence a principal right ideal domain.* ■

In particular, for a skew field K the skew polynomial ring $K[x; \alpha, \delta]$ is right Noetherian and hence also right Ore, by Proposition 7.1.9, so we can form its field of fractions. This is denoted by $K(x; \alpha, \delta)$; its elements are fg^{-1}, where f, g are polynomials (7.3.1) with coefficients in K.

Proposition 7.3.4. *Any skew polynomial ring over a right Ore domain is again right Ore.*

Proof. Let R be a right Ore domain and K its field of fractions. Since α is an injective endomorphism of R, it can by Corollary 7.1.2 be extended to an endomorphism of K, again denoted by α. Further, δ gives rise to an R-inverting homomorphism from R to K_2, and this gives an α-derivation of K, again written δ. Now we have the inclusions

$$R[x; \alpha, \delta] \subseteq K[x : \alpha, \delta] \subseteq K(x; \alpha, \delta).$$

Any element $u \in K(x; \alpha, \delta)$ has the form fg, where $f, g \in K[x; \alpha, \delta]$. By Proposition 7.1.7 we can bring the finite set of coefficients of f, g to a common denominator, say $f = fc$, $g = gc$, where $f, g \in R[x; \alpha, \delta]$, $c \in R^\times$. Now $u = fc(gc)^{-1} = fg^{-1}$, so every element of $K(x; \alpha, \delta)$ can be written as a right fraction of elements of $R[x; \alpha, \delta]$, and hence the latter is right Ore, by Corollary 7.1.6. ■

The Hilbert basis theorem extends to skew polynomial rings relative to an automorphism (for endomorphisms it need not hold, see Exercise 2).

Theorem 7.3.5. *Let R be a right Noetherian domain, α an automorphism and δ an α-derivation of R. Then the skew polynomial ring $R[x; \alpha, \delta]$ is again a right Noetherian domain.*

Proof. This is essentially the same as in the commutative case (BA, Theorem 10.4.1) and will be left to the reader. Here it will be found more convenient to write the coefficients on the left of x; since α is an automorphism, this is possible. ∎

We list some examples of skew polynomial rings. When the derivation is 0, we write $R[x; \alpha]$ in place of $R[x; \alpha, 0]$.

1. $\alpha = 1, \delta = 0$. This is the ordinary polynomial ring $R[x]$ in a central indeterminate (although R need not be commutative).
2. The *complex-skew* polynomial ring $\mathbf{C}[x; \bar{}\ ]$ is the ring of polynomials with complex coefficients and commutation rule

$$ax = x\bar{a}, \quad \text{where } \bar{a} \text{ is the complex conjugate of } a.$$

The centre of this ring is the ring $\mathbf{R}[x^2]$ of all real polynomials in x, and $\mathbf{C}[x, \bar{}\ ]/(x^2 + 1)$ is the division algebra of real quaternions. More generally, let k be a field of characteristic not 2 with a quadratic extension $K = k(\sqrt{b})$; this has an automorphism α given by $(r + s\sqrt{b})^\alpha = r - s\sqrt{b}$. For any $a \in k^\times$, $K[x; \alpha]/(x^2 - a)$ is the quaternion algebra $(a, b; k)$ (see Section 5.4).

3. Let K be any commutative ring and denote by $A_1[K]$ the K-algebra generated by u, v over K with the relation

$$uv - vu = 1. \tag{7.3.9}$$

This ring is called the *Weyl algebra* on u, v over K. It may also be defined as the skew polynomial ring $R[v; 1,']$, where $R = K[u]$ and $'$ denotes differentiation with respect to u. We observe that when K is a Noetherian domain, then so is $A_1[K]$.

From (7.3.9) we obtain by induction on n,

$$u^n v - vu^n = nu^{n-1},$$

hence $u^m v^n . v - v . u^m v^n = mu^{m-1} v^n = \partial(u^m v^n)/\partial u$. A similar formula holds for commutation by u and by linearity it follows that for any $f \in A_1[K]$,

$$fv - vf = \frac{\partial f}{\partial u}, \quad uf - fu = \frac{\partial f}{\partial v}. \tag{7.3.10}$$

From these formulae it is easy to show that for a field k of characteristic 0, $A_1[k]$ is a simple ring. For if $\mathfrak{a}$ is a non-zero ideal in $A_1[k]$, pick an element $f(u, v) \neq 0$ in $\mathfrak{a}$ of least possible degree in u. Then $\partial f/\partial u = fv - vf \in \mathfrak{a}$, but this has lower degree and so must be 0. Hence $f = f(v)$ is a polynomial in v alone. If its v-degree is taken minimal, then $\partial f/\partial v = uf - fu = 0$ and so $f = c \in k$. Thus $\mathfrak{a}$ contains a non-zero element of k and so must be the whole ring, i.e. $A_1[k]$ is simple, as claimed.

We observe that $A_1[k]$ is an example of a simple Noetherian domain, not a field. For a field k of finite characteristic p, $A_1[k]$ is no longer simple, since it has the centre $k[u^p, v^p]$.

4. The *translation ring* $k\langle x, y | xy = y(x+1)\rangle$ may be described as $R = A[y; \sigma]$, where A is the polynomial ring $k[x]$ with the shift automorphism $\sigma : x \mapsto x + 1$.

5. Let k be a field of prime characteristic p and $F : a \mapsto a^p$ the Frobenius endomorphism. Then $k[x; F]$ is a skew polynomial ring whose field of fractions $k(x; F)$ has an inner automorphism inducing F, namely conjugation by x.

More generally, if k is any field, even skew, with an endomorphism α, then $k(x; \alpha)$ is an extension with an inner automorphism inducing α on k, because (7.3.3) now reads $ax = xa^\alpha$. Similarly, if δ is an α-derivation, then $k[x; \alpha, \delta]$ is a ring with an inner α-derivation inducing δ, as we see by writing (7.3.3) in the form $a^\delta = ax - xa^\alpha$.

6. Let K be a commutative field with an automorphism α of finite order n, and consider the field of fractions $E = K(x; \alpha)$. If k is the fixed field of α, then $F = k(x^n)$ is contained in the centre of E, as is easily checked. Moreover, $K(x^n)$ is a commutative subfield, a Galois extension of F of degree n, and provided that K contains a primitive n-th root of 1, the structure of E is given by the equations

$$ax^i = x^i a^{\alpha^i} \quad \text{for all } a \in K, i = 0, 1, \ldots, n-1.$$

It follows that $k(x^n)$ is the precise centre of E and E is of dimension n^2 over its centre, in fact a crossed product (see Section 5.5).

7. If D is a skew field not algebraic over its centre k, then $D \otimes k(t)$ is a simple Noetherian domain (by Theorem 5.1.2), but not Artinian, hence not a skew field.

8. Let R be an integral domain with an automorphism α. In the skew polynomial ring $R[x; \alpha]$ the powers of x form an Ore set, and the ring of fractions consists of all polynomials $\sum_{-r}^{s} x^i a_i$ involving negative as well as positive powers of x. Such an expression is called a *skew Laurent polynomial* and the resulting ring may be written $R[x, x^{-1}; \alpha]$.

For each polynomial $f \in P = R[x; \alpha, \delta]$ of the form (7.3.1) we can define its *order* $o(f)$ as the lowest power of x occurring with a non-zero coefficient:

$$o(f) = \min\{i | a_i \neq 0\}.$$

This function has the properties of a valuation on P:

O.1 $o(f) \geq 0$ for all $f \in P$, $o(0) = \infty$,
O.2 $o(f - g) \geq \min\{o(f), o(g)\}$,
O.3 $o(fg) = o(f) + o(g)$.

Taking first the case $\delta = 0$, we can form the ring $R[[x; \alpha]]$ of formal power series over R as the set of all infinite series

$$f = a_0 + xa_1 + x^2a_2 + \ldots, \tag{7.3.11}$$

with componentwise addition and with multiplication based on the commutation rule $ax = xa^\alpha$. There is of course no question of convergence here; we regard (7.3.11) as a series in the purely formal sense. We can describe f equally well as an infinite sequence $(a_i) = (a_0, a_1, \ldots)$, with addition $(a_i) + (b_i) = (a_i + b_i)$ and with multiplication

$$(a_i)(b_j) = (c_k), \quad \text{where } c_k = \sum_j a_{k-j}^{\alpha^j} b_j. \tag{7.3.12}$$

Alternatively we can regard $R[[x; \alpha]]$ as the completion of the skew polynomial ring $R[x; \alpha]$ with respect to the powers of the ideal generated by x; these powers define a topology called the *x-adic* topology. This topological viewpoint is not essential for the construction, but it helps in understanding the situation.

Let R be a ring and α an automorphism of R. The powers of x form an Ore set in $R[[x; \alpha]]$ and by taking fractions we obtain the ring $R((x; \alpha))$ of all *formal Laurent series*

$$\sum_{-r}^{\infty} x^i a_i = x^{-r} a_{-r} + \ldots + x^{-1} a_{-1} + a_0 + x a_1 + x^2 a_2 + \ldots. \tag{7.3.13}$$

This is again a ring, with the same multiplication (7.3.12); here the restriction to finitely many negative powers is necessary to ensure that the multiplication rule (7.3.12) has a sense. This is also the reason for taking α to be an automorphism, since now j may take negative values in (7.3.12).

Let us now consider a skew polynomial ring $R[x; \alpha, \delta]$, where δ may be non-zero, but α is still an automorphism, and ask whether a power series ring can be formed. If we attempt to define the multiplication of power series by means of the commutation formula (7.3.3), we shall find that (apart from a more complicated form for the coefficients of the product), the product cf, where $c \in R$, cannot always be expressed as a power series, because there will in general be contributions to the coefficient of a given power x^r from each term $cx^n a_n$ $(n \geq r)$ and so we may have infinitely many such terms to consider. In terms of the x-adic topology we can express this by saying that left multiplication by $c \in R$ is not continuous; this follows from (7.3.3), because when $a \neq 0$, we have $o(ax) < o(x)$.

One way to overcome this difficulty is to introduce $y = x^{-1}$ and rewrite (7.3.3) in terms of y. We find

$$ya = a^\alpha y + ya^\delta y = a^\alpha y + a^{\delta\alpha} + ya^{\delta^2} y^2 = \ldots,$$

hence by induction we obtain

$$ya = a^\alpha y + a^{\delta\alpha} y^2 + ya^{\delta^2} y^2 + \ldots. \tag{7.3.14}$$

With the help of this commutation formula we can multiply power series in y and even Laurent series. We observe that in passing from x to $y = x^{-1}$ we have also had to change the side on which the coefficients are put; of course this is immaterial as long as α is an automorphism. To be precise, from any skew polynomial ring $R[x; \alpha, \delta]$ we can form a skew power series ring in x^{-1}, with coefficients on the left; in order to define Laurent series in x^{-1} we need to assume that α is an automorphism. We shall not pursue this point but note one result which illustrates the usefulness of power series.

Theorem 7.3.6. *Let K be a skew field and a an automorphism of K. Further denote the centre of K by C and the subfield of C fixed under α by C_0. If no positive power of α is an inner automorphism of K, then the centre of $K(x;\alpha)$ is C_0.*

Proof. Every element of $K(x;\alpha)$ can be written as a Laurent series $f=\sum x^i a_i$. If this lies in the centre, then $fc=cf$ for all $c\in K$, i.e. $\sum x^i(a_ic-c^{\alpha^i}a_i)=0$, hence

$$a_ic=c^{\alpha^i}a_i \quad \text{for all } i\in \mathbf{Z} \text{ and all } c\in K. \tag{7.3.15}$$

If $a^i\neq 0$ for some $i>0$, then α^i is inner, by (7.3.15); in case $i<0$, α^{-i} is inner, but this contradicts the hypothesis. Hence $a^i=0$ for $i\neq 0$ and $f=a_0\in K$. Now (7.3.15) reads $a_0c=ca_0$, so $a_0\in C$, and since $xa_0=a_0x=xa_0^{\alpha}$, we have $f=a_0\in C_0$. Conversely, every element of C_0 commutes with every element of $K(x;\alpha)$. ∎

Finally we note a rationality criterion for power series, which applies also in the skew case.

Proposition 7.3.7. *Let K be a skew field with an automorphism α, and consider the natural embedding of $K(x;\alpha)$ in the skew field $K((x:\alpha))$ of formal Laurent series. A given series $f=\sum x^ia_i$ lies in $K(x;\alpha)$ if and only if there exist integers r, n_0 and elements $c_1,\dots,c_r\in K$ such that*

$$a_n=a_{n-1}^{\alpha}c_1+a_{a-2}^{\alpha^2}c_2+\dots+a_{n-r}^{\alpha^r}a_r \quad \text{for all } n>n_0. \tag{7.3.16}$$

Proof. The series f lies in $K(x;\alpha)$ iff there is a polynomial g with constant term 1 such that fg is a Laurent polynomial. Writing $g=1-xc_1-\dots-x^rc_r$, we require that in the product $(\sum x^ia_i)(1-\sum x^jc_j)$ all coefficients of powers beyond a given one, say x^{n_0} vanish. On equating the coefficient of x^n to 0 we just obtain (7.3.16), and the conclusion follows. ∎

Exercises

1. Supply the details of the proof of Theorem 7.3.5, and point out where the fact that α is an automorphism is used.
2. (Jategaonkar, Koshevoi) Let K be a skew field and α an endomorphism of K. Show that $K[x;\alpha]$ is a left Ore domain iff α is surjective (and hence an automorphism). Using Proposition 7.1.8, obtain an embedding of the free algebra $K\langle x,y\rangle$ in a right Ore domain, and hence an embedding of $K\langle x,y\rangle$ in a skew field.
3. Find a localization of the Weyl algebra over a field of finite characteristic, which is a crossed product.
4. Show that for any ring K, a Weyl K-ring $A_1[K]$ on u, v may be defined by (7.3.9), where u, v commute with the elements of K. Show that if K is a simple ring of characteristic 0, then so is $A_1[K]$.
5. Show that if R is a Noetherian ring with an automorphism α, then $R[x,x^{-1};\alpha]$ is again Noetherian.

6. Let K be a ring with an injective endomorphism α. Put $K_0 = K$ and take an ascending chain of K-rings K_n, all isomorphic to K, such that K_{n-1} is identified with the image of K_n under α. Show that their union is a ring $K^{[\alpha]}$ to which α can be extended as an automorphism. Verify that with the Laurent polynomial ring $K[x, x^{-1}; \alpha]$ the subring $\cup x^m K x^{-m}$ is isomorphic to $K^{[\alpha]}$.
7. Let $F \subseteq E$ be any fields and α an automorphism of E mapping F into itself. Show that $E(x; \alpha) \cap F((x; \alpha)) = F(x; \alpha)$.
8. Let K be a skew field with centre C, α an automorphism of K and C_0 the subfield of C left fixed by α. Suppose further that $\alpha^n = 1$, but no power α^i $(0 < i < n)$ is inner. Show that the centre of $K((x; \alpha))$ is $C_0((x^n))$ and the centre of $K(x; \alpha)$ is $C_0(x^n)$.
9. (P. Draxl) Let k be a field of characteristic not 2 and $K = k(u_1, u_2, u_3, u_4)$, where the u_i are independent (central) indeterminates. Put $K_1 = k((x_1)), K_2 = K_1((x_2; \alpha)), K_3 = K_2((x_3)), K_4 = K_3((x_4; \beta))$, where $\alpha : x_1 \mapsto -x_1$, $\beta : x_3 \mapsto -x_3$, and by identifying u_i with x_i^2 show that the K-subalgebra D of K_4 generated by $x_1, \ldots, x_4$ is a central division algebra of degree 4 over K, of the form $D = (u_1, u_2; K) \otimes (u_3, u_4; K)$. Using the representation as Laurent series, show that any multiplicative commutator on D has the form $1 + f$, where f involves only positive powers of the x's. Deduce that if k contains a primitive 4th root of 1, then $\mathbf{SK}_1(D)$ is non-trivial. (Hint. Use Exercise 8 to show that D has centre K and verify that the 4th root of 1 has reduced norm 1. For more details of this solution of the Tannaka–Artin problem see Draxl (1983)).
10. Show that the trace group, defined in Section 4.5, vanishes on the Weyl algebra over a field of characteristic 0.

7.4 Goldie's theorem

The main structure theorem for Artinian rings states that a semisimple ring is a direct product of simple rings and each of the latter is a full matrix ring over a skew field (Wedderburn theorems, BA, Section 5.2). Such precise information is not to be expected for Noetherian rings, but in this case there is a reduction theorem due to Alfred Goldie, which implies that a Noetherian semiprime ring has a quotient ring which is semisimple (hence Artinian), and here prime Noetherian rings correspond to simple Artinian rings. Our aim in this section is a proof of these results, but some preparation will be necessary. We recall that a ring R is called *prime* if $R \neq 0$ and $aRb = 0$ implies $a = 0$ or $b = 0$; a ring R in which $aRa = 0$ implies a $= 0$ is called *semiprime*. The precise relation between these rings will be described in Chapter 8.

We shall also need a substitute for the dimension of a vector space which can be applied to general modules; this is the notion of uniform rank. Let M be an R-module (over any ring R); we recall that a submodule M' of M is said to be *large* in M and M is an *essential extension* of M' if

$$N \neq 0 \Rightarrow N \cap M' \neq 0 \quad \text{for any submodule } N \text{ of } M. \tag{7.4.1}$$

In particular, R itself may be considered as a left or right R-module; we then obtain large left or right ideals, also called *left large* or *right large*, respectively.

We list some properties of large submodules used later:

L.1 *If M' is large in M and M'' is large in M' then M'' is large in M.*
For if $0 \neq N \subseteq M$, then $N \cap M' \neq 0$, hence $N \cap M'' = (N \cap M') \cap M'' \neq 0$. ■

L.2 *Let M be a right R-module and M' a large submodule of M. For any $m \in M$, the set $\mathfrak{c} = \{x \in R \mid mx \in M'\}$ is a large right ideal in R, and if $m \neq 0$, then $0 \neq m\mathfrak{c} \subseteq M'$.*
Clearly $\mathfrak{c}$ is a right ideal; for any non-zero right ideal $\mathfrak{a}$ in R, either $m\mathfrak{a} = 0$, in which case $\mathfrak{a} \subseteq \mathfrak{c}$, or $m\mathfrak{a} \neq 0$; then $mR \cap M' \neq 0$ and for any $a \in \mathfrak{a}$ such that $0 \neq ma \in M'$ we have $0 \neq aR \subseteq \mathfrak{c} \cap \mathfrak{a}$, which shows $\mathfrak{c}$ to be right large. If moreover $ma \in M'$, then $a \in \mathfrak{c}$, so $0 \neq m\mathfrak{c} \subseteq M'$. ■

L.3 *For any nilpotent ideal $\mathfrak{n}$ of R, the left annihilator $(\mathfrak{n})_l = \{x \in R \mid x\mathfrak{n} = 0\}$ is a right large ideal of R.*
For, given $c \neq 0$, there exists $s \geq 1$ such that $c\mathfrak{n}^{s-1} = 0$, $c\mathfrak{n}^s = 0$, and $c\mathfrak{n}^{s-1} \subseteq cR \cap (\mathfrak{n})_l$. ■

L.4 *A module is semisimple if and only if it has no proper large submodules.*

For in a semisimple module every proper submodule has a non-zero complement and so cannot be large. Conversely, if M has no proper large submodules and $M' \subseteq M$ is given, we can by Zorn's lemma find a submodule N which is maximal subject to $N \cap M' = 0$. If P is a non-zero submodule such that $P \cap (N + M') = 0$, then the sum $P + N + M'$ is direct and so $(P + N) \cap M' = 0$, but this contradicts the maximality of N. Hence $P \cap (N + M') \neq 0$ for all $P \neq 0$, so $N + M'$ is large and therefore equal to M. Hence N is a complement of M' in M and this shows M to be semisimple. ■

We recall that a module M is called *uniform* if $M \neq 0$ and every non-zero submodule of M is large. For example, an integral domain R is right uniform (i.e. uniform as right R-module) iff it is a right Ore domain. With the help of uniform modules we can define a form of dependence relation which leads to a notion of rank in general modules. Let M be an R-module (for any ring R) and denote by $\mathscr{U}$ the collection of all its uniform submodules. On $\mathscr{U}$ we introduce a dependence relation as follows. If $N, P_1, \ldots, P_r \in \mathscr{U}$, then N is said to be *dependent* on $P_1, \ldots, P_r$ if $N \cap \sum P_i \neq 0$. Generally N is said to be *dependent* on a (possibly infinite) family of uniform submodules if it is dependent on a finite subfamily. A set of uniform modules is *independent* if no member is dependent on the rest. This dependence relation satisfies the following conditions:

D.0 *In any family $\mathscr{F}$ of uniform submodules of M, each member of $\mathscr{F}$ is dependent on $\mathscr{F}$.*

D.1′ *(Transitivity) If N is dependent on the independent family $\mathscr{F}$ and each member of $\mathscr{F}$ is dependent on $\mathscr{G}$, then N is dependent on $\mathscr{G}$.*

D.2 *(Exchange property) If N is dependent on $\mathscr{F} \cup \{M'\}$ but not on $\mathscr{F}$, then M' is dependent on $\mathscr{F} \cup \{N\}$.*

We note that these conditions are like those listed in Section 11.1 of BA (except that **D**.1′ is a weaker form of **D**.1 listed there).

D.0 is clear; to prove **D**.1′ we may take the families to be finite, say N is dependent on the independent family $\{P_1, \ldots, P_r\}$ and each P_i is dependent on $\{Q_1, \ldots, Q_s\}$. By hypothesis, $N \cap \sum P_i \neq 0$, so there exists $n \in N$, $n \neq 0$, such that

$$n = p_1 + \ldots + p_r, \quad \text{where } p_i \in P_i. \tag{7.4.2}$$

Writing $Q = \sum Q_j$, we have to show that $N \cap Q \neq 0$. If $p_i \in Q$ for all i, the conclusion follows from (7.4.2); otherwise we choose an equation (7.4.2) with $n \neq 0$ such that the least number of p_i are not in Q. If $p_1 \notin Q$ say, then since P_1 is uniform, $p_1R \cap (P_1 \cap Q) \neq 0$, say $0 \neq p_1c \in Q$. Now in

$$nc = p_1c + \ldots + p_rc, \tag{7.4.3}$$

there are fewer terms outside Q than in (7.4.2) and $nc \neq 0$, because $p_1c \neq 0$ and the sum $\sum P_i$ is direct. This contradiction shows that $N \cap Q \neq 0$, as claimed.

The exchange axiom **D**.2 follows easily: if N is dependent on $P_1, \ldots, P_r$ but not on $P_2, \ldots, P_r$ then there exists $n \in N$, $n \neq 0$, such that (7.4.2) holds with $p_1 \neq 0$, by hypothesis. If we now rewrite (7.4.2) as $p_1 = n - p_2 - \ldots - p_r$, we have a relation showing P_1 to be dependent on $N, P_2, \ldots, P_r$, so **D**.2 holds. ■

Let us also note that if N is dependent on a family $\mathscr{F}$ of uniform submodules, then so is any non-zero submodule of N. For if $Q \cap P \neq 0$, where P is a sum of terms in $\mathscr{F}$, and $0 \neq N' \subseteq N$, then $N' \cap P = N' \cap (N \cap P) \neq 0$, because N is uniform.

A set of uniform submodules of M on which every uniform submodule depends will be called a *spanning set*, and an independent spanning set is a *basis*; we recall from BA, Section 11.1 that the bases are just the independent spanning sets and every independent set is contained in a basis. We also have the exchange lemma (BA, Lemma 11.1.2, which used **D**.1 only in the weaker form **D**.1′), which states that for an independent set $\mathscr{F}$ and a spanning set $\mathscr{G}$, it is possible to complete $\mathscr{F}$ to a basis by adjoining a subset $\mathscr{G}'$ of $\mathscr{G}$, and if $\mathscr{G}$ is finite, then so is $\mathscr{F}$, and $|\mathscr{F} \cup \mathscr{G}'| \leq |\mathscr{G}|$.

We remark that such bases need not exist (see Exercise 7), but if every non-zero submodule of M contains a uniform submodule, then we can by Zorn's lemma find a direct sum of uniform submodules which is large in M, and hence a basis. We shall mainly be interested in modules with a finite basis and we shall find conditions for this to exist in Proposition 7.4.1 below.

From the exchange lemma we can deduce in the usual fashion that if M has a finite basis, then any two bases have the same number of terms (see BA, Corollary 11.1.6). This number is called the *uniform rank* or simply the *rank* of M, and is written rk M. It is clear that rk $M = 1$ iff M is uniform, and in a module of rank n, any direct sum of non-zero submodules has at most n terms; it has exactly n terms iff each term is uniform and the sum is large in M. The conditions for a module to have finite rank are easily stated:

Proposition 7.4.1. *Let R be any ring and M an R-module which contains no infinite direct sums of non-zero submodules. Then there is a direct sum of uniform submodules which is large in M, so that M has a rank, and in fact the rank is then finite. Conversely, a module of finite rank contains uniform submodules, but no infinite direct sum of submodules.*

Proof. We begin by showing that every non-zero submodule N of M contains a uniform submodule. For if N is not itself uniform, then it contains a direct sum $N_1 \oplus N_1'$; now either N_1' is uniform or it contains a direct sum $N_2 \oplus N_2'$ and continuing in this way, we obtain in M the direct sum

$$N_1 \oplus N_2 \oplus \dots .$$

Since M contains no infinite direct sums, this process must break off, which can happen only when we reach an N_i' which is uniform. Hence N contains a uniform submodule.

Now let $\sum_1^r U_i$ be a direct sum of uniform submodules in M; such a sum exists, e.g. for $r = 0$. Either it is large in M or we can find $V \neq 0$ such that $V \cap \sum U_i \neq 0$. By the first part of the proof V contains a uniform submodule U_{r+1} and now $\sum_1^{r+1} U_i$ is a direct sum of $r + 1$ terms. If we continue in this way, the process must break off, because M has no infinite direct sums, and it can end only when we have a direct sum of uniform submodules which is large in M. This shows that M has a rank and that this rank is finite. Conversely, if $\text{rk}\ M = n$, then we know that any direct sum contains part of a basis and so cannot have more than n terms. ∎

This result may be applied to R itself, as left or right R-module, and in this way we obtain the notion of *left rank* and *right rank* of R. For example, an integral domain has right rank 1 iff it is a right Ore domain. In fact, by Proposition 7.1.9 we obtain

Corollary 7.4.2. *An integral domain which has finite right rank, necessarily has right rank 1.* ∎

We remark that for a submodule M' of M we have $\text{rk}\ M' \leq \text{rk}\ M$, with equality iff M' is large in M. On the other hand, going over to a quotient module may well raise the rank, e.g. $\text{rk}\ \mathbf{Z} = 1$, but $\text{rk}(\mathbf{Z}/m) > 1$ unless m is a prime power.

Let R be any ring and M be a right R-module. For any subset S of M we define the *right annihilator* of S in R as

$$(S)_r = \{x \in R \mid Sx = 0\}.$$

It is clear that $(S)_r$ is a right ideal; if S is a submodule, $(S)_r$ is even a two-sided ideal. When $S \neq 0$, $(S)_r$ cannot contain 1 and so will be proper. If $S = \{a\}$, we write $(a)_r$ instead of $(\{a\})_r$. In particular, this defines right annihilators of subsets of R; the *left annihilator* of a subset S of R (or more generally, of a left R-module) is defined similarly. The ring R is said to satisfy the *maximum condition on right annihilators* if every collection of right annihilators in R (of subsets of R) has a maximal member, e.g. any right Noetherian ring satisfies the maximum condition on right annihilators.

Proposition 7.4.3. *Let R be a semiprime ring with maximum condition on right annihilators. Then every nil left or right ideal is zero.*

Proof. It is clearly enough to prove the result for principal ideals, and we need only consider left ideals, for Ra is nil iff $(xa)^n = 0$ for all x and suitable $n = n(x)$. Now $(xa)^n = 0$ implies $(ax)^{n+1} = a(xa)^n x = 0$, hence if Ra is nil, then so is aR.

Thus let Ra be a non-zero nil left ideal and choose a maximal annihilator not equal to R of the form $(xa)_r$. Writing $b = xa$, we choose $y \in R$; if $yb \neq 0$, take $\nu \geq 2$ such that $(yb)^{\nu-1} \neq 0$, $(yb)^\nu = 0$. Then $(b)_r \subseteq ((yb)^{\nu-1})_r$ and by maximality we have equality here; since $yb \in ((yb)^{\nu-1})_r$, we conclude that $byb = 0$. This holds for all $y \in R$, even when $yb = 0$. Hence $bRb = 0$, $b \neq 0$, in contradiction to the fact that R is semiprime. Hence every nil left (or right) ideal is 0. ■

We shall also need the notion of a singular submodule. For any ring R and any right R-module M consider the set

$$Z(M) = \{m \in M \mid (m)_r \text{ is a large right ideal of } R\}.$$

This set is a submodule of M, called the *singular submodule*. To verify the module property, let $u, v \in Z(M)$; then $\mathfrak{a} = (u)_r$ and $\mathfrak{b} = (v)_r$ are right large, hence so is $\mathfrak{a} \cap \mathfrak{b}$ and $(u - v)(\mathfrak{a} \cap \mathfrak{b}) = 0$, therefore $(u - v)_r$ is right large. Further, if $a \in R$, then $(ua)_r$ is right large, by **L**.2, for $x \in (ua)_r \Leftrightarrow ax \in (u)_r$ and the latter is right large by hypothesis. Thus $Z(M)$ is indeed a submodule. In particular, taking $M = R$, we obtain the *right singular ideal* $Z(R)$ of R. By what has been shown, it is a right ideal; in fact it is two-sided, for if $(a)_r$ is right large, then so is $(ba)_r \supseteq (a)_r$.

Although Goldie's theorem is concerned with Noetherian rings, it applies to a somewhat wider class, defined as follows. A ring which is of finite right rank and satisfies the maximum condition on right annihilators is called a *right Goldie ring*. In particular, every right Noetherian ring is right Goldie; of course the converse is false, as the example of commutative integral domains shows. In a right Goldie ring the right singular ideal is nilpotent; we shall only need the special case where the ring is semiprime, when it is a consequence of the next result (see Exercise 9 of Section 8.5 for the general case).

Proposition 7.4.4. *In a right Goldie ring R, for each $a \in R$ there exists $n \geq 0$ such that $a^nR + (a^n)_r$ is right large. Moreover, $(a^\nu)_r = (a^n)_r$ for all $\nu \geq n$ and the sum $a^nR + (a^n)_r$ is direct.*

Proof. The sequence $(a)_r \subseteq (a^2)_r \subseteq \dots$ becomes stationary, say $(a^\nu)_r = (a^n)_r$ for $\nu \geq n$. It follows that $(a^n)_r \cap a^nR = 0$, for if $a^n.a^nx = 0$, then $x \in (a^{2n})_r = (a^n)_r$, hence $a^nx = 0$. If $\mathfrak{c}$ is any right ideal such that $\mathfrak{c} \cap (a^nR + (a^n)_r) = 0$, then the sum $\mathfrak{c} + a^n\mathfrak{c} + a^{2n}\mathfrak{c} + \dots$ is direct, for if $a^{sn}c_s + a^{(s+1)n}c_{s+1} + \dots + a^{tn}c_t = 0$, where $c_i \in \mathfrak{c}$, $c_s \neq 0$, then $c_s \in (a^n)_r + a^nR$, which is a contradiction. Since R has finite rank, $a^{sn}\mathfrak{c} = 0$ for some $s \geq 1$, i.e. $\mathfrak{c} \subseteq (a^{sn})_r = (a^n)_r$ and it follows that $\mathfrak{c} = \mathfrak{c} \cap (a^nR + (a^n)_r) \neq 0$. This proves that $a^nR + (a^n)_r$ is right large, and we have seen that the sum is direct. ■

We note two extreme special cases of this result.

Corollary 7.4.5. *In a right Goldie ring R, if a is left regular, then aR is right large.*

Proof. In this case $(a^n)_r = 0$ for all n and if a^nR is right large, then so is aR. ■

Corollary 7.4.6. *In a semiprime right Goldie ring R, $Z(R) = 0$.*

Proof. By Proposition 7.4.4, if $(a)_r$ is right large, then a is nilpotent. Hence $Z(R)$ is a nil ideal and so must be zero, by Proposition 7.4.3. ■

The next lemma, essentially the converse of Corollary 7.4.5, will be needed for Goldie's theorem, but is also useful elsewhere.

Lemma 7.4.7. *In a semiprime right Goldie ring any large right ideal contains a regular element.*

Proof. By Proposition 7.4.3, any nil right ideal of R is 0, so a non-zero right ideal $\mathfrak{a}$ will contain a non-nilpotent element a. By Proposition 7.4.4 we can find a power a_1 of a such that $(a_1)_r = (a_1^2)_r$ and so $a_1R + (a_1)_r$ is direct. If $(a_1)_r \cap \mathfrak{a} \neq 0$, we choose $a_2 \in (a_1)_r \cap \mathfrak{a}$ such that $a_2 \neq 0$ and $(a_2)_r = (a_2^2)_r$. Then $a_2R + ((a_1)_r \cap (a_2)_r \cap \mathfrak{a})$ is a direct sum contained in $(a_1)_r \cap \mathfrak{a}$, hence the sum $a_1R + a_2R + ((a_1)_r \cap (a_2)_r \cap \mathfrak{a})$ is direct. If $(a_1)_r \cap (a_2)_r \cap \mathfrak{a} \neq 0$, we can continue the process; at the n-th stage we have a direct sum

$$a_1R + \ldots + a_nR + ((a_1)_r \cap \ldots \cap (a_n)_r \cap \mathfrak{a}), \tag{7.4.4}$$

where $a_i \in (a_1)_r \cap \ldots \cap (a_{i-1})_r \cap \mathfrak{a}$ and $(a_i)_r = (a_i^2)_r \neq R$.

Since R has finite rank, the process must stop; if this happens at the n-th stage, we have

$$(a_1)_r \cap \ldots \cap (a_n)_r \cap \mathfrak{a} = 0. \tag{7.4.5}$$

So far $\mathfrak{a}$ was any right ideal $\neq 0$; if we take $\mathfrak{a}$ to be right large, then by (7.4.5) we find

$$(a_1)_r \cap \ldots \cap (a_n)_r = 0. \tag{7.4.6}$$

Put $c = \sum a_i$; by construction $c \in \mathfrak{a}$ and we claim that c is regular. If $cx = \sum a_ix = 0$, then by the directness of (7.4.4), $a_ix = 0$ for all i, hence $x = 0$ by (7.4.6), i.e. $(c)_r = 0$. By Corollary 7.4.5, cR itself is right large, hence $(c)_l \subseteq Z(R)$, but $Z(R) = 0$, by Corollary 7.4.6, so $(c)_l = 0$ and c is regular. ■

In the presence of maximum conditions the definition of a right Ore set can be simplified a little; this is sometimes useful, although it is not actually needed here.

Proposition 7.4.8. *Let R be a ring with maximum condition on right annihilators and let S be a multiplicative subset of R such that*

(i) *for any $a \in R, s \in S, aS \cap sR \neq \emptyset$,*
(ii) *for any $a \in R, s \in S, as = 0 \Rightarrow a = 0$.*

Then S is a right Ore set consisting of regular elements.

Proof. We need only show that for any $a \in R, s \in S, sa = 0 \Rightarrow a = 0$. By the maximum condition the sequence

$$(s)_r \subseteq (s^2)_r \subseteq \dots$$

becomes stationary, say $(s^n)_r = (s^{n+1})_r$. If $sa = 0$, then by (i) there exist s', a' such that $as' = s^n a'$; hence $s^{n+1}a' = sas' = 0$, so $a' \in (s^{n+1})_r = (s^n)_r$, and so $0 = s^n a' = as'$, hence $a = 0$ by (ii). ∎

We now come to the main result of this section.

Theorem 7.4.9 (Goldie's theorem). *A ring R has a total quotient ring Q which is semisimple if and only if R is a semiprime right Goldie ring. Moreover, R is simple Artinian if and only if R is prime right Goldie.*

Proof. Assume that R is semiprime right Goldie and let S denote the set of all regular elements of R. We shall show that S is a right Ore set. Given $a \in R$, $s \in S$, define

$$\mathfrak{c} = \{x \in R \mid ax \in sR\}.$$

By Corollary 7.4.5, sR is right large, hence by **L.2**, so is $\mathfrak{c}$, therefore it contains a regular element (Lemma 7.4.7). This shows that $sR \cap aS \neq \emptyset$, so S is a right Ore set.

Let $Q = R_S$ be the total quotient ring. If $\mathfrak{A}$ is a large right ideal of Q, then $\mathfrak{A} \cap R$ is right large in R and so contains a regular element. This must be a unit in Q, hence $\mathfrak{A} = Q$, i.e. Q has no proper large right ideals, and so is semisimple, by **L.4**.

Conversely, let R be a ring with a semisimple quotient ring Q. We shall show that for any right ideal $\mathfrak{c}$ of R the following conditions are equivalent:

(a) $\mathfrak{c}$ *is right large in R,*
(b) $\mathfrak{c}Q = Q$,
(c) $\mathfrak{c}$ *contains a regular element of R.*

(a) $\Leftrightarrow$ (b). Assume that $\mathfrak{c}$ is right large and let $\mathfrak{A}$ be any non-zero right ideal of Q. Then $\mathfrak{A} \cap R \neq 0$, hence $\mathfrak{A} \cap R \cap \mathfrak{c} \neq 0$, and so $\mathfrak{A} \cap \mathfrak{c}Q \neq 0$. This shows that $\mathfrak{c}Q$ is right large in Q, hence $\mathfrak{c}Q = Q$ by **L.4**. Conversely, if $\mathfrak{c}Q = Q$ and $\mathfrak{a}$ is a non-zero right ideal in R, then $\mathfrak{a}Q \cap \mathfrak{c}Q \neq 0$, hence $\mathfrak{a} \cap \mathfrak{c} \neq 0$, so $\mathfrak{c}$ is right large in R.

(b) $\Leftrightarrow$ (c). If $\mathfrak{c}Q = Q$, then $1 = as^{-1}$, where $a \in \mathfrak{c}$ and s is a regular element in R, hence $a = s$ is regular. Conversely, if c contains a regular element, then clearly $\mathfrak{c}Q = Q$.

We can now complete the proof of the theorem by verifying that R is semiprime right Goldie.

Let $\mathfrak{n}$ be a nilpotent ideal of R. Then $(\mathfrak{n})_l$ is a right large ideal in R, by **L.3**, so it contains a regular element and it follows that $\mathfrak{n} = 0$. Thus R is semiprime.

Next let $\mathfrak{a} = \sum \mathfrak{a}_\lambda$ be a direct sum of non-zero right ideals in R which is right large; then $\mathfrak{a}$ contains a regular element c, say:

$$c = x_1 + \dots + x_n, \quad \text{where } x_i \in \mathfrak{a}_{\lambda_i}.$$

Now cR is right large and is contained in $\mathfrak{a}_{\lambda_1} + \dots + \mathfrak{a}_{\lambda_n}$, hence the latter sum is right large, so the sum $\sum a_\lambda$ was finite.

For any subset I of R, its right annihilator in R is obtained by intersecting its annihilator in Q with R. In an obvious notation we have

$$(I)_r^R = (I)_r^Q \cap R.$$

Now the maximum condition for right annihilators follows for R, because it holds in Q. Finally, if R is prime, then so is Q, by Proposition 7.1.10. It follows that Q is simple. Conversely, if Q is simple, then R must be prime, for if $\mathfrak{a}$, $\mathfrak{b}$ are ideals in R such that $\mathfrak{ab} = 0$, then $\mathfrak{b}Q\mathfrak{a} \cap R$ is an ideal of R whose square is zero; since R is semiprime, we have $\mathfrak{b}Q\mathfrak{a} \cap R = 0$ and hence $\mathfrak{b}Q\mathfrak{a} = 0$. But Q is simple, so it follows that $\mathfrak{a} = 0$ or $\mathfrak{b} = 0$, and this shows R to be prime. ■

For prime rings Theorem 7.4.9 was proved by Goldie [1958] and extended by him to semiprime rings in 1960. The result raises the question of localizing relative to a prime ideal. If R is a Noetherian ring with a prime ideal $\mathfrak{p}$, and $C_{\mathfrak{p}}$ denotes the set of all elements of R that are *regular* mod $\mathfrak{p}$, i.e. whose image in $R/\mathfrak{p}$ is regular, then it is not necessarily the case that $C_{\mathfrak{p}}$ is a right Ore set. Here it is necessary to consider more than one prime (a so-called 'clique' of prime ideals), and to localize at the set of elements that are regular modulo all the prime ideals considered. For a detailed account of this method see Jategaonkar (1986), McConnell and Robson (1987) or Goodearl and Warfield (1989).

Exercises

1. (Kasch–Sandomierski) Show that the socle of a module is the intersection of all its large submodules. (Hint. Show first that every submodule is a direct summand of a large submodule).
2. Let A be the direct sum of an infinite and a finite (non-zero) cyclic group. Show that for the dependence relation defined for A as **Z**-module the form of **D**.1′ without the independence of $\mathscr{F}$ (**D**.1 of BA) does not hold. Show also that this form of **D**.1′ does hold on any torsion-free module.
3. Show that an Artinian semiprime ring is self-injective.
4. Find the rank of $\mathbf{Z}/m$, for a positive integer m.
5. Show that for any ring R and any $n \geq 1$, $\mathrm{rk}(R_n) = n.\mathrm{rk}\, R$.
6. Show that the injective hull of a uniform module is indecomposable. Use this fact and the Krull–Schmidt theorem to give another proof of Proposition 7.4.1.
7. Let F be the ring of all real continuous functions on the unit interval with pointwise operations: $(f+g)(x) = f(x) + g(x)$, $(fg)(x) = f(x)g(x)$. Show that F has no uniform ideals.
8. Show that the maximum condition on left annihilators is equivalent to the minimum condition on right annihilators.
9. Show that a reduced ring ($x^2 = 0 \Rightarrow x = 0$) is non-singular.
10. Show that over an Ore domain every finitely generated flat module is projective. (Hint. Use Further Exercise 16 of Chapter 4.)
11. Show that every left and right self-injective ring is its own quotient ring (R is *right self-injective* if R_R is injective).

12. Show that a non-zero submodule of a direct sum of uniform modules has a uniform submodule. (Hint. First find a non-zero submodule in a finite direct sum.)

7.5 PI-algebras

Let k be a field and $F = k\langle x_1, \ldots, x_d \rangle$ the free k-algebra on $x_1, \ldots, x_d$. The elements of F are called *polynomials* and a k-algebra A is said to satisfy the *polynomial identity*

$$p(x_1, \ldots, x_d) = 0, \tag{7.5.1}$$

if p is an element of F which vanishes for all values of the x's in A. If A satisfies a *non-trivial* polynomial identity, i.e. an identity (7.5.1) where p is not the zero polynomial, then A is called a *PI-algebra*. Many of the results proved for Noetherian rings have their counterparts for PI-algebras. It will simplify matters to assume a field of coefficients, though it is possible to consider more general coefficient rings (see Procesi (1973) or Rowen (1980)).

Examples

1. Every commutative k-algebra satisfies the identity $xy - yx = 0$ and so is a PI-algebra.
2. Every Boolean ring satisfies the identity $x^2 - x = 0$.
3. Every finite-dimensional algebra satisfies an identity. If $\dim A < n$, then A satisfies the identity

$$S_n(x_1, \ldots, x_n) = \sum_\sigma \operatorname{sgn}(\sigma) x_{1\sigma} x_{2\sigma} \ldots x_{n\sigma} = 0, \tag{7.5.2}$$

   where σ runs over all permutations of $1, \ldots, n$ and $\operatorname{sgn}(\sigma)$ is 1 or -1 according as σ is even or odd. S_n is called the *standard polynomial* and (7.5.2) the *standard identity* of degree n. For any $a_1, \ldots, a_n \in A$ it is clear that $S_n(a_1, \ldots, a_n) = 0$ if two of the a's coincide. Now take a k-basis $e_1, \ldots, e_r$ ($r = \dim A < n$) of A; then for any $a_1, \ldots, a_n \in A$, $S_n(a_1, \ldots, a_n)$ can by linearity be written as a linear combination of terms $S_n(e_{i_1}, \ldots e_{i_n})$. Since $r < n$, at least two of the e's must coincide, so all terms vanish. This shows that A satisfies (7.5.2).
4. If A is a commutative k-algebra, then $\mathfrak{M}_n(A)$ satisfies the standard identity of degree $n^2 + 1$, for we have a basis of n^2 elements, so the result follows as in Example 3. In Theorem 7.5.8 below we shall see that $\mathfrak{M}_n(A)$ actually satisfies $S_{2n} = 0$.
5. If every element of A is algebraic over k, of degree at most n, each element of A satisfies an equation

$$x^n + a_1 x^{n-1} + \ldots + a_n = 0, \quad \text{where } a_i \in k. \tag{7.5.3}$$

If the equation for some element of A has degree less than n, we can bring it to the form (7.5.3) by multiplying by a power of x. Writing $[x,y]=xy-yx$, we obtain from (7.5.3),

$$[x^n,y]+a_1[x^{n-1},y]+\ldots+a_{n-1}[x,y]=0.$$

Thus the commutators in this expression are linearly dependent and so A satisfies the identity

$$S_n([x,y],[x^2,y],\ldots,[x^n,y])=0.$$

We collect some elementary facts about PI-algebras.

Proposition 7.5.1. *Any subalgebra or homomorphic image of a PI-algebra is again a PI-algebra.*

Proof. The proof is immediate. ∎

Since the free algebra is clearly not a PI-algebra, we deduce from Proposition 7.1.8,

Corollary 7.5.2. *Every PI-algebra which is also an integral domain is a left and right Ore domain.* ∎

A polynomial p and the corresponding identity $p=0$ is said to be *multilinear* if it is homogeneous of degree 1 in each variable. In a PI-algebra we can always find multilinear identities:

Proposition 7.5.3. *Any algebra A satisfying a polynomial identity of degree n also satisfies a polynomial identity of degree n which is multilinear.*

Proof. Let $p=p(x_1,\ldots,x_d)$ be a polynomial of degree n which vanishes identically on A, and let r be the highest degree to which any variable occurs, say p has degree r in x_1. If $r>1$, replace p by $p(x_1+x_{d+1},x_2,\ldots)-p(x_1,x_2,\ldots)-p(x_{d+1},x_2,\ldots)$; since $(x_1+x_{d+1})^r-x_1^r-x_{d+1}^r\neq 0$ (even in finite characteristic, because $x_1x_{d+1}\neq x_{d+1}x_1$), we get a polynomial in which x_1,x_{d+1} occur, but of degree less than r and the degree of the other variables is not raised. By a double induction, on the highest degree r and on the number of variables occurring to this degree, we reduce p to a polynomial of degree 1 in each variable. The process preserves the total degree, so we get a polynomial q with a term $ax_1x_2\ldots x_n$, say. Now replace q by $q(\ldots,x_i,\ldots)-q(\ldots,0,\ldots)(1\leq i\leq n)$ to get rid of terms not involving x_i. ∎

To illustrate the proposition, suppose we have an algebra A satisfying the identity

$$x^2=0. \tag{7.5.4}$$

Here we do not restrict our algebra to be unital (for otherwise it would have to be trivial, by (7.5.4)). Then A also satisfies $(x+y)^2-x^2-y^2=0$, i.e.

$$xy+yx=0. \tag{7.5.5}$$

and this is multilinear. In characteristic other than 2 the identities (7.5.4) and (7.5.5) are equivalent, for we can get back from (7.5.5) to (7.5.4) by putting $y = x$, which gives $2x^2 = 0$.

A multilinear identity has the advantage that to verify it we need only check the elements of a basis. Moreover, such identities are preserved under extensions:

Corollary 7.5.4. *Let R be a k-algebra with centre C. If R contains a subalgebra A which is a PI-algebra such that $R = AC$, then R is a PI-algebra.*

Proof. By Proposition 7.5.3, A satisfies a multilinear identity $p(x_1, \ldots, x_n) = 0$. Let $\{u_\lambda\}$ be a k-basis for R and put $a_i = \sum \alpha_{i\lambda} u_\lambda$, where $\alpha_{i\lambda} \in C$. Then

$$p(a_1, \ldots, a_n) = \sum \alpha_{1\lambda_1} \ldots \alpha_{n\lambda_n} p(u_{\lambda_1}, \ldots, u_{\lambda_n}) = 0,$$

by multilinearity, thus p vanishes on R. ∎

In special cases we can make the assertion of this corollary more precise.

Proposition 7.5.5. *Let A be a finite-dimensional algebra over an infinite field k. Then any polynomial identity $p(x_1, \ldots, x_d) = 0$ for A also holds for A_E, for any extension field E of k.*

Proof. Let $e_1, \ldots, e_n$ be a k-basis of A, take a field F obtained by adjoining dn independent indeterminates t_{ij} $(i = 1, \ldots, d, j = 1, \ldots, n)$ to k and put $x_i = \sum t_{ij} e_j$. Then in A_F we have

$$p(x_1, \ldots, x_d) = \sum_\nu f_\nu(t_{ij}) e_\nu, \tag{7.5.6}$$

where the f_ν are polynomials in the t_{ij} with coefficients in k. By hypothesis, f_ν vanishes identically on k: if $a_i = \sum \alpha_{ij} e_j$ $(\alpha_{ij} \in k)$, then $\sum f_\nu(\alpha_{ij}) e_j = p(a_1, \ldots, a_d) = 0$; hence $f_\nu(\alpha_{ij}) = 0$ for all α_{ij} in k. Since k is infinite, f_ν vanishes identically, and by (7.5.6), $p = 0$ in A_E for any extension field E of k. ∎

In the opposite direction from Proposition 7.5.3 one can show that every PI-algebra satisfies an identity in two variables. To prove this fact we shall need the fact that every free algebra of rank at least 2 contains a free subalgebra of countable rank. This is easily verified: in $k\langle x, y\rangle$ the elements $z_n = xy^n$ $(n = 1, 2, \ldots)$ are free, because in any equation $f(z_1, \ldots, z_n) = 0$ we can equate homogeneous components, and if $\sum u_i z_i = 0$, then $\sum u_i xy^i = 0$, and it follows that $u_i = 0$, so we reach the conclusion by induction on the degree.

Proposition 7.5.6. *Every PI-algebra satisfies an identity in two variables.*

Proof. If $p(x_1, \ldots, x_d) = 0$ is a non-trivial identity in a k-algebra A, then so is $p(xy, xy^2, \ldots, xy^d) = 0$, and as we have just seen, this is non-trivial. ∎

One of the main results in PI-theory, Kaplansky's theorem, asserts that a primitive PI-algebra is finite-dimensional; this will be proved in Chapter 8, where primitive rings are discussed. For the moment we shall show that the $n \times n$ matrix ring over a commutative ring satisfies the standard identity S_{2n} (Amitsur–Levitzki theorem). The proof uses exterior algebras; we recall that the *exterior algebra* on a vector space V is the algebra generated by V with the defining relations $v^2 = 0\ (v \in V)$. If V has the basis $v_1, \ldots, v_n$ then a basis for the algebra is given by the elements $v_{i_1} \ldots v_{i_r}$ $(i_1 < \ldots < i_r, 1 \le r \le n)$ (see BA, Section 6.4). We shall also need an elementary result on traces.

Lemma 7.5.7. *Let K be a commutative* **Q**-*algebra and $A \in \mathfrak{M}_n(K)$. If $\operatorname{tr}(A^r) = 0$ for $r = 1, \ldots, n$, then $A^n = 0$.*

Proof. Suppose first that K is an algebraically closed field (necessarily of characteristic 0) and let A be an $n \times n$ matrix over K, with eigenvalues $\lambda_1, \ldots, \lambda_n$. The characteristic polynomial of A is

$$\det(\lambda I - A) = x^n + c_1 x^{n-1} + \ldots + c_n, \tag{7.5.7}$$

where the c_i are (except for sign) the elementary symmetric functions of the λ's. Since we are in characteristic 0, we can express $c_1, \ldots, c_n$ as polynomials in the power sums of the λ's, $s_r = \sum \lambda_\nu^r = \operatorname{tr}(A^r)$, $r = 1, \ldots, n$ (Newton's formulae):

$$c_i = f_i(s_1, \ldots, s_i) \tag{7.5.8}$$

where $s_r = \operatorname{tr}(A^r)$ and f_i is of weight i and with rational coefficients.

Now let K be as stated in the lemma, and A the given matrix. Its characteristic polynomial is given by (7.5.7), where the c_i are still given by (7.5.8) in terms of the s_i (since these equations are identities in the entries of A). By hypothesis, $\operatorname{tr}(A^r) = 0$ for $1 \le r \le n$, hence (7.5.7) reduces to x^n, and the Cayley–Hamilton theorem (which clearly holds for any commutative ring) shows that $A^n = 0$. ∎

Theorem 7.5.8 (Amitsur–Levitzki, 1950). *Let K be any commutative ring and $A_1, \ldots, A_{2n} \in K_n$. Then*

$$S_{2n}(A_1, \ldots, A_{2n}) = 0. \tag{7.5.9}$$

Thus $\mathfrak{M}_n(K)$ satisfies a polynomial identity of degree $2n$.

Proof. (S. Rosset) Suppose first that K is a **Q**-algebra and let E be the exterior algebra on the free K-module of rank $2n$, with basis $u_1, \ldots, u_{2n}$, say. In E consider the matrix

$$A = \sum_1^{2n} A_i u_i.$$

For any $r = 1, 2, \ldots$ we have

$$A^r = \sum S_r(A_{i_1}, \ldots, A_{i_r}) u_{i_1} \wedge \ldots \wedge u_{i_r}. \tag{7.5.10}$$

In particular, $A^{2n} = S_{2n}(A_1, \ldots, A_{2n}) u_1 \wedge \ldots \wedge u_{2n}$, so we have to prove that $A^{2n} = 0$. Let E_0 be the subalgebra of E of terms of even degree; then E_0 is commutative and by (7.5.10), $A^2 \in \mathfrak{M}_n(E_0)$, so we need only show that $\operatorname{tr}(A^{2r}) = 0$, for $r = 1, \ldots, n$, by Lemma 7.5.7. By (7.5.10) this will follow if we show

$$\operatorname{tr}(S_m(A_1, \ldots, A_m)) = 0, \quad \text{where } m \text{ is even.} \tag{7.5.11}$$

Let S be the symmetric group on $1, \ldots, m$ and T the stabilizer of 1 in S; then a transversal of T in S is $1, \tau, \ldots, \tau^{m-1}$, where $\tau = (1, 2, \ldots, m)$. Every element of S is uniquely expressible as $\tau^i\sigma$, where $0 \leq i < m$ and $\sigma \in T$; for τ^i brings 1 to the right place and σ then permutes $2, \ldots, m$ as needed. Hence we can write

$$S_m(A_1, \ldots, A_m) = \sum \operatorname{sgn}(\tau^i\sigma) A_{1\tau^i\sigma} \ldots A_{m\tau^i\sigma}.$$

Now τ has the effect of permuting the factors cyclically, which leaves the trace unaffected, hence

$$\operatorname{tr}(S_m(A_1, \ldots, A_m)) = \sum_i \operatorname{sgn}(\tau^i) \left\{ \sum_\sigma \operatorname{sgn}(\sigma) \operatorname{tr}(A_{1\sigma} A_{2\sigma} \ldots A_{m\sigma}) \right\}.$$

Here the second sum is independent of i, and $\sum \operatorname{sgn}(\tau^i) = 0$, because m is even, so (7.5.11) follows, and this proves (7.5.9) when K is a $\mathbf{Q}$-algebra. Therefore it holds for a polynomial ring over $\mathbf{Z}$ (which can be embedded in a $\mathbf{Q}$-algebra), and hence for any commutative ring (which is a homomorphic image of a polynomial ring). ■

We note that the bound $2n$ in this result is best possible:

Lemma 7.5.9 (Staircase lemma). *Let A be a K-algebra with $1 \neq 0$. Then $\mathfrak{M}_n(A)$ satisfies no polynomial identity of degree less than $2n$.*

Proof. If A satisfies an identity of degree $r < 2n$, then it also satisfies a multilinear identity of degree r. Let this be $p = 0$, where each term of p consists of $x_1, \ldots, x_r$ in some order. Thus p has the form

$$p = \alpha x_1 x_2 \ldots x_r + p', \tag{7.5.12}$$

where $\alpha \in k$ and p' is the sum of products of the x's in other orders than that shown. Now the matrix units in A satisfy

$$e_{11}e_{12}e_{22}e_{23} \ldots e_{n-1n}e_{nn} = e_{1n} \neq 0,$$

while the product in any other order is 0, and this applies even if we only take the first r, where $r > 1$. Hence if we put $x_1 = e_{11}$, $x_2 = e_{12}$, $x_3 = e_{22}, \ldots$ then the first term in (7.5.12) is αe_{1i} for some i, while all other terms vanish, so p does not vanish on A_n and we have reached a contradiction. ■

Exercises

1. Show that a polynomial which vanishes identically on a non-trivial algebra must have zero constant term.
2. Let R be a prime PI-algebra, satisfying an identity of degree d. Show that the left (or right) uniform rank of R is $< d$.

3. Show that if a $\mathbf{Q}$-algebra with 1 satisfies the standard identity $S_{2n+1} = 0$, then it also satisfies $S_{2n} = 0$.
4. Show that $S_n([x, y], [x^2, y], \ldots, [x^n, y]) = 0$ holds in $\mathfrak{M}_n(k)$ but not in $\mathfrak{M}_{n+1}(k)$, if k is infinite.
5. Show that $S_{n+1}(x_1, \ldots, x_{n+1}) = \sum(-1)^{i+1}x_i S(x_1, \ldots, x_{i-1}, x_{i+1}, \ldots, x_{n+1})$. Deduce that $S_n = 0$ implies that $S_m = 0$ for all $m > n$.
6. Let R be a central simple algebra of degree n over an infinite field as centre. Show that R satisfies $S_{2n} = 0$ but no identity of degree $< 2n$.
7. Let R, S be such that $\mathfrak{M}_n(R) \cong \mathfrak{M}_n(S)$ and R is commutative. Show that S is also commutative and deduce that $R \cong S$. (Hint. Apply the standard identity with suitable arguments including ae_{11}, be_{11}.)
8. Explain the name of Lemma 7.5.9 (keeping in mind matrix notation).

7.6 Varieties of PI-algebras and Regev's theorem

Let I be any subset of the free algebra $F = k\langle X\rangle$ and denote by $V(I)$ the collection of all k-algebras on which all the members of I vanish identically. It is clear that $V(I)$ is a variety of algebras; in fact the countably generated algebras in $V(I)$ are just the homomorphic images of $F/\mathfrak{t}$, where $\mathfrak{t}$ is the ideal of F generated by all elements obtained by substituting elements of F for the variables in members of I. To obtain another description of $F/\mathfrak{t}$ we need some definitions.

Let A be a k-algebra and Y a generating set of A. The algebra A is said to be *relatively free* on Y if every mapping $Y \to A$ extends to an endomorphism of A (necessarily unique). For example, the free algebra F is relatively free on X, since as we know, every mapping from X to any algebra A extends to a homomorphism $F \to A$, and we need only take the special case $A = F$.

An ideal $\mathfrak{t}$ of the free algebra $F = k\langle X\rangle$ is said to be *fully invariant* or a *T-ideal* if it admits all endomorphisms of F. By the freeness of F this means that if an element $f(x) \in F$ belongs to $\mathfrak{t}$, then $f(a) \in \mathfrak{t}$ for all possible ways of replacing $x_i \in X$ by $a_i \in F$. Given any subset I of F, the ideal generated by all elements obtained from I by replacing $x_i \in X$ by $a_i \in F$ in all possible ways is the least T-ideal containing I; this is also called the T-ideal *generated by I*. Now the ideals defining relatively free algebras are precisely the T-ideals:

Proposition 7.6.1. *A k-algebra A is relatively free on a set Y if and only if* $A \cong F/\mathfrak{t}$, *where F is the free k-algebra on a set X equipotent with Y and* $\mathfrak{t}$ *is a T-ideal.*

Proof. Any k-algebra A with generating set Y can be written as a homomorphic image of a free algebra F on a set X equipotent with Y:

$$A \cong F/\mathfrak{t}, \qquad (7.6.1)$$

where $\mathfrak{t}$ is the ideal of relations in A. Suppose that A is relatively free and that $\lambda : F \to A$ is a surjective homomorphism with kernel $\mathfrak{t}$, whose restriction to X defines a bijection with Y. Given any mapping $\varphi : X \to F$, there is a unique endomorphism of F agreeing with φ on X, which we may again denote by φ. We have to show that φ maps $\mathfrak{t}$ into itself. By hypothesis the mapping $\lambda^{-1}\varphi\lambda : Y \to A$ extends to an endomorphism g of A; thus $x\varphi\lambda = x\lambda g$ for all $x \in X$, hence $\varphi\lambda = \lambda g$ holds on all of F. If $p \in \mathfrak{t}$, then $p\lambda = 0$, hence $p\varphi\lambda = p\lambda g = 0$ and it follows that $p\varphi \in \ker \lambda = \mathfrak{t}$, which shows that $A \cong F/\mathfrak{t}$.

Conversely, assume that $A \cong F/\mathfrak{t}$, where $\mathfrak{t}$ is a T-ideal and let a mapping $g : Y \to A$ be given. We define $\varphi : X \to F$ as follows: given $x \in X$, choose an element u of F such that $u\lambda = x\lambda g$ and put $u = x\varphi$. The mapping φ extends to an endomorphism of F, which will again be denoted by φ. Since $\mathfrak{t}$ is a T-ideal, if $p \in \mathfrak{t}$, then $p\varphi \in \mathfrak{t}$; thus φ can be factored by λ to give an endomorphism h of A such that $\varphi = \lambda h$. For any $x \in X$ we have $x\varphi\lambda = x\lambda h = x\lambda g$, hence h is an endomorphism of A which agrees with g on Y, and this shows A to be relatively free. ■

Our aim in what follows is to prove that the tensor product of two PI-algebras is again a PI-algebra. This is Amitai Regev's theorem; the proof given here is due to Victor Latyshev. Some preparation is necessary.

Consider a permutation σ of $1, 2, \ldots, n$. We use σ to define a partial ordering on the set $\{1, \ldots, n\}$ by writing $i \prec j$ whenever $i < j$ and $i\sigma < j\sigma$. Let us recall that an *antichain* in a partially ordered set is a subset of pairwise incomparable elements, and the *width* of the set is the maximum number of elements in an antichain. For example, for our permutation σ an antichain of d elements is a set of numbers $i_1 < i_2 < \ldots < i_d$ such that $i_1\sigma > i_2\sigma > \ldots > i_d\sigma$. We also recall (BA, Theorem 1.3.1):

Dilworth's theorem. *In any finite partially ordered set S, the minimum number of disjoint chains into which S can be decomposed is the width of S.* ■

Given a permutation σ, suppose that the corresponding partially ordered set can be decomposed into d chains. Then to specify σ we need only give the d chains and their images under σ. For example, the permutation $\begin{pmatrix} 1 & 2 & 3 & 4 & 5 & 6 & 7 & 8 \\ 3 & 5 & 8 & 1 & 4 & 7 & 6 & 2 \end{pmatrix}$ defines a partially ordered set of width 4, and it may be expressed as the unions of the chains $\{1, 2, 3\}, \{4, 5, 6\}, \{7\}, \{8\}$ with images $\{3, 5, 8\}, \{1, 4, 7\}, \{6\}, \{2\}$. Dilworth's theorem allows us to estimate the number of permutations of a given width:

Theorem 7.6.2. *The set of permutations of* $1, 2, \ldots, n$ *for which any set of d numbers* $(2 \le d < n)$ *contains at least one pair in their natural order is at most* $(d-1)^{2n}$.

Proof. Let σ be such a permutation and consider the partial ordering defined by σ (as above) on $\{1, 2, \ldots, n\}$. The hypothesis states that no antichain has d elements, so the set has width less than d, and by Dilworth's theorem it can be written as a disjoint union of at most $d-1$ chains. Let us number these chains from 1 to δ, where $\delta < d$. To specify σ we have to give the distribution of $1, 2, \ldots, n$ and their images under σ over these δ chains. This can be done by defining two mappings from $\{1, \ldots, n\}$ to $\{1, \ldots, \delta\}$. For each mapping there are δ^n choices, hence in all there are δ^{2n} choices and so the number of permutations satisfying the given conditions is at most $(d-1)^{2n}$. ■

Let $F = k\langle X\rangle$ be the free k-algebra on $X = \{x_1, x_2, \dots\}$ and denote by L_n the subspace of all multilinear forms in $x_1, \dots, x_n$; L_n is spanned by the monomials

$$x_\sigma = x_{1\sigma}x_{2\sigma}\dots x_{n\sigma}, \tag{7.6.2}$$

hence its dimension over k is $n!$. We fix an integer d, $2 \le d \le n$, and call a monomial (7.6.2) *good* if the partial ordering defined by σ has width $< d$; by Theorem 7.6.2 the number of good monomials (7.6.2) does not exceed $(d-1)^{2n}$. We shall use this fact to bound the dimension of a relatively free algebra; here it will be convenient to restrict attention to multilinear elements.

Proposition 7.6.3. *Let $\mathfrak{t}$ be a T-ideal in the free algebra $F = k\langle X\rangle$, and put $\mathfrak{t}_n = \mathfrak{t} \cap L_n$, where L_n is the space of multilinear forms of degree n, as above. If $\mathfrak{t}$ contains a polynomial of degree d, where $2 \le d \le n$, then*

$$[L_n/\mathfrak{t}_n : k] \le (d-1)^{2n}. \tag{7.6.3}$$

Proof. Let us take the monomial basis in L_n with the lexicographic ordering. By linearization it follows that $F/\mathfrak{t}$ satisfies a d-linear identity:

$$y_1y_2\dots y_d = \sum_{\sigma\neq 1} \alpha_\sigma y_\sigma \quad \text{where } \alpha_\sigma \in k. \tag{7.6.4}$$

For the proof it will be enough to show that L_n is spanned (mod $\mathfrak{t}$) by the good monomials. Suppose that u is a monomial which is not good. Then we have a factorization

$$u = p(x_{\alpha_1}\dots x_{\beta_1})(x_{\alpha_2}\dots x_{\beta_2})\dots(x_{\alpha_{d-1}}\dots x_{\beta_{d-1}})x_{\alpha_d}q,$$

where $\alpha_1 > \alpha_2 > \dots > \alpha_d$. If in (7.6.4) we put $y_i = x_{\alpha_i}\dots x_{\beta_i}, y_d = x_{\alpha_d}$, we obtain

$$u \equiv \sum \lambda_j v_j \pmod{\mathfrak{t}_n},$$

where v_j comes before u in the lexicographic ordering. Repeating the process if necessary we can after a finite number of steps reduce u to a linear combination of good monomials. Now the conclusion follows by Theorem 7.6.2. ■

In Proposition 7.6.3 we have obtained in (7.6.3) a bound for $[L_n/\mathfrak{t}_n : k]$ of the form $f(d)^n$, and so we have $f(d)^{2n} < n! = [L_n : k]$ for large enough n. This estimate now allows us to achieve our aim:

Theorem 7.6.4 (Regev's theorem, 1972). *The tensor product of two PI-algebras over a field k is again a PI-algebra.*

Proof. (Latyshev) Let A, B be PI-algebras over k; clearly there is a polynomial identity holding in both A and B. Let $\mathfrak{t}$ be the T-ideal of all identities holding in both A and B and suppose that $\mathfrak{t}$ contains an identity of degree $d \geq 2$. From Proposition 7.6.3 we know that $[L_n/\mathfrak{t}_n : k] \leq (d-1)^{2n}$ for all $n \geq d$. Let us fix $n \geq d$ and take a monomial basis $m_1(x), \ldots, m_\nu(x)$ of $L_n/\mathfrak{t}_n$, where $\nu = [L_n/\mathfrak{t}_n : k]$.

For any permutation σ of $1, 2, \ldots, n$ we have

$$x_\sigma = x_{1\sigma}x_{2\sigma}\ldots x_{n\sigma} \equiv \sum_1^\nu \lambda_i(\sigma)m_i(x) \pmod{\mathfrak{t}_n}. \tag{7.6.5}$$

In particular, this relation holds in A and B. Now consider a multilinear polynomial

$$f(x) = \sum \gamma_\sigma x_{1\sigma}\ldots x_{n\sigma}, \tag{7.6.6}$$

with coefficients γ_σ to be determined, in such a way that $f = 0$ is an identity for $A \otimes B$. By multilinearity we need only check f on a spanning set. Given $a_1, \ldots, a_n \in A$, $b_1, \ldots, b_n \in B$, we have

$$\begin{aligned} f(a_1 \otimes b_1, \ldots, a_1 \otimes b_1) &= \sum \gamma_\sigma a_{1\sigma} \otimes b_{1\sigma} \ldots a_{n\sigma} \otimes b_{n\sigma} \\ &= \sum_{i,j=1}^{\nu}\left(\sum_\sigma \lambda_i(\sigma)\lambda_j(\sigma)\gamma_\sigma\right)m_i(a) \otimes m_j(b), \end{aligned}$$

by (7.6.6) and (7.6.5). To ensure that $f = 0$ on $A \otimes B$, we have to solve the equations

$$\sum_\sigma \lambda_i(\sigma)\lambda_j(\sigma)\gamma_\sigma = 0. \tag{7.6.7}$$

These are ν^2 homogeneous linear equations in the $n!$ indeterminates γ_σ and they have a non-trivial solution when $n! > \nu^2$. By Proposition 7.6.3, $\nu \leq (d-1)^{2n}$; now d is given and for large enough n we have

$$\nu^2 \leq (d-1)^{4n} < n!.$$

For such n the equations (7.6.7) have a non-trivial solution γ_σ^0 and then $f = \sum \gamma_\sigma^0 x_\sigma$ is the required polynomial identity for $A \otimes B$. ■

Exercises

1. Show that an algebra A is universal for homomorphisms into some family $\mathscr{C}$ of algebras iff A is relatively free.
2. Show that if two algebras A, B satisfy the same polynomial identity of degree d, then $A \otimes B$ satisfies an identity of degree at most $(d-1)^4$. (Hint. Use Stirling's formula: $(n-1)! \sim n^n e^{-n}\sqrt{(2\pi/n)}$.)
3. Show that an algebra satisfying $x^2 = 0$ also satisfies $xyz = 0$.
4. Show that if A, B satisfy $x^2 = 0$, then $A \otimes B$ satisfies $x^3 = 0$. Find an identity for $A \otimes B$ when A satisfies $x^2 = 0$ and B satisfies $x^3 = 0$.

7.7 Generic matrix rings and central polynomials

As we have seen in Section 7.6, a relatively free algebra may be characterized by the fact that it is universal for homomorphisms to a given class of algebras. We now define the *generic matrix ring* of degree n in $x_1, \ldots, x_d$,

$$F_{(n)} = k\langle x_1, \ldots, x_d\rangle_{(n)}$$

as the k-algebra on $x_1, \ldots, x_n$ which is universal for homomorphisms into $n \times n$ matrix algebras over commutative rings. The elements of $F_{(n)}$ may themselves be thought of as matrices, so that $F_{(n)}$ is a ring of $n \times n$ matrices, generated by $x_1, \ldots, x_d$ and every mapping $x_i \longmapsto a_i \in \mathfrak{M}_n(A)$, where A is commutative, can be extended to a unique k-algebra homomorphism.

An explicit construction of $F_{(n)}$ is obtained as follows: we adjoin dn^2 commuting indeterminates $t_{ij\delta}$ $(\delta = 1, \ldots, d, i, j = 1, \ldots, n)$ to k and in the $n \times n$ matrix ring $\mathfrak{M}_n(k[t_{ij\delta}])$ consider the subalgebra generated by the matrices $x_\delta = (t_{ij\delta})$. Since the $t_{ij\delta}$ are commuting indeterminates, the universal property is easily verified. When $d = 1$, $k\langle x_1\rangle_{(n)}$ is just the polynomial ring in one variable, but for $d > 1$ (and $n > 1$) $k\langle x_1, \ldots, x_d\rangle_{(n)}$ is non-commutative, since $n \times n$ matrices over a non-trivial ring do not all commute. On the other hand, $F_{(n)}$ satisfies certain identities; since it is a subring of $\mathfrak{M}_n(A)$, where A is commutative, it satisfies, for example, the standard identity $S_{2n} = 0$. Thus we may think of $F_{(n)}$ as the k-algebra generated by $x_1, \ldots, x_d$ subject to all the identities holding between $n \times n$ matrices. This is expressed more precisely in

Proposition 7.7.1. *Let $F = k\langle t_1 \ldots, t_d\rangle$ be the free k-algebra, $F = k\langle x_1, \ldots, x_d\rangle_{(n)}$ the generic matrix ring of degree n and $\nu : F \to F_{(n)}$ the k-algebra homomorphism in which $t_\delta \longmapsto x_\delta$. Then $p \in F$ vanishes identically on every $n \times n$ matrix over a commutative k-algebra if and only if $p\nu = 0$.*

Proof. If p vanishes on every $n \times n$ matrix ring, then in particular, $p\nu = 0$. Conversely, if $p\nu = 0$, then since every homomorphism $\varphi : F \to \mathfrak{M}_n(A)$ (A commutative) can be factored by ν, say $\varphi = \nu\varphi'$, we have $p\varphi = p\nu\varphi' = 0$. ■

Thus if $p \in F$ and we want to check whether $p = 0$ holds in all matrix rings, we need only find its image in $F_{(n)}$. Here it is often convenient to embed the coefficient ring $k[t_{ij\delta}]$ in an algebraically closed field K. Then we can transform any matrix over K with distinct eigenvalues to diagonal form. Now the generic matrix $x_\delta = (t_{ij\delta})$ certainly has distinct eigenvalues, since we can specialize it to any other matrix. Hence we can always transform x_1 to diagonal form; of course the same applies to $x_2, \ldots, x_d$, but we cannot transform more than one of the x's simultaneously to diagonal form, because they do not commute.

It turns out that $F_{(n)}$ can be embedded in a skew field; this follows from

Proposition 7.7.2. *The generic matrix ring $k\langle x_1, \ldots, x_d\rangle_{(n)}$ is a left and right Ore domain.*

Proof. We have already seen that $F_{(n)}$ is a PI-algebra; if we can show that it is an integral domain, the desired result will follow by Corollary 7.5.2.

It remains to show that $F_{(n)}$ is an integral domain; the essence of the proof will be to show that any polynomial identity holding in $F_{(n)}$ also holds in a certain division algebra. Let K be the field of fractions of the coefficient ring $k[t_{ij\delta}]$ and let E be an extension field of K with a K-automorphism α of order n, e.g. we may take $E = K(\xi_1, \ldots, \xi_n)$ to be a rational function field and α a cyclic permutation of the ξ's. Let $D = E(z; \alpha)$ be the skew field of fractions of the skew polynomial ring. As we have seen in Example 6 of Section 7.3, this is a central division algebra of degree n over its centre C, so C is infinite. Let L be a splitting field of D; then

$$D \otimes_C L \cong L_n \cong K_n \otimes_K L. \tag{7.7.1}$$

Here the left-hand side contains D, while the right-hand side contains K_n. Now let $f, g \in F_{(n)}$ and suppose that $fg = 0$. Then fg vanishes identically on L_n and hence (by (7.7.1)) on D. Since D is a skew field, it follows that for each choice of arguments either f or g vanishes, so if y is a new indeterminate, then

$$f(x_1, \ldots, x_d)yg(x_1, \ldots, x_d) = 0, \tag{7.7.2}$$

identically in D. By Proposition 7.5.5, this also holds in $D_L \cong L_n$ and hence in K_n, but K_n is simple, hence prime, so either $f = 0$ or $g = 0$, as elements of $F_{(n)}$. This shows $F_{(n)}$ to be an integral domain. ∎

From this result it follows that F has a skew field of fractions, called the *generic division algebra* of degree n over k. We shall see below (in Corollary 7.7.5) that its degree over its centre is n.

These concepts have been used by Shimshon Amitsur [1972] to prove that not every division algebra is a crossed product. It can be shown that if the generic division algebra of degree n were a crossed product, with Galois group Γ, then every division algebra of degree n would be a crossed product with group Γ. Now Amitsur constructs two division algebras of degree n (for a certain n, and in characteristic 0) which cannot be expressed as crossed products with the same Galois group. It follows that the generic division algebra cannot be a crossed product (see Rowen (1988)).

In studying the centre of a PI-algebra it would be useful to have 'centre-valued' or 'central' polynomials, i.e. polynomials p which when evaluated, always yield elements of the centre. By a *central polynomial* for $n \times n$ matrices one understands a polynomial $p \in k\langle x_1, \ldots, x_d\rangle$ which when evaluated in $\mathfrak{M}_n(A)$, where A is a commutative k-algebra, takes values in the centre of $\mathfrak{M}_n(A)$. Of course we are only interested in non-constant polynomials, i.e. polynomials taking at least two values.

As an example consider 2×2 matrices over R. A commutator of 2×2 matrices has the form $\alpha = \begin{pmatrix} a & b \\ c & -a \end{pmatrix}$, for its trace must be 0, and hence

$$\alpha^2 = \begin{pmatrix} a^2 + bc & 0 \\ 0 & a^2 + bc \end{pmatrix}.$$

This is a scalar, hence $(xy - yx)^2$ is a central polynomial for 2×2 matrices. This was essentially the only non-constant central polynomial known for many years, but a large family of central polynomials for all values of n was discovered in 1972 by Edward Formanek. In 1974 Yuri Razmyslov discovered multilinear central polynomials for $n \times n$ matrices, and we shall now describe his construction. We preface the main theorem by two remarks.

Let k be a field; the matrix ring k_n may be considered as an n^2-dimensional vector space with a bilinear form $\mathrm{tr}(xy)$. The usual matrix basis $\{e_{ij}\}$ has the dual basis $\{e_{ji}\}$, for we clearly have

$$\mathrm{tr}(e_{ij}e_{kl}) = \delta_{jk}\delta_{il}.$$

For simplicity we shall index the e_{ij} by a single suffix $\lambda : e_\lambda$ ($\lambda = 1, \dots, n^2$) and denote the dual basis by e^*_λ. We observe that for any matrix $a = \sum a_{ij}e_{ij}$ we have

$$\sum e_\lambda a e^*_\lambda = \sum e_{ij}a_{rs}e_{rs}e_{ji} = \sum a_{jj}e_{ii} = \mathrm{tr}(a).1.$$

If $\{f_\lambda\}, \{f^*_\lambda\}$ is another pair of dual bases for k_n, say $f_\lambda = \sum c_{\lambda\mu}e_\mu$, $e^*_\lambda = \sum f^*_\mu c_{\mu\lambda}$, then $\sum f_\lambda a f^*_\lambda = \sum c_{\lambda\mu}e_\mu a f^*_\lambda = \sum e_\mu a e^*_\mu$, and this proves the formula

$$\sum e_\lambda a e^*_\lambda = \mathrm{tr}(a).1 \quad \text{for any dual bases } \{e_\lambda\}, \{e^*_\lambda\} \text{ of } k_n. \tag{7.7.3}$$

Secondly, put $F = k\langle x_1, \dots, x_d\rangle$ and consider the subspace Φ of elements that are homogeneous of degree 1 in x_1. These elements can be written in the form $\sum a_i x_1 b_i$, where $a_i, b_i \in k\langle x_2, \dots, x_d\rangle$, and the space Φ admits the linear mapping

$$\sum a_i x_1 b_i \longmapsto \sum b_i x_1 a_i,$$

because $\sum a_i \otimes b_i = 0 \Leftrightarrow \sum b_i \otimes a_i = 0$. In general rings there is no reason for this to hold, but it does hold when x is a generic matrix:

Lemma 7.7.3. *Given any k-algebra R, let x_{ij} be n^2 commuting indeterminates over k and write $x = (x_{ij})$. Then the R-bimodule generated by x in $\mathfrak{M}_n(R \otimes k[x_{ij}])$ admits the k-linear mapping*

$$A_x : \sum a_i x b_i \longmapsto \sum b_i x a_i. \tag{7.7.4}$$

Proof. The subspace of matrices linear in the x_{ij} has the R-basis $x_{ij}e_{rs}$. Let us define $(x_{ij}e_{rs})A_x = x_{sr}e_{ji}$; then (7.7.4) holds for $a_1 = e_{ri}$, $b_1 = e_{js}$, for then $a_1 x b_1 = x_{ij}e_{rs}$, $b_1 x a_1 = x_{sr}e_{ji}$. Hence it holds generally, by linearity. ∎

This transformation A_x is called the *Razmyslov transposition*. We observe that if u is linear homogeneous in x, then

$$\mathrm{tr}(y(uA_x|_{x\to 1})) = \mathrm{tr}(u|_{x\to y}), \tag{7.7.5}$$

where $x \to y$ indicates that x is to be replaced by y. For $u = pxq$ the equation reduces to $\mathrm{tr}(yqp) = \mathrm{tr}(pyq)$, which holds by the cyclic symmetry of the trace; hence it holds generally by linearity.

We can now state the main result; instead of A_{x_i} we write A_i.

Theorem 7.7.4 (Razmyslov). *Let*

$$C = C(x_1, \dots, x_{n^2}; y_0, \dots y_{n^2}) = \sum_\sigma \operatorname{sgn}(\sigma) y_0 x_{1\sigma} y_1 \dots y_{n^2-1} x_{n^2\sigma} y_{n^2},$$

where σ *ranges over all permutations of* $1, \dots, n^2$. *Then*

$$D = D(x_1, \dots, x_{n^2}; y_0, \dots y_{n^2}; z) = \sum x_i z(CA_i|_{x_i \to 1})$$

is a non-constant central polynomial on $\mathfrak{M}_n(k)$; *more precisely, on* $\mathfrak{M}_n(k)$,

$$D(x; y; z) = \operatorname{tr}(z)\operatorname{tr}(C).1 \tag{7.7.6}$$

Proof. Let $\{e_\lambda\}$ be a basis for k_n and denote by E the field of fractions of the polynomial ring in the commuting indeterminates $x_i^{(\lambda)}, y_i^{(\lambda)}$ ($\lambda = 1, \dots, n^2, i = 0, 1, \dots, n^2$). We shall prove the theorem by taking $x_i = \sum x_i^{(\lambda)} e_\lambda, y_i = \sum y_i^{(\lambda)} e_\lambda$ and establishing (7.7.6) in E_n.

In the first place we note that $\operatorname{tr}(C) \neq 0$. For if we take the x_i to be the e_{rs} in some order, we can choose the y_j so that

$$y_0 x_1 y_1 x_2 \dots x_{n^2} y_{n^2} = e_{11},$$

while all the other terms in C are 0, and with these values $\operatorname{tr}(C) = 1$.

It is clear from the definition that C vanishes when we put $x_i = x_j$, where $i \neq j$. Hence by (7.7.5),

$$\operatorname{tr}(x_j(CA_i|_{x_i \to 1})) = \operatorname{tr}(C|_{x_i \to x_j}) = \begin{cases} 0 & \text{if } j \neq i, \\ \operatorname{tr}(C) & \text{if } j = i. \end{cases}$$

It follows that up to a scalar (non-zero!) factor $\{(CA_i|x_i \to 1)\}$ is a dual basis for $\{x_i\}$. So by (7.7.3),

$$D = \sum x_i z(CA_i|_{x_i \to 1}) = \operatorname{tr}(z)\operatorname{tr}(C).1,$$

as we wished to show. ■

The polynomial $C = C_n$ is called the *Capelli polynomial* and $D = D_n$ is the *Razmyslov polynomial* for $n \times n$ matrices. By Theorem 7.7.4 it is a non-constant central polynomial on $\mathfrak{M}_n(k)$ whose value is relatively easy to calculate, using the formula (7.7.6).

As Claudio Procesi has observed, any central polynomial φ for $n \times n$ matrices, with zero constant term, vanishes identically on $\mathfrak{M}_m(k)$ for $m < n$, for we can regard any $m \times m$ matrix as an $n \times n$ matrix whose last $n - m$ rows and columns are 0. Now the value of φ must be central, i.e. a scalar, and this scalar is 0, as we see by looking at the (n, n)-entry. More generally, let A be a simple algebra of degree m with centre k. If E is a splitting field, then $A_E = A \otimes E \cong E_m$ and since D_n is multilinear, it vanishes on A whenever $m < n$. This proves

Corollary 7.7.5. *The Razmyslov polynomial D_n is central and non-vanishing for any central simple algebra of degree n, while it vanishes identically on central simple algebras of degree less than n.* ∎

It is clear that D_n does not vanish on the generic division algebra $F_{(n)}$, whereas D_{n+1} does; this shows $F_{(n)}$ to be of degree n. We shall return to this point in Section 8.5, when we come to discuss prime PI-algebras.

Exercises

1. Show that the exterior algebra $A(V)$ on any vector space V is a PI-algebra, but if V is infinite-dimensional, $A(V)$ does not satisfy a standard identity and so cannot be embedded in a matrix algebra over a commutative ring.
2. Show that in a prime PI-algebra every non-zero ideal contains a central regular element.
3. Show that Corollary 7.7.5 holds for any central polynomial with zero constant term. (Hint. Use Proposition 7.5.5 and treat the case of a finite ground field separately.)
4. Show that $[\mathrm{tr}(ax)b]A_x = a.\mathrm{tr}(xb)$, where A_x is the Razmyslov transposition.
5. (Razmyslov)Show that the commutators $[a, b] = ab - ba$ span an $(n^2 - 1)$-dimensional subspace of k_n and use this fact to prove that (in the notation of Theorem 7.7.4) $B = C(x, [u_2, v_2], \dots, [u_{n^2}, v_{n^2}]; y_0, \dots, y_{n^2}) = \mathrm{tr}(x)p$, where p is a matrix depending on u, v, y. Deduce that BA_x is again a central polynomial.
6. Let f be a polynomial in the entries of a square matrix A over a field. Show that if f is unchanged when A is replaced by $P^{-1}AP$, then f is a symmetric function of the eigenvalues of A.

7.8 Generalized polynomial identities

In all the polynomial identities considered so far the variables commute with all the elements of the ground field; in other words, our identities were obtained by equating elements of $k\langle X\rangle$ to zero. However, one may wish to consider situations where the variables do not centralize the ground ring; this leads to identities obtained by equating elements of the tensor ring $A_k\langle X\rangle$ to zero. We recall that the *tensor A-ring* $A_k\langle X\rangle$, where A is any k-algebra, is defined as the ring generated by a set X over A subject to the defining relations $\alpha x = x\alpha$, where $x \in X$, $\alpha \in k$.

Let A be any k-algebra. By a *generalized polynomial identity* (GPI) in A one understands a non-zero element of the tensor A-ring $A_k\langle X\rangle$ which vanishes under all mappings $X \to A$. Shimshon Amitsur [1965] has shown that a primitive ring R satisfies a GPI iff R has a non-zero socle and the endomorphism ring D of a simple R-module is finite-dimensional over its centre. The main difference from Kaplansky's theorem (as regards the conclusion) is that this time the degree of the identity does not provide a bound on the dimension. This can be illustrated by the example of the $n \times n$ matrix ring over a commutative ring, which satisfies a GPI of degree 2:

$$e_{11}xe_{11}ye_{11} - e_{11}ye_{11}xe_{11} - 0,$$

whereas an ordinary identity has degree at least $2n$, as we saw in the staircase lemma (Lemma 7.5.9). An even simpler example is given by a non-prime ring, which always satisfies a GPI of degree 1: $axb = 0$ (for suitable $a, b \neq 0$). We shall confine our attention to the special case of Amitsur's theorem where R is a skew field; thus we shall essentially prove that a skew field satisfying a GPI is finite-dimensional over its centre. This then will also provide an independent proof of Kaplansky's theorem for the case of a skew field. We begin with two lemmas.

Lemma 7.8.1. *Let D be a skew field, Σ a multiplicative subset of D and K its centralizer in D. If $a_i, b_i \in D^\times$ $(i = 1, \dots, n)$ are such that*

$$\sum a_i x b_i = 0 \quad \text{for all } x \in \Sigma, \tag{7.8.1}$$

then $a_1, \dots, a_n$ are right linearly dependent over K.

Proof. Since $a_1, b_1 \neq 0$, we have $a_1 x b_1 \neq 0$ and it follows that $n > 1$ in any relation (7.8.1). Taking n minimal, we may assume that no such relation exists for a proper subfamily of $a_1, \dots, a_n$. Now any element of D can be written in the form $\sum u_j b_1 v_j$ for $u_j \in \Sigma$, $v_j \in D$, $j = 1, \dots, r$, because $\Sigma b_1 D = D$. For $i = 1, \dots, n$ we define a mapping $\gamma_i : D \to D$ by the rule

$$\gamma_i : \sum u_j b_1 v_j \mapsto \sum u_j b_i v_j, \quad \text{where } u_j \in \Sigma, v_j \in D. \tag{7.8.2}$$

To show that γ_i is well-defined, we have to verify that

$$\sum u_j b_1 v_j = 0 \Rightarrow \sum u_j b_i v_j = 0 \quad (i = 2, \dots, n).$$

Suppose then that $\sum u_j b_1 v_j = 0$. For any $x \in \Sigma$ and any $j = 1, \dots, r$ we have $\sum a_i x u_j b_i = 0$ by (7.8.1), hence

$$0 = -\sum_j a_1 x u_j b_1 v_j = \sum_{i=2}^{n} a_i x \left(\sum_j u_j b_i v_j \right).$$

This holds for all $x \in \Sigma$ and is a shorter relation than (7.8.1), hence the coefficient of each a_i must vanish, i.e. $\sum u_j b_i v_j = 0$, for $i = 2, \dots, n$, as we wished to show.

This shows the mapping (7.8.2) to be well-defined. It is right D-linear, i.e. $\gamma_i(zc) = \gamma_i(z)c$ for all $c \in D$. Taking $z = 1$, we find that $\gamma_i(c) = \gamma_i.c$ is left multiplication by an element γ_i. Moreover, γ_i is also left Σ-linear: $\gamma_i w = w\gamma_i$ for all $w \in \Sigma$, hence $\gamma_i \in K$. By definition, $\gamma_i b_1 = b_i$, hence

$$0 = \sum a_i x b_i = \sum a_i x \gamma_i b_1 = \left(\sum a_i \gamma_i\right) x b_1.$$

Since $b_1 \neq 0$, we have $\sum a_i \gamma_i = 0$, and here $\gamma_1 = 1$, so the a's are linearly dependent over K. ∎

Lemma 7.8.2. *Let D, Σ, K be as in Lemma 7.8.1. Suppose that $a_1, \ldots, a_n \in D$ are right linearly independent over K and $b_1, \ldots, b_n \in D$ are such that the set $E = \{\sum a_i x b_i \,|\, x \in \Sigma\}$ is contained in a finite-dimensional left K-space. Then there exists $c \in D^\times$ such that $Kc\Sigma$ is finite-dimensional as left K-space.*

Proof. Let $u_1, \ldots, u_r$ be a left K-basis for a space containing E, so that

$$\sum_1^n a_i x b_i = \sum_1^r \lambda_j(x) u_j \quad \text{for all } x \in \Sigma \text{ and some } \lambda_j(x) \in K.$$

Here we may take $b_1 = 1$ by multiplying by b_1^{-1} on the right. We shall use induction on n; for $n = 1$ we have $a_1 x = \sum \lambda_j(x) u_j$, so $u_1, \ldots, u_r$ is a K-basis for a space containing $Ka\Sigma$ and the conclusion holds with $c = a_1$.

Assume now that $n > 1$. For any $y \in \Sigma$ we have

$$\sum_2^n a_i x(b_i y - y b_i) = \sum \lambda_j(x) u_j y - \sum \lambda_j(xy) u_j.$$

If $b_i y \neq y_i b$ for some $y \in \Sigma$ and some i, we can apply induction on n to reach the conclusion. Otherwise $b_i y = y b_i$ for all $y \in \Sigma$ and so $b_i \in K$; hence $\sum a_i x b_i = (\sum a_i b_i) x$ and we are reduced to the case $n = 1$. ∎

We now come to the main result of this section. Here the restriction to multilinear elements is necessary because Σ may not admit sums.

Theorem 7.8.3 (GPI theorem). *Let D be a skew field, Σ a multiplicative subset of D and K its centralizer in D. If for all $c \in D^\times$, $Kc\Sigma$ is infinite-dimensional as left K-space then any non-zero multilinear element of $D_K\langle X\rangle$ has a non-zero value for some choice of values of X in Σ.*

Proof. Let f be a non-zero multilinear polynomial in $D_K\langle X\rangle$ of degree n, say, which vanishes on Σ. We single out the terms in which x_1 occurs last and write

$$f = \sum_1^r g_i x_1 b_1 + \sum_1^s p_j x_1 q_j, \tag{7.8.3}$$

where $b_i \in D$ and no term in any q_j has zero degree in the x's. We may again take $b_1 - 1$ and suppose f chosen so that r and s are minimal. Then for any $y \in \Sigma$

$$f(x_1, \ldots, x_n)y - f(x_1 y, x_2, \ldots, x_n) = \sum_2^r g_i x_1(b_i y - y b_i) + \sum_1^s p_j x_1(q_j y - y q_j). \tag{7.8.4}$$

Since r was chosen minimal in (7.8.3), $1, b_2, \ldots, b_r$ are linearly independent over K, in particular, $b_i \notin K$ for $i > 1$, so none of the terms $b_i y - y b_i$ in the first sum can vanish identically for $y \in \Sigma$. Choosing $y = y_0 \in \Sigma$ such that $b_2 y_0 \neq y_0 b_2$, we obtain a GPI in D with a smaller value of r, unless $r = 1$, when the first sum on the right of (7.8.4) is absent. In the latter case consider $q_j y - yq_j$; if this vanishes identically for all $y \in \Sigma$, then $q_j \in K$ for all values of the x's in Σ. Write $q_j = \sum c_i x_2 d_i$, where the values of $x_3, \ldots, x_n$ in Σ are chosen so that $q_j \neq 0$ (which is possible, by induction on n). Then the set $\{\sum c_i y d_i | y \in \Sigma\}$ is one-dimensional over K, hence $Kc\Sigma$ is finite-dimensional for some $c \in D^\times$, by Lemma 7.8.2, in contradiction to the hypothesis.

There remains the case where the $q_j y - yq_j$ do not all vanish identically; we shall show that this leads to a contradiction. In this case the left-hand side of (7.8.4), for suitable $y = y_0 \in D$ is a non-zero polynomial f_1, again multilinear in $x_1, \ldots, x_n$, with no term in which x_1 is last. Moreover, each term in f_1 has the x's in the same order as some term in f, so if x_ν does not come last in any term of f, then the same is true of f_1. We apply the same reduction to $x_2, \ldots, x_n$ in turn and finally obtain a polynomial f^* in which no x_n comes last. This is impossible, so this case cannot occur. ∎

If in this theorem we choose $\Sigma = D$, then K is the centre of D and we obtain the skew field case of Amitsur's theorem:

Theorem 7.8.4 (Amitsur). *Let D be a skew field with centre C such that $[D : C] = \infty$. If $f \in D_C\langle X\rangle$ is non-zero, then $f(a) \neq 0$ for some $a \in D^\times$.*

Proof. We need only observe that a non-zero polynomial f leads to a non-zero multilinear polynomial by the linearization process of Proposition 7.5.3, which clearly still applies to generalized polynomials. ∎

The above proof uses simplifications by Wallace Martindale and Yitz Herstein, see Herstein (1976).

Exercises

1. Let R be a simple ring with centre C. Show that Lemma 7.8.1 holds for $D = R$, $\Sigma = R$, $K = C$. Deduce that if $a, b \in R$ satisfy $axb = bxa$ for all $x \in R$, and $a \neq 0$, then $b = \lambda a$ for some $\lambda \in C$.
2. Let k be a field of characteristic 0. Show that in the Weyl algebra $A_1(k)$ with generators u, v every non-trivial multilinear polynomial f is non-zero when the variables in f are replaced by suitable powers of u and v.
3. Let G be a group and $\Delta(G)$ the set of elements of G which have only finitely many conjugates in G. Show that $\Delta(G)$ is a characteristic subgroup of G. Show further that if the group algebra kG has a GPI holding for all arguments in G, then $\Delta(G)$ is of finite index in G.

Further exercises on Chapter 7

1. Let f be a homomorphism from R to a skew field K. Show that f is an epimorphism iff K is the subfield generated by im f. Show also that if f is injective, then K is left R-flat iff R is a right Ore domain.

2. Show that the skew field of fractions of a right Ore domain is unique up to an isomorphism leaving R fixed. (Note that this does not extend to general rings, e.g. a free algebra of rank at least two has many non-isomorphic skew fields of fractions, see Exercise 2 of Section 7.3 and Cohn (1985), (1995).)
3. Let M be a cancellation monoid. Given $a, b \in M$, if $aM \cap bM = \emptyset$, show that the submonoid generated by a and b is free on these generators.
4. Let R be a ring with total quotient ring Q. Show that Q is semisimple whenever every right Q-module is injective as right R-module. (Hint. Take a right ideal in Q, form its complement C as right R-module and verify that C is a Q-module.)
5. Let k be an algebraically closed field and let A be the translation ring over k, generated by x, y with $xy = y(x+1)$ (see Section 7.3, Example 4). Verify that yA is a prime ideal and its complement is an Ore set. Show that any prime ideal other than 0 or yA has the form $\mathfrak{c}_\alpha = yA + (x-\alpha)A$, where $\alpha \in k$. Verify that $y\mathfrak{c}_\alpha = \mathfrak{c}_{\alpha+1}y$ and deduce that the complement of $\mathfrak{c}_\alpha$ is not an Ore set, but the complement of $\bigcap_n \mathfrak{c}_{\alpha+n}$ is an Ore set, for each $\alpha \in k$.
6. In the k-algebra with generators x, y and defining relation $xy = \lambda yx$, where $\lambda \in k^\times$, find the maximal prime ideals and the complements of intersections of prime ideals that are Ore sets. (Hint. Treat the case where λ is a root of 1 separately.)
7. (Ore) Let k be a field containing a finite subfield $\mathbf{F}_q$. Show that the elements of $k[x]$ which as functions on k are linear over $\mathbf{F}_q$ are the *q-polynomials* $\sum a_i x^{q^i}$, and that they form a ring under substitution as multiplication: $fg(x) = f(g(x))$. Verify that this ring is isomorphic to $k[z; \varphi]$, where $\varphi : a \mapsto a^q$.
8. (P. Fatou) A power series over $\mathbf{Z}$ is called *primitive* if no prime divides all its coefficients. Show that the product of primitive power series is primitive. Deduce that if $P, Q \in \mathbf{Z}[x]$ are coprime polynomials such that the power series P/Q has integer coefficients, then $Q(0) = \pm 1$. (Hint. Find polynomials $f, g \in \mathbf{Z}[x]$ such that $fP + gQ = m$ is a positive integer and express m as a product of Q and another series in $\mathbf{Z}[[x]]$.)
9. Let R be a right Bezout domain. Show that any finitely generated torsion-free left R-module is free. Deduce that over a 2-sided Bezout domain every finitely generated module splits over its torsion submodule.
10. Let R be a right Ore domain and K its field of fractions. Show that any ring between R and K is again right Ore.
11. Show that a semiprime right Goldie ring satisfies the minimum condition on right annihilator ideals. (Hint. In any chain the rank becomes stationary; now use (a) $\Rightarrow$ (c) in the proof of Theorem 7.4.9 and **L**.2 to show that essential extensions are trivial.)
12. A module is called *semi-Artinian* if every non-zero quotient contains a simple submodule. Show that a semi-Artinian module is non-singular iff its socle is projective.
13. (A. R. Kemer) If the symmetrizer h_α corresponds to the polynomial $F_\alpha(x_1, \ldots, x_n)$ by the rule $\sum a_\sigma \sigma \mapsto \sum a_\sigma x_{1\sigma} x_{2\sigma} \ldots x_{n\sigma}$, show that F_α is the linearization of $S_{n_1}(x_1, \ldots, x_{n_1}) \ldots S_{n_r}(x_1, \ldots, x_{n_r})$, where the S_i are standard polynomials and α has columns $n_1, \ldots, n_r$.

14. (P. J. Higgins) Write $P(x, y) = \sum_0^n x^i y x^{n-1-i}$. Show that if an algebra A over a field of characteristic prime to $n!$ satisfies $x^n = 0$, then it also satisfies $\sum x_{1\sigma} \ldots x_{n\sigma} = 0$, where σ runs over all permutations of $1, \ldots, n$, and deduce that $P(x, y) = 0$ in A. By evaluating the expression $\sum x^i z y^j x^{n-1-i} y^{n-1-j}$ in two ways, show that A satisfies the identity $x^{n-1} z y^{n-1} = 0$.
15. (P. J. Higgins) Use Exercise 14 to prove the Nagata–Higman theorem: if an algebra of characteristic prime to $n!$ satisfies the identity $x^n = 0$, then $A^{2^n - 1} = 0$. (Hint. Let I be the ideal generated by all elements a^{n-1}, where $a \in A$. Show that $n!IAI = 0$ by Exercise 14 and apply induction on n to A/I.)

8 Rings without finiteness assumption

For general rings there is naturally not as much structure theory as in the Artinian or Noetherian case. It is true that some of the same methods can be used, e.g. the radical can be defined, semiprimitive rings can be expressed as subdirect products of primitive rings etc., but these methods are less precise and they do not lead to a complete classification. For primitive rings a structure theorem can be proved using a general version of the density theorem; this is presented in Section 8.1 and applied in Section 8.2, while Section 8.3 deals with semiprimitive rings. So far we have taken the existence of a unit element for granted, but some work has been done on 'rings without one' and we cast a brief glance at it in Section 8.4; we shall examine the case of simple rings and also see when the existence of a 'one' follows from other assumptions. In Section 8.5 we study semiprime rings, and in Section 8.6 we present an analogue of Goldie's theorem for PI-rings. The final section, Section 8.7, takes a brief look at a natural generalization of principal ideal domains: free ideal rings.

8.1 The density theorem revisited

One of the basic results of ring theory is the Wedderburn–Artin theorem: a simple Artinian ring is a matrix ring over a skew field (BA, Theorem 5.2.2). A related result is the density theorem (Theorem 5.1.1), which for a simple ring A finite-dimensional over its centre k tells us that $A^0 \otimes A$ is a full matrix ring over k. In 1945 Nathan Jacobson (and independently, Claude Chevalley) proved a far-reaching generalization, as part of his theory without finiteness assumptions. Our object is to present this result, but we begin by examining a particular case, the endomorphism ring of a vector space.

Let K be a skew field and V a left K-module; as is well known (see BA, Theorem 11.1.5), V is free as K-module and any two bases of K have the same cardinal, called the *rank* or also the *dimension* of V over K and written $[V : K]$. Let $E = \mathrm{End}_K(V)$ be its endomorphism ring; when $[V : K]$ is finite, equal to n say, then $E \cong \mathrm{End}_K(K^n) \cong \mathfrak{M}_n(K)$, the $n \times n$ matrix ring over K, and this is a simple ring (BA, Theorem 5.2.2). We now ask: what can we say about E when V is infinite-dimensional? In this case E is no longer simple; its ideal structure is described in Theorem 8.1.3 below.

The first step is to find a matrix representation for E. Here it is not necessary for K to be a skew field; we may take any ring R and consider a free R-module of infinite rank ν. Thus we take an index-set I of cardinal ν and take a free left R-module V with basis $\{v_\alpha\}$ $(\alpha \in I)$. In terms of this basis any endomorphism a of V is described by the equations expressing the image of each v_α in terms of the v's:

$$v_\alpha a = \sum_\beta a_{\alpha\beta} v_\beta, \quad \text{where } a_{\alpha\beta} \in R. \tag{8.1.1}$$

Here $(a_{\alpha\beta})$ is a $\nu \times \nu$ matrix, i.e. a square array of elements of R, whose rows and columns are indexed by a set I of cardinal ν. Moreover, for each $\alpha \in I$, there are only finitely many non-zero coefficients $a_{\alpha\beta}$ in (8.1.1), hence each row of the matrix $(a_{\alpha\beta})$ contains only finitely many non-zero entries; we say: it is *row-finite.* Conversely, every row-finite $\nu \times \nu$ matrix over R defines an endomorphism of V relative to the basis $\{v_\alpha\}$. For we can define $v_\alpha a$ by (8.1.1), and for a general element $x = \sum \xi_\alpha v_\alpha$ of V put

$$xa = \sum \xi_\alpha a_{\alpha\beta} v_\beta.$$

It is easily checked that the mapping a so defined is an endomorphism of V, so that we have a bijection between $\mathrm{End}_R(V)$ and the set $\mathfrak{M}_\nu(R)$ of all row-finite $\nu \times \nu$ matrices over R. If we define the addition and multiplication of row-finite matrices, as in the finite case, by the formulae

$$(a_{\alpha\beta}) + (b_{\alpha\beta}) = (a_{\alpha\beta} + b_{\alpha\beta}), \; (a_{\alpha\beta})(b_{\alpha\beta}) = \left(\sum_\lambda a_{\alpha\lambda} b_{\lambda\beta}\right),$$

we find that $\mathfrak{M}_\nu(R)$ is a ring isomorphic to E. We observe that some restriction on the matrices, such as row-finiteness, is essential for the product to be defined. Our conclusion may be stated as

Theorem 8.1.1. *Let R be any ring and V a free left R-module of infinite rank ν. Then* $\mathrm{End}_R(V)$ *is isomorphic to the ring of all row-finite $\nu \times \nu$ matrices over R.* ■

Let us return to the case of a skew field K. When V is a finite-dimensional K-space, say $[V : K] = n$, then $\mathrm{End}_K(V) \cong \mathfrak{M}_n(K)$ is a simple ring, and it can be written as a direct sum of n pairwise isomorphic simple left ideals (BA, Theorem 5.2.2), corresponding to the columns of the matrix. In the infinite case the situation is rather different. We still have the minimal left ideals, corresponding to the columns of the matrix, but we can no longer express the general matrix as the sum of a finite number of such columns, and $\mathrm{End}_K(V)$ is no longer simple.

For any $a \in \mathrm{End}_K(V)$ let us define the *rank* of a, $\rho(a)$, as the K-dimension of the image space:

$$\rho(a) = [\mathrm{im}\, a : K].$$

This agrees with the usual definition in the finite-dimensional case, and we have the following rules:

R.1 $\rho(a)$ *is a cardinal satisfying* $0 \leq \rho(a) \leq [V : K]$,
R.2 $\rho(a) = 0 \Leftrightarrow a = 0$,
R.3 $\rho(a - b) \leq \rho(a) + \rho(b)$,
R.4 $\rho(ab) \leq \min\{\rho(a), \rho(b)\}$.

Of these, **R**.1 and **R**.2 are clear, and **R**.3 follows because $V(a - b) \subseteq Va + Vb$, and so $[V(a - b) : K] \leq [Va : K] + [Vb : K]$. Turning to **R**.4, we clearly have $Vab \subseteq Vb$, hence $\rho(ab) \leq \rho(b)$, but also $[Vb : K] \leq [V : K]$; hence on replacing V by Va we find that $[Vab : K] \leq [Va : K]$ and so $\rho(ab) \leq \rho(a)$.

To elucidate the ideal structure of $\mathrm{End}_K(V)$, we note the following relation between ranks of endomorphisms:

Lemma 8.1.2. *Let K be a skew field and V a left K-space. If $a, b \in \mathrm{End}_K(V)$ and $\rho(a) \geq \rho(b)$, then there exist $p, q \in \mathrm{End}_K(V)$ such that*

$$b = paq, \tag{8.1.2}$$

and $\rho(p) = \rho(q) = \rho(b)$.

Proof. Choose complements N_a, N_b of ker a, ker b in V respectively, so that $V = \ker a \oplus N_a = \ker b \oplus N_b$. Then $N_a \cong \mathrm{im}\ a$, $N_b \cong \mathrm{im}\ b$, so by hypothesis, $[N_a : K] \geq [N_b : K]$; hence there exists $p \in \mathrm{End}_K(V)$ mapping ker b to 0 and embedding N_b in N_a; clearly $\rho(p) = \rho(b)$. If $\{u_\alpha\}$ is a basis of N_b, then the $u_\alpha p$ are linearly independent in N_a and the $u_\alpha pa$ are linearly independent, because the restriction $a|N_a$ is injective. Likewise the $u_\alpha b$ are linearly independent in Vb. Now choose a complement L in V for the subspace spanned by the $u_\alpha pa$, and define q as the endomorphism mapping L to 0 and $u_\alpha pa$ to $u_\alpha b$. Then $\rho(q) = \rho(b)$ and (8.1.2) holds, for both sides map u_α to $u_\alpha b$ and ker b to 0. ∎

With this preparation we can describe the ideal structure of $\mathrm{End}_K(V)$:

Theorem 8.1.3. *Let K be a skew field and V a K-space of infinite dimension ν. For any infinite cardinal μ denote by E_μ the set of all endomorphisms of V of rank $< \mu$. Then the E_μ (for $\mu \leq \nu$) are distinct, each E_μ is an ideal in $E = \mathrm{End}_K(V)$, and these are the only ideals apart from 0 and E.*

Proof. Let $a, b \in E_\mu$, where μ is an infinite cardinal. Then

$$\rho(a - b) \leq \rho(a) + \rho(b) < 2\mu = \mu,$$

(by BA, Proposition 1.2.7), hence $a - b \in E_\mu$. Next, if $a \in E_\mu$ and $c \in E$, then $\rho(ac) < \mu$, $\rho(ca) < \mu$, by **R**.4, and this shows E_μ to be an ideal in E. There are endomorphisms of any rank $\leq \nu$, e.g. projections on subspaces, hence all the E_μ are distinct. It remains to show that there are no other ideals.

Let $\mathfrak{a}$ be a non-zero ideal in E and denote by μ the least cardinal $> \rho(a)$ for all $a \in \mathfrak{a}$. Then μ is infinite; for $\mathfrak{a}$ contains endomorphisms of positive rank, hence (by Lemma 8.1.2) $\mathfrak{a}$ contains all endomorphisms of rank 1. But every endomorphism of finite rank can be written as a sum of endomorphisms of rank 1, so $\mathfrak{a}$ contains all endomorphisms of finite rank and μ must be infinite. By definition, each $a \in \mathfrak{a}$ has rank $< \mu$, hence $\mathfrak{a} \subseteq E_\mu$. Conversely, if $b \in E_\mu$, then $\rho(b) < \mu$, hence $\rho(b) \leq \rho(a)$ for some $a \in \mathfrak{a}$. By Lemma 8.1.2, $b = paq$ for some $p, q \in E$, so $b \in \mathfrak{a}$, and this shows that $\mathfrak{a} = E_\mu$. Thus every non-zero ideal in E is of the form E_μ for some infinite cardinal μ, and clearly if $\mu > \nu$, then $E_\mu = E$. ■

This result shows in particular that (for infinite $[V : K]$) $\mathrm{End}_K(V)$ is never simple. However, it has another property; as we shall see in Section 8.2, $\mathrm{End}_K(V)$ is primitive, and there is a close relation between primitive rings and such endomorphism rings, which is described in Theorem 8.2.3.

We now come to the density theorem in its general form; our presentation follows Nicolas Bourbaki. We begin by describing the bicentral action on a module.

Let R be any ring and M a right R-module; the action of R on M may be described by saying that we have a homomorphism of R into $\mathrm{End}(M)$, each $a \in R$ corresponding to an endomorphism a' of M, qua additive group. The centralizer S of this set $R' = \{a' | a \in R\}$ in $\mathrm{End}(M)$ is the ring of all R-endomorphisms, $S = \mathrm{End}_R(M)$, and we shall regard M as left S-module. Since $(\alpha x)a = \alpha(xa)$ for all $x \in M$, $a \in R$, $\alpha \in S$, by definition of S, we see that M becomes an (S, R)-bimodule in this way. Let T be the centralizer of S; this is again a subring of $\mathrm{End}(M)$, called the *bicentralizer* of R. Clearly for any $a \in R$, a' centralizes S and so lies in T, i.e. $R \subseteq T$. If equality holds, we say that R acts *bicentrally* on M, or also that M_R is *bicentral.*

As an example consider the ring R itself. As is well known and easily proved (see BA, Theorem 5.1.3), the centralizer of ${}_RR$ is the set of all right multiplications; by symmetry the centralizer of R_R is the left of all left multiplications, hence ${}_RR$ as well as R_R is bicentral. For this property the presence of a unit element is of course material.

The definition of 'bicentral' may be restated as follows: Given a right R-module M, R acts *bicentrally* if for each θ in the bicentralizer T there exists $a \in R$ such that

$$x\theta = xa \quad \text{for all } x \in M. \tag{8.1.3}$$

This is a very strong requirement, because a in (8.1.3) is independent of x. To obtain a weaker condition, let us say that R acts *fully* on M if for each θ in the bicentralizer T and each $x \in M$ there exists $a \in R$ such that $x\theta = xa$, where a may depend on the choice of $x \in M$. But the most useful condition is intermediate between these two. We shall say that R acts *densely* on M if for each $\theta \in T$ and each finite family $x_1, \ldots, x_n \in M$ there exists $a \in R$ such that

$$x_i\theta = x_ia \quad \text{for } i = 1, \ldots, n. \tag{8.1.4}$$

This reduces to the definition in Section 5.1 when the centralizer of R is a field k.

It is clear that bicentral $\Rightarrow$ dense $\Rightarrow$ full. The next result gives a useful condition under which dense $\Leftrightarrow$ bicentral.

Proposition 8.1.4. *Let M be an R-module with centralizer S. If M is finitely generated as S-module, then R acts densely on M if and only if it acts bicentrally.*

Proof. Let $M = \sum_1^n Su_i$; given θ in the bicentralizer, choose $a \in R$ such that $u_i\theta = u_ia$ $(i = 1, \ldots, n)$ and consider the set $N = \{x \in M | x\theta = xa\}$. This is clearly an S-submodule of M and it contains the generating set $u_1, \ldots, u_n$ of M, hence $N = M$, i.e. $x\theta = xa$ for all $x \in M$, so R acts bicentrally. The converse is clear. ■

To motivate the terminology we remark that End(M) may be regarded as a set of mappings from M to M, i.e. a subset of M^M, and this set can be topologized as a product, taking the discrete topology on M. The topology so defined on End(M) is called the *topology of pointwise convergence*. Given $f \in M^M$, we obtain a typical neighbourhood of f by taking a finite set $x_1, \ldots, x_n \in M$ and considering all $f' \in M^M$ such that $x_i f' = x_i f (i = 1, \ldots, n)$. Now (8.1.4) may be restated by saying that R is *dense* in its bicentralizer T (i.e. every non-empty open subset of T contains an element of R). In this connexion we note that any centralizer and hence any bicentralizer is closed in the topology defined here on End(M).

Let M be a right R-module, denote the centralizer (acting on the left) by S and the bicentralizer by T. As is easily verified, the centralizer of ${}^nM_R$ is S_n (BA, Corollary 4.4.2), so that we may regard ${}^nM$ as left S_n-module, and the centralizer of this module is T (BA, Theorem 4.4.6). Thus we have

Proposition 8.1.5. *For any R-module M and any $n \geq 1$, M and ${}^nM$ have isomorphic bicentralizers.* ■

To say that R acts densely on M means that for all n and all $x \in {}^nM$, $\theta \in T$, there exists $a \in R$ such that $x\theta = xa$. This just states that R acts fully on ${}^nM$, for all n; thus we have proved

Theorem 8.1.6 (Density theorem). *Let M be any right R-module. Then R acts densely on M if and only if R acts fully on ${}^nM$, for all $n \geq 1$.* ■

We give several applications, which show the power of this result. In the first place, we clearly have

Corollary 8.1.7. *If R acts densely on a module M, then it acts densely on ${}^nM$, for any $n \geq 1$.* ■

An important case of dense action is provided by semisimple modules.

Theorem 8.1.8. *Let M be a semisimple right R-module. Then R acts densely on M, and M is also semisimple over the centralizer of M. Moreover, if R is right Artinian, then M is finitely generated over the centralizer of R and R acts bicentrally on M.*

Proof. Denote by S the centralizer and by T the bicentralizer of R on M. We first show that every R-submodule of M is a T-submodule. Let M_1 be an R-submodule of M; since M is semisimple, we have $M = M_1 \oplus M_2$ for a submodule M_2. Let θ_1 be the projection on M_1; then $\theta_1 \in S$, hence for any $t \in T$ and $x \in M_1$ $xt = (\theta_1 x)t = \theta_1(xt)$, so $xt \in M_1$ as claimed.

Next we show that R acts fully on M, i.e. for each $t \in T$ and $x \in M$ there exists $a \in R$ such that $xt = xa$. This states that for each $x \in M$, $xT \subseteq xR$. But xR is an

R-submodule containing x, hence it admits T, by what has been shown, and so $xT \subseteq xR$, as required.

We apply this result to ${}^nM$. This is again semisimple, hence R acts fully on ${}^nM$ for all n, and so, by the density theorem, R acts densely on M.

To prove that M is semisimple as S-module, we can write $M = \sum Sx$, where x runs over all elements of all simple R-submodules, for every element of M is a sum of such elements. The conclusion will follow if we show that Sx is simple or 0. Take $0 \neq y \in Sx$, say $y = sx$. Then for any $a \in R$, the mapping $xa \mapsto sxa = ya$ is a homomorphism $xR \to yR$, which is surjective and $yR \neq 0$. Hence it is an isomorphism; if $xR = yR$, we can write $M = xR \oplus N$ and find an R-automorphism of M mapping y to x. Otherwise $xR \cap yR = 0$ and we can write $M = xR \oplus yR \oplus N$; now it is clear how the isomorphism from yR to xR can be extended to an R-endomorphism of M. It follows that Sy contains x, hence $Sy = Sx$ and this shows Sx to be simple. Thus M has been expressed as a sum of simple S-modules and by omitting redundant terms we see that ${}_SM$ is semisimple.

Finally, assume that R is right Artinian. Let us write $M = \oplus Su_i$; we claim that this sum is finite, for if not there is a countable subset: $u_1, u_2, \ldots$. If we put $\mathfrak{a}_n = \{a \in R \mid u_1a = \ldots = u_na = 0\}$, then $\mathfrak{a}_n$ is a right ideal of R and

$$\mathfrak{a}_1 \supseteq \mathfrak{a}_2 \supseteq \ldots. \tag{8.1.5}$$

The projection of M on the complement of $Su_1 \oplus \ldots \oplus Su_n$ is an S-endomorphism, i.e. there exists $t \in T$ such that $u_it = 0\ (i = 1, \ldots, n)$, $u_it = u_i\ (i > n)$. By density there exists $a \in R$ such that $u_ia = 0\ (i = 1, \ldots, n)$, $u_{n+1}a = u_{n+1}$. Hence $u_{n+1} \in \mathfrak{a}_n \backslash \mathfrak{a}_{n+1}$ and this shows the inclusions in (8.1.5) to be strict. But this contradicts the fact that R is right Artinian. Therefore $M = Su_1 \oplus \ldots \oplus Su_n$ for some n, and by Proposition 8.1.4, R acts bicentrally. ∎

For a simple module we can say rather more. By combining Corollary 8.1.7 with Schur's lemma (Lemma 6.3.1) we obtain a generalization of Wedderburn's first structure theorem (see BA, Theorem 5.2.2):

Corollary 8.1.9. *Let M be a simple right R-module. Then the centralizer K of R is a skew field and R acts as a dense ring of linear transformations on M. In particular, if R is right Artinian, then the image of R in* End(M) *acts bicentrally, and so is a full matrix ring over K.*

Proof. By Schur's lemma the centralizer K is a skew field. When R is right Artinian, then M is finitely generated over K, by Theorem 8.1.8, and so is finite-dimensional, say $M \cong K^n$ and R acts bicentrally. Hence R as centralizer of K is K_n. ∎

Exercises

1. Let $E = \text{End}(V)$ and E_ν be as in the text, where $[V : K] = \nu$ is infinite. Verify that E/E_ν is simple but not semisimple. Show that for $a, b \in E$ there exists p such that $b = pa$ iff $\text{im}\ b \subseteq \text{im}\ a$ and there exists q such that $b = aq$ iff $\ker b \supseteq \ker a$. Deduce that E is neither Artinian nor Noetherian; prove the

same for E/E_ν. (Hint. Consider the set of endomorphisms with image in a given finite-dimensional subspace, or with kernel containing a given finite-dimensional subspace.)

2. In $E = \mathrm{End}_K(V)$ show that the endomorphisms with image in a given one-dimensional subspace form a minimal left ideal and that all such left ideals are isomorphic. Show further that the sum of all these left ideals is the unique minimal ideal in E (it is the *socle* of E).
3. In the notation of Theorem 8.1.3 show that E_μ, as algebra without 1, has as ideals precisely the E_λ ($\lambda \leq \mu$) and 0.
4. Show that for any vector space V over a field k, $\mathrm{End}_k(V)$ is a regular ring.
5. Show that in any monoid the centralizer of any subset is its own bicentralizer. Deduce that for any R-module M, $\mathrm{End}_R(M)$ acts bicentrally on M.
6. Let A be a finitely generated abelian group, as $\mathbf{Z}$-module. Show that $\mathbf{Z}$ acts bicentrally on A.
7. Show that a simple ring acts bicentrally on any right ideal.

8.2 Primitive rings

Let R be any ring and M a right R-module. We say that R acts *faithfully* on M or that M is a *faithful* R-module if for any $a \in R$, $a \neq 0$, we have $Ma \neq 0$. This means that the standard homomorphism $R \to \mathrm{End}(M)$ is injective. A ring R is called *primitive* if it has a simple faithful right R-module. Strictly speaking this type of ring should be called 'right primitive' and a corresponding notion 'left primitive' should be defined. In fact these concepts are distinct, as examples show, but we shall only be dealing with right primitive rings and so omit the qualifying adjective.

To obtain an internal characterization of primitive rings, we define for any right ideal $\mathfrak{a}$ of R its *core* as the set

$$(\mathfrak{a} : R) = \{x \in R \mid Rx \subseteq \mathfrak{a}\}. \tag{8.2.1}$$

If we regard $M = R/\mathfrak{a}$ as a right R-module, $(\mathfrak{a} : R)$ is the annihilator of M in R. This shows the core to be an ideal in R; moreover $(\mathfrak{a} : R) \subseteq \mathfrak{a}$ and any (two-sided) ideal of R contained in $\mathfrak{a}$ is contained in $(\mathfrak{a} : R)$, by the definition; thus the core of $\mathfrak{a}$ is the largest ideal of R contained in $\mathfrak{a}$. With its help primitive rings can be characterized as follows:

Proposition 8.2.1. *A ring R is (right) primitive if and only if R contains a maximal right ideal whose core is zero.*

Proof. If R is primitive, there is a faithful simple right R-module M. We have $M = R/\mathfrak{a}$, where $\mathfrak{a}$ is a maximal right ideal (by the simplicity of M), and since M is faithful, $(\mathfrak{a} : R) = 0$. Conversely, if $\mathfrak{a}$ is a maximal right ideal with zero core, then $R/\mathfrak{a}$ is a faithful simple right R-module. ∎

Corollary 8.2.2. *A commutative ring is primitive if and only if it is a field.*

Proof. In this case the core of $\mathfrak{a}$ is $\mathfrak{a}$ itself, so 0 must be the maximal ideal of R and this means that R is a field. ■

To give an example, any simple ring (Artinian or not) is primitive, for R has a maximal right ideal by Krull's theorem (see BA, Theorem 4.2.6) and its core is a proper ideal, which must be 0. The converse does not hold: if V is an infinite-dimensional vector space over a field and E is its endomorphism ring, then E acts faithfully on V and V is clearly simple as E-module, so E is primitive, but as we saw in Section 8.1, E is not simple. The next result describes primitive rings more precisely:

Theorem 8.2.3. *Any primitive ring is isomorphic to a dense ring of linear transformations in a vector space over a skew field K. Conversely, any dense subring of* $\mathrm{End}_K(V)$ *is primitive.*

Proof. Given a primitive ring R, let V be a simple R-module on which R acts faithfully. Its centralizer K is a skew field, by Schur's lemma, and R is naturally embedded as a dense subring in $\mathrm{End}(V)$, by Corollary 8.1.9. For the converse we need only observe that any dense subring R of $\mathrm{End}(V)$ acts simply: given $u, v \in V$, $u \neq 0$, there exist $a \in R$ such that $ua = v$. Hence the R-submodule generated by any $u \neq 0$ is V, i.e. V is simple. ■

As we saw, a primitive ring need not be simple, but if R is right Artinian as well as primitive, say it is a dense subring of $\mathrm{End}_K(V)$, then V is finitely generated over K, by Theorem 8.1.8, hence $V \cong K^n$ and R acts bicentrally: $R \cong \mathrm{End}_K(K^n) \cong K_n$. This expression is unique up to isomorphism of K (BA, Theorem 5.2.2). In the general case there is no such uniqueness, but we have the following consequence which is sometimes useful:

Proposition 8.2.4 (O. Litoff). *Let R be a primitive ring which is not right Artinian. Then for every* $n \geq 1$, *R has a subring with a homomorphism onto a full* $n \times n$ *matrix ring over a skew field.*

Proof. We may take R to be a dense subring of $\mathrm{End}_K(V)$, where V is an infinite-dimensional vector space over K. Given $n \geq 1$, take an n-dimensional subspace U of V, with a basis $u_1, \ldots, u_n$. Let R_1 be the subring of R mapping U into itself: every element of R_1 defines by restriction an endomorphism of U, thus we have a homomorphism $R_1 \to \mathrm{End}_K(U) \cong K_n$, and this is surjective, by density, so it is the required homomorphism. ■

Let R be a ring with a minimal right ideal and define the *socle* $\mathfrak{s}$ of R as the sum of all minimal right ideals. This socle is an ideal, for, given any minimal right ideal $\mathfrak{a}$ of R and any $x \in R$, then $x\mathfrak{a}$ is a minimal right ideal or 0 and so is contained in $\mathfrak{s}$. Primitive rings with non-zero socle have a more precise description:

Theorem 8.2.5. *A primitive ring has a non-zero socle if and only if in its representation as a dense ring of linear transformations of a K-space, R contains transformations of* *finite (non-zero) rank. When this is so, all faithful simple right ideals of R are isomorphic and the skew field K is determined up to isomorphism as the centralizer of a faithful simple right R-module.*

Proof. If there is an element of R defining a linear transformation of finite rank, take $c \in R$ such that the rank $\rho(c)$ is the least positive number possible. Then $\ker cx \supseteq \ker c$ for any $x \in R$, and if $cx \neq 0$, then $\rho(cx) = \rho(c)$, hence $\ker cx = \ker c$, and the complement of $\ker c$ is finite-dimensional. By density we can find $y \in R$ such that $cxy = c$; this shows cR to be minimal, hence R has a non-zero socle.

Now assume that R has a non-zero socle, and hence a minimal right ideal $\mathfrak{a}$. If M is any faithful simple right R-module, then $M\mathfrak{a} \neq 0$, so $u\mathfrak{a} \neq 0$ for some $u \in M$. But $u\mathfrak{a}$ is a submodule of M, hence $u\mathfrak{a} = M$ and so the mapping $x \mapsto ux$ $(x \in \mathfrak{a})$ is a surjective homomorphism $\mathfrak{a} \to M$. By the minimality of $\mathfrak{a}, M \cong \mathfrak{a}$ as right R-module, so $\mathfrak{a}$ is also faithful; this shows that every simple faithful right R-module is isomorphic to M. It follows that the isomorphism type of K is uniquely determined as the endomorphism ring of $\mathfrak{a}$. Finally, since $\mathfrak{a}$ is faithful, $\mathfrak{a}^2 \neq 0$, hence $\mathfrak{a}^2 = \mathfrak{a}$, and so $c\mathfrak{a} = \mathfrak{a}$ for some $c \in \mathfrak{a}$. We claim that $\rho(c) = 1$; for if not, then there exist $x, y \in V$, where V is the K-space on which R acts, such that xc, yc are linear independent over K, and by density there exists $b \in R$ such that $xcb \neq 0$, $ycb = 0$. Then $\mathfrak{a} = cR$ meets the annihilator $\mathfrak{n}$ of y in R, and $\mathfrak{n}$ is a right ideal, so $\mathfrak{a} \subseteq \mathfrak{n}$, by the minimality of $\mathfrak{a}$. But this means that $yc = 0$, which is a contradiction, and it shows that R contains elements of rank 1. ■

By contrast, simple rings with minimal right ideals are much more special:

Proposition 8.2.6. *Any simple ring with minimal right ideals is Artinian.*

Proof. Let R be a simple ring with minimal right ideals. The sum of all minimal right ideal is the socle, a two-sided ideal, which coincides with R, by simplicity. Thus R is a sum of simple right R-modules, hence a direct such sum, and since R is finitely generated (by 1), this direct sum is finite. Thus R is right Artinian. Now R is also semisimple as left R-module, hence it is also left Artinian, and so it is an Artinian ring. ■

Exercises

1. For any ring R and any $n \geq 1$, show that R_n is primitive iff R is. More generally, show that being primitive is a Morita invariant.
2. Show that every minimal right ideal in a primitive ring has an idempotent generator. (Hint. Use the primitivity to show that the right ideal is not nilpotent.)
3. Show that if e is a non-zero idempotent in a primitive ring R, then eRe is primitive.
4. Show that for any idempotent e in a ring R, eR is a minimal right ideal iff Re is a minimal left ideal. Deduce that in a primitive ring with non-zero socle, the socle coincides with the left socle (defined correspondingly).
5. Show that the socle of a primitive ring is a minimal two-sided ideal. Deduce that a simple ring with non-zero socle is Artinian (Proposition 8.2.6).

6. Show that a left Artinian primitive ring is simple. What can be said about a (right) primitive ring with minimal left ideals?
7. Show that the centre of a primitive ring is an integral domain. If A is any commutative integral domain with field of fractions K, show that the set of infinite matrices over K which are equal to a scalar in A outside a finite square is a primitive ring with centre A.

8.3 Semiprimitive rings and the Jacobson radical

In the last section we defined primitive rings as rings with a faithful simple module. Often one needs to consider a wider class of rings, and we define a ring to be *semi-primitive* if it has a faithful semisimple module; clearly it is equivalent to require that for each $a \in R^{\times}$ there exists a simple module M such that $Ma \neq 0$; for if such a module M_a is chosen for each a, then $\sum M_a$ is faithful and semisimple, and conversely, if M is faithful and semisimple, then each $a \neq 0$ acts non-zero on at least one simple summand.

We recall from BA (Lemma 5.3.2) that $\mathrm{J}(R)$, the Jacobson radical of R, is defined in the following equivalent ways:

$\mathrm{J}(R)$ is the set of all $a \in R$ such that
(a) $Ma = 0$ for each simple right R-module M,
(b) a belongs to each maximal right ideal,
(c) $1 - ay$ has a right inverse for each $y \in R$,
(d) $1 - xay$ has an inverse for all $x, y \in R$,
(a^0)–(d^0) the left-hand analogues of (a)–(d).

It is clear from (a) of this definition that a ring R is semiprimitive precisely when $\mathrm{J}(R) = 0$. Moreover, the symmetry of the definition shows that the notion 'semi-primitive' is left–right symmetric. This is in contrast to the situation for primitive rings, where one-sided examples exist, first found by George Bergman in 1964 and Arun Jategaonkar in 1968.

We also recall that a *quasi-inverse* of an element c is an element c' such that

$$c + c' = cc' = c'c. \tag{8.3.1}$$

The quasi-inverse, when it exists, is unique, because $1 - c'$ is the inverse of $1 - c$. An element which has a quasi-inverse is sometimes called *quasi-regular*; for example, any nilpotent element is quasi-regular. Now $\mathrm{J}(R)$ may also be defined as the largest ideal consisting entirely of quasi-regular elements; this is an easy consequence of (d) above.

Semiprimitive rings admit a subdirect product representation which is sometimes useful; however, it should be borne in mind that a given product may contain many different subdirect products, and the relation to the direct product is not very close.

Theorem 8.3.1. *Every semiprimitive ring R is a subdirect product of primitive rings which are homomorphic images of R. Conversely, every subdirect product of primitive rings is semiprimitive.*

Proof. Let $\{\mathfrak{p}_\lambda\}$ be the family of all maximal right ideals of the semiprimitive ring R, and denote the core of $\mathfrak{p}_\lambda$ by $\mathfrak{c}_\lambda$. Since R is semiprimitive, we have $\cap\mathfrak{p}_\lambda = 0$, and since $\mathfrak{c}_\lambda \subseteq \mathfrak{p}_\lambda$, it follows that $\cap\mathfrak{c}_\lambda = 0$. If we put $R_\lambda = R/\mathfrak{c}_\lambda$, then R_λ is primitive, for it is represented faithfully on the simple module $R/\mathfrak{p}_\lambda$. Now the natural maps $f_\lambda : R \to R_\lambda$ can be combined to a homomorphism into the direct product

$$f : R \to P = \prod R_\lambda, \tag{8.3.2}$$

and $\ker f = \cap \ker f_\lambda = \cap\,\mathfrak{c}_\lambda = 0$. Thus f is injective and if $\varepsilon_\lambda : P \to R_\lambda$ denotes the canonical projection, then $f\varepsilon_\lambda = f_\lambda$ is surjective, by the definition of R_λ, so (8.3.2) is the required direct product representation. Conversely, if $\{R_\lambda\}$ is a family of primitive rings and M_λ is a faithful simple R_λ-module, then for any subdirect product R of the R_λ and any $a \in R^\times$ we have $a\varepsilon_\lambda \neq 0$ for some λ, hence a has a non-zero action on M_λ and this shows R to be semiprimitive. ∎

In particular, when R is commutative, we have by Corollary 8.2.2,

Corollary 8.3.2. *Any commutative semiprimitive ring is a subdirect product of fields, and conversely, such a subdirect product is primitive.* ∎

For example, $\mathbf{Z}$ is semiprimitive because 0, 2 are the only quasi-regular elements and so $\mathbf{J}(\mathbf{Z}) = 0$. Hence $\mathbf{Z}$ is a subdirect product of fields; in fact $\mathbf{Z}$ is a subdirect product of the fields $\mathbf{F}_p$, where p ranges over all primes.

The definition of $\mathbf{J}(R)$ shows that it measures how far R is from being semiprimitive. It is a pleasant (and by no means self-evident) property that $R/\mathbf{J}(R)$ is semiprimitive. This follows from the next result, itself more general.

Proposition 8.3.3. *Let R be a ring and $\mathfrak{a}$ an ideal such that $\mathfrak{a} \subseteq \mathbf{J}(R)$. Then*

$$\mathbf{J}(R/\mathfrak{a}) = \mathbf{J}(R)/\mathfrak{a}. \tag{8.3.3}$$

Proof. The natural homomorphism $R \to R/\mathfrak{a}$ induces a lattice-isomorphism between the lattice of right ideals of $R/\mathfrak{a}$ and that of all right ideals of R which contain $\mathfrak{a}$. Each maximal right ideal of $R/\mathfrak{a}$ corresponds to a maximal right ideal of R; the converse also holds because $\mathfrak{a} \subseteq \mathbf{J}(R) = \cap\{\text{max. right ideals}\}$. Taking intersections of these sets of maximal right ideals we obtain (8.3.3). ∎

If in (8.3.3) we put $\mathfrak{a} = \mathbf{J}(R)$, the right-hand side reduces to 0 and we deduce

Corollary 8.3.4. *For any ring R, $R/\mathbf{J}(R)$ is semiprimitive.* ∎

We note that (8.3.3) may not hold without restriction on $\mathfrak{a}$, for the mapping $\mathbf{J}(R) \to \mathbf{J}(R)/\mathfrak{a}$ is not generally surjective, e.g. if $R = \mathbf{Z}$, $\mathfrak{a} = (4)$, then $\mathbf{J}(\mathbf{Z}) = 0$, but $\mathbf{J}(\mathbf{Z}/4) \neq 0$.

It is well known (see BA, Section 5.3) and easily checked that the Jacobson radical contains all nil ideals and for an Artinian ring R, $\mathbf{J}(R)$ is nilpotent, although in general $\mathbf{J}(R)$ need not even be nil, e.g. in the power series ring $k[[x]]$ the Jacobson radical is (x). However, in the absence of nil ideals we obtain a semiprimitive ring by adjoining an indeterminate.

Proposition 8.3.5 (Amitsur). *If R is a ring with no non-zero nil ideals, then $R[t]$ is semiprimitive, where t is an indeterminate.*

Proof. We have to show that the radical J of $R[t]$ is 0. If $J \neq 0$, let $\mathfrak{a}$ be the set consisting of 0 and all leading coefficients of elements in J. It is clear that $\mathfrak{a}$ is an ideal in R and the conclusion will follow if we prove that $\mathfrak{a} = 0$. Let $a_1 \in \mathfrak{a}$, say $f = a_1 x^n + \ldots \in J$. Then $ft \in J$ and so there exists $g \in R[t]$ such that $(1+g)(1-ft) = 1$, i.e.

$$g = ft + gft = ft + f^2t^2 + gf^2t^2 = \ldots = ft + f^2t^2 + \ldots + f^rt^r + gf^rt^r,$$

for all $r \geq 1$. Hence we obtain

$$(1+g)(1 - f^rt^r) = 1 + ft + \ldots + f^{r-1}t^{r-1}.$$

Let us take $r > \deg g$ and equate the coefficients of terms of degree $r(n+1)$. On the right there is no contribution, while on the left we have $(1+g)a_1^r$, therefore $a_1^r = 0$. Thus $\mathfrak{a}$ is a nil ideal, hence $\mathfrak{a} = 0$ and so $J = 0$, as we had to show. ■

We can now also prove the basic theorem on PI-algebras:

Theorem 8.3.6 (Kaplansky's theorem, 1948). *Let R be a primitive PI-algebra, with a polynomial identity of degree d. Then R is a simple algebra of finite dimension n over its centre, where $n \leq d/2$. More precisely, if V is a simple faithful R-module and $D = \mathrm{End}_R(V)$, then $R \cong \mathfrak{M}_m(D)$, where $m = [V : D]$.*

Proof. Let p be a polynomial of degree d which vanishes on R; by Proposition 7.5.3 we may take p to be multilinear. Since R is primitive, there is a simple faithful R-module V; we identify R with its image in $\mathrm{End}(V)$ and put $D = \mathrm{End}_R(V)$. By Schur's lemma D is a skew field, and by the density theorem (Theorem 8.1.6) R is dense in $\mathrm{End}_D(V)$. If $[V : D]$ is finite, we have $R = \mathrm{End}_D(V)$ and the result follows; otherwise, by Proposition 8.2.4 we can for any m, find a subring of R mapping onto D_m, so D_m again satisfies $p = 0$. By the staircase lemma (Lemma 7.5.9) it follows that $d \geq 2m$, so we have a bound on m. Hence $[V : D] = m \leq d/2$ and $R = \mathrm{End}_D(V) = D_m$.

In particular, this shows that R is simple and its centre is the centre of D, a field C. Let K be a maximal commutative subfield of D containing C; we have a natural homomorphism $R \otimes_C K \to RK$, but the left-hand side is simple, hence $R \otimes_C K \cong RK$. Now RK is a K-algebra, its centre is K and it acts densely on V, therefore $RK \cong K_n$, where again $n \leq d/2$. Moreover, $[R : C] = [RK : K] = n^2$. ■

We record separately the special case of a skew field:

Corollary 8.3.7. *A skew field which satisfies a polynomial identity of degree d is of finite dimension $\leq [d/2]^2$ over its centre.* ■

The result can be extended to semiprimitive rings as follows. We remark that any ring R can be embedded in the full matrix ring R_d (e.g. as scalar matrices), hence R_m can be embedded in R_n whenever $m|n$.

Corollary 8.3.8. *Any semiprimitive PI-algebra satisfying an identity of degree d can be embedded in a matrix algebra $\mathfrak{M}_r(A)$ over a commutative ring A, where $r \leq [d/2]$.*

Proof. By Theorem 8.3.1, R is a subdirect product of primitive rings R_λ, where R_λ is a homomorphic image of R and hence again satisfies an identity of degree d. By Theorem 8.3.6, R_λ is a simple algebra of degree n over its centre, where $n \leq d/2$. By taking a splitting field E_λ we can thus embed R_λ in a matrix algebra $\mathfrak{M}_{n_\lambda}(E_\lambda)$. The least common multiple of the degrees n_λ which occur is $r \leq [d/2]!$ and R_λ can also be embedded in $\mathfrak{M}_r(E_\lambda)$. Now $\prod \mathfrak{M}_r(E_\lambda) \cong \mathfrak{M}_r(A)$, where $A = \prod E_\lambda$ and so we have an embedding of R in $\mathfrak{M}_r(E)$. ■

If R is a PI-algebra without non-zero nil ideals, then $R[t]$ is semiprimitive, by Proposition 8.3.5 and it is again a PI-algebra, by Corollary 7.5.4, hence we obtain

Theorem 8.3.9. *Let R be a PI-algebra without non-zero nil ideals. Then R can be embedded in $\mathfrak{M}_n(A)$, where A is a commutative ring, and if R satisfies an identity of degree d, then $n \leq [d/2]!$.* ■

For Artinian rings 'semiprimitive' reduces to 'semisimple'; this follows from the fact that for an Artinian ring R, $R/\mathbf{J}(R)$ is semisimple (see BA, Theorem 5.3.5), and it will also be derived in a more general context below in Section 8.4. In the Artinian case the radical may be described as the intersection of all maximal two-sided ideals. This does not hold generally: we clearly have, for any ring R,

$$\cap\{\text{max. left ideals}\} = \cap\{\text{max. right ideals}\} \subseteq \cap\{\text{max. ideals}\}, \tag{8.3.4}$$

where the first equality holds by the characterization of $\mathbf{J}(R)$. For an example where the inclusion is strict, take $R = \mathrm{End}_K(V)$, where V is an infinite-dimensional vector space over a field K. We have seen in Section 8.2 that R is primitive, hence $\mathbf{J}(R) = 0$, but the ideals in R form a chain, so the intersection on the right is the unique maximal ideal of R. Let us denote by $\mathfrak{s}_0$ the socle of R; in the representation on V this corresponds to the set of elements of finite rank. Assuming $[V : K]$ to be countable, with basis $\{e_i\}$, let us write $\mathfrak{a}_i$ for the set of elements of R mapping e_i to 0. Then $\mathfrak{a}_i$ is a maximal right ideal and $\mathfrak{s}_0 \not\subset \mathfrak{a}_i$, $\cap \mathfrak{a}_i = 0$. Similarly, if $\mathfrak{b}_i$ is the set of elements of R mapping V into $\sum_{j \neq i} Ke_j$, then b_i is a maximal left ideal such that $\mathfrak{s}_0 \not\subset \mathfrak{b}_i$, $\cap \mathfrak{b}_i = 0$. By Krull's theorem R also has maximal right ideals (and maximal left ideals) containing $\mathfrak{s}_0$, but the proof is non-constructive, and there is no obvious procedure for finding them.

Exercises

1. Show that a subdirect product of semiprimitive rings is semiprimitive.
2. Show that the Jacobson radical of a ring contains no non-zero idempotent. Deduce that every regular ring is semiprimitive.
3. Verify that in an Artinian ring the intersection of all maximal two-sided ideals is just the radical.
4. Let R be a ring and $\mathfrak{a}$ an ideal in R. Show that if $\mathbf{J}(R/\mathfrak{a}) = 0$, then $\mathbf{J}(R) \subseteq \mathfrak{a}$.
5. Show that a subdirect product of a finite number of simple rings is a direct product of simple rings.

8.4 Non-unital algebras

So far we have taken the existence of a unit-element or 'one' as part of the definition of a ring, but there are some occasions when a 'non-unital' ring arises naturally. For example, the algebra $C(X)$ of all continuous functions on a topological space X has a one precisely when X is compact. We shall maintain the convention that a ring necessarily has a one, and allow for the case where a one is lacking by speaking of an *algebra*. The algebra is called *unital* if it has a one. The coefficient ring K (with 1) may be any commutative ring; this is no restriction since every ring may be regarded as a $\mathbf{Z}$-algebra.

Let A be a K-algebra; by a right A-module M we understand a (K, A)-bimodule with the rule $\alpha(xa) = x(\alpha a)$ for all $x \in M$, $\alpha \in K$, $a \in A$. Even if A has a one, e say, this need not define the identity mapping on M. If it does, i.e. if

$$xe = x \quad \text{for all } x \in M,$$

the module M is said to be *unital*.

Our first observation is that a K-algebra may always be embedded in a unital K-algebra. This may be done in many ways; we shall single out one which uses the notion of an augmented algebra. A unital K-algebra A is said to be *augmented* if there exists a K-algebra homomorphism, called the *augmentation mapping*:

$$\varepsilon : A \to K.$$

This means that ε is a ring homomorphism such that $(\alpha a)\varepsilon = \alpha(a\varepsilon)$ for all $\alpha \in K$, $a \in A$. In particular, $(\alpha e)\varepsilon = \alpha$, hence $\alpha e \mapsto \alpha$ is a bijection between $K.e$ (where e is the one of A) and K, and we may embed K in A by identifying a with αe. The kernel of ε is the augmentation ideal of A and we have the direct sum decomposition

$$A = \ker \varepsilon \oplus K, \tag{8.4.1}$$

corresponding to the decomposition for any $x \in A$:

$$x = (x - x\varepsilon) + x\varepsilon.$$

For any K-algebra A we can form the augmented algebra $A^1 = A \oplus K$ by defining the multiplication

$$(a, \alpha)(b, \beta) = (ab + a\beta + \alpha b, \alpha\beta),$$

and this algebra has the unit-element $(0, 1)$ and augmentation ideal A. Conversely, if C is an augmented K-algebra with augmentation ideal A, then $A^1 \cong C$, as augmented K-algebras. Moreover, the category of A-modules is equivalent to the category of unital A^1-modules.

For K-algebras the notion of simple module has to be modified. A right module M over a K-algebra A is called *simple* if $MA \neq 0$ and M has no submodules other than 0 and M. When A has a one and acts unitally, this reduces to the previous definition. Our object will be to study simple A-modules in terms of A^1 (where the results of Section 8.3 can be used).

Any simple A-module M can again be represented as A/I for a maximal right ideal I, but we must also have $A^2 \not\subseteq I$, to ensure that $MA \neq 0$. Further, we can no longer use Krull's theorem to find maximal right ideals because A may not be finitely generated as right A-module. To overcome these difficulties, let us look more closely at the correspondence between right ideals in A and in A^1.

Let I be a right ideal of A and I' a right ideal of A^1 such that $I' \supseteq I$. Then $I \subseteq I' \cap A$, so there is a natural homomorphism of A-modules

$$A/I \to A/(I' \cap A) \cong (A + I')/I' \subseteq A^1/I'.$$

This is an isomorphism iff $I' \cap A = I$ and $I' + A = A^1$. Given any right ideal I' of A^1 such that $A + I' = A^1$, we can put $I = I' \cap A$; then by what has just been said, $A/I \cong A^1/I'$. In particular, this holds for any maximal right ideal I' of A^1 which does not contain A. If we start from I in A, the next lemma shows under what conditions we can find a suitable I'.

Lemma 8.4.1. *Let A be a K-algebra with a right ideal I. Then there is a right ideal I' of A^1 such that*

$$I' + A = A^1, \quad I' \cap A = I, \tag{8.4.2}$$

if and only if A contains an element e such that

$$(1 - e)A \subseteq I. \tag{8.4.3}$$

In (8.4.3) 1 is the one of A^1, but we can express (8.4.3) entirely within A by writing $a - ea \in I$ for all $a \in A$. A right ideal I satisfying (8.4.3) for some $e \in A$ is called a *modular* right ideal.

Proof. If (8.4.2) holds, we can write $1 = u + e$, where $u \in I'$, $e \in A$. Then $(1 - e)\mathrm{A} = \mathrm{uA} \subseteq \mathrm{I}' \cap \mathrm{A} = \mathrm{I}$, and (8.4.3) follows. Conversely, given (8.4.3), we put $I' = I + (1 - e)K$. Then I' is a right ideal in A^1, by (8.4.3) and $I' + A$ is a right ideal containing $(1 - e) + 1 = 1$, hence $I' + A = A^1$. Moreover, if $x = a + (1 - e)\alpha \in A$, where $\alpha \in K$, $a \in I$, then $\alpha = 0$, hence $x \in I$, so $I' \cap A = I$. ■

This lemma shows in particular that for any maximal right ideal I' of A^1 which does not contain A, $I' \cap A$ is a modular right ideal of A.

By a *maximal modular right ideal* of A we shall understand a maximal member of the set of all proper modular right ideals. Any right ideal containing a modular right ideal is again modular (by the definition), hence a maximal modular right ideal is also maximal in the set of all proper right ideals. Moreover, any proper modular right ideal is contained in a maximal modular right ideal for, given $I \supseteq (1-e)A$, we can by Krull's theorem find a right ideal containing I but not e, and maximal with these properties, and this is easily seen to be modular. In fact the notion of a modular right ideal may be regarded as a device for producing maximal right ideals in non-unital algebras; the corresponding quotients are simple modules, as we shall see below.

In any K-algebra A let us define the *Jacobson radical* $\mathbf{J}(A)$ as the set of all elements of A represented by 0 in any simple A-module. If $\mathbf{J}(A) = 0$, A is said to be *semi-primitive.* For unital algebras these definitions reduce to the earlier ones, by the characterization quoted in Section 8.3. Now $\mathbf{J}(A)$ can be described as follows in terms of the maximal modular right ideals of A:

Theorem 8.4.2. *Let A be a K-algebra, where K is a semiprimitive coefficient ring, and denote by A^1 the corresponding augmented K-algebra. Then* $\mathbf{J}(A) = \mathbf{J}(A^1)$, and $\mathbf{J}(A)$ is the intersection of all maximal modular right ideals of A.

Proof. It is clear that a right A-module M such that $MA \neq 0$ is simple iff it is simple as A^1-module. Thus for each $a \in A$,

$$\begin{aligned} a \in \mathbf{J}(A) &\Leftrightarrow a \text{ is represented by 0 in every simple } A\text{-module} \\ &\Leftrightarrow a \text{ is represented by 0 in every simple } A^1 - \text{module} \\ &\Leftrightarrow a \in \mathbf{J}(A^1) \cap A. \end{aligned}$$

It follows that $\mathbf{J}(A) = \mathbf{J}(A^1) \cap A$. Now assume that $x \in \mathbf{J}(A^1) \backslash A$; then

$$x = a + \alpha.1, \quad \text{where } a \in A, \alpha \in K, \alpha \neq 0. \tag{8.4.4}$$

Since K is semiprimitive, there is a maximal ideal $\mathfrak{m}$ of K not containing α. The residue class field $K/\mathfrak{m}$ has a natural K-module structure (by multiplication) and we can define an A^1-module structure by the rule $u.y = u(y\varepsilon)$, for $u \in K/\mathfrak{m}$, $y \in A^1$. Then $K/\mathfrak{m}$ is a simple A^1-module, and if $\bar{1}$ is the residue class of 1, then by applying the element (8.4.4) we get $\bar{1}.x = \bar{1}.a \neq 0$. This contradicts the fact that $x \in \mathbf{J}(A^1)$; hence $\mathbf{J}(A^1) \subseteq A$ and it follows that $\mathbf{J}(A^1) = \mathbf{J}(A)$. Now

$$\begin{aligned} a \in J(A^1) &\Leftrightarrow a \in \text{each maximal right ideal of } A^1 \\ &\Leftrightarrow a \in \text{each maximal right ideal of } A^1 \text{ which does not contain } A \\ &\Leftrightarrow a \in \text{each maximal modular right ideal of } A, \end{aligned}$$

by the remarks preceding the theorem. ■

From the definition of the Jacobson radical quoted in Section 8.3 we now obtain

Corollary 8.4.3. *Let A be a K-algebra, where K is semiprimitive. Then $\mathbf{J}(A)$ consists of all elements $a \in A$ such that ay has a quasi-inverse, for all $y \in A$.* ■

We note that the intersection of all the maximal right ideals of A is in general different from $\mathbf{J}(A)$. For example, let k be a field and A the algebra of formal power series in x with zero constant term. Then $\mathbf{J}(A) = A$, but the intersection of all maximal ideals is xA. More generally, simple algebras have been constructed which coincide with their Jacobson radical, by Edward Saşiada in 1961 (see Saşiada and Cohn [1967]).

We can now prove the Wedderburn structure theorem for semisimple rings under weaker hypotheses. First an auxiliary result on modular right ideals; we recall that two right ideals $\mathfrak{a}$, $\mathfrak{b}$ of A are called *comaximal* if $\mathfrak{a} + \mathfrak{b} = A$.

Lemma 8.4.4. *Let A be any K-algebra and $\mathfrak{a}$, $\mathfrak{b}$ modular right ideals which are comaximal. Then $\mathfrak{a} \cap \mathfrak{b}$ is again modular.*

Proof. By hypothesis there exist $e, f \in A$ such that $(1-e)A \subseteq \mathfrak{a}$, $(1-f)A \subseteq \mathfrak{b}$. Since $\mathfrak{a} + \mathfrak{b} = A$, by comaximality, there exist $a_i \in \mathfrak{a}$, $b_i \in \mathfrak{b}$ $(i = 1, 2)$ such that

$$e = a_1 + b_1, f = a_2 + b_2.$$

Hence for any $x \in A$,

$$(a_2 + b_1)x \equiv b_1 x \equiv ex \equiv x \pmod{\mathfrak{a}},$$

$$(a_2 + b_1)x \equiv a_2 x \equiv fx \equiv x \pmod{\mathfrak{b}}.$$

Therefore $(1 - (a_2 + b_1))A \subseteq \mathfrak{a} \cap \mathfrak{b}$ and the conclusion follows. ■

Theorem 8.4.5 (Wedderburn decomposition theorem). *Let A be a semiprimitive K-algebra over a semiprimitive coefficient ring K, and assume that A is right Artinian. The A is unital and semisimple, as a ring.*

Proof. Let $\{I_\lambda\}$ be the family of all maximal modular right ideals of A. Since A is semiprimitive, $\cap I_\lambda = 0$, and since A is right Artinian, the chain $I_1 \supset I_1 \cap I_2 \supset \ldots$ breaks off, hence for some r,

$$I_1 \cap \ldots \cap I_r = 0. \tag{8.4.5}$$

By omitting superfluous terms, we can choose this representation to be irredundant, i.e. such that $I_1 \cap \ldots \cap I_{i-1} \not\subseteq I_i$. Since I_i is maximal modular, it is comaximal with $I_1 \cap \ldots \cap I_{i-1}$, so by Lemma 8.4.4, 0 is modular, i.e. $(1-e)A = 0$ for some $e \in A$. Hence $A(1-e)$ is a nilpotent left ideal, which must be 0, by semiprimitivity, so $x = ex = xe$ for all $x \in A$ and e is the one of A.

Since each I is a maximal right ideal, $P_i = A/I_i$ is a simple right A-module and A is a submodule of the direct sum $\oplus P_i$. As submodule of a semisimple module, A itself is semisimple as right A-module, hence it is semisimple as a ring. ■

In Section 8.2 we saw that any simple ring with minimal right ideals is Artinian; for non-unital algebras this is no longer always so, and it is of some interest to examine the form such algebras take. To begin with we need to clarify what is to be understood by a simple algebra; for example, a 1-dimensional vector space A over a field with multiplication $xy = 0$ for all $x, y \in A$ has no ideals apart from 0 and A, but such trivial cases should clearly be excluded. We therefore define an algebra A to be *simple* if $A^2 \neq 0$ and A has no ideals other than 0, A. Now a simple algebra with minimal right ideals is again equal to its socle, but the latter need not be finitely generated. To describe such algebras more precisely we need the notion of a Rees matrix algebra.

In the rest of this section we shall take all our algebras to be bimodules over a skew field K such that $x(yz) = (xy)z$ for any x, y, z in K or in the algebra. Thus they could be described as K-rings, were it not for the fact that they will in general lack 1. We shall regard them as algebras over the centre of K; an alternative would be to suspend the convention that rings have a 1, but we shall not take that course to avoid confusion. In practical terms this makes no difference, since we shall not have occasion to consider the augmented algebra.

Let C be any algebra and I, Λ any sets; we shall write ${}^{\Lambda}C^{I}$ for the set of all matrices over C with rows indexed by Λ and columns indexed by I, briefly, $\Lambda \times I$ matrices. Fix a matrix P in ${}^{\Lambda}C^{I}$ which is *regular*, i.e. whose rows are left linearly independent and whose columns are right linearly independent over C. By the *Rees matrix algebra* over C with *sandwich matrix* P one understands the set M of all $I \times \Lambda$ matrices $A = (a_{i\lambda})$ over C, with almost all entries zero, with componentwise addition:

$$(a_{i\lambda}) + (b_{i\lambda}) = (a_{i\lambda} + b_{i\lambda}),$$

and with the multiplication

$$A^*B = APB, \quad \text{where } A = (a_{i\lambda}),\ B = (b_{i\lambda}).$$

This is well-defined, since A, B are both $I \times \Lambda$ and zero almost everywhere, while P is $\Lambda \times I$. With these definitions we have the following characterization of simple algebras with minimal right ideals.

Theorem 8.4.6. *For any algebra A the following conditions are equivalent:*

(a) *A is a simple algebra with a minimal right ideal,*
(b) *A is a prime algebra which coincides with its socle,*
(c) *A is isomorphic to a Rees matrix algebra over a skew field,*
(d) *A is isomorphic to a dense algebra of linear transformations of finite rank on a vector space over a skew field,*
(a^0)–(d^0) *the left–right analogues of* (a)–(d).

Proof. The equivalence (a) $\Leftrightarrow$ (d) follows as in Section 8.2, so it only remains to prove (a), (b), (c) equivalent; the equivalence to (a^0)–(d^0) then follows by the symmetry of (c).

(a) $\Rightarrow$ (b). By hypothesis the socle of A is not zero, so it equals A, and being simple, A is prime.

(b) $\Rightarrow$ (c). Let $\mathfrak{r}$ be a minimal right ideal. Since A is prime, $\mathfrak{r}^2 \neq 0$, so $\mathfrak{r}^2 = \mathfrak{r}$ and hence $\mathfrak{r} = a\mathfrak{r}$ for some $a \in \mathfrak{r}$. Choose $e \in \mathfrak{r}$ such that $ae = a$; if $e^2 \neq e$, then $(e^2 - e)\mathfrak{r} = \mathfrak{r}$ and we have $\mathfrak{r} = a\mathfrak{r} = a(e^2 - e)\mathfrak{r} = 0$, a contradiction. Hence $e^2 = e$ and of course $e \neq 0$, so $\mathfrak{r} = e\mathfrak{r}$ and e is an idempotent generator. By Lemma 4.3.8 and Schur's lemma, $\mathrm{End}_A(eA) = eAe$ is a skew field, K say. Now A, being equal to its socle, is semisimple as right A-module. Let $\mathfrak{s}$ be the sum of all right ideals isomorphic to $\mathfrak{r}$. This is a two-sided ideal; if $\mathfrak{s} \neq A$, then we have $A = \mathfrak{s} \oplus \mathfrak{s}'$ for some non-zero right ideal $\mathfrak{s}'$, hence $\mathfrak{s}'\mathfrak{s} = 0$, and so $\mathfrak{s}' = 0$, because A is prime. Thus $A = \mathfrak{s}$ is a direct sum of right ideals isomorphic to $\mathfrak{r}$. Now $\mathfrak{r} = eA$ is a left K-space and we can take a basis u_λ $(\lambda \in \Lambda)$; likewise Ae is a right K-space with basis v_i $(i \in I)$, say; further we note that $u_\lambda v_i \in eAe = K$. We define a $\Lambda \times I$ matrix $P = (p_{\lambda i})$ over K by

$$p_{\lambda i} = u_\lambda v_i,$$

and claim that P is regular. For if (c_λ) is a family in K, almost all zero, such that $\sum c_\lambda p_{\lambda i} = 0$ for all i, then $\sum c_\lambda u_\lambda v_i = 0$, hence $\sum c_\lambda u_\lambda$ annihilates Ae and so must be 0, and now we have $c_\lambda = 0$ by the linear independence of the u_λ. Similarly $\sum p_{\lambda i} d_i = 0$ for all λ implies $d_i = 0$.

Let M be the Rees matrix algebra over K with sandwich matrix P and define a mapping $f : M \to A$ by

$$(a_{i\lambda})f = \sum v_i a_{i\lambda} u_l. \tag{8.4.6}$$

Since almost all the $a_{i\lambda}$ vanish, this is well-defined. We shall establish (c) by proving that f is an isomorphism. It is clearly additive, and we have

$$\begin{aligned}[(a_{i\lambda})^*(b_{i\lambda})]f &= \left[\sum a_{i\lambda} u_\lambda v_j b_{j\mu}\right] f \\ &= \sum v_i a_{i\lambda} u_\lambda v_j b_{j\mu} u_\mu \\ &= (a_{i\lambda})f.(b_{j\mu})f;\end{aligned}$$

thus f is a homomorphism. It is surjective because AeA is the socle and so equal to A, and f is injective because P is regular, for if $\sum v_i a_{i\lambda} u_\lambda = 0$, then $\sum p_{\mu i} a_{i\lambda} p_{\lambda j} = 0$, hence $a_{i\lambda} = 0$. Thus f is an isomorphism and (c) follows.

(c) $\Rightarrow$ (a). Given a Rees matrix algebra M over K with sandwich matrix P, take any non-zero matrix $C = (c_{i\lambda})$ in M. For any $(\mu, j) \in \Lambda \times I$ consider

$$b_{\mu j} = \sum p_{\mu i} c_{i\lambda} p_{\lambda j}.$$

If $b_{\mu j} = 0$ for all μ, j, then by the regularity of P, $c_{i\lambda} = 0$ for all i, λ, a contradiction. Hence $b_{\mu j} \neq 0$ for some $(\mu, j) \in \Lambda \times I$. Let us write $E_{i\lambda}$ for the matrix unit with (i, λ)-entry 1 and the rest 0. For any $d \in K$ we have

$$\mathrm{d}E^*_{i\mu} C^* E_{j\lambda} = \mathrm{d}E_{i\mu} PCPE_{j\lambda} = (f_{k\rho}),$$

where

$$f_{k\rho} = \sum d\delta_{ki} p_{\mu l} c_{l\sigma} p_{\sigma j} \delta_{\lambda\rho} = \delta_{ki}\delta_{\lambda\rho}\, db_{\mu j}.$$

Since d was arbitrary in K, $f_{k\rho}$ ranges over K, and every matrix of M is a finite sum of terms $dE_{i\lambda}$, it follows that every matrix of M lies in the ideal generated by C, hence M is simple.

It remains to find a minimal right ideal. By the definition of P, $p_{\lambda i} \neq 0$ for some pair $(\lambda, i) \in \Lambda \times I$. Writing $p_{\lambda i} = p$, we have $p^{-1}E^{*}_{i\lambda}p^{-1}E_{i\lambda} = p^{-1}E_{i\lambda}PE_{i\lambda}p^{-1} = p^{-1}E_{i\lambda} \neq 0$, hence $e = p^{-1}E_{i\lambda}$ is a non-zero idempotent in M. Now the mapping $\varphi : K \to M$ defined by $c \mapsto p^{-1}E_{i\lambda}c$ is an injective homomorphism with image eMe, as is easily verified. Hence eMe is a skew field and so eM is a minimal right ideal of M. ∎

This result is closely analogous to the structure theorem on 0-simple semigroups proved by David Rees in 1940. The equivalence of (a) and (c) is due to Eckehart Hotzel in 1970; the equivalence of (a) and (d) was proved by Nathan Jacobson in 1945, building on work by Jean Dieudonné in 1942.

The generalization of the Wedderburn–Artin theory set forth in Sections 8.2–8.4 was developed by Jacobson in 1945. The density theorem generalizes a theorem of William Burnside to the effect that a monoid acting irreducibly on an n-dimensional vector space over an algebraically closed field contains n linearly independent endomorphisms; this is the essential content of Corollary 8.1.9.

Exercises

1. Show that a non-zero ideal in a primitive ring is again primitive (as a non-unital algebra).
2. Let A be an algebra without nilpotent (two-sided) non-zero ideal. Show that A has no nilpotent non-zero left or right ideals. Deduce that for any non-zero idempotent e in A, eA is a minimal right ideal iff eAe is a skew field.
3. Let A be a nil algebra (i.e. every element is nilpotent). Show that every maximal right ideal is two-sided and contains A^2.
4. Let A be an algebra over a field k; show that if A has no zero-divisors, then A is an integral domain iff 0 is not modular, as right ideal. By considering End(A), show that A can then be embedded in an integral domain.
5. Let k be a field and A the k-subalgebra of the free algebra $k\langle x, y\rangle$ consisting of all polynomials with zero constant term. Find two modular right ideals of A whose intersection is not modular.
6. Show that if a $\Lambda \times I$ sandwich matrix is row-finite, then $|I| \leq |\Lambda|$.
7. Let A be a Rees matrix algebra over a skew field K with $\Lambda \times I$ sandwich matrix P. Show that $|\Lambda| \leq 2^{|I|}$ and give an example where equality occurs. (Hint. If V is a K-space of dimension $|I|$, then its dual V^* has dimension $2^{|I|}$; interpret the rows of P as vectors in V^*.)

8.5 Semiprime rings and nilradicals

We have already briefly met prime and semiprime rings and now come to examine the relation between them. We recall that a *prime ring* is a ring $R \neq 0$ such that for any two ideals $\mathfrak{a}$, $\mathfrak{b}$ of R,

$$\mathfrak{ab} = 0 \Rightarrow \mathfrak{a} = 0 \quad \text{or} \quad \mathfrak{b} = 0.$$

Equivalently, for any $a, b \in R$, $aRb = 0$ implies $a = 0$ or $b = 0$. In the commutative case a prime ring is just an integral domain. In general every integral domain is prime, but not conversely. More generally we have

Proposition 8.5.1. *Every primitive ring is prime.*

Proof. Let R be primitive and take a simple faithful right R-module M. If $\mathfrak{a}$, $\mathfrak{b}$ are non-zero ideals in R, then $M\mathfrak{a} \neq 0$, hence $M\mathfrak{a} = M$ by the simplicity of M; likewise $M\mathfrak{b} = M$, so $M\mathfrak{ab} = M\mathfrak{b} = M$. This shows that $\mathfrak{ab} \neq 0$.

Since the notion 'prime' is left–right symmetric, a similar result holds for left primitive rings. Of course the converse of Proposition 8.5.1 is false, as we see already in the commutative case, when primitive rings are fields, whereas prime rings are integral domains.

By a *prime ideal* in a ring R we understand a two-sided ideal $\mathfrak{p}$ such that $R/\mathfrak{p}$ is a prime ring; for commutative rings this reduces to the definition given in BA, Section 10.2. We observe that a prime ideal must always be proper.

There is a method of constructing prime ideals, rather as in the commutative case; the notion of multiplicative set has to be replaced here by that of an m-system. By this term one understands a subset M of R such that $1 \in M$ and if $a, b \in M$ then $axb \in M$ for some $x \in R$. We note that an ideal is prime iff its complement in R is an m-system. Now there is an analogue of BA, Theorem 10.2.6:

Theorem 8.5.2. *Let R be a ring, M an m-system in R and $\mathfrak{a}$ an ideal in R disjoint from M. Then there exists an ideal $\mathfrak{p}$ in R which contains $\mathfrak{a}$, is disjoint from M and is maximal with respect to these properties. Any such ideal $\mathfrak{p}$ is prime.*

Proof. Let $\mathscr{A}$ be the set of all ideals $\mathfrak{a}'$ of R such that $a' \supseteq \mathfrak{a}$, $\mathfrak{a}' \cap M = \emptyset$. Then $\mathfrak{a} \in \mathscr{A}$, so $\mathscr{A}$ is not empty. It is easily seen to be inductive and so by Zorn's lemma, it contains a maximal member $\mathfrak{p}$, which is an ideal with the required properties.

Now let $\mathfrak{p}$ be an ideal with the properties stated and assume that $\mathfrak{a}, \mathfrak{b} \not\supseteq \mathfrak{p}$, $\mathfrak{ab} \subseteq \mathfrak{p}$. Then $\mathfrak{a} + \mathfrak{p}$, $\mathfrak{b} + \mathfrak{p}$ are strictly larger than $\mathfrak{p}$ and so must meet M, say $s = p + a$, $t = q + b \in M$, where $p, q \in \mathfrak{p}$, $a \in \mathfrak{a}$, $b \in \mathfrak{b}$. By hypothesis there exists $x \in R$ such that $sxt \in M$; thus M contains

$$(p + a)x(q + b) = px(q + b) + axq + axb \in \mathfrak{p} + \mathfrak{ab} = \mathfrak{p},$$

a contradiction. Hence $\mathfrak{a}, \mathfrak{b} \not\subseteq \mathfrak{p}$ implies $\mathfrak{ab} \not\subseteq \mathfrak{p}$ and clearly $1 \notin \mathfrak{p}$, so $\mathfrak{p}$ is indeed prime. ∎

In the commutative case we found that the intersection of all prime ideals of R is the set of all nilpotent elements of R. Theorem 8.5.2 can be used to obtain a corresponding result for general rings; however, the situation is a little more complicated here because the ideal generated by a nilpotent element need not be nilpotent. Let us recall that an ideal $\mathfrak{a}$ is called *nilpotent* if $\mathfrak{a}^r = 0$ for some $r \geq 1$, i.e. $x_1 \ldots x_r = 0$ for any $x_i \in \mathfrak{a}$, and $\mathfrak{a}$ is *nil* if it consists of nilpotent elements. Clearly every nilpotent ideal is nil, but the converse does not hold generally.

We also recall that a ring R is called *semiprime* if for any two-sided ideal $\mathfrak{a}$ of R,

$$\mathfrak{a}^2 = 0 \Rightarrow \mathfrak{a} = 0. \tag{8.5.1}$$

In terms of elements this states that $aRa = 0$ implies $a = 0$, for all $a \in R$. We observe that a ring is semiprime iff it has no non-zero nilpotent ideals. For when this holds, (8.5.1) is clearly satisfied. Conversely, assume (8.5.1) and let $\mathfrak{a}$ be a nilpotent ideal, say $\mathfrak{a}^{r-1} \neq 0$, $\mathfrak{a}^r = 0$, where $r \geq 2$. Then $2r - 2 \geq r$, hence $0 = \mathfrak{a}^{2r-2} = (\mathfrak{a}^{r-1})^2 \neq 0$, by (8.5.1), which is a contradiction.

In the commutative case the semiprime rings are the *reduced* rings, i.e. rings in which 0 is the only nilpotent element; in the general case every reduced ring is semiprime, but not conversely.

An ideal $\mathfrak{a}$ in a ring R is called *semiprime* if $R/\mathfrak{a}$ is a semiprime ring. We note that R itself, as an ideal of R, is semiprime but not prime.

We first elucidate the relation between prime and semiprime ideals.

Proposition 8.5.3. *Let R be a ring. Then every intersection of prime ideals is semiprime; conversely every semiprime ideal is an intersection of prime ideals.*

Proof. Let $\mathfrak{c} = \cap \mathfrak{p}_\lambda$, where the $\mathfrak{p}_\lambda$ are prime ideals, and suppose that $aRa \subseteq \mathfrak{c}$. Then $aRa \subseteq \mathfrak{p}_\lambda$, hence $a \in \mathfrak{p}_\lambda$ for all λ, and so $a \in \cap \mathfrak{p}_\lambda = \mathfrak{c}$, and this shows $\mathfrak{c}$ to be semiprime. Conversely, let $\mathfrak{c}$ be a semiprime ideal in R; by passing to the residue class ring $R/\mathfrak{c}$, we may take $\mathfrak{c} = 0$. We have to show that in a semiprime ring the intersection of all prime ideals is 0. Take any $a \in R$; then $aRa \neq 0$, so there exists b_0 such that $a_1 = ab_0a \neq 0$. Generally, if we have constructed $a_1, \ldots, a_n$ such that $a_{i+1} \in a_iRa_i$ and $a_n \neq 0$, then there exists b_n such that $a_{n+1} = a_nb_na_n \neq 0$. The set $M = \{1, a_0 = a, a_1, a_2, \ldots\}$ is an m-system, for given a_r, a_s, choose $n > r, s$; then a_r, a_s are factors of a_n and $a_nb_na_n \neq 0$, hence $a_rua_s \neq 0$ for some $u \in R$. Thus M is an m-system and $a \in M$, $0 \notin M$; hence we can find a prime ideal $\mathfrak{p}$ disjoint from M, by Theorem 8.5.2. It follows that $a \notin \mathfrak{p}$; since a was any non-zero element of R, we see that the intersection of all prime ideals of R is zero. ∎

Corollary 8.5.4. *A ring is semiprime if and only if the intersection of all its prime ideals is zero. Hence any semiprime ring R can be written as a subdirect product of prime rings, which are homomorphic images of R. In particular, every semiprimitive ring is semiprime.*

Proof. The first part follows by applying Proposition 8.5.3 to the zero ideal, and the second part now follows as in the proof of Theorem 8.3.1. ∎

Just as semiprimitivity leads to the Jacobson radical, so there is a type of radical arising from semiprimeness, but it is not so clear cut and in general there is more than one radical. Let us define a *nilradical* in a ring R as an ideal N which is nil and such that R/N is semiprime. A ring may have more than one nilradical; in what follows we shall describe the greatest and least such radical. We begin with a lemma.

Lemma 8.5.5. *The sum of any family of nil ideals in a ring is a nil ideal.*

Proof. Consider first two nil ideals $\mathfrak{a}_1, \mathfrak{a}_2$ and write $\mathfrak{a} = \mathfrak{a}_1 + \mathfrak{a}_2$. Then the ideal $\mathfrak{a}/\mathfrak{a}_2$ of the ring $R/\mathfrak{a}_2$ is nil, because $\mathfrak{a}/\mathfrak{a}_2 \cong \mathfrak{a}_1/(\mathfrak{a}_1 \cap \mathfrak{a}_2)$ and the latter is a homomorphic image of $\mathfrak{a}_1$ and hence is nil. Thus any element of $\mathfrak{a}$ has a power in $\mathfrak{a}_2$ and a power of this is 0, therefore $\mathfrak{a}$ is nil. Now an induction shows that the sum of any finite number of nil ideals is nil. In the general case let $\mathfrak{a} = \sum \mathfrak{a}_\lambda$, where each $\mathfrak{a}_\lambda$ is a nil ideal. Any element of $\mathfrak{a}$ lies in the sum of a finite number of the $\mathfrak{a}_\lambda$ and so is nilpotent, therefore $\mathfrak{a}$ is nil. ∎

By this lemma the sum of all nil ideals in a ring R is a nil ideal; it is denoted by $U = U(R)$ and is called the (*Baer*) *upper nil radical* or also the *Köthe nilradical.* It is indeed a nilradical, for R/U cannot have any non-zero nil ideals, by the maximality of U; a fortiori U must be semiprime. Since U contains all nil ideals of R, it contains all nil radicals.

To obtain the least nilradical we need another definition. An element c of R is called *strongly nilpotent* if any sequence $c_1 = c, c_2, c_3, \ldots$ such that $c_{n+1} \in c_nRc_n$ is ultimately zero. It is clear that such an element is nilpotent and that any element of a nilpotent (left or right) ideal is strongly nilpotent. Moreover, in a semiprime ring R, the only strongly nilpotent element is 0. For if $c \neq 0$, then $cRc \neq 0$, say $c_2 = cac \neq 0$, now $c_3 = c_2a'c_2 \neq 0$ for some $a' \in R$ and continuing in this way we obtain a sequence $c_1 = c, c_2, c_3, \ldots$ such that $c_{n+1} \in c_nRc_n$ and no c_n is zero, so c is not strongly nilpotent. Conversely, if 0 is the only strongly nilpotent element in R, then R is semiprime. For if not, then $cRc = 0$ for some $c \neq 0$, so c is strongly nilpotent. Thus we have

Proposition 8.5.6. *Any ring R is semiprime if and only if the only strongly nilpotent element is 0.* ∎

Now the least nilradical may be described as follows:

Theorem 8.5.7. *In any ring R the set $L(R)$ of all strongly nilpotent elements is the least nilradical, and is equal to the intersection of all prime ideals in R:*

$$L(R) = \cap\{\mathfrak{p} \mid \mathfrak{p} \text{ prime in } R\}. \tag{8.5.2}$$

Proof. If x is strongly nilpotent in R, so is its residue class $\bar{x}$ (mod $\mathfrak{p}$) for any prime ideal $\mathfrak{p}$, so $\bar{x} = 0$, which means that $x \in \mathfrak{p}$. Thus $L(R)$ is contained in the right-hand side of (8.5.2). Now suppose that $c \notin L(R)$; then c is not strongly nilpotent, so there exists a sequence $\{c_n\}$ such that $c_1 = c, 0 \neq c_{n+1} \in c_nRc_n$. The set $S = \{1, c_1, c_2, \ldots\}$ is an m-system, for given c_r, c_s, where $r \leq s$ say, we have $c_{s+1} \in c_rRc_r \cap c_sRc_s$. By Theorem 8.5.2 there is an ideal $\mathfrak{p}$ which is maximal disjoint from S, and $\mathfrak{p}$ is prime, thus $c \notin \mathfrak{p}$ and this shows that equality holds in (8.5.2), and incidentally, that $L(R)$ is indeed an ideal.

Now $L(R)$ is clearly a nil ideal and $R/L(R)$ is semiprime, hence $L(R)$ is a nilradical, and by (8.5.2) it is the least ideal with semiprime quotient, hence it is the least nilradical. ■

$L(R)$ is called the *prime radical* or the (*Baer*) *lower nilradical*.

If a ring R has a non-zero nilpotent right ideal $\mathfrak{a}$, then $R\mathfrak{a}$ is a two-sided ideal, again nilpotent, since $(R\mathfrak{a})^n \subseteq R\mathfrak{a}^n$. The following conjecture, raised in the 1930s, is still unanswered:

Köthe's conjecture. *If a ring has a non-zero nil right ideal, then it has a non-zero nil ideal.*

Equivalently, the prime radical contains every nil right ideal. We remark that in Noetherian rings every nil right ideal is nilpotent (Proposition 7.4.3), so the conjecture is valid in that case.

In general rings the upper and lower radical may well be distinct; to give an example it is enough to construct a semiprime ring with a non-zero nil ideal. Let A be the (non-unital) k-algebra generated by $x_1, x_2, \ldots$ with the following defining relations: for any element a involving only $x_1, x_2, \ldots, x_n$ we have $a^n = 0$. Then every element of A is nilpotent, and in the augmented algebra $R = A^1$ we have the nil ideal A. However, R is semiprime, for, given $a \in R^{\times}$, let a be of degree m in the x's and put $N = 2m + 2$. Any relation involving x_N consists of terms of degree at least N, hence $ax_Na \neq 0$, because this expression has only terms of degree at most $2m + 1$. This shows R to be semiprime.

For another example, this time finitely generated, take the finitely generated non-nilpotent nil algebra A constructed by Golod (see BA, Exercise 5 of Section 6.3). It can be verified that A^1 is semiprime but it has the nil ideal A. More generally, a simple nil algebra has recently been constructed by Agata Smoktunowicz [2002].

However, in a Noetherian ring the upper and lower nilradicals coincide. For as we saw in Proposition 7.4.3, in a semiprime ring with maximum condition on right annihilators every nil left or right ideal is 0. Hence if R is right Noetherian, then $R/L(R)$ has no non-zero nil ideals and so $U(R) = L(R)$; by symmetry the same holds for left Noetherian rings, so we have

Proposition 8.5.8. *In a right (or left) Noetherian ring the upper and lower nilradical coincide. Thus in a Noetherian ring the prime radical is a nilpotent ideal containing all nilpotent ideals.* ■

There is a radical intermediate between U and L which is sometimes considered; it is defined as follows. An algebra A is said to be *locally nilpotent* if the subalgebra generated by any finite subset is nilpotent. It is clear that for any ideal we have the implications: nilpotent ⇒ locally nilpotent ⇒ nil. Moreover, it is not hard to show that the sum of any number of locally nilpotent ideals is again locally nilpotent. Now the *Levitzki radical* N of a ring R is defined as the sum of all locally nilpotent ideals of R. By what has been said, N is locally nilpotent, hence nil and it can be shown that

R/N has zero Levitzki radical, so it is semiprime, and this shows N to be indeed a nilradical. In any ring we have

$$U \supseteq N \supseteq L,$$

and in the first example given above $N \supset L$, at least in characteristic 0, by the Nagata–Higman theorem (see Further Exercise 15 of Chapter 7), while $U \supset N$ in Golod's example.

Exercises

1. Show that in any right Noetherian ring 0 can be expressed as a product of prime ideals.
2. Show that a prime ring with a minimal right ideal is primitive.
3. Show that in a semiprime ring the socle and the left socle coincide. (Hint. Use Exercise 2 of Section 8.4.) Give an example to show that this does not hold generally.
4. Let R be the subring of $\mathfrak{M}_2(\mathbf{Z})$ consisting of all matrices $\begin{pmatrix} a & b \\ c & d \end{pmatrix}$ such that $a \equiv d$, $b \equiv c \pmod 2$. Show that R is a prime ring, but not an integral domain; find all its idempotents.
5. Give an example of a commutative ring with a nil ideal which is not nilpotent.
6. Show that the sum of any family of locally nilpotent ideals is locally nilpotent. (Hint. Imitate the proof of Lemma 8.5.5.)
7. In any ring with Levitzki radical N, verify that R/N has zero Levitzki radical.
8. In any ring R, define N_1 as the sum of all nilpotent ideals and define N_α for any ordinal α recursively by $N_{\alpha+1}/N_\alpha = N_1(R/N_\alpha)$, while at a limit ordinal, $N_\alpha = \cup_{\beta<\alpha} N_\alpha$. Show that the union of all the N_α is the lower nilradical of R.
9. Show that in a left or right Noetherian ring any nilpotent element is strongly nilpotent. Deduce that every nil (left or right) ideal is nilpotent.
10. Show that a reduced prime ring is an integral domain.
11. Show that any ideal of a semiprime ring is semiprime, qua non-unital algebra.
12. Show that in a reduced ring R, if a product of n elements in a certain order is 0, then the product in any order is 0. (Hint. Show that if $a_1 \dots a_n = 0$, then $a_1 x_1 a_2 x_2 \dots a_n x_n = 0$ for all $x_i \in R$.)
13. Show that in any ring R, every prime ideal contains a minimal prime ideal. (Hint. Take a maximal m-system containing the complement of the given prime ideal.)
14. Let R be a reduced ring and $\mathfrak{p}$ a minimal prime ideal. Show that the monoid M generated by the complement of $\mathfrak{p}$ does not contain 0. Deduce that M does not meet $\mathfrak{p}$ and hence that $R/\mathfrak{p}$ is an integral domain.
15. Show that every reduced ring is a subdirect product of integral domains. (Hint. Use Exercises 10–14. This is a theorem of Andrunakievich–Ryabukhin; the proof is due to Herstein.)

8.6 Prime PI-algebras

In some respects the presence of a polynomial identity has effects similar to the maximum condition and we shall in this section prove Posner's theorem, which is an analogue of Goldie's theorem for PI-rings. The first step is to form fractions with respect to a central Ore set.

Theorem 8.6.1. *Let R be a semiprimitive PI-algebra with centre C. Then every non-zero ideal $\mathfrak{a}$ of R meets C non-trivially: $\mathfrak{a} \cap C \neq 0$.*

Proof. By hypothesis R has a family of primitive ideals $\mathfrak{t}_\lambda$ $(\lambda \in \Lambda)$ whose intersection is 0. Let $f = 0$ be a polynomial identity for R; then $R_\lambda = R/\mathfrak{t}_\lambda$ is a primitive PI-algebra satisfying $f = 0$ and we have an embedding

$$R \to \prod R_\lambda,$$

by Theorem 8.3.1. If $\deg f = d$, then by Theorem 8.3.6, R_λ is simple of degree at most $d/2$ over its centre. The canonical homomorphism $\varepsilon_\lambda : R \to R_\lambda$ is surjective and R_λ contains $\mathfrak{a}\varepsilon_\lambda$ as an ideal, which is therefore 0 or R_λ, by simplicity. Put

$$\Lambda_0 = \{\lambda \in \Lambda | \mathfrak{a}\varepsilon_\lambda \neq 0\},$$

and choose $\mu \in \Lambda$ such that the degree m of R_μ is maximal. Then the Razmyslov polynomial $D = D_\mu$ is central and non-zero on R_μ and since $\mathfrak{a}\varepsilon_\mu = R_\mu$, there exist $a_1, \ldots, a_r \in \mathfrak{a}$ such that $x = D(a_1\varepsilon_\mu, \ldots, a_r\varepsilon_\mu)$ is a non-zero element of the centre of R_μ. It follows that $y = D(a_1, \ldots, a_r) \neq 0$; moreover, $y\varepsilon_\lambda = 0$ for $\lambda \notin \Lambda_0$, while for $\lambda \in \Lambda_0$, $y\varepsilon_\lambda$ is in the centre of R_λ; therefore $y \in \mathfrak{a} \cap C$. ∎

Corollary 8.6.2. *Any semiprimitive PI-algebra whose centre is a field is simple.*

Proof. For if $\mathfrak{a} \neq 0$, then $\mathfrak{a} \cap C \neq 0$, so $\mathfrak{a}$ contains a unit and so $\mathfrak{a} = R$. ∎

Here none of the conditions can be omitted, for the polynomial ring $k[x]$ is a semiprimitive PI-algebra which is not simple, and for an infinite-dimensional k-space V, $\mathrm{End}_k(V)$ is a primitive k-algebra whose centre is a field, but which is not simple (and of course not a PI-algebra).

Theorem 8.6.1 has a useful generalization. To prove it we first note a PI-analogue of Proposition 7.4.3.

Proposition 8.6.3. *Any prime PI-algebra A satisfies the maximum condition on left (or right) annihilators; hence every nil left or right ideal of A is zero.*

Proof. Let $0 \subset \mathfrak{l}_1 \subset \mathfrak{l}_2 \subset \ldots$ be a strictly ascending chain of left annihilators and put $\mathfrak{r}_n = (\mathfrak{l}_n)_r$, so that $\mathfrak{r}_1 \supset \mathfrak{r}_2 \supset \ldots$. By Proposition 7.5.3 we may take the polynomial identity to be multilinear; denote by d the least degree for which there is a homogeneous multilinear polynomial f of degree d:

$$f = \sum a_\sigma x_{1\sigma} \ldots x_{d\sigma},$$

such that f vanishes for all $x \in \mathfrak{l}_i$ $(i = 1, \ldots, d)$. Here $d > 1$, for if $a_1\mathfrak{l}_1 = 0$, then $\mathfrak{l}_1 = 0$, because A is prime. We thus have

$$\sum a_\sigma x_{1\sigma} \ldots x_{d\sigma}\mathfrak{r}_{d-1} = 0 \quad \text{for all } x_i \in \mathfrak{l}_i. \tag{8.6.1}$$

Now when $d\sigma \neq d$, then $x_{d\sigma} \in \mathfrak{l}_{d-1}$ and the corresponding term in (8.6.1) is 0; the remaining terms have the form $a_\sigma x_{1\sigma} \ldots x_{(d-1)_\sigma} x_d \mathfrak{r}_{d-1}$. Hence

$$\sum a_\sigma x_{1\sigma} \ldots x_{(d-1)_\sigma} \mathfrak{l}_d \mathfrak{r}_{d-1} = 0, \tag{8.6.2}$$

where σ ranges over all permutations of $1, \ldots, d-1$. But $\mathfrak{l}_d\mathfrak{r}_{d-1}$ is a non-zero two-sided ideal and A is prime, so

$$\sum a_\sigma x_{1\sigma} \ldots x_{(d-1)_\sigma} = 0 \quad \text{for } x_i \in \mathfrak{l}_i,$$

and this contradicts the definition of d. Hence the chain of $\mathfrak{l}_i$ breaks off and the maximum condition holds. By symmetry the same is true for right annihilators and now the rest follows by Proposition 7.4.3. ■

Corollary 8.6.4. *In a semiprime PI-algebra every nil left or right ideal is zero.*

Proof. If R is semiprime, it can be embedded in $\prod R_\lambda$, where $R_\lambda = R/\mathfrak{c}_\lambda$ is a prime PI-algebra, $\cap \mathfrak{c}_\lambda = 0$. If $\mathfrak{n}$ is a nil ideal in R and $\varepsilon_\lambda : R \to R_\lambda$ is the canonical projection, then $\mathfrak{n}\varepsilon_\lambda$ is a nil ideal in R_λ, hence equal to 0 by Proposition 8.6.3, and so $\mathfrak{n} \subseteq \cap \ker \varepsilon_\lambda = 0$.

We deduce

Theorem 8.6.5 (Rowen). *In a semiprime PI-algebra, every non-zero ideal meets the centre nontrivially. In particular, a semiprime PI-algebra whose centre is a field is simple.*

Proof. Let R be a semiprime PI-algebra with centre C. By Corollary 8.6.4, R has no non-zero nil ideals, hence by Proposition 8.3.5, the polynomial ring $R[t]$ is semi-primitive, and its centre is clearly $C[t]$, so by Theorem 8.6.1, for any non-zero ideal $\mathfrak{a}$ in R we have $\mathfrak{a}[t] \cap C[t] \neq 0$. On comparing coefficients we find that $\mathfrak{a} \cap C \neq 0$. Now the rest follows as in Corollary 8.6.2. ■

It is now an easy matter to deduce Edward Posner's theorem, in a rather strong form, following Louis Rowen, 1973 (see Rowen (1980)):

Theorem 8.6.6 (Posner, 1960). *Let R be a prime PI-algebra with centre C. Then C is an integral domain, and if K is its field of fractions, then the natural mapping $R \to Q = R \otimes_C K$ is an embedding and Q is a finite-dimensional simple K-algebra.*

Proof. The first part follows by Corollary 7.1.11; now any multilinear identity in R also holds in Q, so by Theorem 8.6.5, Q is simple. Hence by Kaplansky's theorem (Theorem 8.3.6), Q is finite-dimensional over K. ■

Let R be a prime PI-algebra and Q its quotient ring by fractions of the centre. By Theorem 8.6.6, Q is of finite dimension over its centre and this dimension is a square, n^2 say (Proposition 5.2.2); the number n is called the *PI-degree* of R. We can now determine the PI-degree of the generic division algebra:

Corollary 8.6.7. *The generic division algebra of degree n has PI-degree n.*

Proof. The generic matrix ring $F_{(n)}$ is a prime PI-algebra, hence Theorem 8.6.6 applies, and its skew field of fractions D is a finite-dimensional central simple algebra. Moreover, $F_{(n)}$ satisfies the standard identity S_{2n} but no identity of lower degree, hence the same is true of D. It follows that on passing to a splitting field we obtain an $n \times n$ matrix ring, hence D is of degree n over its centre. ■

The PI-degree has been used to give a simple characterization of separable algebras, by Michael Artin and Claudio Procesi.

For any ring R with centre C, the mapping

$$\varphi_{a,b} : x \longmapsto axb \quad (a, b \in R)$$

is an endomorphism of R as C-module and it is easily checked that the correspondence $a \otimes b \longmapsto \varphi_{a,b}$ defines a homomorphism

$$\lambda : R^0 \otimes_C R \to \mathrm{End}_C(R). \tag{8.6.3}$$

If R is a finitely generated projective C-module and (8.6.3) is an isomorphism, R is called an *Azumaya algebra* or also *central separable* (it can be shown that this is equivalent to R being separable as C-algebra, see Section 4.7). The proof of the Artin–Procesi theorem given below uses the Razmyslov polynomial D. We recall that $D = D_n$ is multilinear alternating in n^2 variables $x_1, \ldots, x_{n^2}$ (and others which we shall not name explicitly). If we put $D_{n\nu}(x) = D_n(x, x_1, \ldots, x_{\nu-1}, x_{\nu+1}, \ldots, x_{n^2})$, then the expression $x_0 D_n - \sum x_\nu D_{n\nu}(x_0)$ is an alternating multilinear function of the $n^2 + 1$ variables $x_0, \ldots, x_{n^2}$ and so vanishes on any prime ring of PI-degree n, because such a ring can be embedded in its quotient ring which is spanned by n^2 elements over C.

Theorem 8.6.8 (Artin–Procesi). *Let R be a prime PI-algebra of PI-degree n. Then R is Azumaya provided that each simple homomorphic image of R has PI-degree n.*

Proof. (W. Schelter) Consider the Razmyslov polynomial D_n; if its arguments lie in R, its values will be in the centre C of R. If they all lie in some maximal ideal of R, then D_n will vanish on some homomorphic image of R, which therefore has PI-degree less than n, against the hypothesis. Hence the left ideal generated by the values must be the whole of R:

$$1 = \sum b_i D_{n\nu}(a_{\nu i}) \quad \text{where } a_{\nu i}, b_i \in R, \tag{8.6.4}$$

and the other arguments of $D_{n\nu}$ also lie in R. Denote by $D_{n\nu i}$ the function obtained from $D_{n\nu}$ (defined above) by replacing x_μ by $a_{\mu i}$ for $\mu \neq \nu$. By the remark preceding the theorem, we have $cD_{n\nu}(a_{\nu i}) - \sum a_{\nu i}D_{n\nu i}(c) = 0$ for any $c \in R$; hence by (8.6.4),

$$c = \sum b_i c D_n(a_{\nu i}) = \sum b_i a_{\nu i} D_{n\nu i}(c).$$

Thus we have a (finite) projective coordinate system for R, showing that R is finitely generated projective over C.

It remains to show that (8.6.3) is an isomorphism. We note that $\mathrm{End}_C(R)$ is generated by elements of the form $cD_{n\nu i}$ and if $D_{n\nu i}(x) = \sum_j u_{\nu ij} x v_{\nu ij}$, then $(\sum cu_{\nu ij} \otimes v_{\nu ij})\lambda = cD_{n\nu i}$, so λ is surjective. Now

$$\begin{aligned}\sum p_j \otimes q_j &= \sum_{j\nu i} p_j \otimes b_i a_{\nu i} D_{n\nu i}(q_j)\\ &= \sum p_j D_{n\nu i}(q_j) \otimes b_i a_{\nu i}\\ &= \sum_{\nu ir} \left\{ \sum_j p_j u_{\nu ir} q_j \right\} v_{\nu ir} \otimes b_i a_{\nu i}.\end{aligned}$$

If $\sum p_j \otimes q_j \in \ker \lambda$, then $\sum p_j x q_j = 0$ for all $x \in R$, hence $\sum p_j \otimes q_j = 0$ by (8.6.5); this shows λ to be injective, so it is indeed an isomorphism. ■

It can be shown that the sufficient condition of this theorem is also necessary (see Rowen (1980), (1988)).

Exercises

1. Let D be a skew field with centre k. Show that for any (skew) subfield E of D, E and Ek have the same PI-degree.
2. Show that in any PI-algebra the upper and lower nilradicals coincide.
3. (Amitsur) Let R be a PI-algebra and $\mathfrak{N}$ the sum of all its nil ideals. Use Corollary 8.3.8 to show that $R/\mathfrak{N}$ can be embedded in a matrix ring $\mathfrak{M}_n(E)$, where E is a commutative ring. Deduce that R satisfies an identity $S_{2n}^m = 0$, where S_{2n} is the standard polynomial and $m \geq 1$. (By Exercise 1 of Section 7.7, m cannot always be taken to be 1. Hint. Express R as homomorphic image of a relatively free algebra satisfying a given polynomial identity holding in R.)

8.7 Firs and semifirs

Principal ideal domains (PIDs) include several important types of commutative ring such as the ring of integers, polynomial rings in one variable over a field and certain rings of algebraic integers, more specifically, any Dedekind domain with unique factorization (see BA, Section 10.5). In the non-commutative case there are no such striking examples of PIDs, but there is a wider class of rings which reduce to PIDs when commutativity is imposed.

Definition. By a *free right ideal ring* or *right fir* for short, we understand a ring R in which each right ideal is free, of uniquely determined rank. *Left firs* are defined similarly and a left and right fir is called a *fir*.

It follows that each right (or left) fir R has invariant basis number (IBN), for this is so when R is right Noetherian, and when this is not the case, it will contain free right ideals of arbitrary rank, by Proposition 7.1.9. We shall meet firs again in Section 11.5, where it will be shown that the free algebra $k\langle X\rangle$ on any set X over a field k is a fir; this may be regarded as a generalization of the fact that the polynomial ring $k[x]$ in a single variable over a field is a PID.

Often one meets an even wider class than firs; to describe it we shall need to look at general relations in rings. A relation

$$x_1y_1 + \ldots + x_ny_n = 0 \tag{8.7.1}$$

or in terms of vectors, $x.y = 0$, in a ring R is said to be *trivial* if for each $i = 1, \ldots, n$, either $x_i = 0$ or $y_i = 0$. Every non-zero ring has non-trivial relations, for example, if $x = (1, 1)$, $y = (-1, 1)^T$, then we have the non-trivial relation

$$x.y = \begin{pmatrix} 1 & 1 \end{pmatrix} \begin{pmatrix} -1 \\ 1 \end{pmatrix} = 0. \tag{8.7.2}$$

However, this relation can be transformed into a trivial relation by replacing x, y by x', y', given by

$$x' = \begin{pmatrix} 1 & 1 \end{pmatrix} \begin{pmatrix} 1 & -1 \\ 1 & 1 \end{pmatrix} = \begin{pmatrix} 2 & 0 \end{pmatrix},\ y' = \frac{1}{2} \begin{pmatrix} 1 & 1 \\ -1 & 1 \end{pmatrix} \begin{pmatrix} -1 \\ 1 \end{pmatrix} = \begin{pmatrix} 0 \\ 1 \end{pmatrix}. \tag{8.7.3}$$

Let us write the relation (8.7.1) as $x.y = 0$, where $x = (x_1, \ldots, x_n)$, $y = (y_1, \ldots, y_n)^T$. Then (8.7.1) is said to be *trivializable* if there is an invertible matrix P over R such that the relation $xP^{-1}.Py = 0$ is trivial; we also say that (8.7.1) is *trivialized* by P. More generally, a matrix relation $XY = 0$, where X is $r \times n$ and Y is $n \times s$, is *trivial* if for each $i = 1, \ldots, n$, either the i-th column of X or the i-th row of Y is 0, and $XY = 0$ is *trivializable* if $XP^{-1}.PY = 0$ is trivial for some $P \in \mathbf{GL}_n(R)$. In the above example, (8.7.3) shows the relation (8.7.2) to be trivializable.

We begin by describing the rings in which every relation is trivializable:

Theorem 8.7.1. *Let R be a non-zero ring. Then the following conditions are equivalent:*

(a) *Every relation in R can be trivialized.*
(b) *Every finitely generated right ideal in R is free, as right R-module, of unique rank.*
(c) *R has IBN and every finitely generated submodule of a free right R-module is free.*
(d) *Every matrix relation in R can be trivialized.*
(a^0)–(d^0) *the leftright analogues of* (a)–(d).

Further, any such ring is an integral domain.

Proof. (a) $\Rightarrow$ (b). Let $\mathfrak{a}$ be a finitely generated right ideal of R and let n be the least integer such that $\mathfrak{a}$ has an n-element generating set, $u_1, \ldots, u_n$ say. Then $\mathfrak{a}$ is free on

$u_1, \ldots, u_n$, for if not, take a non-trivial relation $u.a = 0$. By (a) this can be trivialized, say $u' = uP^{-1}$, $a' = Pa$. Since $a \neq 0$, we have $a' \neq 0$, say $a'_n \neq 0$. But $u'.a' = 0$ is trivial, so $u'_n = 0$ and it follows that $\mathfrak{a}$ is generated by $u'_1, \ldots, u'_{n-1}$, which contradicts the choice of n; hence $\mathfrak{a}$ is free on $u_1, \ldots, u_n$. If $\mathfrak{a}$ has another basis $v_1, \ldots, v_m$ and $m \neq n$, then $m > n$ and $R^m \cong R^n$; this yields an endomorphism f of $\mathfrak{a}$ which is surjective but not injective. Thus $fu_1, \ldots, fu_n$ generate $\mathfrak{a}$ but not freely; by the first part we see that $\mathfrak{a}$ can be generated by fewer than n elements, which is again a contradiction, hence $m = n$ and $\mathfrak{a}$ has unique rank.

(b) $\Rightarrow$ (c). Let F be a free right R-module and G a finitely generated submodule. The finite generating set of F involves only finitely many generators of F; by ignoring the rest we may take F to be finitely generated. Let λ_1 be the projection of F on the first factor R and denote by F' the kernel of λ_1. Then we have the exact sequence

$$0 \to F' \cap G \to G \to \mathfrak{a} \to 0, \tag{8.7.4}$$

where $\mathfrak{a}$ is the image of G under λ_1, a finitely generated right ideal of R. By (b), $\mathfrak{a}$ is free, hence (8.7.2) splits and $G \cong (F' \cap G) \oplus \mathfrak{a}$. By induction on the rank of F, $F' \cap G$ as finitely generated submodule of F' is free and it follows that G is free.

We next show that R is an integral domain. Let $a \in R$ and denote its right annihilator by $\mathfrak{n}$. Then $aR \cong R/\mathfrak{n}$; since aR is free, we have $R = \mathfrak{n} \oplus \mathfrak{a}$, where $\mathfrak{a} \cong aR$. Both $\mathfrak{n}$ and $\mathfrak{a}$ are free; by the uniqueness of the rank, either $\mathfrak{a} = 0$ or $\mathfrak{n} = 0$, so a is either 0 or right regular. This holds for all $a \in R$, hence R is an integral domain. Now if R is right Ore, then it is embeddable in a skew field and so has IBN (BA, Theorem 4.6.7); otherwise by Proposition 7.1.9 it has free right ideals of any finite rank and this rank is unique by (b); hence R has IBN.

(c) $\Rightarrow$ (d). Let $XY = 0$ be a matrix relation, where $X \in {}^rR^n$, $Y \in {}^nR^s$. The matrix X defines a linear map $\varphi : {}^nR \to {}^rR$ by left multiplication and we have an exact sequence

$$0 \to \ker \varphi \to {}^nR \to \operatorname{im} \varphi \to 0. \tag{8.7.5}$$

As a finitely generated submodule of ${}^rR$, $\operatorname{im} \varphi$ is free, so the exact sequence (8.7.5) splits, and by changing the basis in ${}^nR$ we obtain a basis adapted to the submodule $\ker \varphi : {}^nR = \ker \varphi \oplus F$, where $F \cong \operatorname{im} \varphi$, $\dim \ker \varphi = t$, say. If this change of basis is described by $P \in \mathbf{GL}_n(R)$, we put $X' = XP^{-1}$, $Y' = PY$. Since $X'Y' = XY = 0$, the columns of Y' lie in $\ker \varphi$; thus the rows of Y' after the first t are 0, while the first t columns of X' are 0. Hence the matrix relation $X'Y' = 0$ is indeed trivial.

Now (a) is a special case of (d) and (a^0)–(d^0) follow by the evident symmetry of (a). ∎

A non-zero ring satisfying the conditions of this theorem is called a *semifir*. By (b) every left or right fir is a semifir; however there are right firs that are not left firs (see Exercise 5). We also remark that a commutative semifir is just a Bezout domain, while a commutative fir is a PID.

As already remarked, the free algebra $k\langle X\rangle$ on any set X over a field k is a fir; more generally, this holds for the tensor D-ring $D_k\langle X\rangle$, where D is any skew field and k a subfield. This is proved by means of the weak algorithm, which we shall not define here (see Cohn (1985), Chapter 2); it is easier to prove the weaker statement that $D_k\langle X\rangle$ is a semifir and we shall do so now (but see also Section 11.5).

Theorem 8.7.2. *Let D be a skew field and F a subfield. Then the tensor D-ring $D_K\langle X\rangle$ on any set X is a semifir.*

Proof. Let $\{u_\nu\}$ be a right F-basis of D and $X=\{x_i\}$; then every element of $D_F\langle X\rangle$ can be uniquely written in the form $c+\sum u_\nu x_i f_{\nu i}$, where $c\in D$ and $f_{\nu i}\in D_F\langle X\rangle$. This follows because we can express every element of D as $a=\sum u_\nu a_\nu$, where $a_\nu\in F$ and $ax_i=\sum u_\nu a_\nu x_i=\sum u_\nu x_i a_\nu$.

Suppose now that we have a relation in $D_F\langle X\rangle$:

$$\sum_1^n a_i b_i = 0. \tag{8.7.6}$$

In order to show that this relation can be trivialized it is enough to do this in a given degree; thus we can assume that the a_i, b_i are homogeneous and that $\deg a_i+\deg b_i=r>0$. We shall use double induction, on n and r. If each a_i has positive degree, we can write $a_i=\sum u_\nu x_t a_{\nu t i}$; equating cofactors of $u_\nu x_t$ we find

$$\sum_i a_{\nu t i} b_i = 0,$$

and now the result follows by induction on r, for we can make a transformation reducing one of the b's to 0. There remains the case where some a_i, say a_1 has degree 0. Then we can replace a_2 by $a_2-a_1.a_1^{-1}a_2=0$ and b_1 by $b_1'=b_1+a_1a_1^{-1}b$; we thus obtain

$$\sum a_i b_i = a_1 b_1' + a_3 b_3 + \ldots + a_n b_n = 0,$$

and we have now diminished n, and so can apply induction on n to complete the proof. ∎

We conclude this section with a result that is often useful, the inertia lemma for semifirs (see also Cohn (1985), Lemma 4.6.3). We recall that from any ring R we can form the polynomial ring $R[t]$ in a central indeterminate t; its completion by power series is the formal power series ring $R[[t]]$, and if we localize now at the powers of t we obtain the formal Laurent series ring $R((t))$. Since t is regular in $R[[t]]$, the latter ring is embedded in the Laurent series ring.

Let B be any ring and A a subring. Then A is said to be *finitely inert* in B if for any matrix $Z\in {}^rA^s$, if $Z=XY$ over B, where X is $r\times n$ and Y is $n\times s$, there exists $P\in \mathbf{GL}_n(B)$ such that XP^{-1}, PY have their entries in A.

Lemma 8.7.3 (Inertia lemma). *Let R be a semifir. Then for any central indeterminate t, the formal power series ring $R[[t]]$ is finitely inert in $R((t))$.*

Proof. Put $S=R[[t]]$, $T=R((t))$ and indicate the natural homomorphism $S\to R$ by $f\mapsto (f)_0$; it amounts to putting $t=0$. We take $A\in {}^rS^s$ and suppose that over T:

$$A=PQ,\ \text{where } P \text{ is } r\times n \text{ and } Q \text{ is } n\times s. \tag{8.7.7}$$

If P or Q is 0, there is nothing to prove, so we may suppose $P, Q \neq 0$. Over T every non-zero matrix C can be written in the form $t^{\nu}C'$, where $\nu \in \mathbf{Z}$, C' has entries in S and $(C')_0 \neq 0$. Let $P = t^{\mu}P'$, $Q = t^{\nu}Q'$, where $(P')_0$, $(Q')_0 \neq 0$. Dropping the dashes and writing $\mu + \nu = -\lambda$, we can rewrite (8.7.7) as

$$A = t^{-\lambda}PQ, \text{ where } P \in {}^rS^n,\ Q \in {}^nS^s \text{ and } (P)_0, (Q)_0 \neq 0. \tag{8.7.8}$$

If $\lambda \leq 0$, there is nothing to prove, so assume that $\lambda > 0$. Then $(PQ)_0 = 0$; since R is a semifir, we can find a matrix $U \in \mathbf{GL}_n(R)$ trivializing this relation, and on replacing P, Q by PU^{-1}, UQ we find that for some h $(0 \leq h \leq n)$ all the columns in $(P)_0$ after the first h are 0, while the first h rows of $(Q)_0$ are 0. If we multiply P on the right by $V = tI_h \oplus I_{n-h}$ and Q on the left by V^{-1}, then P becomes divisible by t while Q still has all its entries in S. In this way we can, by cancelling a factor t, replace $t^{-\lambda}$ by $t^{1-\lambda}$ in (8.7.8) and after λ steps obtain the same equation with $\lambda = 0$. This proves the finite inertia. ∎

We remark that there is a stronger notion, total inertia, and the inertia theorem asserts that an inversely filtered fir with inverse weak algorithm is totally inert in its completion (see Cohn (1985), Theorem 2.9.15).

Exercises

1. Show that a right Ore domain is a right fir iff it is right principal.
2. Show that every semifir is weakly finite.
3. Let R be a semifir and A, B finitely generated submodules of a free R-module. Show that $A \cap B$, $A + B$ are again free and that $\text{rk}(A + B) + \text{rk}(A \cap B) = \text{rk}\, A + \text{rk}\, B$.
4. Let R be a right fir. Show that any submodule of a free right R-module is free.
5. In the group algebra over a field k of the free group on x, y let R be the subalgebra generated by $x, y, x^{-1}y, x^{-2}y, \ldots$. Verify that R is a semifir but not a left fir (it can be shown that R is a right fir, see Cohn (1985), Section 2.10).

Further exercises on Chapter 8

1. Show that if V is a vector space of infinite dimension ν over a skew field K of cardinal α, then V has the cardinal $\nu\alpha$ and $V^* = \text{Hom}_K(V, K)$ has cardinal α^{ν}. Show that if K is commutative and $\dim V^* = \nu^*$, then $\nu^* \geq \alpha$, and deduce that $\nu^* = \alpha^{\nu}$ (this is true even when K is skew, see Jacobson (1953)).
2. Let K be a skew field and for $n \geq 1$ embed $\mathfrak{M}_{2^n}(K)$ in $\mathfrak{M}_{2^{n+1}}(K)$ by mapping A to $A \oplus A$. Show that the direct limit of the rings $\mathfrak{M}_{2^n}(K)$ is simple. What are its projective modules?
3. Let K be a skew field. Describe the ideal structure of the following infinite matrix rings over K: (i) the ring of all row-finite and column-finite matrices, (ii) the ring of all matrices which are equal to a scalar outside a finite square.

4. Let V be a (K, R)-bimodule, where K is a skew field and R is a ring which acts fully on V with centralizer K. Show that V_R is simple; if moreover R acts fully on ${}^2V$, then R acts densely on V.
5. (Jacobson) Let k be a field and R the (unital) k-algebra generated by u, v subject to $uv = 1$. Show that R is primitive by representing it by the linear transformations $e_iu = e_{i+1}$, $e_iv = e_{i-1}$ if $i > 1$, $e_1v = 0$.
6. (P. Samuel) Let k be a field and $k\langle x, y\rangle$ the free k-algebra on x, y. Show that this is primitive by representing it as follows: $e_ix = e_{i^2+1}$, $e_iy = e_{i-1}$ if $i > 1$, $e_1y = 0$.
7. Show that in any ring R the ideals with primitive quotient are just the cores of maximal right ideals. Show further that the intersection of these ideals is $\mathbf{J}(R)$.
8. Show that in a prime ring R every non-zero right ideal is a faithful R-module.
9. Show that every maximal ideal in a ring is a prime ideal.
10. Show that the least ideal $\mathfrak{N}$ in a ring R such that $R/\mathfrak{N}$ is semiprime is a nil ideal and deduce that $\mathfrak{N}$ is the lower nilradical.
11. Find the socle of the ring of all upper triangular matrices over a field, and show that it may differ from the left socle.
12. Let K be a commutative ring and t an indeterminate. Show that if $a_0 + a_1t + \ldots + a_nt^n \in K[t]$ is a unit, then a_0 is a unit and $a_1, \ldots, a_n$ are nilpotent. Deduce that the Jacobson radical of $K[t]$ is its nilradical.
13. Show that the left (or right) ideal generated by a strongly nilpotent element is nilpotent.
14. Let R be a prime ring. Show that any non-zero central element of R is a non-zero-divisor; deduce that the centre of R is an integral domain.
15. (Kaplansky) With any square matrix A over a ring K we can associate an 'infinite periodic' matrix by taking the diagonal sum of countably many copies of A. Let R be the ring of all upper triangular infinite periodic matrices over a field k. Show that R is prime, with Levitzki nilradical equal to the Jacobson radical.
16. Let R be any K-algebra and define the *centroid* of R as $\mathrm{End}({}_RR_R)$. Show that the centroid Γ is a commutative K-algebra and that R has a natural Γ-algebra structure. If R is primitive and $R^2 = R$, show that Γ is an integral domain.
17. Show that the group algebra of the additive group of rationals is a Bezout domain. For real numbers α, β show that the monoid M of all positive numbers of the form $m\alpha + n\beta$ is such that the monoid algebra kM is Bezout iff α/β is rational.
18. Over a semifir R show that if a matrix product AB ($A \in {}^rR^n$, $B \in {}^nR^s$) has a $u \times v$ block of zeros, then for some t ($0 \le t \le n$) and a suitable $P \in \mathbf{GL}_n(R)$, AP^{-1} has a $u \times t$ block of zeros and PB has an $n - t \times v$ block of zeros (this is the partition lemma, see Cohn (1985), Lemma 1.1.4).
19. Show that the group algebra of a free group is a semifir.

9 Skew fields

Skew fields arise quite naturally in the application of Schur's lemma and elsewhere, but most of the known theory deals with the case of skew fields finite-dimensional over their centres (see Chapter 5). In the general case only isolated results are known, much of it depending on the coproduct construction (see Cohn (1995), Schofield (1985)). This lies outside our framework and we shall confine ourselves to presenting some of the highlights which do not require special prerequisites.

After some general remarks in Section 9.1, including the Cartan–Brauer–Hua theorem, we shall give an account in Section 9.2 of determinants over skew fields. This is followed in Section 9.3 by a proof of the existence of free fields, based on the specialization lemma whose proof has been much simplified. Many of the concepts (though fewer of the actual results) of valuation theory carry over to skew fields and in Section 9.4 we shall examine the situation and pursue some of the results that continue to hold in the general case. The final section, Section 9.5, is concerned with the question when the left and right dimensions of a field extension are equal. We shall also meet examples of skew field extensions whose left and right dimensions are different (Emil Artin's problem); they are most easily obtained as pseudo-linear extensions, a more tractable class of finite-dimensional extensions.

Throughout this chapter we shall use the term 'field' to mean 'not necessarily commutative division ring'; the prefix 'skew' is sometimes added for emphasis.

9.1 Generalities

Let K be a skew field. Its centre C is a commutative field, and the characteristic of C is also called the *characteristic* of K. We have already seen in Theorem 5.1.14 that every finite field is commutative. However, an infinite field may well have a finite centre, in fact, for every commutative field k there is a skew field with centre k. In characteristic 0 we can adjoin variables u, v with the relation $uv - vu = 1$ to obtain such a field, but there is another construction which applies quite generally; a third construction, generally valid, is used in the proof of Theorem 9.3.3 below.

Proposition 9.1.1. *Let k be any commutative field. Then there is a skew field D whose centre is k.*

Proof. Consider the group algebra of the additive group of rationals over $k : k[x^{\lambda} | \lambda \in \mathbf{Q}]$. Clearly this is a commutative k-algebra; we take the subalgebra generated by $x^{2^{-n}}, n = 0, 1, \ldots$ and form its field of fractions

$$E = k(x, x^{1/2}, x^{1/4}, \ldots).$$

This field has the automorphism $\alpha : f(x) \mapsto f(x^2)$, which is of infinite order. Moreover, k is the precise subfield fixed by α, for if f involves x, let 2^n $(n \geq 0)$ be the largest denominator in any power of x occurring in f. Then $x^{r/2^n}$ occurs in f for some odd r, but it does not occur in f^{α}, so f is not fixed under α. Now form the skew polynomial ring $E[y; \alpha]$ and let D be its field of fractions. By Theorem 7.3.6, D has the precise centre k. ∎

Much of linear algebra can be done over a skew field (see BA, Chapter 4); this rests on the fact that every module over a field is free. It is well known (BA, Theorem 4.6.8) that this property actually characterizes skew fields.

We go on to prove two general results, which although not used here, are of importance, with many applications. The first concerns normal subgroups of the multiplicative group of a field; the proof is based on an idea of Jan Treur.

Theorem 9.1.2 (Cartan–Brauer–Hua theorem). *Let $K \subseteq L$ be skew fields, and assume that $K^{\times}$ is a normal subgroup of $L^{\times}$. Then either $K = L$ or K is contained in the centre of L.*

Proof. For any $c \in L^{\times}$ the mapping $\alpha_c : x \mapsto c^{-1}xc$ satisfies

$$(x + y)\alpha_c = x\alpha_c + y\alpha_c.$$

Further, we have $x\alpha_c = x.(x, c)$, where $(x, c) = x^{-1}c^{-1}xc \in K$, whenever $c \in K$. If $K \neq L$, take $a \in L\backslash K$; for any $c \in K^{\times}$, since $(1, c) = 1$, we have

$$(a + 1)(a + 1, c) = a(a, c) + 1.$$

By the linear independence of 1, a over K, we have $(a, c) = (a + 1, c) = 1$; thus $ac = ca$ for all $c \in K$, $a \notin K$. But if $b \in K$, then $a + b \notin K$, therefore $bc - cb = (a + b)c - c(a + b) - [ac - ca] = 0$, hence c commutes with every element of L, i.e. K is contained in the centre of L. ∎

This result was obtained in the case of division algebras by Henri Cartan in 1947 as a consequence of his Galois theory of skew fields. Richard Brauer and independently Hua Loo-Keng observed in 1949 that the general case could be proved directly.

Our second result concerns additive mappings preserving inversion:

Theorem 9.1.3 (Hua's theorem). *Let $\sigma : K \to L$ be a mapping between two skew fields such that*

$$(a + b)^{\sigma} = a^{\sigma} + b^{\sigma}, \ 1^{\sigma} = 1, \ (a^{-1})\sigma = (a^{\sigma})^{-1}. \tag{9.1.1}$$

Then σ is either a homomorphism or an antihomomorphism.

Proof. (E. Artin) We must show that for all $a, b \in K$, either $(ab)^\sigma = a^\sigma b^\sigma$ or for all $a, b \in K$, $(ab)^\sigma = b^\sigma a^\sigma$. We observe that $a^\sigma (a^{-1})^\sigma = 1$ by (9.1.1), so $a \neq 0 \Rightarrow a^\sigma \neq 0$ and it follows that σ is injective. We start from the following identity (Hua's identity):

$$a - (a^{-1} + (b^{-1} - a)^{-1})^{-1} = aba, \tag{9.1.2}$$

valid whenever all inversions are defined, i.e. $ab \neq 0, 1$. To prove (9.1.2), we observe that for any $x \neq 0, 1$,

$$(x^{-1} - 1)^{-1} = (1 - x)^{-1} - 1, \tag{9.1.3}$$

as we see by multiplying out. Let $ab \neq 0, 1$; then $a^{-1}(b^{-1} - a) = (ba)^{-1} - 1$, hence on taking $x = ba$ in (9.1.3), we find

$$(b^{-1} - a)^{-1} a = ((ba)^{-1} - 1)^{-1} = (1 - ba)^{-1} - 1,$$

$$(b^{-1} - a)^{-1} = (1 - ba)^{-1} a^{-1} - a^{-1} = (a - aba)^{-1} - a^{-1}.$$

Hence

$$a - aba = (a^{-1} + (b^{-1} - a)^{-1})^{-1},$$

and (9.1.2) follows from this on rearranging the terms.

Now $(a^{-1})^\sigma = (a^\sigma)^{-1}$ and we may denote both sides by $a^{-\sigma}$. If we apply σ to (9.1.2) and observe that σ is compatible with all the operations on the left, we obtain

$$(aba)^\sigma = a^\sigma b^\sigma a^\sigma. \tag{9.1.4}$$

Clearly this still holds if a or b is 0; if $b = a^{-1}$, then $b^\sigma = a^{-\sigma}$ and both sides reduce to a^σ, so (9.1.4) holds in all cases. Put $b = 1$ in (9.1.4):

$$(a^2)^\sigma = (a^\sigma)^2. \tag{9.1.5}$$

Next replace a by $a + b$ in (9.1.5) and simplify, using (9.1.5) again:

$$(ab + ba)^\sigma = a^\sigma b^\sigma + b^\sigma a^\sigma. \tag{9.1.6}$$

Now consider $(c^\sigma - a^\sigma b^\sigma) c^{-\sigma} (c^\sigma - b^\sigma a^\sigma)$ for any $c \neq 0$. This equals

$$\begin{aligned} c^\sigma - a^\sigma b^\sigma - b^\sigma a^\sigma + a^\sigma b^\sigma c^{-\sigma} b^\sigma a^\sigma &= c^\sigma - (ab + ba^\sigma) + (abc^{-1}ba^\sigma) \\ &= (c - ab - ba + abc^{-1}ba)^\sigma, \end{aligned}$$

by an application of (9.1.6), (9.1.4) and (9.1.1). Thus

$$(c^\sigma - a^\sigma b^\sigma) c^{-\sigma} (c^\sigma - b^\sigma a^\sigma) = (c - ab - ba + abc^{-1}ba)^\sigma, \tag{9.1.7}$$

for all a, b, c such that $c \neq 0$. For $c = ab$ the right-hand side reduces to $ab - ab - ba + ba = 0$, hence the left-hand side of (9.1.7) vanishes for $c = ab$. Thus $(ab)^\sigma$ is either $a^\sigma b^\sigma$ or $b^\sigma a^\sigma$; it only remains to show that the same alternative holds for all pairs.

Fix $a \in K$ and put $U_\sigma = \{b \in K \mid (ab)^\sigma = a^\sigma b^\sigma\}$, $V_\sigma = \{b \in K \mid (ab)^\sigma = b^\sigma a^\sigma\}$. They are clearly subgroups of the additive group of K whose union is K, hence for each $a \in K$, one of them must be all of K (see Exercise 2). Now put $U = \{a \in K | U_\sigma = K\}$, $V = \{a \in K | V_\sigma = K\}$; then U, V are again subgroups whose union is K, so one of them must be all of K, i.e. either $(ab)^\sigma = a^\sigma b^\sigma$ for all $a, b \in K$ or $(ab)^\sigma = b^\sigma a^\sigma$ for all $a, b \in K$. ■

This result is used in projective geometry to show that a bijective transformation of the line which preserves harmonic ranges necessarily has the form $x \mapsto ax^\sigma + b$, where $a \neq 0$ and σ is an automorphism or an antiautomorphism (for $K = \mathbf{R}$ this is von Staudt's theorem, see E. Artin (1957) p. 37).

Exercises

1. Let k be a commutative field of characteristic 0. Verify that the field of fractions of the Weyl algebra $A_1(k)$ has centre k. What is the centre when k has prime characteristic?
2. In the proof of Theorem 9.1.3 the fact was used that a group G cannot be the union of two proper subgroups H, K. By considering the product of two elements, one not in H and one not in K, prove this fact.
3. Let K be a skew field. Show that for any $c \in K$ the centralizer of the set of all conjugates of c is either K or the centre of K. Deduce that no non-central element c can satisfy the identity $cx^{-1}cx = x^{-1}cxc$ for all $x \in K^\times$.
4. Let $\sigma : K \to L$ be a mapping between fields such that $(x+y)^\sigma = x^\sigma + y^\sigma$, $1^\sigma = \lambda$, $(x^\sigma)^{-1} = \lambda^{-1}(x^{-1})^\sigma\lambda^{-1}$. Show that $x^\sigma = x^\tau\lambda$, where τ is a homomorphism or an antihomomorphism.
5. Show that any non-central element in a skew field has infinitely many conjugates.
6. Let K be a skew field. Show that any conjugacy class of elements of K outside the centre generates K as a field. Show that any subfield containing $K^{\times\prime}$ coincides with K.
7. (Herstein) Let K be a skew field of prime characteristic and G a finite subgroup of $K^\times$. Denoting by P the prime subfield of K, show that the P-space spanned by G is an algebra, hence a finite field. Deduce that G must be cyclic.
8. Let K be a skew field. Show that any abelian normal subgroup of $K^\times$ is contained in the centre of K.

9.2 The Dieudonné determinant

The determinant is a fundamental concept, going back much further than the notion of matrix, on which it is based (matrices were introduced by Arthur Cayley in the middle of the 19th century, whereas determinants had been used at the end of the 17th century by Gottfried Wilhelm von Leibniz). Today we regard the determinant of a square matrix as an alternating multilinear function of the columns of the matrix; its most important property is that its vanishing characterizes the linear dependence of its columns. Here the entries of the matrix are assumed to lie in a (commutative) field, but it is clear that the definition is unchanged when the entries are taken from any commutative ring. So it is natural to try to extend the definition to the non-commutative case; we have seen in Section 5.3 how this can be done for a finite-dimensional algebra by means of the norm. The general definition is due to Jean Dieudonné [1943]; before presenting it let us examine the simplest case, that of a 2×2 matrix.

The columns of the matrix $\begin{pmatrix} a & b \\ c & d \end{pmatrix}$ over a skew field K say, are linearly dependent iff the equations

$$ax + by = 0, \quad cx + dy = 0, \tag{9.2.1}$$

have a non-trivial solution (x, y). If $a = 0$, such a solution exists precisely when $b = 0$ or $c = 0$, so let us assume that $a \neq 0$. Then in any non-trivial solution $y \neq 0$ and by eliminating x from (9.2.1) we obtain $(d - ca^{-1}b)y = 0$, hence the condition for linear dependence is

$$d - ca^{-1}b = 0. \tag{9.2.2}$$

Depending on which entries of A are zero, we can find various expressions whose vanishing characterizes the linear dependence of the columns of A, but a few trials make it clear that there is no polynomial in a, b, c, d with this property. Thus we must expect a determinant function (if one exists) to be a rational function. A second point is that under any reasonable definition one would expect $\begin{pmatrix} a & 0 \\ 0 & b \end{pmatrix}$ and $\begin{pmatrix} b & 0 \\ 0 & a \end{pmatrix}$ to have the same determinant. This suggests that even for a skew field the values of the determinant must lie in an abelian group. For any field K we define the abelianized group K^{ab} as

$$K^{ab} = K^{\times}/K^{\times\prime},$$

where $K^{\times\prime}$ is the derived group of the multiplicative group $K^{\times}$. Thus for a commutative field K^{ab} reduces to $K^{\times}$. It is clear that K^{ab} is universal for homomorphisms of $K^{\times}$ into abelian groups.

As usual we write $\mathbf{GL}_n(K)$ for the group of all invertible $n \times n$ matrices over K. Let us recall that any $m \times n$ matrix A over K may be interpreted as the matrix of a linear mapping from an m-dimensional to an n-dimensional vector space over K. By choosing suitable bases in these spaces we can ensure that A takes the form $I_r \oplus 0$, where r is the rank of A. Thus there exist $P \in \mathbf{GL}_m(K)$, $Q \in \mathbf{GL}_n(K)$ such that

$$PAQ = \begin{pmatrix} I_r & 0 \\ 0 & 0 \end{pmatrix}. \tag{9.2.3}$$

This was proved in the remark after Proposition 7.2.3. In particular, (9.2.3) shows that a square matrix over any field is invertible iff it is left (or equivalently, right) regular (see also Corollary 9.2.3 below). Such a matrix is also called *non-singular* and a non-invertible matrix is called *singular*.

Without a determinant we cannot define SL_n, but we have the group $\mathbf{E}_n(K)$ generated by all elementary matrices $B_{ij}(c) = I + cE_{ij}$ (see Section 3.5). We observe that multiplication by an elementary matrix corresponds to an elementary operation on matrices; more precisely, left multiplication by $B_{ij}(c)$ amounts to adding c times the j-th row to the i-th row and right multiplication by $B_{ij}(c)$ means adding the i-th column multiplied by c to the j-th column.

As a first result we show that we can pass from AB to BA by elementary transformations, provided that the matrices are 'enlarged' by forming their diagonal sum with a unit matrix. Here it is not necessary to assume that the coefficients lie in a field.

Lemma 9.2.1 (Whitehead's lemma). *Let R be any ring and $n \geq 1$. For any $A, B \in \mathbf{GL}_n(R)$, $AB \oplus I$ and $BA \oplus I$ lie in the same (left or right) coset of $\mathbf{E}_{2n}(R)$, a fact which may be expressed by writing*

$$\begin{pmatrix} AB & 0 \\ 0 & I \end{pmatrix} \equiv \begin{pmatrix} BA & 0 \\ 0 & I \end{pmatrix} \quad (\mathrm{mod}\ \mathbf{E}_{2n}(R)). \tag{9.2.4}$$

Proof. We must show that

$$A^{-1}B^{-1}AB \oplus I \in \mathbf{E}_{2n}(R). \tag{9.2.5}$$

In the first place we note that for any $C \in \mathbf{GL}_n(R)$,

$$\begin{pmatrix} 0 & C \\ -C^{-1} & 0 \end{pmatrix} = \begin{pmatrix} I & C \\ 0 & I \end{pmatrix}\begin{pmatrix} I & 0 \\ -C^{-1} & I \end{pmatrix}\begin{pmatrix} I & C \\ 0 & I \end{pmatrix} \in \mathbf{E}_{2n}(R), \tag{9.2.6}$$

for each matrix on the right can be written as a product of n^2 elementary matrices. Hence we have

$$\begin{pmatrix} C^{-1} & 0 \\ 0 & C \end{pmatrix} = \begin{pmatrix} 0 & -I \\ I & 0 \end{pmatrix}\begin{pmatrix} 0 & C \\ -C^{-1} & 0 \end{pmatrix} \in \mathbf{E}_{2n}(R), \tag{9.2.7}$$

for the matrices on the right are instances of (9.2.6). Now we have

$$\begin{pmatrix} A^{-1}B^{-1}AB & 0 \\ 0 & I \end{pmatrix} = \begin{pmatrix} A^{-1} & 0 \\ 0 & A \end{pmatrix}\begin{pmatrix} B^{-1} & 0 \\ 0 & B \end{pmatrix}\begin{pmatrix} AB & 0 \\ 0 & (AB)^{-1} \end{pmatrix} \in \mathbf{E}_{2n}(R). \qquad \blacksquare$$

As we shall soon see, in the case of a skew field $\mathbf{E}_n(K)$ is the derived group $\mathbf{GL}_n(K)'$ (except when $n = 2 = |K|$), in particular, E_n is a normal subgroup of GL_n; for a commutative field k, $\mathbf{E}_n(k) = \mathbf{SL}_n(k)$ and the result was proved in Proposition 3.5.2.

We can embed $\mathbf{GL}_n(K)$ in $\mathbf{GL}_{n+1}(K)$ by mapping A to $A \oplus 1$. In this way we obtain an ascending chain

$$\mathbf{GL}_1(K) \subset \mathbf{GL}_2(K) \subset \ldots,$$

whose union is again a group, written $\mathbf{GL}(K)$ and called the *stable* general linear group. Its elements may be thought of as infinite matrices which differ from the unit matrix only in a finite square. Similarly the union of the groups $\mathbf{E}_n(K)$ is a group $\mathbf{E}(K)$.

In order to obtain a definition for the determinant we shall need to refine the expression (9.2.3) for a matrix. A square matrix is called *lower unitriangular* if all the entries on the main diagonal are 1 and those above it are 0; it is clear from the definition that such a matrix is a product of elementary matrices and hence is invertible. Moreover, the lower unitriangular matrices form a group under multiplication. An *upper triangular matrix* is defined similarly. We observe that left multiplication by a lower unitriangular matrix amounts to adding left multiples of certain rows to later rows, and right multiplication by an upper unitriangular matrix comes to adding right multiples of certain columns to later columns.

We can now describe a decomposition which applies to any matrix over a skew field. Our account follows that of Peter Draxl (1983), with some simplifications.

Theorem 9.2.2 (Bruhat normal form). *Let K be a skew field and $A \in {}^mK^n$. Then A can be written in the form*

$$A = LMU, \tag{9.2.8}$$

where L is an $m \times m$ lower unitriangular matrix, U an $n \times n$ upper unitriangular matrix and M is an $m \times n$ matrix with at most one non-zero entry in any row or column. Moreover, any other such decomposition of A leads to the same matrix M.

The matrix M is called the *core* of A and (9.2.8) is the *Bruhat decomposition*. For example, when $A = \begin{pmatrix} a & b \\ c & d \end{pmatrix}$ then the Bruhat decomposition is

$$\begin{pmatrix} 1 & 0 \\ ca^{-1} & 1 \end{pmatrix}\begin{pmatrix} a & 0 \\ 0 & d - ca^{-1}b \end{pmatrix}\begin{pmatrix} 1 & a^{-1}b \\ 0 & 1 \end{pmatrix} \quad \text{if } a \neq 0,$$

$$\begin{pmatrix} 1 & 0 \\ db^{-1} & 1 \end{pmatrix}\begin{pmatrix} 0 & b \\ c & 0 \end{pmatrix} \quad \text{if } a = 0 \neq b, \qquad \begin{pmatrix} 0 & 0 \\ c & 0 \end{pmatrix}\begin{pmatrix} 1 & c^{-1}d \\ 0 & 1 \end{pmatrix} \quad \text{if } a = b = 0 \neq c,$$

$$\begin{pmatrix} 0 & 0 \\ 0 & d \end{pmatrix} \quad \text{if } a = b = c = 0.$$

Proof. Suppose that the first non-zero row of A is the i-th row and that its first non-zero entry is a_{ij}. By adding left multiples of the i-th row to each succeeding row we can reduce every entry in the j-th column except a_{ij} to 0. These operations correspond to left multiplication by a certain lower unitriangular matrix. Next we add right multiples of the j-th column to succeeding columns to reduce all entries in the i-th row except a_{ij} to 0; these operations will correspond to right multiplication by an upper unitriangular matrix. As a result a_{ij} is the only non-zero element in its row and column, and all the rows above the i-th are zero. Next we take the first non-zero row after the i-th, say the k-th row and with the first non-zero entry a_{kl} in this row we continue the process. After at most m steps A has been reduced to the form M where each row and each column has at most one non-zero entry, with a left factor which is lower and a right factor which is upper unitriangular. This is the required decomposition (9.2.8).

If $A = L'M'U'$ is another such decomposition, then on writing $P = L^{-1}L'$, $Q = U'U^{-1}$, we have

$$PM'Q = M,$$

where P is again lower and Q upper unitriangular. This tells us that we can pass from M' to M by adding left multiples of rows to later rows and right multiples of columns to later columns. If m'_{rs} is a non-zero entry of M', the application of these operations only affects entries in the r-th row or s-th column and leaves m'_{rs} itself unchanged. Hence in any operation on M', the (i, j)-entry is affected only if M' has either a non-zero entry in the i-th row before the j-th column, or a non-zero entry in the j-th column above the i-th row. In either case $m'_{ij} = 0$, while m'_{rs} remains unchanged. Hence $m_{rs} = m'_{rs}$ and it follows that $m_{ij} = 0$. Therefore $M' = M$, as we wished to show. ∎

We remark that the matrices L, U in (9.2.8) are not generally unique. For example, we have

$$\begin{pmatrix} 1 & 0 \\ a & 1 \end{pmatrix}\begin{pmatrix} 0 & u \\ v & 0 \end{pmatrix}\begin{pmatrix} 1 & -v^{-1}au \\ 0 & 1 \end{pmatrix} = \begin{pmatrix} 0 & u \\ v & 0 \end{pmatrix}.$$

For a truly unique form see Exercise 4.

Corollary 9.2.3. *For any $m \times n$ matrix A over a skew field K the following conditions are equivalent:*

(a) *A is left regular: $XA = 0 \Rightarrow X = 0$,*
(b) *A has a right inverse: $AB = I$ for some $B \in {}^nK^m$.*

Moreover, when (a), (b) *hold, then $m \leq n$, with equality if and only if* (a), (b) *are equivalent to their left–right analogues.*

Proof. Either of (a), (b) clearly holds for A precisely when it holds for its core and it holds for the core iff each row has a non-zero entry. ∎

Let us define a *monomial matrix* as a square matrix with precisely one non-zero entry in each row and each column, e.g. the core of an invertible matrix is a monomial matrix. If all the non-zero entries in a monomial matrix are 1, we have a *permutation matrix*; this may also be defined as the matrix obtained by permuting the rows (or equivalently, the columns) of the unit matrix. The determinant of a permutation matrix is 1 or -1 according as the permutation is even or odd. Sometimes it is more convenient to have matrices of determinant 1; this can be accomplished by using a *signed* permutation matrix, i.e. a matrix obtained from the unit matrix by a series of operations which consist in interchanging two columns and changing the sign of one of them. By (9.2.6) such a matrix is a product of elementary matrices.

Any monomial matrix M may be written in the form

$$M = DP, \tag{9.2.9}$$

where D is a diagonal matrix and P is a signed permutation matrix. By Corollary 9.2.3 any matrix over a skew field is invertible iff its core is a monomial matrix.

With the help of Theorem 9.2.2 we can also identify E_n:

Proposition 9.2.4. *For any skew field K and any $n \geq 1$, we have*

$$\mathbf{GL}_n(K) = \mathbf{D}_n(K).\mathbf{E}_n(K), \tag{9.2.10}$$

where $\mathbf{D}_n(K)$ is the group of all diagonal matrices in $\mathbf{GL}_n(K)$. Moreover, for any $n \geq 2$,

$$\mathbf{E}_n(K) = \mathbf{GL}_n(K)', \tag{9.2.11}$$

except when $n = 2$, $K = \mathbf{F}_2$, when $\mathbf{E}_2(\mathbf{F}_2) = \mathbf{GL}_2(\mathbf{F}_2)$.

Proof. By Theorem 9.2.2, any invertible matrix can be written as $LDPU$, where L is lower, U is upper unitriangular, D is diagonal and P is a signed permutation matrix. Now P can be written as a product of elementary matrices, using (9.2.6); moreover, $LD = DL'$, where L' is again lower unitriangular, hence our matrix takes the form $D.F$, where $F \in \mathbf{E}_n(K)$, and this proves (9.2.10).

To establish (9.2.11), let us write $(A, B) = A^{-1}B^{-1}AB$ and (H, K) for the subgroup generated by all (A, B), $A \in H$, $B \in K$. We shall also write G_n for $\mathbf{GL}_n(K)$ and similarly for D_n, E_n. We first show that $(G_n, G_n) \subseteq E_n$; by (9.2.10) this will follow if we show that $(D_n, D_n) \subseteq E_n$ and $(D_n, E_n) \subseteq E_n$. The first inclusion follows from Lemma 9.2.1 (because $n \geq 2$), while the second results from the formula

$$\begin{pmatrix} u^{-1} & 0 \\ 0 & v^{-1} \end{pmatrix}\begin{pmatrix} 1 & -a \\ 0 & 1 \end{pmatrix}\begin{pmatrix} u & 0 \\ 0 & v \end{pmatrix}\begin{pmatrix} 1 & a \\ 0 & 1 \end{pmatrix} = \begin{pmatrix} 1 & a - u^{-1}av \\ 0 & 1 \end{pmatrix}.$$

In the other direction we have

$$\begin{pmatrix} 1 & a \\ 0 & 1 \end{pmatrix} = \left(\begin{pmatrix} 1 & 0 \\ 0 & v \end{pmatrix}, \begin{pmatrix} 1 & b \\ 0 & 1 \end{pmatrix}\right) = \begin{pmatrix} 1 & b(1-v) \\ 0 & 1 \end{pmatrix}, \tag{9.2.12}$$

provided that $a = b(1 - v)$. If $K \neq \mathbf{F}_2$, there is an element $v \neq 0, 1$ and putting $b = a(1 - v)^{-1}$, we can use (9.2.12) to express $B_{ij}(a)$ as a commutator. Hence $E_n \subseteq G'_n$ whenever $K \neq \mathbf{F}_2$. If $n \geq 3$, we have

$$\begin{pmatrix} 1 & a & 0 \\ 0 & 1 & 0 \\ 0 & 0 & 1 \end{pmatrix} = \left(\begin{pmatrix} 1 & 0 & 1 \\ 0 & 1 & 0 \\ 0 & 0 & 1 \end{pmatrix}, \begin{pmatrix} 1 & 0 & 0 \\ 0 & 1 & 0 \\ 0 & a & 1 \end{pmatrix}\right),$$

so we again have $E_n \subseteq G'_n$ and (9.2.11) follows. When $K = \mathbf{F}_2$ and $n = 2$, then $\mathbf{GL}_2(\mathbf{F}_2)$ is the symmetric group of degree 3 and is equal to $\mathbf{E}_2(\mathbf{F}_2)$, as is easily verified. ∎

We now define for any skew field K, a mapping $\delta : \mathbf{GL}(K) \to K^{ab}$ from the stable linear group to the abelianized group of K, as follows: for $A \in \mathbf{GL}_n(K)\delta(A) = \prod_{i=1}^n \bar{d}_i$, where the d_i are the diagonal elements of D in the expression (9.2.9) for the core of A and $\bar{d}_i$ is its residue class mod $K^{\times\prime}$. The value $\delta(A)$ is called the *Dieudonné determinant*, or simply the *determinant* of A. To obtain its properties we shall need to find how it is affected by permutations of its rows; for simplicity we consider the effect of signed permutations.

Lemma 9.2.5. *For any $A \in \mathbf{GL}_n(K)$ and any signed permutation matrix P,*

$$\delta(PA) = \delta(A).$$

Proof. By induction it will be enough to prove the result when P is the signed permutation matrix corresponding to a transposition, (r, s) say, where $r < s$. Denote this matrix by P_0 and take a Bruhat decomposition (9.2.8) of A, where the core is factorized as in (9.2.9):

$$A = LDPU,$$

and denote the (s, r)-entry of L by b. We have

$$\begin{aligned} P_0A = P_0LDPU &= L'B_{rs}(-b)P_0DPU \\ &= L'D'B_{rs}(c)P_0PU, \end{aligned} \tag{9.2.13}$$

where L', D' are again lower unitriangular and diagonal respectively, and D' differs from D by an interchange of the r-th and s-th diagonal elements. If $c = 0$ (which is the case iff $b = 0$) or if the permutation corresponding to P_0P preserves the order of r, s, then the matrix on the right takes the form $L'D'P_0PB_{rs}(c)U$; this is again in Bruhat normal form and so we have $\delta(P_0A) = \delta(A)$ in this case.

Suppose now that $c \neq 0$ and that P_0P inverts the order r, s. Then the formula

$$\begin{pmatrix} 1 & c \\ 0 & 1 \end{pmatrix}\begin{pmatrix} 0 & -1 \\ 1 & 0 \end{pmatrix} = \begin{pmatrix} 1 & 0 \\ c^{-1} & 1 \end{pmatrix}\begin{pmatrix} c & 0 \\ 0 & c^{-1} \end{pmatrix}\begin{pmatrix} 1 & -c^{-1} \\ 0 & 1 \end{pmatrix}$$

shows that, on writing $D_i(c)$ for the matrix differing from the unit matrix only in the (i, i)-entry, which is c, we have

$$B_{rs}(c)P_0 = B_{sr}(c^{-1})D_r(c)D_s(c^{-1})B_{rs}(-c^{-1}).$$

Inserting this in the expression (9.2.13) for P_0A, we find

$$\delta(P_0A) = \bar{c}.\bar{c}^{-1}\delta(A) = \delta(A),$$

and this proves the assertion in all cases. The conclusion follows by induction. ∎

We can now establish the main property of the determinant:

Theorem 9.2.6. *For any skew field K, the determinant function δ is a homomorphism giving rise to an exact sequence*

$$1 \to \mathbf{E}(K) \to \mathbf{GL}(K) \xrightarrow{\delta} K^{ab} \to 1.$$

In particular, the determinant is unchanged by elementary row or column operations.

Proof. Let $A \in GL(K)$; we first show that $\delta(BA) = \delta(A)$ for any $B \in E(K)$, and by induction it is enough to prove this for $B = B_{ij}(a)$. If $i > j$, this matrix is lower unitriangular, so BA and A have the same core and hence the same determinant. If $i < j$, let P_0 be the signed permutation matrix corresponding to the transposition (i, j). Using Lemma 9.2.5 and the case just proved, we have

$$\delta(B_{ij}(a)A) = \delta(P_0 B_{ij}(a)A) = \delta(B_{ji}(a)P_0 A) = \delta(P_0 A) = \delta(A),$$

hence the result holds generally. The same argument applies for multiplying by an elementary matrix on the right. Now Lemma 9.2.1 shows that

$$\delta(AB) = \delta(BA) \quad \text{for any } A, B \in \mathbf{GL}(K). \tag{9.2.14}$$

Here it is of course necessary to consider δ as being defined on the stable linear group.

Now take any two matrices A, B with Bruhat decompositions $A = L_1 M_1 U_1$, $B = L_2 M_2 U_2$. We have, by what has been proved and (9.2.14),

$$\delta(AB) = \delta(L_1 M_1 U_1 L_2 M_2 U_2) = \delta(M_1 U_1 L_2 M_2) = \delta(M_2 M_1 U_1 L_2) = \delta(M_2 M_1).$$

Further it is clear from the definition of δ that $\delta(M_2 M_1) = \delta(M_1 M_2) = \delta(M_1)\delta(M_2)$, so we obtain

$$\delta(AB) = \delta(A)\delta(B). \tag{9.2.15}$$

This shows δ to be a homomorphism. Clearly its image is K^{ab}; its kernel includes $\mathbf{GL}(K)' = \mathbf{E}(K)$ because K^{ab} is abelian. Conversely, if $\delta(A) = 1$, then the core of A has the form DP, where P is a signed permutation matrix and so $1 = \delta(A) = \delta(D)$. By (9.2.7) we can apply elementary operations to reduce D to the form $D_1(c)$, where c is the product of the diagonal elements of D. But by hypothesis $\bar{c} = 1$, hence D has been reduced to 1 and so $A \in \mathbf{E}_n(K)$, as we wished to show. ■

Recently another more general form, the quasideterminant, has been defined by Izrail Gelfand and Vladimir Retakh [1997], which is essentially a rational expression defined recursively in terms of the $n-1 \times n-1$ submatrices.

Exercises

1. Show that the transpose of every invertible matrix over a field K is invertible iff K is commutative. (Hint. Try a 2×2 matrix with $(1, 1)$-entry 1.)
2. Use Theorem 9.2.2 to show that $\mathbf{GL}_n(R) = \mathbf{D}_n(R)\mathbf{E}_n(R)$ for any local ring R.
3. Show that $\mathbf{GL}_2(\mathbf{F}_2) = \mathbf{E}_2(\mathbf{F}_2) = \mathrm{Sym}_3$.
4. Let K be a skew field. Show that $A \in \mathbf{GL}_n(K)$ can be written as $A = LDPU$, where L is lower, U is upper unitriangular, D is diagonal, P is a permutation matrix and PUP^{-1} is also upper triangular. Moreover, such a representation is unique (Draxl (1983); this is known as the *strict* Bruhat normal form).
5. Show that a homomorphism $\mathbf{GL}_n(K) \to \mathrm{Sym}_n$ can be defined by associating with $A \in \mathbf{GL}_n(K)$ the permutation matrix P from the representation $A = LDPU$.
6. Show that if P is the permutation matrix obtained by applying a permutation σ to the rows of I, then it can also be obtained by applying σ^{-1} to the columns of I.

7. Let K be a skew field with centre C. Show that δ restricted to C reduces to the usual determinant, provided that no element of C other than I is a product of commutators.

9.3 Free fields

We have seen that free rings and free algebras may be defined by a universal property; this is not to be expected for fields, since fields do not form a variety (see Theorem 1.3.7). Nevertheless, in the commutative case the rational function field $k(x_1, \ldots, x_d)$ may be regarded as a 'free' field in the sense that all other fields generated by d elements over k can be obtained from it by specialization. Moreover, it is the field of fractions of the polynomial ring $k[x_1, \ldots, x_d]$ and as such it is uniquely determined. By contrast, the free algebra $k\langle x_1, \ldots, x_d\rangle$ has more than one field of fractions (see Exercise 2), but this leaves the question whether in the general case there exists a field that is universal for specializations. A full study of these questions is beyond the scope of this book (see Cohn (1985), Chapter 7), but it is possible to prove the existence of free fields in a relatively straightforward way, and this will now be done.

For a general theory of fields it is convenient to invert matrices rather than elements. We shall not enter into details, but we have to consider which matrices can become invertible under a homomorphism to a field. Clearly we can confine ourselves to square matrices. If A is an $n \times n$ matrix over a ring R, and A can be written in the form

$$A = PQ, \quad \text{where } P \in {}^nR^r,\ Q \in {}^rR^n, \tag{9.3.1}$$

then it is clear that under any homomorphism from R to a field we again have a factorization as in (9.3.1), hence the image of A cannot have a rank greater than r, and so cannot be invertible when $r < n$. The least possible value of r in a factorization of A as in (9.3.1) is called the *inner rank* of A over R and is denoted by ρA. It is easily verified that over a field (even skew) the inner rank reduces to the usual rank. Thus an $n \times n$ matrix over any ring cannot become invertible under a homomorphism to a field, unless its inner rank is n.

A square matrix over any ring R is said to be *full* if its inner rank equals the number of rows. The above remarks show that in studying matrices that can be inverted under a homomorphism to a field, we can confine our attention to full matrices. Our aim in this section is to show that for the tensor ring F there exists a field U containing F and generated by it as field, such that any full matrix over F can be inverted over U. This field U is a *universal field of fractions* of F, in the sense that there is a specialization from U to any other field which is obtained by a homomorphism from F (Cohn (1985), Chapter 7).

We begin with some remarks on the inner rank. We recall from BA, Section 4.6, that a ring R is called *weakly finite* if for any square matrices of the same size $A, B, AB = I$ implies $BA = I$. If R is a weakly finite ring which is non-trivial, then the unit matrix in R must be full, i.e. if A is $r \times s$ and B is $s \times r$ and $AB = I$, then

$r \leq s$. For if $r > s$, we can adjoin zero columns to A and zero rows to B to obtain square matrices. Now we have

$$(A \quad 0)\begin{pmatrix} B \\ 0 \end{pmatrix} = I, \quad \text{hence} \quad \begin{pmatrix} B \\ 0 \end{pmatrix}(A \quad 0) = I.$$

Comparing (r, r)-entries, we obtain $0 = 1$, which contradicts the fact that R is non-trivial.

Lemma 9.3.1. *Let R be a non-trivial weakly finite ring and consider a partitioned matrix over R:*

$$A = \begin{pmatrix} A_1 & A_2 \\ A_3 & A_4 \end{pmatrix}, \quad \text{where } A_1 \text{ is } r \times r.$$

If A_1 is invertible, then $\rho A \geq r$, with equality if and only if $A_3 A_1^{-1} A_2 = A_4$.

Proof. Clearly the inner rank is unchanged on passing to an associated matrix. Hence we can make the transformation

$$A \to \begin{pmatrix} I & A_1^{-1}A_2 \\ A_3 & A_4 \end{pmatrix} \to \begin{pmatrix} I & A_1^{-1}A_2 \\ 0 & A_4 - A_3A_1^{-1}A_2 \end{pmatrix} \to \begin{pmatrix} I & 0 \\ 0 & A_4 - A_3A_1^{-1}A_2 \end{pmatrix},$$

and these transformations leave the inner rank unchanged. If $\rho A = s$, this matrix can be written as

$$PQ = \begin{pmatrix} P_1 \\ P_2 \end{pmatrix}(Q_1 \quad Q_2),$$

where P_1 is $r \times s$ and Q_1 is $s \times r$. Thus $I = P_1 Q_1$, hence $r \leq s$ by weak finiteness, and this shows that $\rho A \geq r$. When equality holds, we have $Q_1 P_1 = I$, but $P_1 Q_2 = 0$, so $Q_2 = 0$. Similarly $P_2 = 0$ and hence $A_4 - A_3 A_1^{-1} A_2 = P_2 Q_2 = 0$. The converse is clear. ■

The existence proof for free fields is based on a lemma of independent interest, the specialization lemma. This may be regarded as an analogue of the GPI-theorem (Theorem 7.8.3), which is used in the proof; its commutative counterpart is the elementary result that a polynomial vanishing for all values in an infinite field must be the zero polynomial. In the proof we shall need the result that any full matrix over the tensor ring $D\langle X\rangle$ remains full over the formal power series ring $D\langle\langle X\rangle\rangle$ (Lemma 5.9.4 of Cohn (1985) or Proposition 6.2.2 of Cohn (1995)).

Lemma 9.3.2 (Specialization lemma). *Let D be a skew field with infinite centre C and such that $[D : C]$ is infinite. Then any full matrix over the tensor ring $D_C\langle X\rangle$ is invertible for some choice of X in D.*

Proof. Let $A = A(x)$ be any full $n \times n$ matrix over $D_C\langle X\rangle$ and denote by r the supremum of its ranks as its arguments range over D. We have to show that

$r = n$, so let us assume that $r < n$. By a translation $x \mapsto x + a$ ($x \in X$, $a \in D$) we may assume that the maximum rank is assumed at the point $x = 0$, and by an elementary transformation we may take the principal $r \times r$ minor to be invertible. Thus if

$$A(x) = \begin{pmatrix} A_1(x) & A_2(x) \\ A_3(x) & A_4(x) \end{pmatrix},$$

where A_1 is $r \times r$, then $A_1(0)$ is invertible. Given $a \in D^X$ and any $\tau \in C$, we have $\rho A(\tau a) \leq r$, hence by Lemma 7.2.4, the rank of $A(ta)$ over $D(t)$ is at most r, and the same holds over $D((t))$, the field of formal Laurent series in t as central indeterminate. Now $A_1(ta)$ is a polynomial in t with matrix coefficients and constant term $A_1(0)$, a unit, hence $A_1(ta)$ is invertible over the power series ring $D[[t]]$. By Lemma 9.3.1, the equation

$$A_4(ta) = A_3(ta)A_1(ta)^{-1}A_2(ta) \tag{9.3.2}$$

holds over $D[[t]]$, for all $a \in D^X$. This means that the matrix

$$A_4(tx) - A_3(tx)A_1(tx)^{-1}A_2(tx) \tag{9.3.3}$$

vanishes when the elements of X are replaced by any values in D. Now (9.3.3) is a power series in t with coefficients that are matrices over $D_C\langle X\rangle$. Thus the coefficients are generalized polynomial identities (or identically 0), so by Amitsur's GPI-theorem (Theorem 7.8.4), the expression (9.3.3) vanishes as a matrix over $D_C\langle X\rangle[[t]]$. It follows that for $r < n$, $A(tx)$ is non-full over $D_C\langle X\rangle[[t]]$. Hence we can write $A(tx)$ as a product PQ, where P is $n \times r$ and Q is $r \times n$ and P, Q have entries from $D_C\langle X\rangle[[t]]$; putting $t = 1$, we obtain a corresponding factorization

$$A(x) = PQ \tag{9.3.4}$$

over $D_C\langle\langle X\rangle\rangle$ which by the result quoted can be taken over $D_C\langle X\rangle$. Thus $A(x)$ is non-full over $D_C\langle X\rangle$, a contradiction, which proves the result. ■

In this lemma the condition $[D : C] = \infty$ is clearly necessary; whether the condition that C be infinite is needed is not known.

We can now prove the existence of free fields:

Theorem 9.3.3. *Let D be a skew field with centre C and X any set. Then $D_C\langle X\rangle$ can be embedded in a field U, generated by $D_C\langle X\rangle$, such that every full matrix over $D_C\langle X\rangle$ becomes invertible over U.*

Proof. Suppose first that $[D : C] = \infty$, $|C| = \infty$, and consider the mapping

$$D_C\langle X\rangle \to D^{D^X},$$

where $p \in D_C\langle X\rangle$ is mapped to (p_f), with $p_f = p(xf)$, for any $f \in D^X$. With each square matrix A over $D_C\langle X\rangle$ we associate a subset $\mathcal{D}(A)$ of D^X defined by

$$\mathcal{D}(A) = \{f \in D^X | A(xf) \text{ is invertible}\}.$$

$\mathscr{D}(A)$ is called the *singularity support* of A. Of course $\mathscr{D}(A) = \emptyset$ unless A is full, but by Lemma 9.3.2, $\mathscr{D}(A) \neq \emptyset$ whenever A is full. If P, Q are any invertible matrices, then $P \oplus Q$ is invertible, hence $A(x) \oplus B(x)$ becomes singular precisely when $A(x)$ or $B(x)$ becomes singular, thus

$$\mathscr{D}(A) \cap \mathscr{D}(B) = \mathscr{D}(A \oplus B).$$

It follows that the family of sets $\mathscr{D}(A)$, where A is full, is closed under finite intersections. Hence it is contained in an ultrafilter $\mathscr{F}$ on D^X (see Section 1.5), and we have a homomorphism to an ultrapower

$$D_C\langle X\rangle \to D^{D^X}/\mathscr{F}, \tag{9.3.5}$$

where by definition, every full matrix A over $D_C\langle X\rangle$ is invertible on $\mathscr{D}(A)$ and so is invertible in the ultrapower. Hence the subfield of the ultrapower generated by $D_C\langle X\rangle$ is the required field U.

In the general case we take indeterminates r, s, t and define $D_1 = D(r)$, $D_2 = D_1(s)$. On D_2 we have an automorphism $\alpha : f(s) \mapsto f(rs)$, with fixed field D_1. We now form $E = D_2(t; \alpha)$; the centre of E is the centre of D_1, namely $C(r)$ (see Theorem 7.3.6). This is infinite and E has infinite dimension over $C(r)$, because the powers s^n are linearly independent. It is clear that we have an embedding $D_C\langle X\rangle \to E_{C(r)}\langle X\rangle$, and it follows from the inertia lemma (Lemma 8.7.3) that this embedding is *honest*, i.e. full matrices are mapped to full matrices. Hence on taking the field U constructed earlier for $E_{C(r)}\langle X\rangle$, we obtain a field over which every full matrix over $D\langle X\rangle$ becomes invertible. ∎

The field U whose existence has been established in Theorem 9.3.3 is denoted by $D_C(\langle X\rangle)$ and is called the *universal field of fractions* of $D_C\langle X\rangle$ or also the *free field* over D with centre C (its centre can be shown to be C). The existence proof for free fields goes back to Shimshon Amitsur [1966], who used his results on generalized rational identities. The existence of such a universal field of fractions can be proved more generally for any tensor ring $K_L\langle X\rangle$, where K is any skew field and L any subfield of K. This is a special case of the fact that every semifir has a universal field of fractions over which every full matrix can be inverted (see Cohn (1985), Chapter 7).

We remark that any automorphism of $D_C\langle X\rangle$ is honest and therefore extends to an automorphism of U. Further, by representing derivations as homomorphisms from $D_C\langle X\rangle$ to U_2 we see that for any automorphisms α, β of U, any (α, β)-derivation of $D_C\langle X\rangle$ extends to one of U.

Exercises

1. Show that the $n \times n$ unit matrix over a ring R is full iff R^n cannot be generated by less than n elements. Show also that a non-trivial weakly finite ring has IBN.
2. Let $E = k(t)$ be the field of rational functions in one variable t, with the endomorphism $\alpha_r : f(t) \mapsto f(t^r)$ $(r > 1)$. Show that the subalgebra of $E[x; \alpha_r]$ generated by x and $y = xt$ is free on x, y. Using Exercise 2 of Section 7.3,

obtain for each $r > 1$ a field of fractions L_r of the free algebra F. Show that these fields are non-isomorphic as F-rings (J. L. Fisher).

3. Show that an endomorphism θ of $D_C\langle X\rangle$ can be extended to an endomorphism of the free field iff θ is honest.
4. Show that every honest endomorphism is injective; give an example of an endomorphism of the free algebra $k\langle X\rangle$ which is injective but not honest.
5. Verify that over a skew field the inner rank agrees with the rank.
6. Let K be a skew field with infinite centre. Show that for any square matrix A over K there is an element α in K such that $A - \alpha I$ is non-singular. For a finite field F find a matrix A such that $A - xI$ is singular for all values of x in F (for infinite fields with finite centre the question remains open).
7. Show that over a PID, a square matrix is regular iff it is full. Give an example of a square matrix over a free algebra which is regular but not full. (Hint. Try a 3×3 matrix with a 2×2 block of zeros.)

9.4 Valuations on skew fields

Valuations may be defined on skew fields as in the commutative case, but there have so far been fewer applications. This is no doubt due to the inherent difficulties in handling general valuations; however in special cases they become more tractable and offer the prospect of a means of gaining information on skew fields. Here we present a part of the general theory that runs parallel to the commutative case, together with some illustrations.

Let K be a skew field. A subring V of K is said to be *total* if for every $a \in K^\times$, either a or a^{-1} lies in V. If for every $a \in K^\times$, $a^{-1}Va = V$, then V is called *invariant*. Now a *valuation ring* of K is an invariant total subring of K. In any valuation ring V in K the set $\mathfrak{m}$ of all non-units is easily seen to be an ideal, hence V is a local ring with $\mathfrak{m}$ as maximal ideal. The set U of all units in V is a normal subgroup of $K^\times$; we shall denote the quotient $K^\times/U$ by Γ and call it the *value group* of V, with natural homomorphism $v : K^\times \to \Gamma$. We shall use additive notation for Γ; our main concern will be the case when Γ is abelian. Given $a, b \in K^\times$, we shall write $v(a) \geq v(b)$ iff $ab^{-1} \in V$, or equivalently (because V is invariant), $b^{-1}a \in V$. This relation is a total ordering on Γ, for if $v(a) \geq v(b)$, $v(b) \geq v(c)$, then ab^{-1}, $bc^{-1} \in V$, hence $ac^{-1} \in V$ and so $v(a) \geq v(c)$. Clearly $v(a) \geq v(a)$ and if $v(a) \geq v(b)$ and $v(b) \geq v(a)$, then ab^{-1}, $ba^{-1} \in V$ hence $ab^{-1} \in U$ and so $v(ab^{-1}) = 0$, hence $v(a) = v(b)$. Under this ordering Γ becomes an ordered group, for if $v(a) \geq v(b)$, then $ab^{-1} \in V$, hence for any $c \in K$, $ac(bc)^{-1} = ab^{-1} \in V, ca(cb)^{-1} = c^{-1}ab^{-1}c^{-1} \in V$, hence $v(ac) \geq v(bc)$, $v(ca) \geq v(cb)$. This is a total ordering, because V is a total subring. Moreover, since v is a homomorphism, we have $v(ab) = v(a) + v(b)$, and the fact that $ab^{-1} \in V \Rightarrow ab^{-1} + 1 \in V$ implies that $v(a) \geq v(b) \Rightarrow v(a + b) \geq v(b)$. Thus v obeys the following rules:

V.1 $v(x) \in \Gamma$ *for* $x \in K$,
V.2 $v(x + y) \geq \min\{v(x), v(y)\}$,
V.3 $v(xy) = v(x) + v(y)$.

Here we had to exclude the values $x, y = 0$, but it is more convenient to allow $x = 0$ and define $v(0) = \infty$. Then **V**.1–**V**.3 continue to hold if we define (as usual) $\infty + \alpha = \alpha + \infty = \infty + \infty = \infty$, $\alpha < \infty$ for all $\alpha \in \Gamma$. As in the commutative case we have equality in **V**.2 whenever $v(x) \neq v(y)$ ('all triangles are isosceles').

A function v from K to an ordered group Γ (with $v(0) = \infty$), satisfying **V**.1–**V**.3 is called a *valuation* on K. Given such a valuation, we can define

$$V = \{x \in K \mid v(x) \geq 0\},$$

and it is easily verified that V is a valuation ring on K. In this way valuation rings on K and valuations correspond to each other; to make the correspondence bijective we define two valuations v, v' on K with value groups Γ, Γ' to be *equivalent* if there is an order-preserving isomorphism $\varphi : \Gamma \to \Gamma'$ such that

$$v(x)\varphi = v'(x) \quad \text{for all } x \in K^{\times}.$$

With this definition we have

Theorem 9.4.1. *On any field K there is a natural bijection between valuation rings and equivalence classes of valuations on K.*

Proof. This is an easy consequence of the above remarks and may be left to the reader to prove. ∎

Valuations on skew fields were introduced by Otto F. G. Schilling in 1945. In the commutative case there is a third notion, equivalent to the above two, namely that of a *place*; this can also be defined for skew fields, but will not be needed here (see Exercise 2).

We note that K itself is a valuation ring in K; it corresponds to the *trivial* valuation, defined by $v(x) = 0$ for all $x \neq 0$, with trivial value group. Of course we shall mainly be interested in non-trivial valuations.

The simplest (non-trivial) type of ordered group is the infinite cyclic group. It can be shown that a valuation has the infinite cyclic group as value group precisely when its valuation ring is a principal ideal domain; such a valuation is called *principal*. For example, the usual p-adic valuation on $\mathbf{Q}$ is principal, and in Chapter 9 of BA we saw that every valuation on a rational function field $k(t)$ which is trivial on k is principal.

Let K be any field with a valuation v and write V for its valuation ring, $\mathfrak{m}$ for its maximal ideal and U for the group of units in V. It is clear from **V**.2 that every element of the form $1 + x$, where $x \in \mathfrak{m}$, is a unit; such a unit is called a *1-unit* (Einseinheit). Thus u is a 1-unit whenever $v(u - 1) > 0$. The group of all 1-units is written $1 + \mathfrak{m}$ or U_1. Let us denote $V/\mathfrak{m}$, the residue class field of V, by k. Then we have a group isomorphism

$$k^{\times} \cong U/U_1, \tag{9.4.1}$$

while the value group of v is given by

$$\Gamma \cong K^{\times}/U. \tag{9.4.2}$$

These isomorphisms may be combined into the following commutative diagram with exact rows and columns:

$$\begin{array}{ccccccc} & & 1 & & 1 & & \\ & & \downarrow & & \downarrow & & \\ 1 & \to & U_1 & \to & U_1 & \to & 1 \\ & & \downarrow & & \downarrow & & \downarrow \\ 1 & \to & U & \to & K^\times & \to & \Gamma \to 1 \\ & & \downarrow & & \downarrow & & \downarrow \\ 1 & \to & k^\times & \to & K^\times/U_1 & \to & \Gamma \to 1 \\ & & \downarrow & & \downarrow & & \downarrow \\ & & 1 & & 1 & & 1 \end{array} \tag{9.4.3}$$

When V is principal, Γ is infinite cyclic, so then the horizontal sequences split and

$$K^\times/U_1 \cong k^\times \times \Gamma. \tag{9.4.4}$$

For this reason the rows of (9.4.3) add nothing to our knowledge in the cases usually encountered, but in general, especially with a non-abelian value group, (9.4.3) provides more information about $K^\times$.

In constructing valuations it is helpful to know that any valuation on an Ore domain has a unique extension to its field of fractions.

Proposition 9.4.2. *Let R be a right Ore domain with field of fractions K. If v is a valuation on R satisfying* **V**.1–**V**.3, *then v has a unique extension to K.*

Proof. If v is to have an extension to K, then for $p = as^{-1} \in K$ we must have

$$v(p) = v(a) - v(s). \tag{9.4.5}$$

Suppose that $as^{-1} = a_1 s_1^{-1}$. Then there exist $u, u_1 \in R$ such that $su_1 = s_1 u \neq 0$, $au_1 = a_1 u$; hence $a = 0 \Leftrightarrow a_1 = 0$, and when $a, a_1 \neq 0$, then $-v(a_1) + v(a) = v(u) - v(u_1) = -v(s_1) + v(s)$ and so $v(a) - v(s) = v(a_1) - v(s_1)$. This shows the definition (9.4.5) to be independent of the particular representation as^{-1} of p. Now it is easily verified that v so defined satisfies **V**.1–**V**.3 on K. ■

Examples of valuations

1. Let K be any field with a valuation v. We can extend v to the rational function field $K(x)$ by defining v on any polynomial $f = \sum x^i a_i$ $(a_i \in K)$ by the rule

$$v(f) = \min\{v(a_i)\}, \tag{9.4.6}$$

and using Proposition 9.4.2 to extend v to $K(x)$. This is called the *Gaussian extension* of v; the value group is unchanged while the residue class field undergoes a simple transcendental extension. The same construction works if instead of $K(x)$ we use $K(x; \alpha)$, the skew function field with respect to an automorphism α of K, provided that $v(a^\alpha) = v(a)$ for all $a \in K$.

If we enlarge the value group Γ by an element δ having no non-zero multiple in Γ (e.g. by forming the direct product of Γ and the infinite cyclic group on Δ, with the lexicographic ordering), and instead of (9.4.6) define

$$v(f) = \min\{v(a_i) + i\delta\}, \tag{9.4.7}$$

we obtain an extension, called the *x-adic extension* (provided that $\delta > 0$), with the same residue class field and enlarged group.

2. Consider the free field $k(\langle x, y\rangle)$; let E be the subfield generated over k by $y_i = x^{-i}yx^i$ $(i \in \mathbf{Z})$. The conjugation by x defines an automorphism α which maps E into itself, the 'shift automorphism' $y_i \mapsto y_{i+1}$. Moreover, $k(\langle x, y\rangle)$ may be obtained as the skew function field $E(x; \alpha)$. Taking the x-adic extension of the trivial valuation on E, we obtain a principal valuation on $k(\langle x, y\rangle)$ with residue class field E. We note that whereas the general aim of valuation theory is to obtain a simpler residue class field, E is actually more complicated than the original field. As we shall see, in order to simplify the residue class field we must allow a more complicated value group.
3. In a skew field it may happen that an element is conjugate to its inverse: $y^{-1}xy = x^{-1}$. For example, let α be the automorphism of the rational function field $F = k(x)$ defined by $f(x) \mapsto f(x^{-1})$ and put $E = F(y; \alpha)$. If v is any valuation on E, then $v(x) = 0$, for if $v(x) \neq 0$, say $v(x) > 0$, then $v(x^{-1}) < 0$, but $x^{-1} = y^{-1}xy$, hence $v(x^{-1}) = -v(y) + v(x) + v(y) > 0$, a contradiction. In fact fields exist in which every element outside the centre is conjugate to its inverse (e.g. the existentially closed fields, see Cohn (1995)). It is clear that such a field can have no non-trivial valuation.

One of the main tools of the commutative theory is Chevalley's extension theorem (see BA, Section 9.5), which allows one to construct extensions for any valuations defined on a subfield. Such a result is not to be expected in general, but an analogue exists when the value group is abelian, and it is no harder to prove.

A valuation on a field K is said to be *abelian* if its value group is abelian. For any field K we denote the derived group of $K^\times$ by K^c. It is clear that any abelian valuation on K is trivial on K^c. As an almost immediate consequence we have

Lemma 9.4.3. *Let K be a skew field with a valuation v and valuation ring V. Then v is abelian if and only if $V \supseteq K^c$ or equivalently, $v(a) = 0$ for all $a \in K^c$. Moreover, any subring A of K such that $A \supseteq K^c$ is invariant, and any ideal in A is invariant.*

Proof. The second sentence follows from the fact that v is abelian iff the unit group of V contains K^c. To prove the last sentence, take any $a \in A^\times$, $b \in K^\times$; then $b^{-1}ab = a.a^{-1}b^{-1}ab \in A$, and similarly for any ideal of A. ∎

To state the analogue of Chevalley's theorem (Lemma 9.4.4) we require the notion of domination. On any field K we consider the pairs $(R, \mathfrak{a})$ consisting of a subring R of K and a proper ideal $\mathfrak{a}$ of R. Given two such pairs $P_i = (R_i, \mathfrak{a}_i)$ $(i = 1, 2)$, we say that P_1 *dominates* P_2, in symbols $P_1 \geq P_2$ if $R_1 \supseteq R_2$ and $\mathfrak{a}_1 \supseteq \mathfrak{a}_2$ and write $P_1 > P_2$ (as usual) for proper domination, i.e. to exclude equality. If the pair $(R, \mathfrak{a})$ is such that $R \supseteq K^c$, then every element of K^c is a unit in R (because K^c is a group), and so $K^c \cap \mathfrak{a} = \emptyset$. The essential step in our construction is the following

Lemma 9.4.4. *Let K be a skew field, R a subring containing K^c and $\mathfrak{a}$ a proper ideal in R. Then there is a subring V with a proper ideal $\mathfrak{m}$ such that $(V, \mathfrak{m})$ is maximal among pairs dominating $(R, \mathfrak{a})$, and any such maximal pair $(V, \mathfrak{m})$ consists of a valuation ring and its maximal ideal.*

Proof. This is quite similar to the commutative case (BA, Lemma 9.4.3); we briefly recall it to show where changes are needed.

The pairs dominating $(R, \mathfrak{a})$ form an inductive family, so a maximal pair exists by Zorn's lemma. If $(V, \mathfrak{m})$ is a maximal pair, then $\mathfrak{m}$ is a maximal ideal in V, and since $V \supseteq K^c$, V and $\mathfrak{m}$ are invariant. To show that V is a total subring in K, take $c \in K$; if $c \notin V$, then $V[c] \supset V$, so if the ideal $\mathfrak{m}'$ generated by $\mathfrak{m}$ in $V[c]$ is proper, we have $(V[c], \mathfrak{m}') > (V, \mathfrak{m})$, contradicting the maximality of the latter. Hence $\mathfrak{m}' = V[c]$ and we have an equation

$$a_0 + a_1 c + \ldots + a_m c^m = 1, \quad a_i \in \mathfrak{m}. \tag{9.4.8}$$

Here we were able to collect powers of c on the right of each term because of the invariance of $\mathfrak{m}$, using the equation $cb = cbc^{-1}.c$.

Similarly if $c^{-1} \notin V$, we have

$$b_0 + b_1 c^{-1} + \ldots + b_n c^{-n} = 1, \quad b_j \in \mathfrak{m}. \tag{9.4.9}$$

We assume that m, n are chosen as small as possible and suppose that $m \geq n$, say. Multiplying (9.4.9) on the right by c^m, we obtain

$$(1 - b_0)c^m = b_1 c^{m-1} + \ldots + b_n c^{m-n}. \tag{9.4.10}$$

By the invariance of V, $xc = c.x\gamma$ for all $x \in V$, where $\gamma = \gamma(c)$ is an automorphism of V which maps $\mathfrak{m}$ into itself. If we multiply (9.4.8) by $1 - b_0$ on the left and (9.4.10) by $a_m \gamma^m$ on the right and substitute into (9.4.8), we obtain an equation of the same form as (9.4.8) but of degree less than m, a contradiction. This proves V to be total, and hence a valuation ring; from the maximality it is clear that $\mathfrak{m}$ is the maximal ideal on V. ■

We now have the following form of the extension theorem:

Theorem 9.4.5. *Let $K \subseteq L$ be an extension of skew fields. Given an abelian valuation v on K, there is an extension of v to L if and only if there is no equation*

$$\sum a_i c_i = 1, \quad \text{where } a_i \in K, v(a_i) > 0 \text{ and } c_i \in L^c. \tag{9.4.11}$$

Proof. If there is an equation (9.4.11), then any abelian extension w of v to L must satisfy $w(a_i c_i) = w(a_i) = v(a_i) > 0$, hence $w(1) \geq \min\{w(a_i c_i)\} > 0$, a contradiction. Conversely, if no equation (9.4.11) holds, this means that if V is the valuation ring of

v, with maximal ideal $\mathfrak{m}$, then $\mathfrak{m}L^c$ is a proper ideal in VL^c, and by Lemma 9.4.4 there is a maximal pair $(W, \mathfrak{n})$ dominating $(VL^c, \mathfrak{m}L^c)$. Now W is a valuation ring satisfying $W \cap K \supseteq V$, $\mathfrak{n} \cap K \supseteq \mathfrak{m}$, hence $W \cap K = V$ and so W defines the desired extension. ■

To make valuations more tractable we shall require an abelian value group and commutative residue class field. It is convenient to impose an even stronger condition, as in the next result:

Proposition 9.4.6. *Let K be a skew field with a valuation v, having valuation ring V, maximal ideal $\mathfrak{m}$ and group of 1-units U_1. Then the following conditions are equivalent:*

(a) *$K^\times/U_1$ is abelian,*
(b) *$K^c \subseteq 1 + \mathfrak{m} = U_1$,*
(c) *$v(1 - a) > 0$ for all $a \in K^c$.*

Moreover, when (a)–(c) hold, then the value group and residue class field are commutative.

Proof. This is an almost immediate consequence of the definitions, and the last sentence is clear from a glance at the diagram (9.4.3). ■

A valuation satisfying the conditions of Proposition 9.4.6 will be called *quasi-commutative*. The following condition for extensions of quasi-commutative valuations is an easy consequence:

Theorem 9.4.7. *Let K be a skew field with a quasi-commutative valuation v, and let L be an extension field of K. Then v can be extended to a quasi-commutative valuation of L if and only if there is no equation in L of the form*

$$\sum a_i p_i + \sum b_j(q_j - 1) = 1, \tag{9.4.12}$$

where $a_i, b_j \in K$, $v(a_i) > 0$, $v(b_j) \geq 0$, $p_i, q_j \in L^c$.

Proof. If there is a quasi-commutative extension w of v to L, then for a_i, b_j, p_i, q_j as above we have

$$w\Big(\sum a_i p_i + \sum b_j(q_j - 1)\Big) \geq \min\{v(a_i) + w(p_i), v(b_j) + w(q_j - 1)\} > 0,$$

because $v(a_i) > 0$, $w(q_j - 1) > 0$. It follows that no equation of the form (9.4.12) can hold. Conversely, assume that there is no equation (9.4.12) and consider the set $\mathfrak{q}$ of all expressions $\sum a_i p_i + \sum b_j(q_j - 1)$, where a_i, b_j, p_i, q_j are as before. It is clear that $\mathfrak{q}$ is closed under addition and contains the maximal ideal corresponding to the valuation v. Moreover, $\mathfrak{q}$ is invariant in L, i.e. $u^{-1}\mathfrak{q}u = \mathfrak{q}$ for all $u \in L^\times$, because $u^{-1}a_i p_i u = a_i.a_i^{-1}u^{-1}a_i u.u^{-1}p_i u \in VL^c$, and similarly for the other terms. In the same way we verify that $\mathfrak{q}$ admits multiplication. We now define

$$T = \{c \in L \mid c\mathfrak{q} \subseteq \mathfrak{q}\}. \tag{9.4.13}$$

It is clear that T is a subring of L containing L^c, and it also contains the valuation ring V of v, for if $c \in V$ and we multiply the expression on the left of (9.4.12) by c, we obtain $\sum ca_i p_i + \sum cb_j(q_j - 1)$; this is of the same form, because $v(ca_i) = v(c) + v(a_i) > 0$ and $v(cb_j) \geq 0$. Moreover, $\mathfrak{q}$ is an ideal in T, for we have $c\mathfrak{q} \subseteq \mathfrak{q}$ for all $c \in T$, by the definition of T, and $\mathfrak{q}c = c.c^{-1}\mathfrak{q}c = c\mathfrak{q} \subseteq \mathfrak{q}$. Since $1 \notin \mathfrak{q}$ by hypothesis, $\mathfrak{q}$ is a proper ideal in T. Thus T is a subring of L containing L and the valuation ring of v. By Lemma 9.4.4 we can find a maximal pair $(W, \mathfrak{p})$ dominating $(T, \mathfrak{q})$ and W is a valuation ring such that $W \supseteq T \supseteq L^c$, while $1 + \mathfrak{p} \supseteq 1 + \mathfrak{q} \supseteq L^c$. Hence the valuation w defined by W extends v and is quasi-commutative, by Proposition 9.4.6. ∎

Let D be a skew field with centre C and let X be any set. If D has a quasi-commutative valuation v, one can use Theorem 9.4.7 to extend v to a quasi-commutative valuation of the free field $D(\langle X \rangle)$, but this requires more detail on how free fields are formed. In essence one uses the specialization lemma to show that if there is an equation (9.4.12), then X can be specialized to values in D so as to yield an equation (9.4.12) in D, which is a contradiction (see Cohn [1987], [1989]).

Exercises

1. Show that any total subring of a field is a local ring.
2. A *place* of a field K in another, L, is defined as a mapping $f : K \to L \cup \{\infty\}$ such that $f^{-1}(L) = V$ is an invariant subring of K, the restriction $f|V$ is a homomorphism and $xf = \infty$ implies $x \neq 0$ and $(x^{-1})f = 0$. Show that V is a valuation ring and that conversely, every valuation ring on K leads to a place of K in the residue class field of V. Define a notion of equivalence of places and show that there is a natural bijection between valuation rings on K and equivalence classes of places.
3. Verify that a valuation on a field K has value group $\mathbf{Z}$ iff its valuation ring is a PID.
4. Let $F = D_C\langle X \rangle$ and denote by U the free field $D_C(\langle X \rangle)$. Form $U(t)$ with a central indeterminate t and define a homomorphism $\lambda : U \to U(t)$ as the identity on D and mapping $x \in X$ to xt. Let v_0 be the t-adic valuation on $U(t)$ over U (i.e. trivial on U) and put $v(p) = v_0(p\lambda)$ for $p \in U$. Verify that v is a valuation on U; find its value group and residue class field.
5. Let K be a field with a valuation v whose value group Γ is a subgroup of $\mathbf{R}$. Define an extension of v to the rational function field $K(x)$ by (9.4.7), where $\delta \in \mathbf{R}$, $\delta > 0$, and find the new value group and residue class field. Distinguish the cases $\Gamma \cap \delta\mathbf{Z} = \{0\}$, $\Gamma \cap \delta\mathbf{Z} \neq \{0\}$.
6. Let E be the field of fractions of the Weyl algebra on k (generated by u, v with $uv - vu = 1$), where char $k = 0$. Writing $t = u^{-1}$, verify that $vt = t(v + t)$. Show that the t-adic valuation on E is quasi-commutative.

9.5 Pseudo-linear extensions

We recall that (for any ring K) a *K-ring* A is essentially a ring A with a homomorphism $K \to A$. If K is a field, this means that A contains a copy of K except when $A = 0$. Every K-ring A, for a field K, may be regarded as a left or right K-module; we shall denote the corresponding dimensions by $[A : K]_L$ and $[A : K]_R$ and note that they need not be equal, even when A is itself a field. Below we shall give an example of a field extension in which the right dimension is two and the left dimension is infinite (Artin's problem, see Cohn [1961]). For examples in which both dimensions are finite but different see Schofield [1985] or also Cohn (1995).

We note the product formula for dimensions, familiar from the commutative case. The proof is essentially the same as in that case (BA, Proposition 7.1.2) and so will not be repeated.

Proposition 9.5.1. *Let $D \subseteq E$ be skew fields and V a left E-space. Then*

$$[V : D] = [V : E][E : D]_L,$$

whenever either side is finite. ■

There are a number of cases where the left and right dimensions of a field extension are equal. In the first case equality holds for an extension E/D if E is finite-dimensional over its centre.

Proposition 9.5.2. *Let E/D be a skew field extension and assume that E is finite-dimensional over its centre. Then*

$$[E : D]_L = [E : D]_R, \tag{9.5.1}$$

whenever either side is finite.

Proof. By hypothesis E is finite-dimensional over its centre C. Write $A = DC = \{\sum x_i y_i | x_i \in D, y_i \in C\}$; then A is a subring containing C in its centre, hence a C-algebra and also a D-ring. Since it is generated by C over D, we may choose a basis of A, as left D-space, consisting of elements of C. This is also a basis of A as right D-space, because C is the centre of E, so

$$[A : D]_L = [A : D]_R. \tag{9.5.2}$$

Now A is a subalgebra of the division algebra E, so A is also a skew field and by Proposition 9.5.1,

$$[E : C] = [E : A]_L[A : C] = [E : A]_R[A : C]. \tag{9.5.3}$$

Since $[E : C]$ is finite, so is $[A : C]$; dividing (9.5.3) by $[A : C]$, we find that $[E : A]_L = [E : A]_R$. If we now multiply by (9.5.2) and use Proposition 9.5.1 and its left–right dual, we obtain the required formula (9.5.1). ■

We also have equality when D is commutative:

Proposition 9.5.3. *Let E be a skew field and D a commutative subfield. Then*

$$[E:D]_L = [E:D]_R, \tag{9.5.4}$$

whenever either side is finite.

Proof. Suppose that $[E:D]_L = n$, and denote the centre of E by C. Then $E^0 \otimes_C E$ is simple, by Corollary 5.1.3, so if $M(E)$ denotes the multiplication algebra of E (generated by the left and right multiplications), then the natural mapping $E^0 \otimes_C E \to M(E)$ is injective. Now E is an n-dimensional left D-space, so $[\mathrm{End}({}_D E) : D] = n^2$ and by restriction we obtain an injective mapping $D \otimes_C E \to \mathrm{End}({}_D E)$. It follows that $[E:C] = [D \otimes_C E : D] \leq n^2$, and now (9.5.4) is a consequence of Proposition 9.5.2. ∎

By combining these results we obtain

Theorem 9.5.4. *If E/D is a skew field extension, then (9.5.4) holds whenever either side is finite, provided that either* (i) *D is commutative or* (ii) *E or D is finite-dimensional over its centre.*

Proof. It only remains to treat the case where D is finite-dimensional over its centre. Let Z be this centre, and denote the centre of E by C; further assume that $[E:D]_L$ is finite, so

$$[E:Z]_L = [E:D]_L[D:Z], \tag{9.5.5}$$

and this is also finite. Denote by K the subfield generated by C and Z; clearly K is commutative and $[E:Z]_L = [E:K]_L[K:Z]_L$, so $[E:K]_L$ is finite, hence by Proposition 9.5.3, $[E:K]_L = [E:K]_R$; now $[K:Z]_L = [K:Z]_R$ because K is commutative, and by combining these equalities, we find that $[E:Z]_L = [E:Z]_R$. Now (9.5.4) follows from this equation, combined with (9.5.5) and its right-hand analogue. ∎

The study of finite-dimensional field extensions is greatly complicated by the lack of commutativity. Let us consider the simplest case, of a quadratic extension E/D, $[E:D]_R = 2$. For any $u \in E \backslash D$ the pair $1, u$ is a right D-basis of E. Thus every element of E can be uniquely expressed in the form $ua + b$, where $a, b \in D$. In particular, we have

$$cu = uc^\alpha + c^\delta \quad \text{for all } c \in D, \tag{9.5.6}$$

and

$$u^2 + u\lambda + \mu = 0 \quad \text{for certain } \lambda, \mu \in D. \tag{9.5.7}$$

Here c^α, c^δ are uniquely determined by c and a calculation as in Section 7.3 shows α to be an endomorphism of D and δ an α-derivation. Moreover, the structure of E is completely determined by D and (9.5.6), (9.5.7).

Conversely, if D is any field with an endomorphism α and an α-derivation δ, then for given $\lambda, \mu \in D$ it is possible to write down necessary and sufficient conditions for a quadratic extension of D to be defined by (9.5.6) and (9.5.7) (see Exercise 3).

Generalizing the above discussion, we may define a *pseudo-linear extension* of right dimension n, with generator u, as an extension field E of D with right D-basis $1, u, u^2, \ldots, u^{n-1}$ such that (9.5.6) holds, and in place of (9.5.7),

$$u^n + u^{n-1}\lambda_1 + \ldots + \lambda_n = 0 \quad \text{for certain } \lambda_i \in D. \tag{9.5.8}$$

What has been said shows that every extension of right dimension 2 is pseudo-linear; in higher dimensions the pseudo-linear extensions form a special class. We note the following formula for the left dimension:

Proposition 9.5.5. *Let E/D be a pseudo-linear extension of right dimension n, with commutation formula (9.5.6). Then*

$$[E:D]_L = 1 + [D:D^\alpha]_L + [D:D^\alpha]_L^2 + \ldots + [D:D^\alpha]_L^{n-1}. \tag{9.5.9}$$

In particular, any pseudo-linear extension E/D satisfies

$$[E:D]_L \geq [E:D]_R,$$

with equality if and only if α is an automorphism of D.

Proof. Take a generator u of E/D and write $E_0 = D$, $E_i = uE_{i-1} + D$ $(i \geq 1)$. Then by induction on r we have

$$E_r = D + uD + \ldots + u^r D.$$

Moreover, each E_i is a left D-module, by (9.5.6), so we have a tower of left D-modules

$$D = E_0 \subset E_1 \subset \ldots \subset E_{n-1} = E,$$

and (9.5.9) will follow if we prove

$$[E_r/E_{r-1} : D]_L = [D:D^\alpha]_L^r. \tag{9.5.10}$$

Let $\{e_\lambda | \lambda \in I\}$ be a left D-basis for D. We claim that the elements

$$u^r e_{\lambda_{r-1}}^{\alpha^{r-1}} \ldots e_{\lambda_1}^{\alpha} e_{\lambda_0} \tag{9.5.11}$$

where $\lambda_0, \lambda_1, \ldots, \lambda_{r-1}$ range independently over I, form a basis of $E_r \pmod{E_{r-1}}$. This will prove (9.5.10) and hence (9.5.9).

Any $c \in D$ can be written as a linear combination of the e_λ with coefficients in D^α, say $c = \sum c_{\lambda_0}^\alpha e_{\lambda_0}$. If we repeat the process on c_{λ_0} we obtain $c_{\lambda_0} = \sum c_{\lambda_0\lambda_1}^\alpha e_{\lambda_1}$. Hence

$$c = \sum c_{\lambda_0\lambda_1}^{\alpha^2} e_{\lambda_1}^\alpha e_{\lambda_0},$$

and after r steps we find

$$c = \sum c_{\lambda_0 \ldots \lambda_{r-1}}^{\alpha^r} e_{\lambda_{r-1}}^{\alpha^{r-1}} \ldots e_{\lambda_0}.$$

Therefore

$$u^r c \equiv \sum u^r c^{\alpha^r}_{\lambda_0 \dots \lambda_{r-1}} e^{\alpha^r - 1}_{\lambda_{r-1}} \dots e_{\lambda_0}$$
$$\equiv \sum c_{\lambda_0 \dots \lambda_{r-1}} u^r e^{\alpha^r - 1}_{\lambda_{r-1}} \dots e_{\lambda_0} \pmod{E_{r-1}}.$$

Hence the elements (9.5.11) span $E_r \pmod{E_{r-1}}$. To prove their linear independence, assume that

$$\sum c_{\lambda_0 \dots \lambda_{r-1}} u^r e^{\alpha^r - 1}_{\lambda_{r-1}} \dots e_{\lambda_0} \equiv 0 \pmod{E_{r-1}}$$

By applying the rule (9.5.6) repeatedly, we obtain

$$\sum u^r c^{\alpha^r}_{\lambda_0 \dots \lambda_{r-1}} e^{\alpha^r - 1}_{\lambda_{r-1}} \dots e_{\lambda_0} \equiv 0 \pmod{E_{r-1}}.$$

Since the e_λ are left linearly independent over D^α, we can equate the coefficients of e_{λ_0} to 0 and using induction on r we find that $c_{\lambda_0 \dots \lambda_{r-1}} = 0$. ■

In order to obtain a quadratic extension satisfying (9.5.6) and (9.5.7) let us assume that $\lambda = 0$, $\alpha\delta + \delta\alpha = 0$, and that δ^2 is the inner α^2-derivation induced by $-\mu$. Further assume that $\mu^\alpha = \mu, \mu^\delta = 0$ and that D contains no element a satisfying

$$a.a^\alpha + a^\delta + \mu = 0. \tag{9.5.12}$$

We form the skew polynomial ring $R = D[t; \alpha, \delta]$ and consider $f = t^2 + \mu$. For any $c \in D$ we have

$$\begin{aligned} cf &= c(t^2 + \mu) = (tc^\alpha + c^\delta)t + c\mu \\ &= t^2 c^{\alpha^2} + tc^{\alpha\delta} + tc^{\delta\alpha} + c^{\delta^2} + c\mu \\ &= t^2 c^{\alpha^2} + \mu c^{\alpha^2} - c^{\delta^2} + c^{\delta^2} \\ &= fc^{\alpha^2}. \end{aligned}$$

Hence fR is a two-sided ideal. Moreover, f is irreducible, for if f could be factorized, we would have a product of two linear factors which may both be taken monic, without loss of generality. Then

$$t^2 + \mu = (t - a)(t - b) = t^2 - t(a^\alpha + b) - a^\delta + ab.$$

Therefore $b = -a^\alpha$ and $aa^\alpha + a^\delta + \mu = 0$, but this contradicts the fact that (9.5.12) has no solution in D. It follows that $E = R/fR$ is an integral domain of right dimension 2 over D, hence a field. It has the right D-basis 1, u, where u is the residue class of t, while the left dimension is $1 + [D : D^\alpha]_L$, and this will be greater than 2 provided that α is not an automorphism.

As an example let us take any commutative field k, put $E = k(\langle x, y \rangle)$, the free field on x and y over k, take D to be the subfield generated over k by x, y^2 and $xy - yx^2$. On E we have a k-linear endomorphism $\alpha : x \mapsto x^2, y \mapsto -y$. To show this one has either to verify directly that α as endomorphism of $k\langle x, y \rangle$ is honest, or show that E admits an extension in which x has a square root. For then, by iterating the process one obtains a field P containing E in which x has a 2^n-th root for all $n \geq 0$, so the mapping $x \mapsto x^2$, $y \mapsto -y$ is an automorphism of P and the restriction to E is the required endomorphism. This is fairly plausible, but we shall not give a formal proof here (see Cohn (1995), Section 5.9).

It is easily checked that α maps D into itself; moreover D admits the inner α-derivation $\delta : a \mapsto ay - ya^\alpha$ induced by y. If we can show that $y \notin D$ and that α is not an automorphism of D, we have a quadratic extension E/D with left dimension > 2; in fact we shall find that $[E : D]_L = \infty$.

Consider the $(\alpha, 1)$-derivation γ on E such that $x^\gamma = 0$, $y^\gamma = 1$. We have $(ab)^\gamma = a^\gamma b + a^\alpha b^\gamma$, by definition, hence $x^\gamma = 0$, $(y^2)^\gamma = y + y^\alpha = 0$, $(xy - yx^2)^\gamma = x^2 - x^2 = 0$. Hence γ vanishes on D, but $y^\gamma = 1$, so $y \notin D$.

Finally to show that $[E : D]_L = \infty$, we first note that if $[D : D^\alpha]_L = n$, then $[E : E^\alpha]_L$ is finite. For let $u_1, \dots, u_n$ be a left D^α-basis of D. We claim that u_i, u_iu_j, u_iyu_j span E as left E^α-space. Any $p \in E$ has the form $p = a + yb$, where $a, b \in D$, say $a = \sum a_i^\alpha u_i$, $b = \sum b_i^\alpha u_i$; hence $p = \sum a_i^\alpha u_i + y \sum b_i^\alpha u_i = \sum a_i^\alpha u + \sum b_i y u_i - \sum b_i^\delta u_i = \sum a_i^\alpha u + \sum b_{ij}^\alpha u_i y u_j - \sum c_{ij}^\alpha u_i u_j$, for suitable b_{ij}, $c_{ij} \in D$. This proves our claim; in particular, it shows that $[E : E^\alpha]_L \leq 2n^2 + n$, if $[D : D^\alpha]_L = n$. Now $E^\alpha = k(\langle x^2, y\rangle)$ and so the elements xy^r $(r = 1, 2, \dots)$ are left E^α-linearly independent. This is intuitively plausible and can be proved with the methods of Cohn (1977), Lemma 5.5.5. It follows that $[E : E^\alpha]_L = \infty$ and by Proposition 9.5.5, $[E : D]_L = \infty$.

Exercises

1. Show that every cyclic division algebra may be described as a pseudo-linear extension of a maximal commutative subfield.
2. Use the methods of this section to construct a field extension of right dimension n and infinite left dimension.
3. Show that (9.5.6) and (9.5.7) define an algebra of right dimension 2 over D iff $c^{\alpha\delta} + c^{\delta\alpha} = \lambda c^{\alpha^2} - c^\alpha \lambda$, $c^{\delta^2} + c^\delta \lambda = \mu c^{\alpha^2} - c\mu$, $\lambda^\delta = \mu - \mu^\alpha - \lambda(\lambda - \lambda^\alpha)$, $\mu^\delta = \mu(\lambda^\alpha - \lambda)$, and this extension is a field iff $c.c^\alpha + c\lambda + \mu \neq 0$ for all $c \in D$.

Further exercises on Chapter 9

1. Let D be a field with centre C and k a subfield of C such that C is algebraic over k. Show that if D is finite-dimensional over C and finitely generated as k-algebra, then $[D : k]$ is finite.
2. Show that over a local ring any matrix A can be written as $A = LUP$, where L is lower unitriangular, U is upper triangular and P is a permutation matrix.
3. Let $P = \begin{pmatrix} A & B \\ C & D \end{pmatrix}$ be a square matrix written in block form, where A is invertible.

   Show that $\delta(P) = \delta(A).\delta(D - CA^{-1}B)$.
4. Show that the core of a matrix A (over a field) has diagonal form iff each principal minor of A is invertible, and when this is so, then the factors L and U in (9.2.8) (as well as M) are unique.

5. Show that in a skew field of characteristic not 2, $[(x+y-2)^{-1} - (x+y+2)^{-1}] - [(x-y-2)^{-1} - (x-y+2)^{-1}] = 1/2(xy+yx)$.
6. Let A be a full matrix over $k\langle X\rangle$. Show that there exists $n = n(A)$ such that for every central division k-algebra D of degree at least n, A is non-singular for some set of values of X in A.
7. Let K be a field with a valuation v. Given any matrix A over K, show that $A = PLDUQ$, where P, Q are signed permutation matrices, L is lower and U upper unitriangular and $D = \text{diag}(d_1, \ldots, d_r)$ is a diagonal matrix with $v(d_1) \leq v(d_2) \leq \ldots$. Show that if A is square and v is abelian, then $v(\delta(A)) = \sum v(d_i)$.
8. Show that in a quadratic extension E/D, subject to (9.5.6) and (9.5.7), α may be extended to an endomorphism α' of E by $u^{\alpha'} = -u - \lambda$, and δ is then the inner α'-derivation induced by u. Verify that α' is an automorphism of E iff α is an automorphism of D.

10 Coding theory

The theory of error-correcting codes deals with the design of codes which will detect, and if possible correct, any errors that occur in transmission. Codes should be distinguished from cyphers, which form the subject of cryptography. The subject of codes dates from Claude Shannon's classic paper on information theory (Shannon [1948]) and Section 10.1 provides a sketch of the background, leading up to the statement (but no proof) of Shannon's theorem. Most of the codes dealt with are block codes which are described in Section 10.2, with a more detailed account of special cases in Sections 10.3–10.5; much of this is an application of the theory of finite fields (see BA, Section 7.8).

10.1 The transmission of information

Coding theory is concerned with the transmission of information. For example, when a spacecraft wishes to send pictures back to Earth, these pictures are converted into electrical impulses, essentially representing strings of 0's and 1's, which are transmitted back to Earth, but the message sent may be distorted by 'noise' in space, and one has to build in some redundancy to overcome these errors (provided they are not too numerous). An important everyday application is to digital recording and transmission, for example the compact disc, on which a piece of music is stored by means of tiny pits representing 0's and 1's according to a code.

To transmit our messages we need an *alphabet* Q consisting of q symbols, where $q \geq 2$; we also speak of a *q-ary code*. A finite string of letters from Q is called a *word*. The information to be transmitted is *encoded* by words of Q; during transmission the coded message may be slightly changed (due to a 'noisy' channel) and, as a result, a slightly different message is received. However, if the code is appropriately chosen, it is nevertheless possible to decode the result so as to recover the original message.

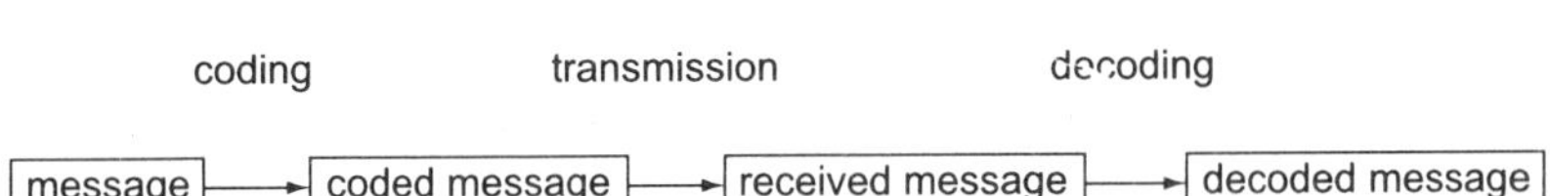

Many of our codes are binary: $q = 2$. For example, in the game of 'twenty questions' an object has to be guessed by asking 20 questions which can be answered 'yes' or 'no'. This allows one to pick out one object in a million (since $2^{20} \sim 10^6$). Usually a binary code will have the alphabet $\{0, 1\}$; our coded message will then be a string of 0's and 1's. As a simple check we can add 1 when the number of 1's in the message is odd and 0 when it is even. If the received message contains seven 1's we know that a mistake has occurred and we can ask for the message to be repeated (if this is possible). This is a *parity check*; it will show us when an odd number of errors occurs, but it does not enable us to correct errors, as is possible by means of more elaborate checks. Before describing ways of doing this we shall briefly discuss the question of information content, although strictly speaking this falls outside our topic. The rest of this section will not be used in the sequel and can be omitted without loss of continuity.

It is intuitively clear that the probability of error can be made arbitrarily small by adding sufficiently many checks to our message, and one might think that this will make the transmission rate also quite small. However, a remarkable theorem due to Shannon asserts that every transmission channel has a capacity C, usually a positive number, and for any transmission rate less than C the probability of error can be made arbitrarily small. Let us briefly explain these terms.

The information content of a message is determined by the likelihood of the event it describes. Thus a message describing a highly probable event (e.g. 'the cat is on the mat') has a low information content, while for an unlikely message ('the cow jumped over the moon') the information content is large. If the probability of the event described by the message is p, where $0 \leq p \leq 1$, we shall assign as a measure of information $-\log_2 p$. Here the minus sign is included to make the information positive and the logarithm is chosen to ensure that the resulting function is additive. If independent messages occur with probabilities p_1, p_2 then the probability that both occur is $p_1 p_2$ and here the information content is

$$-\log p_1 p_2 = -\log p_1 - \log p_2.$$

All logs are taken to the base 2 and the unit of information is the bit (binary digit). Thus if we use a binary code and 0, 1 are equally likely, then each digit carries the information $-\log(1/2) = 1$, i.e. one bit of information.

Suppose we have a channel transmitting our binary code in which a given message, consisting of blocks of k bits, is encoded into blocks of n bits; the *information rate* of this system is defined as

$$R = k/n. \tag{10.1.1}$$

We assume further that the probability of error, x say, is the same for each digit; this is the binary symmetric channel. When an error occurs, the amount of information lost is $-\log x$, so on average the information lost per digit is $-x.\log x$. But when no error occurs, there is also some loss of information (because we do not know that no error occurred); this is $-\log(1-x)$. The total amount of information lost per digit is therefore

$$H(x) = -x.\log x - (1-x).\log(1-x).$$

This is also called the *entropy*, e.g. $H(0.1) = 0.469$, $H(0.01) = 0.0808$. The *channel capacity*, in bits per digit, is the amount of information passed, i.e.

$$C(x) = 1 - H(x).$$

Thus $C(0.1) = 0.531$, $C(0.01) = 0.9192$. We note that $C(0) = 1$; this means that for $x = 0$ there is no loss of information. By contrast, $C(1/2) = 0$; thus when there is an even chance of error, no information can be sent. The fundamental theorem of coding theory, proved by Shannon in 1948, states that for any $\delta, \varepsilon > 0$, there exist codes with information rate R greater than $C(x) - \varepsilon$, for which the probability of error is less than δ. In other words, information flows through the channel at nearly the rate $C(x)$ with a probability of error that can be made arbitrarily small. Here the information rate of the code is represented by (10.1.1). More generally, if there are M different code words, all of length n, then $R = (\log M)/n$. For a binary code there are 2^k different words of length k, so $\log M = k$ and the rate reduces to k/n.

Exercises

1. How many questions need to be asked to determine one object in 10^9 if the reply is one of three alternatives?
2. In the binary symmetric channel with probability of error 1, no information is lost, i.e. $C(1) = 1$. How is this to be interpreted?
3. If n symbols are transmitted and the probability of error in each of them is x, show that the probability of exactly k errors is $\binom{n}{k} x^k (1-x)^{n-k}$.

10.2 Block codes

Most of our codes in this chapter will be *block codes*, that is codes in which all code words have the same number of letters. This number, n say, is called the *length* of the code. Thus a block code of length n in an alphabet Q may be thought of as a sequence of words chosen from Q^n. For any $x, y \in Q^n$ we define the *distance* (also called the *Hamming distance*) between x and y, written $d(x, y)$, as the number of positions in which x and y differ, e.g. $d(\text{pea, pod}) = 2$. We note that this function satisfies the usual axioms of a metric space:

M.1 $d(x, y) \geq 0$ with equality iff $x = y$,
M.2 $d(x, y) = d(y, x)$,
M.3 $d(x, y) + d(y, z) \geq d(x, z)$ (triangle inequality).

M.1–**M**.2 are clear and **M**.3 follows because if y differs in r places from x and in s places from z, then x and z can differ in at most $r + s$ places.

Let C be a block code which is q-ary of length n. The least distance $d = d(C)$ between code words of C is called the *minimum distance* of C. If the number of code words in C is M, then C is called a q-ary (n, M, d)-code, or an $(n, {}^*, d)$-code if we do not wish to specify M. For successful decoding we have to ensure that the code words are not too close together; if $d(x, y)$ is large, this means that x and y differ in many of the n places, and x is unlikely to suffer so many changes in transmission that y is received. Thus our aim will be to find codes for which d is large. Our first result tells us how a large value of d allows us to detect and correct errors. A code is said to be *r-error-detecting* (*correcting*) if for any word differing from a code word u in at most r places we can tell that an error has occurred (resp. find the correct code word u).

We shall define the *r-sphere* about a word $x \in Q$ as the sphere of radius r with centre x:

$$B_r(x) = \{y \in Q^n | d(x, y) \leq r\}. \tag{10.2.1}$$

Clearly it represents the set of all words differing from x in at most r places.

Proposition 10.2.1. *A code with minimum distance d can* (i) *detect up to $d - 1$ errors, and* (ii) *correct up to $[(d - 1)/2]$ errors.*

Here $[\xi]$ denotes the greatest integer $\leq \xi$.

Proof. (i) If x is a code word and s errors occur in transmission, then the received word x' will be such that $d(x, x') = s$. Hence if $0 < s < d$, x' cannot be a code word and it will be noticed that an error has occurred.

(ii) If $e \leq [(d - 1)/2]$, then $2e + 1 \leq d$ and it follows that the e-spheres about the different code words are disjoint. For if x, y are code words and $u \in B(x) \cap B(y)$, then

$$2e \geq d(x, u) + d(u, y) \geq d(x, y) \geq d \geq 2e + 1,$$

which is a contradiction. Thus for any word differing from a code word x in at most e places there is a unique nearest code word, namely x. ■

For example, the parity check mentioned in Section 10.1 has minimum distance 2 and it will detect single errors, but will not correct errors.

Proposition 10.2.1 puts limits on the number of code words in an error-correcting code. Given n, d, we denote by $A(n, d)$ or $A_q(n, d)$ the largest number of code words in a q-ary code of length n and minimum distance d; thus $A_q(n, d)$ is the largest M for which a q-ary (n, M, d)-code exists. A code for which this maximum is attained is also called an *optimal code*; any optimal (n, M, d)-code is necessarily *maximal*, i.e. it is not contained in an $(n, M + 1, d)$-code.

To obtain estimates for $A_q(n, d)$ we need a formula for the number of elements in $B_r(x)$. This number depends on q, n, r but not on x; it is usually denoted by $V_q(n, r)$. To find its value let us count the number of words at distance i from x. These words differ from x in i places, and the values at these places can be any one of $q - 1$ letters, so there are $\binom{n}{i}(q - 1)^i$ ways of forming such words. If we do this for $i = 0, 1, \ldots, r$ and add the results, we obtain

$$V(n,r) = V_q(n,r) = 1 + \binom{n}{1}(q-1) + \binom{n}{2}(q-1)^2 + \ldots + \binom{n}{r}(q-1)^r. \tag{10.2.2}$$

A set of spheres is said to *cover* Q or form a *covering* if every point of Q lies in at least one sphere. It is called a *packing* if every point of Q lies in at most one sphere, i.e. the spheres are non-overlapping. We note that for $0 \le r \le n$,

$$q^r = V_q(r,r) \le V_q(n,r) \le V_q(n,n) = q^n.$$

Theorem 10.2.2. *Given integers $q \ge 2$, n, d, put $e = [(d-1)/2]$. Then the number $A_q(n,d)$ of code words in an optimal $(n, {}^*, d)$-code satisfies*

$$\frac{q^n}{V_q(n,d-1)} \le A_q(n,d) \le \frac{q^n}{V_q(n,e)}. \tag{10.2.3}$$

Proof. Let C be an optimal (n,M,d)-code; then C is maximal, and it follows that no word in Q^n has distance $\ge d$ from all the words of C, for such a word would allow us to enlarge the code, and so increase M. Hence every word of Q^n is within distance at most $d-1$ of some word of C; thus the $(d-1)$-spheres about the code words as centres cover Q^n and so $M.V(n,d-1) \ge q^n$, which gives the first inequality in (10.2.3).

On the other hand, we have $2e+1 \le d$, so the spheres $B_e(x)$, as x runs over an (n,M,d)-code, are disjoint and hence form a packing of Q^n. Therefore $M.V(n,e) \le q^n$, and the second inequality in (10.2.3) follows. ∎

The above proof actually shows that a code with $q^n/V(n,d-1)$ code words and minimum distance d can always be constructed; we shall not carry out the construction yet, since we shall see in Section 10.3 that it can always be realized by a linear code.

The first inequality in (10.2.3) is called the *Gilbert–Varshamov bound*, and the second is the *sphere-packing* or *Hamming bound*. A code is said to be *perfect* if, for some $e \ge 1$, *the* e-spheres with centres at the code words form both a packing and a covering of Q^n. Such a code is an $(n,M,2e+1)$-code for which $M.V(n,e) = q^n$, so it is certainly optimal. It is characterized by the property that every word of Q^n is nearer to one code word than to any of the others. To give an example, any code consisting of a single code word, or of the whole of Q^n is perfect. For $q = 2$ and odd n (with alphabet $\{0,1\}$) the binary repetition code $\{0^n, 1^n\}$ is also perfect. These are the trivial examples; we shall soon meet non-trivial ones.

There are several ways of modifying a code to produce others, possibly with better properties. Methods of extending codes will be discussed in Section 10.3, when we come to linear codes. For the moment we observe that from any (n,M,d)-code C we obtain an $(n-1,M,d')$-code, where $d' = d$ or $d-1$, by deleting the last symbol of each word. This is called *puncturing* the code C. If we consider all words of C ending in a given symbol and take this set of words with the last symbol omitted, we obtain an $(n-1, M', d')$-code, where $M' \leq M$ and $d' \geq d$. This is called *shortening* the code C. Of course we can also puncture or shorten a given code by operating on any position other than the last one.

Exercises

1. Show that $A_q(n, 1) = q^n, A_q(n, n) = q$.
2. Use the proof of Theorem 10.2.2 to construct an $(n, q^n/V_q(n, d-1), d)$-code, for any $q \geq 2$, n and d.
3. Show that there is a binary $(8, 4, 5)$-code, and that this is optimal.
4. (The Singleton bound) Prove that $A_q(n, d) \leq q^{n-d+1}$. (Hint. Take an optimal (n, M, d)-code and puncture it repeatedly; for a linear $[n, d]$-code this gives $k \leq n - d + 1$.)

10.3 Linear codes

By a *linear* code one understands a code with a finite field $\mathbf{F}_q$ as alphabet, such that the set of code words forms a subspace. A linear code which is a k-dimensional subspace of an n-dimensional space over $\mathbf{F}_q$ will be called an $[n, k]$-code over $\mathbf{F}_q$. Thus any $[n, k]$-code over $\mathbf{F}_q$ is a linear (n, q^k, d)-code for some d. The number of non-zero entries of a vector x is called its *weight* $w(x)$; hence for a linear code we can write

$$d(x, y) = w(x - y). \tag{10.3.1}$$

A linear (n, M, d)-code has the advantage that to find d we need not check all the $M(M-1)/2$ distances between code words but only the M weights. Moreover, to describe an $[n, k]$-code C we need not list all q^n code words but just give a basis of the subspace defining C. The $k \times n$ matrix whose rows form the basis vectors is called a *generator matrix* of C; clearly its rank is k.

A matrix over a field is called *left full* if its rows are linearly independent, *right full* if its columns are linearly independent. For a square matrix these concepts are of course equivalent and we then speak of a *full* matrix. Thus a generator matrix of a linear code is left full, and any left full $k \times n$ matrix over $\mathbf{F}_q$ forms a generator matrix of an $[n, k]$-code.

Let C be an $[n, k]$-code with generator matrix G. Any code word is a linear combination of the rows of G, since these rows form a basis for the code. Thus to encode a message $u \in \mathbf{F}_q^k$ we multiply it by G:

$$u \longmapsto uG.$$

For example, for the simple parity check code with generator matrix $\begin{pmatrix} 1 & 0 & 1 \\ 0 & 1 & 0 \end{pmatrix}$ this takes the form

$$(u_1, u_2) \longmapsto (u_1, u_2, u_1 + u_2).$$

For any $[n, k]$-code C we define the *dual code* $C^{\perp}$ as

$$C^{\perp} = \{y \in \mathbf{F}_q^n | xy^T = 0 \quad \text{for all } x \in C\}.$$

Since any $x \in C$ has the form $x = uG(u \in \mathbf{F}_q^k)$, the vectors y of $C^{\perp}$ are obtained as the solutions of the system $Gy^T = 0$. Here G has rank k, therefore $C^{\perp}$ is an $(n-k)$-dimensional subspace of $\mathbf{F}_q^n$; thus $C^{\perp}$ is an $[n, n-k]$-code. It is clear from the definition that $C^{\perp\perp} = C$. When n is even, it may happen that $C^{\perp} = C$; in that case C is said to be *self-dual*. For example, the binary code with generator matrix $\begin{pmatrix} 1 & 0 & 0 & 1 \\ 0 & 1 & 1 & 0 \end{pmatrix}$ is self-dual.

Generally, if G, H are generator matrices for codes C, $C^{\perp}$ that are mutually dual, we have

$$GH^T = 0; \tag{10.3.2}$$

since G, H are both left full, it follows that the sum of their ranks is n and by the theory of linear equations (see e.g. Cohn (1994), Chapter 4),

$$x = uG \text{ for some } u \in \mathbf{F}_q^k \Leftrightarrow xH^T = 0, \tag{10.3.3}$$

$$y = vH \text{ for some } v \in \mathbf{F}_q^{n-k} \Leftrightarrow yG^T = 0. \tag{10.3.4}$$

A generator matrix H for $C^{\perp}$ is called a *parity check matrix* for C. For example, when $G = \begin{pmatrix} 1 & 0 & 1 \\ 0 & 1 & 1 \end{pmatrix}$, then $H = (1\ 1\ 1)$. For any code word x we form $xH^T = x_1 + x_2 + x_3$; if this is non-zero, an error has occurred. Our code detects one error, but no more (as we have already seen). Before introducing more elaborate codes we shall describe a normal form for the generator and parity check matrices.

Two block codes of length n are said to be *equivalent* if one can be obtained from the other by permuting the n places of the code symbols and (in the case of a linear code) multiplying the symbols in a given place by a non-zero scalar. For a generator matrix of a linear code these operations amount to (i) permuting the columns and (ii) multiplying a column by a non-zero scalar. We can of course also change the basis and this will not affect the code. This amounts to performing elementary operations on the rows of the generator matrix (and so may affect the encoding rules). We recall that any matrix over a field may be reduced to the form

$$\begin{pmatrix} I & P \\ 0 & 0 \end{pmatrix} \tag{10.3.5}$$

by elementary row operations and column permutations (see Cohn (1994), p. 59), and for a left full matrix the zero rows are of course absent. Hence we obtain

Theorem 10.3.1. *Any $[n, k]$-code is equivalent to a code with generator matrix of the form*

$$G = (I \;\; P), \tag{10.3.6}$$

where P is a $k \times (n-k)$ matrix. ■

It should be emphasized that whereas the row operations change merely the basis, the column operations may change the code (to an equivalent code). More precisely, an $[n, k]$-code has a generator matrix of the form (10.3.6) iff the first k columns of its generator matrix are linearly independent. This condition is satisfied in most practical cases. If we use a generator matrix G in the standard form (10.3.6) for encoding, the code word uG will consist of the message symbols $u_1, \dots, u_k$ followed by $n-k$ check symbols.

The standard form (10.3.6) for the generator matrix of C makes it easy to write down the parity check matrix; its standard form is

$$H = (-P^T \;\; I_{n-k}). \tag{10.3.7}$$

For we have $GH^T = P - P = 0$, and since H is left full, H^T is right full, thus of rank $n-k$, and it follows that the rows of H form a basis for the dual code $C^\perp$.

The process of decoding just consists in finding the code word nearest to the received word. Let us see how the parity check matrix may be used here. We have an $[n, k]$-code C with parity check matrix H. For any vector $x \in \mathbf{F}_q^n$ the vector $xH^T \in \mathbf{F}_q^{n-q}$ is called a *syndrome* of x. By (10.3.3), the syndrome of x is 0 precisely when $x \in C$. More generally, two vectors $x, x' \in \mathbf{F}_q^n$ are in the same coset of C iff $xH^T = x'H^T$. Thus the syndrome determines the coset: if a vector x in C is transmitted and the received word y is $x + e$, then y and e have the same syndrome. To 'decode' y, i.e. to find the nearest code word, we choose a vector f of minimum weight in the coset of C containing y and then replace y *by* $y - f$. Such a vector f of minimum weight in its coset need not be unique; we choose one such f in each coset and call it the *coset leader*. The process described above is called *syndrome decoding*.

For example consider a binary $[4, 3]$-code with generator matrix

$$G = \begin{pmatrix} 1 & 0 & 0 & 1 \\ 0 & 1 & 0 & 1 \\ 0 & 0 & 1 & 1 \end{pmatrix}.$$

To encode a vector we have

$$(u_1, u_2, u_3) \longmapsto (u_1, u_2, u_3, u_1 + u_2 + u_3).$$

The parity check matrix is $H = (1 \;\; 1 \;\; 1 \;\; 1)$. The possible syndromes are 0 and 1. We arrange the 16 vectors of $\mathbf{F}_2^4$ as a 2×8 array with cosets as rows, headed by the syndrome and coset leaders:

0	0000	0001	0101	0011	1100	1010	0110	1111
1	1000	0001	1101	1011	0100	0010	1110	0111

This is called a *standard array*. To decode x we form its syndrome $x_1 + x_2 + x_3 + x_4$. If this is 0, we can take the first three coordinates as our answer. If it is 1, we subtract the coset leader 1000 before taking the first three coordinates. We note that in this case there are four possible coset leaders for the syndrome 1; this suggests that the code is not very effective; in fact $d = 2$, so the code is 1-error-detecting.

Next take the $[4, 2]$-code with generator and parity check matrices

$$G = \begin{pmatrix} 1 & 0 & 1 & 1 \\ 0 & 1 & 0 & 1 \end{pmatrix}, \quad H = \begin{pmatrix} 1 & 0 & 1 & 0 \\ 1 & 1 & 0 & 1 \end{pmatrix}.$$

A standard array is

00	0000	1011	0101	1110
01	0100	1111	0001	1010
10	0010	1001	0111	1100
11	1000	0011	1101	0110

To decode $x = (1101)$ we form $xH^T = (11)$ and then subtract from x the coset leader for the syndrome (11), giving $(1101) - (1000) = (0101)$. The minimum distance is 2, so the code again detects single errors.

It is not necessary for decoding to write down the complete standard array, but merely the first column, consisting of the coset leaders. We note that this method of decoding assumes that all errors are equally likely, i.e. that we have a symmetric channel. In more general cases one has to modify the weight function by taking the probability of error into account; we shall not enter into the details.

We now turn to the construction of linear codes with a large value of M for given n and d. By the Gilbert–Varshamov bound in Theorem 10.2.2 we have $A_q(n, d) \geq q^k$, provided that $q^{n-k} \geq V_q(n, d-1)$. However, this result does not guarantee the construction of linear codes. In fact we can construct linear codes with a rather better bound, as the next result shows.

Theorem 10.3.2. *There exists an $[n, k]$-code over $\mathbf{F}_q$ with minimum distance at least d, provided that*

$$V_q(n-1, d-2) < q^{n-k}. \tag{10.3.8}$$

For comparison we note that Theorem 10.2.2 gives $V_q(n, d-1) \leq q^{n-k}$, so (10.3.8) is a weaker condition.

Proof. Let C be any $[n, k]$-code over $\mathbf{F}_q$ with parity check matrix H. Each vector x in C satisfies $xH^T = 0$; this equation means that the entries of x define a linear dependence between the rows of H^T, i.e. the columns of H. We require a code for which the minimum distance is at least d, i.e. no vector of C has weight less than d; this will follow if no $d - 1$ columns of H are linearly dependent.

To construct such a matrix H we need only choose successively n vectors in ${}^{n-k}\mathbf{F}_q$ such that none is a linear combination of $d - 2$ of the preceding ones. In choosing the r-th column we have to avoid the vectors that are linear combinations of at most

$d-2$ of the preceding $r-1$ columns. We count the vectors to be avoided by picking $\delta \le d-2$ columns in $\binom{r-1}{\delta}$ ways and choosing the coefficients in $(q-1)^\delta$ ways. Hence the number of vectors to be avoided is

$$1+\binom{r-1}{1}(q-1)+\binom{r-1}{2}(q-1)^2+\ldots+\binom{r-1}{d-2}(q-1)^{d-2}$$
$$= V_q(r-1, d-2).$$

Thus we can adjoin an r-th column, provided that $V_q(r-1, d-2) < q^{n-k}$. By (10.3.8) this holds for $r = 0, 1, \ldots, n$, so we can form the required parity check matrix H. This proves the existence of a code with the required properties, since it is completely determined by H. ∎

Let us examine the case $d = 3$. If n, q are such that

$$V_q(n, 1) = q^{n-k}, \tag{10.3.9}$$

then (10.3.8) holds for $d = 3$, because V_q is an increasing function of its arguments; so when (10.3.9) holds, we can construct an $[n, k]$-code C with minimum distance 3. This means that the 1-spheres about the code words are disjoint, and since by (10.3.9), $q^k.V_q(n, 1) = q^n$, it follows that our code is perfect. These codes are known as *Hamming codes*. The equation (10.3.9) in this case reduces to $1 + n(q-1) = q^{n-k}$, i.e.

$$n = \frac{q^{n-k}-1}{q-1}. \tag{10.3.10}$$

Thus a Hamming code has the property that any two columns of its parity check matrix are linearly independent. In any code with odd minimum distance $d = 2e+1$ every error pattern of weight at most e is the unique coset leader in its coset, because two vectors of weight $\le e$ have distance $\le 2e$ and so are in different cosets. For a perfect code all coset leaders are of this form. Thus in a Hamming code each vector of weight 1 is the unique vector of least weight in its coset. Now the number of cosets is $q^n/q^k = q^{n-k}$. Omitting the zero coset we see from (10.3.10) that we have just $n(q-1)$ non-zero cosets and these are represented by taking as coset leaders the $n(q-1)$ vectors of weight 1. This makes a Hamming code particularly easy to decode: given $x \in \mathbf{F}_q^n$, we calculate xH^T. If x has a single error (which is all we can detect), then $xH^T = \gamma H_j^T$, where H_j is the j-th column of H and $\gamma \in \mathbf{F}_q$. Now the error can be corrected by subtracting γ from the j-th coordinate of x.

The simplest non-trivial case is the binary $[3, 1]$-code with generator and parity check matrices

$$G = (1 \quad 1 \quad 1), \quad H = \begin{pmatrix} 1 & 1 & 0 \\ 1 & 0 & 1 \end{pmatrix}.$$

It consists in repeating each code word three times. The information rate is $1/3 = 0.33$.

The next case of the Hamming code, the binary $[7, 4]$-code, is one of the best-known codes and one of the first to be discovered (in 1947). Its generator and parity check matrices are

$$G = \begin{pmatrix} 1 & 0 & 0 & 0 & 0 & 1 & 1 \\ 0 & 1 & 0 & 0 & 1 & 0 & 1 \\ 0 & 0 & 1 & 0 & 1 & 1 & 0 \\ 0 & 0 & 0 & 1 & 1 & 1 & 1 \end{pmatrix}, \quad H = \begin{pmatrix} 0 & 1 & 1 & 1 & 1 & 0 & 0 \\ 1 & 0 & 1 & 1 & 0 & 1 & 0 \\ 1 & 1 & 0 & 1 & 0 & 0 & 1 \end{pmatrix}.$$

Here the information rate is $4/7 = 0.57$. The minimum distance is 3, so the code will correct 1 and detect 2 errors.

From any q-ary $[n, k]$-code C we can form another code

$$\bar{C} = \left\{ \left(x_1, x_2, \ldots, x_n, -\sum x_i \right) \middle| x \in C \right\},$$

called the *extension* of C by parity check. If C is binary with odd minimum distance d, then $\bar{C}$ has minimum distance $d + 1$ and its parity check matrix $\bar{H}$ is obtained from the of C by bordering it first with a zero column and then a row of 1's. From an (n, M, d)-code we thus obtain an $(n + 1, M, d + 1)$-code, and we can get C back by puncturing $\bar{C}$ (in the last column).

Theorem 10.3.2 implicitly gives a lower bound for d in terms of q, n, k, but it does not seem easy to make this explicit. However, we do have the following upper bound for d:

Proposition 10.3.3 (Plotkin bound). *Let C be a linear $[n, k]$-code over $\mathbf{F}_q$. Then the minimum distance d of C satisfies*

$$d \leq \frac{n(q-1)q^{k-1}}{q^k - 1}. \tag{10.3.11}$$

Proof. C contains $q^k - 1$ non-zero vectors; their minimum weight is d, hence the sum of their weights is at least $d(q^k - 1)$. Consider the contribution made by the different components to this sum. If all the vectors in C have zero first component, this contribution is 0; otherwise we write C_1 for the subspace of vectors in C whose first component is zero. Then $C/C_1 \cong \mathbf{F}_q$ and so $|C_1| = q^{k-1}$. Thus there are $q^k - q^{k-1}$ vectors with non-zero first component. In all there are n components, and their total contribution to the sum of the weights is at most $n(q^k - q^{k-1})$. Hence $d(q^k - 1) \leq n(q^k - q^{k-1})$ and (10.3.11) follows. ∎

Sometimes a more precise measure than the minimum distance is needed. This is provided by the *weight enumerator* of a code C, defined in terms of the weights of the code words as

$$A(z) = \sum_{a \in C} z^{w(a)} = \sum A_i z^i,$$

where A_i is the number of code words of weight i in C. A basic result, the *MacWilliams identity*, relates the weight enumerator of a code to that of its dual. This is useful for finding the weight enumerator of an $[n, k]$-code when k is close to n, so that $n - k$ is small. We begin with a lemma on characters in fields. Here we understand by a *character* on a field F a homomorphism from the additive group of F to the multiplicative group of complex numbers, non-trivial if it takes values other than 1. By the duality of abelian groups (BA, Section 4.9) every finite field has non-trivial characters χ, and $\sum \chi(a) = 0$ by orthogonality to the trivial character.

Lemma 10.3.4. *Let χ be a non-trivial character of $(\mathbf{F}_q, +)$ and define*

$$f(u) = \sum_{v \in \mathbf{F}_q^n} \chi(uv^T) z^{w(v)}, \quad \text{where } u \in \mathbf{F}_q^n. \tag{10.3.12}$$

Then for any code C,

$$\sum_{u \in C} f(u) = |C|.B(z), \tag{10.3.13}$$

where $B(z)$ is the weight enumerator of the dual code $C^{\perp}$ and $|C|$ is the number of code words in C.

Proof. We have

$$\sum_{u \in C} f(u) = \sum_{u \in C} \sum_{v \in \mathbf{F}_q^n} \chi(uv^T) z^{w(v)} = \sum_{v \in \mathbf{F}_q^n} z^{w(v)} \sum_{u \in C} \chi(uv^T).$$

For $v \in C^{\perp}$ the second sum on the right is $|C|$. If $v \notin C^{\perp}$, then uv^T takes every value in $\mathbf{F}_q$ the same number of times, say N times, and we have $\sum \chi(uv^T) = N.\sum \chi(a) = 0$, because χ is non-trivial. Hence the right-hand side reduces to $|C|.B(z)$. ∎

We can now derive a formula for the weight enumerator of the dual code.

Theorem 10.3.5 (MacWilliams identity). *Let C be an $[n, k]$-code over $\mathbf{F}_q$ with weight enumerator $A(z)$ and let $B(z)$ be the weight enumerator of the dual code $C^{\perp}$. Then*

$$B(z) = q^{-k}[1 + (q-1)z]^n.A\left(\frac{1-z}{1+(q-1)z}\right). \tag{10.3.14}$$

Proof. Let us extend the weight to $\mathbf{F}_q$ by treating it as a one-dimensional vector space; thus $w(a) = 1$ for $a \in F_q^{\times}$ and $w(0) = 0$. Next, defining u as in (10.3.12), we have

$$f(u) = \sum_{v \in \mathbf{F}_q^n} z^{\sum w(v_i)} \chi\left(\sum u_i v_i\right)$$

$$= \sum_{v \in \mathbf{F}_q^n} \prod_{i=1}^{n} z^{w(v_i)} \chi(u_i v_i)$$

$$= \prod_{i=1}^{n} \sum_{t \in F_q^\times} z^{w(t)} \chi(u_i t).$$

If $u_i = 0$, the sum in this expression is $1 + (q-1)z$, while for $u_i \neq 0$ it is

$$1 + z\left(\sum \chi(a)\right) = 1 - z.$$

Hence we obtain

$$f(u) = (1-z)^{w(u)} [1 + (q-1)z]^{n - w(u)}.$$

Substituting into (10.3.13) and remembering that $|C| = q^k$, we obtain (10.3.14). ■

Sometimes it is more convenient to use $A(z)$ in its homogeneous form, defined by $A(x, y) = A(yx^{-1})x^n = \sum A_i x^{n-i} y^i$. Then (10.3.14) takes the form

$$B(x, y) = q^{-k} A(x + (q-1)y, x - y). \tag{10.3.15}$$

To illustrate Theorem 10.3.5, consider the binary Hamming code C of length $n = 2^k - 1$ and dimension $n - k$ over $\mathbf{F}_2$. Its dual code has as generator matrix the parity check matrix H of C, whose columns are all the non-zero vectors in $\mathbf{F}_2^k$. Hence any non-zero linear combination of the rows of $H = (h_{ij})$ has the i-th coordinate

$$a_1 h_{1i} + a_2 h_{2i} + \ldots + a_k h_{ki}$$

This vanishes for $2^{k-1} - 1$ columns (forming with 0 a $(k-1)$-dimensional subspace), and so is non-zero for the remaining 2^{k-1} columns. Hence every non-zero vector in the dual code $C^\perp$ has exactly 2^{k-1} non-zero components, and so $B(z) = 1 + nz^{(n+1)/2}$. By Theorem 10.3.5, the weight enumerator of the Hamming code is

$$A(z) = 2^{-k}\{[1 + (q-1)z]^n + n[1 + (q-1)z]^{(n-1)/2}(1-z)^{(n+1)/2}\}.$$

The weight enumerator is used in computing probabilities of transmission error. In any code C, an error, changing a code word x to $y = x + e$, will be detected provided that y is not a code word. Thus the error will go undetected if $y \in C$ or, equivalently, if $e \in C$. If the channel is binary symmetric, with probability p of symbol error, then the probability of the error vector e being of weight i is $p^i(1-p)^{n-i}$. Thus the probability $P_{un}(C)$ that an incorrect code word will be received is independent of the code word sent and is

$$P_{un}(C) = \sum A_i p^i (1-p)^{n-i} = A\left(\frac{p}{1-p}\right)(1-p)^n.$$

Exercises

1. Show that a code C can correct t errors and detect a further s errors if $d(C) > 2t + s + 1$.
2. Let C be a block code of length n over an alphabet Q and let $\lambda \in Q$. Show that C is equivalent to a code which includes the word λ^n.
3. Show that for odd d, $A_2(n, d) = A_2(n + 1, d + 1)$. (Hint. Use extension by parity check and puncturing.)
4. Construct a table of syndromes and coset leaders for the ternary $[4, 2]$-Hamming code.
5. Show that the binary $[7, 4]$-Hamming code extended by parity check is self-dual.
6. Show that for linear codes $A_q(n, d) \geq q^k$, where k is the largest integer satisfying $q^k.V_q(n - 1, d - 2) < q^n$.
7. Show that for an $[n, k]$-code the MacWilliams identity can be written

$$\sum_{i=0}^{n} \binom{i}{r} A_i = q^{k-r}.\sum_{i=0}^{n} (-1)^i \binom{n-i}{n-r} B_i, \quad \text{for } 0 \leq r \leq n.$$

8. Verify that formula (10.3.14) is consistent with the corresponding formula for the dual code. (Hint. Use the form (10.3.15).)

10.4 Cyclic codes

A code C is said to be *cyclic* if the set of code words is unchanged by permuting the coordinates cyclically: if $c = (c_0, c_1, \ldots, c_{n-1}) \in C$, then $(c_{n-1}, c_0, c_1, \ldots, c_{n-2}) \in C$. We shall here assume all our cyclic codes to be linear. For cyclic codes it is convenient to number the coordinates from 0 to $n - 1$. We shall identify any code word c with the corresponding polynomial

$$c(x) = c_0 + c_1 x + \ldots + c_{n-1} x^{n-1},$$

in the ring $A_n = \mathbf{F}_q[x]/(x^{n-1})$. In this sense we can interpret any linear code as a subset of A_n and the cyclic permutation corresponds to multiplication by x. Clearly a subspace of A_n admits multiplication by x iff it is an ideal, and this proves

Theorem 10.4.1. *A linear code in F is cyclic if and only if it is an ideal in $A_n = \mathbf{F}_q[x]/(x^n - 1)$.* ∎

We note that the ring A_n has q^n elements; as a homomorphic image of $\mathbf{F}_q[x]$ it is a principal ideal ring, but it is not an integral domain, since $x^n - 1$ is reducible for $n > 1$.

Henceforth we shall assume that $(n, q) = 1$. Then $x^n - 1$ splits into distinct irreducible factors over $\mathbf{F}_q$ and hence A_n is a direct product of extension fields of $\mathbf{F}_q$ (see BA, Corollary 11.7.4). By Theorem 10.4.1 every cyclic code over $\mathbf{F}_q$ can be generated by a polynomial g. Here g can be taken to be a factor of $x^n - 1$, since we can replace it by $ug - v(x^n - 1)$ without affecting the code. As a monic factor of $x^n - 1$ the generator of a cyclic code is uniquely determined. We record an expression for the generator matrix in terms of the generator polynomial:

Theorem 10.4.2. *Let C be a cyclic code of length n with generator polynomial*

$$g = g_0 + g_1 x + \ldots + g_r x^r \quad (g_r = 1).$$

Then $\dim(C) = n - r$ *and a generator matrix for C is given by*

$$G = \begin{pmatrix} g_0 & g_1 & \ldots & g_r & 0 & 0 & \ldots & 0 \\ 0 & g_0 & \ldots & g_{r-1} & g_r & 0 & \ldots & 0 \\ \ldots & \ldots & & \ldots & \ldots & \ldots & & \ldots \\ 0 & 0 & \ldots & g_0 & g_1 & g_2 & \ldots & g_r \end{pmatrix}.$$

Proof. We have $g_r = 1$ and by considering the last r columns we see that G is left full. The $n - r$ rows represent the code words $g, xg, \ldots, x^{n-r-1}g$ and we have to show that the linear combinations are just the code words. This is clear since the code words are of the form fg, where f is a polynomial of degree $< n - r$. ■

We next derive a parity check matrix for the cyclic code C. This is done most easily in terms of an appropriate polynomial. Let C be a cyclic $[n, k]$-code with generator polynomial g. Then g is a divisor of $x^n - 1$, so there is a unique polynomial h satisfying

$$g(x)h(x) = x^n - 1. \tag{10.4.1}$$

h is called the *check polynomial* of C. Clearly it is monic and its degree is $n - \deg g = n - (n - k) = k$. A cyclic code is said to be *maximal* if its generator polynomial is irreducible; if its dual is a maximal cyclic code, it is called *minimal* or *irreducible*. It is now an easy matter to describe a parity check matrix; to simplify the notation we shall use $\equiv$ to indicate congruence $\bmod (x^n - 1)$, i.e. equality in the ring A_n.

Theorem 10.4.3. *Let C be a cyclic $[n, k]$-code with check polynomial*

$$h = h_0 + h_1 x + \ldots + h_k x^k$$

Then

(i) *$c \in C$ if and only if $ch \equiv 0$,*
(ii) *a parity check matrix for C is*

$$H = \begin{pmatrix} h_k & h_{k-1} & \ldots & h_0 & 0 & 0 & \ldots & 0 \\ 0 & h_k & \ldots & h_1 & h_0 & 0 & \ldots & 0 \\ \ldots & \ldots & & \ldots & \ldots & \ldots & & \ldots \\ 0 & 0 & \ldots & h_k & h_{k-1} & h_{k-2} & \ldots & h_0 \end{pmatrix}.$$

(iii) *the dual code* $C^\perp$ *is cyclic, generated by the reciprocal of the check polynomial for C:*

$$\bar{h} = h_k + h_{k-1}x + \ldots + h_0 x^k.$$

Proof. (i) By definition, $c \in C$ iff $c \equiv ag$. By (10.4.1) $gh \equiv 0$, hence if $c \in C$, then $ch \equiv agh \equiv 0$. Conversely, if $ch \equiv 0$, then ch is divisible by $x^n - 1 = gh$, hence c is divisible by g.

(ii) On multiplying the i-th row of G by the j-th row of H, we obtain

$$g_0 h_{k-i+j} + g_1 h_{k-i+j-1} + \ldots + g_{k-i+j} h_0, \tag{10.4.2}$$

and this vanishes, as the coefficient of x^{k-i+j} in gh. Thus $GH^T = 0$, and since H is a left full $r \times n$ matrix, it is indeed a parity check matrix for C.

(iii) By comparing the form of H with that for G, we see that $\bar{h}$ is a generator polynomial for $C^\perp$. ∎

We go on to describe how generator and check polynomials are used for coding and decoding a cyclic code. Let C be a cyclic $[n, k]$-code with generator polynomial g of degree $r = n - k$ and check polynomial h of degree k. Given a message $a = a_0 a_1 \ldots a_{k-1} \in \mathbf{F}_q^k$, we regard this as a polynomial $a = \sum a_i x^i$ of degree $< k$ over $\mathbf{F}_q$. We encode a by multiplying it by g and obtain a polynomial $u = ag$ of degree $< n$. We note that any code word $\neq 0$ has degree at least $r = \deg g$.

For any polynomial f of degree $< n$ we calculate its syndrome $S(f)$ by multiplying the coefficients of f by the rows of the parity check matrix H. The result is

$$(fh)_k, (fh)_{k+1}, \ldots, (fh)_{n-1}, \tag{10.4.3}$$

where for any polynomial φ in x, φ_i denotes the coefficient of x^i. To represent (10.4.3) as a polynomial, we take the polynomial part of $x^{-k}(fh)$, ignoring powers beyond x^{r-1}. This can also be achieved by reducing $fh \pmod{x^n - 1}$ to a polynomial of degree $< n$ and then taking the quotient of the division by x^k:

$$fh = x^k S(f) + p, \quad \text{where } \deg p < k. \tag{10.4.4}$$

Since $\deg f < n$, the highest possible power in fh is x^{n+k-1}. When reduced this becomes x^{k-1}, and so does not affect the quotient in (10.4.4). Therefore $S(f)$ is indeed the syndrome of f, and as before $S(f) = 0$ precisely when f has the form ag. By reducing $fh \pmod{x^n - 1}$, we obtain a representative of degree $< n$; hence $S(f)$ is of degree $< n - k = r$, as one would expect.

Now we choose for each possible syndrome u a coset leader $L(u)$ of least weight. To decode a word f we compute its syndrome $S(f)$ and subtract the corresponding coset leader: $f - LS(f)$ is a code word, so we have $(f - LS(f))h = a(x^n - 1)$, for some a, and this a is the required decoding of f. For example, $x^7 - 1 = (x-1)(x^3 + x^2 + 1)(x^3 + x + 1)$ is a complete factorization over $\mathbf{F}_2$. Let us take $g = x^3 + x + 1$, $h = x^4 + x^2 + x + 1$, so $r = 3$, $k = 4$. Suppose we encode $x^2 + x$, obtaining the code word $(x^2 + x)(x^3 + x + 1) = x^5 + x^4 + x^3 + x$. Owing to errors in transmission this is received as $x^5 + x^4 + x$. We have $x^5 + x^4 + x =$

$(x^2+x)(x^3+x+1)+x^3$, so the coset leader is x^3, and adding this to the received word we get $x^5+x^4+x^3+x$. Now $(x^5+x^4+x^3+x)(x^4+x^2+x+1) = (x^2+x)(x^3+x+1)$, so our message is (correctly) decoded as x^2+x.

Sometimes it is useful to choose the generator for a cyclic code in a different way. In BA, Corollary 11.7.4 we saw that $A_n = \mathbf{F}_q[x]/(x^n-1)$ is a direct product of fields, say

$$A_n = K_1 \times \ldots \times K_t,$$

where K_i corresponds to the irreducible factor f_i of $x^n - 1$. The generator polynomial of a cyclic code C is the product of certain of the f_i, say (in suitable numbering) $f_1, \ldots, f_s$. If e_i denotes the unit element of K_i, then $e = e_1 + \ldots + e_s$ is an element of A_n which is idempotent and which can also be used for the code. For the polynomials corresponding to the code words are the elements of $K_1 \times \ldots \times K_s$ and these are the elements $c \in A_n$ such that $c = ce$. Thus every cyclic code has an idempotent generator; of course this will in general no longer be a factor of $x^n - 1$. To find the idempotent generator, suppose that C is a cyclic code with generator polynomial g and check polynomial h, so that $gh = x^n - 1$. Since g, h are coprime, there exist polynomials u, v such that $ug + vh = 1$. It follows that ug is the idempotent generator, for we have $(ug)^2 = ug(1 - vh) \equiv ug$. Thus in the above example the idempotent generator is $xg = x^4 + x^2 + x$.

For examples of cyclic codes consider again the binary $[n, n-k]$-Hamming code. Its parity check matrix is $k \times n$ and its columns are all the non-zero vectors of ${}^k\mathbf{F}_2$, for they are distinct; hence any two are linearly independent, and there are $n = 2^k - 1$ of them. Let us write $q = 2^k$ and consider $\mathbf{F}_q$ as a vector space over $\mathbf{F}_2$; this is a k-dimensional space, so the columns of the above parity check matrix are represented by the non-zero elements of $\mathbf{F}_q$. If these elements are $\alpha_1, \ldots, \alpha_{q-1}$, we can regard

$$(\alpha_1, \ldots, \alpha_{q-1}) \tag{10.4.5}$$

as a parity check matrix for the Hamming code. This will correct one error, and it seems plausible that we can correct more errors by including further rows, independent of (10.4.5).

To obtain such rows we recall that for any n distinct elements $c_1, \ldots, c_n$ over a field, the Vandermonde matrix

$$V(c_1, c_2, \ldots, c_n) = \begin{pmatrix} 1 & 1 & \ldots & 1 \\ c_1 & c_2 & \ldots & c_n \\ c_1^2 & c_2^2 & \ldots & c_n^2 \\ \ldots & \ldots & \ldots & \ldots \\ c_1^{n-1} & c_2^{n-1} & \ldots & c_n^{n-1} \end{pmatrix}$$

is non-singular; as is well known (and easily checked) its determinant is $\prod_{i>j}(c_i - c_j)$.

Theorem 10.4.4. *Let $q = 2^m$ and denote by $\alpha_1, \ldots, \alpha_{q-1}$ the non-zero elements of $\mathbf{F}_q$. Then for any integer $t < q/2$ there is a $[q, k]$-code over $\mathbf{F}_2$ with $k \geq q - mt$ and minimum distance at least $2t + 1$ and with parity check matrix*

$$H = \begin{pmatrix} \alpha_1 & \alpha_2 & \ldots & \alpha_{q-1} \\ \alpha_1^3 & \alpha_2^3 & \ldots & \alpha_{q-1}^3 \\ \alpha_1^5 & \alpha_2^5 & \ldots & \alpha_{q-1}^5 \\ \ldots & \ldots & \ldots & \ldots \\ \alpha_1^{2t-1} & \alpha_2^{2t-1} & \ldots & \alpha_{q-1}^{2t-1} \end{pmatrix}. \tag{10.4.6}$$

Proof. A vector $c \in \mathbf{F}_2^q$ is a code word iff $cH^T = 0$, i.e.

$$\sum_i c_i \alpha_i^j = 0 \quad \text{for } j = 1, 3, 5, \ldots 2t - 1. \tag{10.4.7}$$

On squaring the j-th equation we find $(\sum c_i \alpha_i^j)^2 = \sum c_i \alpha_i^{2j}$, because $\alpha_i \in \mathbf{F}_q$ and $c_i \in \mathbf{F}_2$. Hence (10.4.7) holds for all $j = 1, 2, \ldots, 2t$; if we insert the corresponding rows in the matrix (10.4.6), we see that the square matrix formed by the first $2t$ columns has determinant $\alpha_1 \ldots \alpha_{2t} V(\alpha_1, \ldots, \alpha_{2t})$, and this is non-zero since $0, \alpha_1, \ldots, \alpha_{2t}$ are distinct. Thus the first $2t$ columns of the new matrix are linearly independent, and similarly for any other set of $2t$ columns. This means that none of the vectors c in (10.4.7) can have weight $\leq 2t$, so the minimum distance of our code is at least $2t + 1$. ∎

The binary code with parity check matrix (10.4.6) is called a BCH-code, after its discoverers R. C. Bose, D. K. Ray-Chaudhuri and A. Hocquenghem.

Exercises

1. The *zero code* of length n is the subspace 0 of $\mathbf{F}_q^n$. Find its generator polynomial (as cyclic code) and describe its dual, the *universal code.*
2. The *repetition code* is the $[n, 1]$-code consisting of all code words $(\gamma, \gamma, \ldots, \gamma)$, $\gamma \in F_q$. Find its generator polynomial and describe its dual, the *zero-sum* code.
3. Describe the cyclic code with generator polynomial $x + 1$ and its dual.
4. Verify that the $[7, 4]$-Hamming code is cyclic and find its generator polynomial.
5. Show that the $[8, 4]$-code obtained by extending the [7,4]-Hamming code by parity check is self-dual and has weight enumerator $z^8 + 14z^4 + 1$.
6. Show that a binary cyclic code contains a vector of odd weight iff $x - 1$ does not divide the generator polynomial. Deduce that such a code contains the repetition code.
7. The *weight* of a polynomial f is defined as the number $w(f)$ of its non-zero coefficients. Show that for polynomials over $\mathbf{F}_2$, $w(fg) \leq w(f)w(g)$.

10.5 Other codes

There are many other codes adapted to various purposes, and it is neither possible nor appropriate to include all the details here, but it may be of interest to make a brief mention of some of them.

(i) Goppa codes. In (10.4.7) we can take the elements of $\mathbf{F}_q$ to be $\alpha_1^{-1}, \ldots, \alpha_{q-1}^{-1}$. Then the defining equations for the BCH-code take the form $\sum_i c_i \alpha_i^{-j} = 0$ $(i, j = 1, 2, \ldots, 2t)$, or equivalently, $\sum_{ij} c_i x^j \alpha_i^{-j} = 0$. This can also be written

$$\sum_i \frac{c_i}{x - \alpha_i} \equiv 0 \pmod{x^{2t}}, \tag{10.5.1}$$

and it leads to the following generalization.

Definition. Let g be a polynomial of degree t over $\mathbf{F}_{q^m}$ and let $\alpha_0, \ldots, \alpha_{n-1} \in \mathbf{F}_{q^m}$ be such that $g(\alpha_i) \neq 0$ $(i = 0, \ldots, n-1)$. The *Goppa code* with *Goppa polynomial* g is defined as the set of all vectors $c = (c_0, c_1, \ldots, c_{n-1})$ satisfying

$$\sum_i \frac{c_i}{x - \alpha_i} \equiv 0 \pmod{g(x)}. \tag{10.5.2}$$

We see that Goppa codes are linear but not necessarily cyclic. In order to find a parity check matrix for the Goppa code we recall that in the special case of the BCH-code this was obtained by writing

$$(x - \alpha_i)^{-1} = -\sum x^j \alpha_i^{-j-1}(1 - x^{2t}\alpha_i^{-2t}).$$

The coefficients of x^j (taken mod x^{2t}) form the entries of the $(j+1)$-th row in the parity check matrix. We have

$$\frac{1}{x - \alpha} \equiv -g(\alpha)^{-1} . \frac{g(x) - g(\alpha)}{x - \alpha} \pmod{g(x)}, \tag{10.5.3}$$

and here the right-hand side is a polynomial in x; it is therefore the unique polynomial congruent (mod g) to $(x - \alpha)^{-1}$. So in order to express (10.5.2) as a polynomial in x we can proceed as follows: if $g(x) = \sum g_i x^i$, then

$$\frac{g(x) - g(y)}{x - y} = \sum g_{i+j+1} x^i y^j;$$

further write $g(\alpha_i)^{-1} = h_i$. Then (10.5.2) becomes $\sum c_i h_{ij} = 0$, where $h_{ij} = h_i \sum g_{\nu+j+1} x^j \alpha_i^{\nu}$. Thus the matrix (h_{ij}) is

$$\begin{pmatrix} h_0 g_t & h_1 g_t & \cdots & h_{n-1} g_t \\ h_0(g_{t-1} + g_t\alpha_0) & \cdots & \cdots & h_{n-1}(g_{t-1} + g_t\alpha_{n-1}) \\ \cdots & \cdots & \cdots & \cdots \\ h_0(g_1 + g_2\alpha_0 + \ldots + g_t\alpha_0^{t-1}) & \cdots & \cdots & h_{n-1}(g_1 + g_2\alpha_{n-1} + \ldots + g_t\alpha_{n-1}^{t-1}) \end{pmatrix}.$$

By elementary row transformations (remembering that $g_t \neq 0$) we find that

$$H = \begin{pmatrix} h_0 & h_1 & \dots & h_{n-1} \\ h_0\alpha_0 & h_1\alpha_1 & \dots & h_{n-1}\alpha_{n-1} \\ \dots & \dots & \dots & \dots \\ h_0\alpha_0^{t-1} & h_1\alpha_1^{t-1} & \dots & h_{n-1}\alpha_{n-1}^{t-1} \end{pmatrix}.$$

We see again that any t columns are linearly independent, hence the Goppa code has minimum distance $> t$ and its dimension is $\geq n - mt$. There are several methods of decoding Goppa codes, based on the Euclidean algorithm (see McEliece (1977)) and the work of Ramanujan (see Hill (1985).

(ii) Let C be any block code of length n. We obtain another code, possibly the same, by permuting the n places in any way. The permutations which do not change C form a subgroup of Sym_n, the *group* of C, which may be denoted by $G(C)$. For example, the group of a cyclic code contains all translations $i \mapsto i + r \pmod{n}$. If s is prime to n, we have the permutation

$$\mu_s : i \mapsto si \pmod{n}. \tag{10.5.4}$$

We remark that μ_s is an automorphism of $A_n = \mathbf{F}_q[x]/(x^n - 1)$. For if $a(x) = \sum a_i x^i$, then $a\mu_s = \sum a_i x^{si} = a(x^s)$, and the operation $a(x) \mapsto a(x^s)$ is clearly an endomorphism; since s is prime to n, μ_s has finite order dividing $\varphi(n)$ and so is an automorphism.

A *QR-code* is a cyclic code of length n, an odd prime, which admits the permutation (10.5.4) of its places, where s is a quadratic residue mod n. We shall examine a particular case of QR-codes, following essentially van Lint (1982). Let n again be an odd prime and q a prime power such that q is a non-zero quadratic residue mod n. We write Q for the set of all non-zero quadratic residues mod n, N for the set of all quadratic non-residues and let α be a primitive n-th root of 1 in an extension of $\mathbf{F}_q$. Write $E = \mathbf{F}_q(\alpha)$ and put

$$g_0(x) = \prod_{r \in Q} (x - \alpha^r), \quad g_1(x) = \prod_{r \in N} (x - \alpha^r).$$

Then

$$x^n - 1 = (x - 1)g_0(x)g_1(x).$$

The Galois group of $E/\mathbf{F}_q$ is generated by the map $x \mapsto x^q$. Since $q \in Q$, this operation permutes the zeros of g_0 as well as those of g_1; therefore g_0 and g_1 have their coefficients in $\mathbf{F}_q$. We note that μ_s interchanges g_0 and g_1 if $s \in N$, hence the codes with generators g_0 and g_1 are equivalent. We shall be particularly interested in the QR-code generated by g_0; our aim will be to find restrictions on the maximum distance d. We recall from number theory (see e.g. BA, Further Exercise 24 of Chapter 7) that 2 is a quadratic residue mod n iff $n \equiv \pm 1 \pmod{8}$, and that -1 is a quadratic residue mod n iff $n \equiv 1 \pmod{4}$. We shall also need a lemma on weights; for a polynomial f (regarded as a code word) the weight $w(f)$ is of course just the number of non-zero coefficients.

Lemma 10.5.1 *Let f be a polynomial over $\mathbf{F}_q$ such that $f(1) \neq 0$. Then $(1+x+\ldots+x^{n-1})f$ has weight at least n. If $q=2$ and $\deg f < n$, then $(1+x+\ldots+x^{n-1})f$ has weight exactly n.*

Proof. By the division algorithm, $f=(x-1)u+c$, where $c=f(1)\neq 0$. It follows that

$$(1+x+\ldots+x^{n-1})f=(x^n-1)u+(1+x+\ldots+x^{n-1})c. \tag{10.5.5}$$

Suppose that $w(u)=r$; then the right-hand side has at least r terms of degree $\geq n$, while the terms in u can cancel at most r terms in $(1+x+\ldots+x^{n-1})c$. So the total weight is $\geq r+(n-r)=n$.

If $q=2$ and $\deg f < n$, we again have (10.5.5), where now $\deg u < n-1$. Each non-zero term in u will cancel a term in $1+x+\ldots+x^{n-1}$, and this is exactly compensated by the corresponding term in $x^n u$. Hence there are exactly n terms on the right of (10.5.5). ∎

Proposition 10.5.2. *Let C be a QR-code with generator g_0 and let $c=c(x)$ be a code word in C such that $c(1)\neq 0$. Then*

$$w(c)^2 \geq n. \tag{10.5.6}$$

Moreover, if $n \equiv -1 \pmod 4$, then

$$w(c)^2-w(c)+1 \geq n. \tag{10.5.7}$$

If further, $q=2$ and $n\equiv -1 \pmod 8$, then

$$w(c)\equiv -1 \pmod 4. \tag{10.5.8}$$

Proof. The polynomial $c(x)$ is divisible by g_0 but not by $x-1$, because $c(1)\neq 0$. For suitable s, μ_s will transform $c(x)$ into a polynomial $c^*(x)$ divisible by g_1 and again not by $x-1$. This means that cc^* is a multiple of $g_0g_1 = 1+x+\ldots+x^{n-1}$, and so, by Lemma 10.5.1, $w(cc^*)\geq n$. Now (10.5.6) follows because $w(cc^*)\leq w(c)w(c^*) = w(c)^2$.

If $n\equiv -1 \pmod 4$, then $-1\in N$ and so the operation $x \longmapsto x^{-1}$ transforms g_0 into g_1; thus $c(x)c(x^{-1})$ is divisible by g_0g_1. Now corresponding terms in $c(x)$ and $c(x^{-1})$ give rise to a term of degree zero in $c(x)c(x^{-1})$, so there are at most $w(c)^2-w(c)+1$ terms in all and (10.5.7) follows.

Finally assume that $n\equiv -1 \pmod 8$ and $q=2$; then (10.5.7) applies. Further, any code word c has degree $< n$; hence on writing $r=\deg c$, we have $x^r c(x^{-1})c(x) = fg_0g_1$, where f is a polynomial of degree $< n$. By Lemma 10.5.1, this product has weight exactly n, and writing $d=w(c)$, we have $d^2-d+1\geq n$. Now consider how terms in $c(x)c(x^{-1})$ can cancel. We have $c=\sum x^{r_i}$, $c(x^{-1})=\sum x^{-r_i}$ and a pair of terms in the product will cancel if $r_i-r_j=r_k-r_l$. But in this case $r_j-r_i=r_l-r_k$ and another pair will cancel, so that terms cancel in fours. Hence we have $d^2-d+1-4t=n$; therefore $d^2-d\equiv 2 \pmod 4$, and so $d\equiv -1 \pmod 4$. ∎

Let us now take $q = 2$. Then the condition that q is a quadratic residue mod n gives $n \equiv \pm 1 \pmod 8$. Consider the polynomial

$$\theta(x) = \sum_{r \in Q} x^r.$$

Since $\theta(x)^2 = \sum x^{2r} = \theta(x)$, it follows that θ is idempotent. In particular, for the primitive element α of E we have $\theta(\alpha)^2 = \theta(\alpha)$, so $\theta(\alpha)$ is 0 or 1. For any $r \in Q$ we have $\theta(\alpha^r) = \theta(\alpha)$, while for $r \in N$, $\theta(\alpha^r) + \theta(\alpha) = \sum_1^{n-1} \alpha^r = 1$. If $\theta(\alpha) = 1$, replace α by α^s, where $s \in N$; since $(s, n) = 1$, α^s is again a primitive n-th root of 1 and $\theta(\alpha^s) = 0$. Thus for a suitable choice of α we have $\theta(\alpha) = 0$. It follows that

$$\theta(\alpha^i) = \begin{cases} 0 & \text{if } i \in Q, \\ 1 & \text{if } i \in N, \\ (n-1)/2 & \text{if } i = 0. \end{cases}$$

If $n \equiv 1 \pmod 8$, then $\theta(\alpha^i)$ vanishes exactly when $i \in Q \cup \{0\}$, so the code is then generated by $(x-1)g_0$. Similarly if $n \equiv -1 \pmod 8$, then $\theta(\alpha^i)$ vanishes when $i \in Q$, so in this case the generator is g_0.

Let $C(n)$ be the binary code defined in this way and $C(n)^+$ its extension by parity check. It can be shown that the group of $C(n)^+$ is transitive on the $n+1$ places. (These places may be interpreted as the points on the projective line over $\mathbf{F}_n$ and the group is then the projective special linear group, see van Lint (1982), p. 88.) Consider a word $c \in C$ of least weight d. Since the group is transitive, we may assume that the last coordinate in C^+ (the parity check) is 1. This means that c has odd weight d, say, and so $c(1) = 1$. Hence by Proposition 10.5.2, $d \equiv -1 \pmod 4$ and $d^2 - d + 1 \geq n$.

For example, for $n = 7$ we obtain the [7, 4]-Hamming code; here $d = 3$. A second (and important) example is the case $n = 23$. Here

$$g_0 = x^{11} + x^9 + x^7 + x^6 + x^5 + x + 1.$$

Since $23 \equiv -1 \pmod 8$, g_0 is a generator for this code, which is known as the [23, 12]-*Golay code*. Since $d^2 - d + 1 \geq 23$ and $d \equiv -1 \pmod 4$, it follows that $d \geq 7$, so the 3-spheres about the code words form a packing. On checking their size, we note the remarkable fact that

$$V_2(23, 3) = 1 + \binom{23}{1} + \binom{23}{2} + \binom{23}{3} = 2^{11}.$$

This shows that $C(23)$ is a perfect code, with minimum distance $d = 7$. The extended code C^+ is of length 24, with minimum distance 8, giving rise to the Leech lattice, a particularly close sphere packing in 24 dimensions. The symmetry group of a point in this lattice in R is the first Conway group $.O$ ('dotto') of order $2^{22}.3^9.5^4.7^2.11.13.23 \sim 8.3 \times 10^{18}$. The quotient by its centre (which has order 2) is the sporadic simple group known as .1, discovered in 1968 by John Conway (see Conway and Sloane (1988)).

Exercises

1. Construct the ternary $[11, 6]$-Golay code and verify that it is perfect. Find its weight enumerator.
2. Construct the extended ternary $[12, 6]$-Golay code and find its weight enumerator. Is it self-dual?
3. A binary self-dual code is called *doubly even* if all weights of code words are divisible by 4. Show that the extended $[8, 4]$-Hamming code is doubly even. Show that if there is a $[2k, k]$-code which is doubly even, then $k \equiv 0 \pmod 4$.
4. Show that the extended binary $[24, 12]$-Golay code is doubly even. Find its weight enumerator.

Further exercises on Chapter 10

1. (The Plotkin bound) Prove that for any q-ary block code, if $d > \theta n$, where $\theta = 1 - q^{-1}$, then $A_q(n, d) \leq d/(d - \theta n)$. (Hint. Write the words of a maximal (n, M, d)-code as an $M \times n$ matrix and compute the sums of the distances between distinct words in two ways, along rows and along columns.)
2. Deduce the Plotkin bound of Proposition 10.3.3 from the general Plotkin bound of Exercise 1.
3. Show that the binary $[n, k]$-Hamming code has a parity check matrix whose columns are the numbers 1 to $2^n - 1$ in binary notation.
4. Show that the weight enumerator $A(z)$ of the Hamming code satisfies the differential equation $(1 - z^2)A'' + (1 + nz)A = (1 + z)^n$.
5. Examine the linear q-ary codes with the property that for any code word $c = (c_0, c_1, \ldots, c_{n-1})$, $L(c) = (\lambda c_{n-1}, c_0, \ldots, c_{n-2})$ is again a code word, where λ is a fixed element of $\mathbf{F}_q$. In particular consider the case $\lambda^n = 1$.
6. Let q be 2 or 3. Show that for a self-dual code the homogeneous weight enumerator is invariant under the transformations

$$(x, y) \mapsto ([x + (q-1)y]/\sqrt{q},\ (x - y)/\sqrt{q}),\quad (x, y) \mapsto (x, \omega y),$$

   where $\omega^q = 1$. Show that for $q = 2$ the group generated by these transformations has order 16. What is the order for general q?
7. Show that the weight enumerator of any binary self-dual code is a combination of $g_1 = z^2 + 1$ and $g_2 = z^8 + 14z^4 + 1$ (the Gleason polynomials). (Hint. Apply Molien's theorem, see Exercise 8 of Section 6.4, to Exercise 6.)
8. In any ISBN book number $\alpha_1 \ldots \alpha_{10}$ the final digit is chosen so that $\sum k\alpha_k \equiv 0 \pmod{11}$ (using the digits $0, 1, \ldots, 9, X$). Show that this allows single errors and transpositions to be detected, but not necessarily a total reversal (writing the number back to front).

11 Languages and automata

Many problems in mathematics consist in the calculation of a number or function, and our task may be to classify the different types of calculation that can arise. This can be done very effectively by describing simple machines which could carry out these calculations. Of course the discussion is entirely theoretical (we are not concerned with building the machines), but it is no accident that this way of thinking became current in the age of computers. Alan Turing, one of the pioneers of digital computers, used just this method in 1936 to attack decision problems in logic, by introducing the class of 'computable functions', i.e. functions that could be computed on a Turing machine. This development has had many consequences, most of them outside our scope. However, the simplest machines, automata, have an immediate algebraic interpretation. In the algebraic study of languages one uses simple sets of rules ('grammars') to derive certain types of languages, not mirroring all the complexities of natural languages, but more akin to programming languages. It turns out that these languages can also be described in terms of the machines needed to generate them, and in this chapter we give a brief introduction to algebraic languages and automata.

The natural mathematical concept to describe these formal languages is the free monoid, and in Section 11.1 we discuss monoids and their actions. Languages form the subject of Section 11.2, while Section 11.3 introduces automata. The monoid ring of a free monoid is a free associative algebra, an object of independent interest, which in turn can be used to study languages, and Section 11.5 provides a brief account of free algebras and their completions (free power series rings). This is also the natural place to study variable-length codes (Section 11.4), which in their turn have influenced the development of free monoids and free algebras.

11.1 Monoids and monoid actions

We recall from BA, Section 2.1, that a *monoid* is a set M with a binary operation $(x, y) \mapsto xy$ and a distinguished element 1, the *neutral element* or also *unit element*, such that

M.1 $x(yz) = (xy)z$ for all $x, y, z \in M$ (associative law),
M.2 $x1 = 1x = x$.

Groups form the particular case where every element has an inverse. As an example of a monoid other than a group we may take, for any set A, the set $\mathrm{Map}(A) = A^A$ of all mappings of A into itself, with composition of mappings as multiplication and the identity mapping as neutral. Many of the concepts defined for groups have a natural analogue for monoids, e.g. a *submonoid* of a monoid M is a subset of M containing 1 and admitting multiplication. A *homomorphism* between monoids M, N is a mapping $f : M \to N$ such that $(xy)f = xf.yf$, $1_M f = 1_N$ for $x, y \in M$. Here we had to assume explicitly that the unit element is preserved by f; for groups this followed from the other conditions. A *generating set* of a monoid M is a subset X such that every element of M can be written as a product of a number of elements of X. For example, the set $\mathbf{N}$ of all natural numbers is a monoid under multiplication, with neutral element the number 1; here a generating set is given by the set of all prime numbers, for every positive integer can be written as a product of prime numbers, with 1 expressed as the empty product. Likewise the set $\mathbf{N}_0 = \mathbf{N} \cup \{0\}$ is a monoid under addition, with neutral element 0 and generating set $\{1\}$.

An example of particular importance for us in the sequel is the following monoid. Let X be any set, called the *alphabet*, and denote by X^* the set of all finite sequences of elements of X:

$$w = x_1 x_2 \dots x_r, x_i \in X \quad (i = 1, \dots, r), r \geq 0. \tag{11.1.1}$$

Here we include the empty sequence, written as 1. We define multiplication in X^* by juxtaposition:

$$(x_1 \dots x_r)(y_1 \dots y_s) = x_1 \dots x_r y_1 \dots y_s. \tag{11.1.2}$$

The associative law is easily verified, and it is also seen that the empty sequence 1 is the neutral element. X^* is called the *free monoid* on X. We remark that when $X = \emptyset$, X^* reduces to the trivial monoid consisting of 1 alone. This case will usually be excluded in what follows. Apart from this trivial case the simplest free monoid is that on a one-element set, $\{x\}$ say. The elements are $1, x, x^2, x^3, \dots,$ with the usual multiplication. We see that $\{x\}^*$ is isomorphic to $\mathbf{N}_0$, the monoid of non-negative integers under addition, by the rule $n \leftrightarrow x^n$. Since the expression (11.1.1) for an element w of a free monoid is unique, the number r of factors on the right is an invariant of w, called its *length* and written $|w|$.

The name 'free monoid' is justified by the following result:

Theorem 11.1.1. *Every monoid is a homomorphic image of a free monoid.*

Proof. Let M be any monoid and A a generating set. Take a set A' in bijective correspondence with A and write F for the free monoid on A'. We have a mapping $f : F \to M$ defined by

$$(a'_1 \dots a'_r)f = a_1 \dots a_r, \tag{11.1.3}$$

where $a' \leftrightarrow a$ is the given correspondence between A' and A. Since every element of F can be written as a product $a'_1 \dots a'_r$ in just one way, f is well-defined by (11.1.3). It is surjective because A generates M, and f is easily seen to be a homomorphism by (11.1.2). ∎

Just as groups can be represented by permutations, so can monoids be realized by means of mappings. If M is any monoid, then by an *M-set* or a set with an *M-action* we understand a set S, with a mapping from $S \times M$ to S, written $(s, x) \mapsto sx$, such that

S.1 $s(xy) = (sx)y$ for all $s \in S, x, y \in M$,
S.2 $s1 = s$.

Writing for the moment R_x for the mapping $s \mapsto sx$ of S into itself, we can express S.1, S.2 as

$$R_{xy} = R_x R_y, \quad R_1 = 1. \tag{11.1.4}$$

This just amounts to saying that the mapping $R : x \mapsto R_x$ is a monoid homomorphism of M into $\mathrm{Map}(S)$. For example, M itself is an M-set, taking the multiplication in M as M-action. This is sometimes called the *regular representation* of M. We can use it to obtain the following analogue of Cayley's theorem for groups (BA, Theorem 2.2.1):

Theorem 11.1.2. *Every monoid can be faithfully represented as a monoid of mappings.*

Proof. Given a monoid M, we take the regular representation of M. If this is $x \mapsto \rho_x$, then ρ is a homomorphism from M to $\mathrm{Map}(M)$, by what has been said, and if $\rho_x = \rho_y$, then $x = 1.\rho_x = 1.\rho_y = y$, hence the homomorphism is injective. ■

Let us return to a general monoid M and an M-set S. By Theorem 11.1.1 we can write M as a homomorphic image of a free monoid X^*, for some set X. Thus we have a homomorphism

$$X^* \to M \to \mathrm{Map}(S);$$

this shows that any set S with an M-action can also be regarded as a set with an X^*-action, where X corresponds to a generating set of M.

A free monoid has several remarkable properties, which can also be used to characterize it. A monoid M is called *conical* if $xy = 1 \Rightarrow x = y = 1$; M is said to have *cancellation* or be a *cancellation monoid* if for all $x, y \in M$, $xu = yu$ or $ux = uy$ for some $u \in M$ implies $x = y$. Further, M is *rigid* if it has cancellation and whenever $ac = bd$, there exists $z \in M$ such that either $a = bz$ or $b = az$. We observe that any free monoid is conical and rigid; the first property is clear from (11.1.2), because the product in (11.1.2) cannot be 1 unless $r = s = 0$. To prove cancellation we note that in any element $x_1 \dots x_r \neq 1$ the leftmost factor x_1 is unique, as is the rightmost factor x_r. Thus $x_1 \dots x_r = y_1 \dots y_s$ can hold only if $r = s$ and $x_i = y_i$ $(i = 1, \dots, r)$. It follows that when $xu = yu$, say

$$x_1 \dots x_r u_1 \dots u_t = y_1 \dots y_2 u_1 \dots u_t,$$

then both sides have the same length and $x_i = y_i$ $(i = 1, \dots, r = s)$, therefore $x = x_1 \dots x_r = y_1 \dots y_r = y$; a similar argument applies when $ux = uy$. To prove rigidity, let $ac = bd$, say $a = x_1 \dots x_r$, $b = y_1 \dots y_s$, $c = u_1 \dots u_h$, $d = v_1 \dots v_k$. Then we have

$$x_1 \dots x_r u_1 \dots u_h = y_1 \dots y_s v_1 \dots v_k.$$

By symmetry we may assume that $r \le s$; then $x_1 = y_1, \dots, x_r = y_r$, and hence $b = x_1 \dots x_r y_{r+1} \dots y_s = az$, where $z = y_{r+1} \dots y_s$. This shows a free monoid to be rigid. We remark that when $ac = bd$, then $a = bz$ or $b = az$ according as $|a|$ is $\ge$ or $\le |b|$.

By a *unit* in a monoid M we understand an element u such that v exists in M satisfying $uv = 1$, $vu = 1$. For example, in a conical monoid the only unit is 1. When M has cancellation, it is enough to assume one of these equations, say $uv = 1$; for then $(vu)v = v(uv) = v1 = 1v$, hence $vu = 1$ by cancellation, and similarly if $vu = 1$. Let us define an *atom* as a non-unit which cannot be expressed as a product of two non-units (as in rings). For example, in a free monoid the atoms are just the elements of length 1. This shows incidentally that in a free monoid the free generating set is uniquely determined as the set of all atoms. We now have the following characterization of free monoids:

Theorem 11.1.3. *Let F be a monoid and X the set of all its atoms. Then F is free, on X as free generating set if and only if F is conical and rigid, and is generated by X.*

Proof. We have seen that in a free monoid these conditions are satisfied. Conversely, assume that they hold; we shall show that every element of F can be written in just one way as a product of elements of X. Any $a \in F$ can be expressed as such a product in at least one way, because X generates F. If we have

$$a = x_1 \dots x_r = y_1 \dots y_s, x_i, \quad y_j \in X,$$

then by rigidity, $x_1 = y_1 b$ or $y_1 = x_1 b$ for some $b \in F$, say the former holds. Since x_1, y_1 are atoms, b must be a unit and so $b = 1$ because F is conical. Thus $x_1 = y_1$ and we can cancel this factor and obtain $x_2 \dots x_r = y_2 \dots y_s$. By induction on $\max(r, s)$ we find $r - 1 = s - 1$, i.e. $r = s$ and $x_2 = y_2, \dots, x_r = y_r$. Thus F is indeed free on X, as we had to show. ∎

Exercises

1. Show that every finite cancellation monoid is a group.
2. Let a, b be any elements of a monoid M. Show that if ab and ba are invertible in M, then so are a and b, but this does not follow if we only know that ab is invertible. What can we say if aba is invertible?
3. Show that every finitely generated monoid which is conical and rigid is free.
4. Show that the additive monoid of non-negative rational numbers is conical and rigid but not free.
5. Show that a submonoid of a free monoid is free iff it is rigid. Give examples of submonoids of free monoids that are not free. (Hint. Consider first the 1-generator case.)

6. A set with an associative multiplication is called a *semigroup*. Show that any semigroup S may be embedded in a monoid by defining $S^1 = S \cup \{1\}$ with multiplication $x1 = 1x = x$ for all $x \in S$.
7. A *zero* in a monoid M is an element 0 such that $0x = x0 = 0$ for all $x \in M$. Verify that (i) a monoid has at most one zero, (ii) every monoid M can be embedded in a monoid M_0 with zero. If M already has a zero, how can the presence of these two zeros be reconciled with (i)?

11.2 Languages and grammars

Algebraic language theory arose from the attempt by Noam Chomsky to analyse and make precise the process of forming sentences in natural languages. The first point to notice is that whereas one may often test a sentence of a natural language like English for its grammatical correctness by checking its meaning, this is really irrelevant. This is well illustrated by Chomsky's example of a meaningless sentence which is grammatically correct:

Colourless green ideas dream furiously.

To emphasize the point, he confronts it with a sentence which is not correct:

Furiously ideas green colourless dream.

In principle the analysis ('parsing') of a sentence consists in determining its constituents and checking that they have been put together correctly according to prescribed rules:

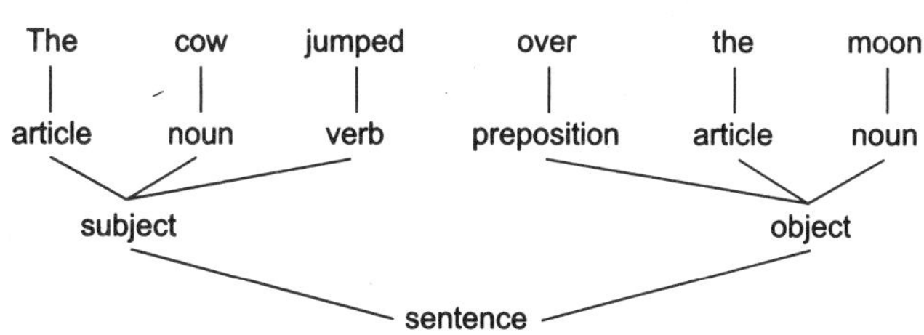

The mathematical model consists of a set of rules of the form: sentence → {subject, verb}, noun → cow, etc., which will lead to all the sentences of the language and no others. This amounts to reading the above diagram from the bottom upwards.

In order to write our sentences we need a finite (non-empty) set X, our *alphabet*. As we have seen, the free monoid on X is the set X^* of all strings of letters from X, also called *words* in X (with a multiplication which we ignore for the moment). By a *language* on X we understand any subset of X^*. Here we do not distinguish between words and sentences; we can think of an element of X^* as a message, with a particular symbol of X as a blank space, to separate the words of the message.

We single out a particular language by prescribing a set of rules according to which its sentences are to be formed. These rules constitute the grammar of the language and are formally defined as follows:A *phrase structure grammar* or simply *grammar* G consists of three sets of data:

(i) An *alphabet* X; its letters are also called *terminal* letters.
(ii) A set V of *clause-indicators* or *variables*, including a symbol σ for a complete sentence. We write $A = X \cup V$.
(iii) A finite set of *rewriting rules*: $u \to v$, where $u, v \in A^*$ and u contains at least one variable, i.e. $u \notin X^*$.

A string of letters from A, i.e. a member of A^* is called *terminal* if it lies in X^*, non-terminal otherwise. The rewriting rule $u \to v$ is applied by replacing a string fug over A by fvg. To obtain a sentence in our language we start from σ and apply the rewriting rules until no variables (clause-indicators) are left. If the resulting terminal string is f, we write $\sigma \to\to f$ and call the sequence of rules applied a *derivation* of f, while f itself is a *sentence* of the language. In this way we obtain the language $L(G)$ *generated* by the given grammar; it consists of all the strings on X that are sentences of the language. A language is called *proper* if it does not include the empty word 1.

In giving example we shall use latin letters for the terminal letters and greek letters for the variables. With this convention it is not necessary to mention the alphabets X, V separately.

Examples

1. $L = \{x^n | n \geq 1\}$. Rules $\sigma \to x, \sigma \to \sigma x$. We shall write this more briefly as $\sigma \to x$; σx. A typical derivation is $\sigma \to \sigma x \to \sigma x^2 \to \sigma x^3 \to x^4$. Similarly the language $\{x^{mr+ns} | m, n \geq 0\}$ is generated by the rules $\sigma \to \sigma x^r$; σx^s; 1.
2. $L = \{xy^n | n \geq 0\}$, also written xy^*. Rules: $\sigma \to x$; $\sigma \to \sigma y$.
3. $L = \{x^m y^n | m, n \geq 0\}$. Rules: $\sigma \to x\sigma$; σy; 1.
4. $L = \{x^m y^n | 0 \leq m \leq n\}$. Rules: $\sigma \to x\sigma y$; σy; 1.
5. $L = \{x^n z y^n | n \geq 0\}$. Rules: $\sigma \to x\sigma y$; z.
6. The *empty* language $L = \emptyset$ has the rule $\sigma \to \sigma$.
7. The *universal* language X^* has the rules $\sigma \to \sigma x$; 1 $(x \in X)$.

This concept of a language is of course too wide to be of use and one singles out certain classes of languages by imposing conditions on the generating grammar, as follows. The classification below is known as the *Chomsky hierarchy.*

0. By a language of *type 0* or a *phrase structure language* we understand any language generated by a phrase structure grammar. By no means every language is of type 0; in fact, since the alphabet and the set of rewriting rules are finite, the set of all languages of type 0 is countable, whereas there are uncountably many languages, because an infinite set has uncountably many subsets. It can be shown that the languages of type 0 are precisely the recursively enumerable subsets of X^* (see e.g. M. Davis (1958)).

1. A language is said to be of *type 1*, or *context-sensitive*, or a *CS-language* if it can be generated by a grammar in which all the rewriting rules are of the form

$$f\alpha g \to fug, \text{ where } \alpha \in V, u \in A^+, f, g \in A^*, \ (A^+ = A^* \backslash \{1\}). \tag{11.2.1}$$

The grammar is then also called a CS-grammar. The rule (11.2.1) can be taken to mean: α is replaced by u in the context $f\alpha g$.

2. A language is said to be of *type 2*, or *context-free*, or a *CF-language* if it can be generated by a grammar with rewriting rules of the form

$$\alpha \to u, \quad \text{where } \alpha \in V, u \in A^*. \tag{11.2.2}$$

The grammar is then also called a CF-grammar. The rule (11.2.2) means that α is replaced by u independently of the context in which it occurs.

3. A language is said to be of *type 3*, or *regular*, or *finite-state* if it can be generated by a grammar with rules of the form

$$\alpha \to x\beta, \alpha \to 1, \quad \text{where } x \in X, \alpha, \beta \in V. \tag{11.2.3}$$

Again the term *regular* is also used for the grammar. Here α is replaced by a variable following a letter or by 1. Instead of writing the variable β on the right of the terminal letter we can also restrict the rules so as to have β on the left of the terminal letter throughout. It can be shown that this leads to the same class of languages (see Exercise 4 of Section 11.3).

If $\mathcal{L}_i$ $(i = 0, 1, 2, 3)$ denotes the class of all proper languages of type i, then it is clear that

$$\mathcal{L}_0 \supseteq \mathcal{L}_1 \supseteq \mathcal{L}_2 \supseteq \mathcal{L}_3; \tag{11.2.4}$$

in fact the inclusion can all be shown to be strict, but in general it may not be easy to tell where a given language belongs, since there are usually many grammars generating it. Thus to show that a given language is context-free we need only find a CF-grammar generating it, but to show that a language is *not* context-free we must show that none of the grammars generating it is CF.

We note the following alternative definition of grammars of type 1, 2:

Proposition 11.2.1. *Let G be a grammar with alphabets X, V. Then*

(i) *If G is a CS-grammar, then for every rule $u \to v$ in G,*

$$|u| \leq |v|. \tag{11.2.5}$$

(ii) *If G is a CF-grammar, then for every rule $u \to v$ in G,*

$$|u| = 1. \tag{11.2.6}$$

Conversely, if G satisfies (11.2.5), (11.2.6) resp., then there is a CS-grammar resp. a CF-grammar generating $L(G)$.

Proof. (i) Let G be a CS-grammar; any rule in G has the form $f\alpha g \to fug$, where $u \neq 1$, hence $|u| \geq 1$ and so $|fug| \geq |f| + 1 + |g| = |f\alpha g|$, and (11.2.5) follows. Conversely, when (11.2.5) holds for every rule, we can achieve the effect of $u \to v$

by replacing the letters in u one at a time by a letter in v, taking care to leave a (new) variable until last. To give a typical example, if $u = u_1\alpha u_2u_3$, $v = v_1 \dots v_5$, we replace $u \to v$ by the rules $u_1\alpha u_2u_3 \to v_1\beta u_2u_3 \to v_1\beta u_2v_5 \to v_1\beta v_4v_5 \to v$, where β does not occur elsewhere.

(ii) It is clear from the definition of a CF-language that its rules are characterized by (11.2.6); the details may be left to the reader. ■

Sometimes one may wish to include the empty word in a proper language. This is most easily done by replacing any occurrence of σ on the right of a rule by a new variable λ, say, for each rule with σ on the left add the same rule with σ replaced by 1 on the left and adding the rule $\sigma \to 1$. For example, to generate the language $\{x^nzy^n, 1|n \geq 0\}$ we modify the example (11.2.5) above: $\sigma \to x\lambda y$; $1, \lambda \to x\lambda y$; z. If we just added $\sigma \to 1$ to the rules of 5, we would also get xy.

From any improper CF-language L we can obtain the proper CF-language $L\backslash\{1\}$ by replacing in any CF-grammar for L, any rule $\alpha \to 1$ by $\beta \to \bar{u}$, where $\bar{u}$ runs over all words obtained from derivations of the form $\beta \to\to u$, where u contains α, by replacing α in u by 1. For example, the language $\{x^my^n|m+n > 0\}$ is generated by $\sigma \to x\sigma$; σy; y; x.

Looking at the examples given earlier, we see that Examples 1, 2 and 3 are regular, as well as Examples 6 and 7. Examples 4 and 5 are context-free but not regular, as we shall see in Section 11.3. We conclude with an example of a CS-language which is not context-free, as Proposition 11.3.6 will show.

Example

8. $\{x^nz^ny^n|n \geq 0\}$ has the generating grammar $\sigma \to x\sigma\lambda\mu$; $xz\mu$; $1, \mu\lambda \to \lambda\mu$, $z\lambda \to z^2$, $\mu \to y$. The first two rules generate all the words $x^nz\mu(\lambda\mu)^{n-1}$, the fourth moves the λ's past the μ's next to z and the next replaces each λ by z. Finally each μ is replaced by y. To obtain the same language without 1 we simply omit the rule $\sigma \to 1$.

Exercises

1. Find a regular grammar to generate the set of all words of even length in X.
2. Show that each finite language is regular.
3. Show that if L, L' are any languages of type i ($= 0, 1, 2$ or 3), then so are $L \cup L'$, $LL' = \{uv|u \in L, v \in L'\}$ and L^0 obtained from L by writing each word in reverse order.
4. Show that if L is regular, then so is L^*, the language whose words are all the finite strings of words from L.
5. Show that regular languages form the smallest class containing all finite languages and closed under union, product and *.
6. Show that every context-free language can be generated by a CF-grammar G with the property: for each non-terminal variable α there is a derivation $\alpha \to u\,(u \in X^*)$ and for each terminal letter x there is a derivation $\alpha \to u$, where x occurs in u.

7. Show that for any CF-grammar $G = (X, V)$ there is a CF-grammar G' producing the same language as G such that (i) G' contains no rule $\alpha \to \beta$, where $\alpha, \beta \in V$, (ii) if $L(G)$ is improper, then G' contains the rule $\alpha \to 1$ but no other rules with 1 on the right-hand side and (iii) no rule of G' has σ occurring on the right. Thus all rules of G' have the form $\sigma \to 1$, $\alpha \to x$ or $\alpha \to f$, where $f \in (X \cup V\backslash\{\sigma\})^+$, $|f| \geq 2$.
8. Show that for a given CF-grammar G there exists a CF-grammar G' producing the same language as G, with rules $\alpha \to xf$, $f \in V^*$ and possibly $\sigma \to 1$ (Greibach normal form).

11.3 Automata

Logical machines form a convenient means of studying recursive functions. In particular, Turing machines lead precisely to recursively enumerable sets, and so correspond to grammars of type 0, as mentioned earlier. These machines are outside the scope of this book and will not be discussed further, but we would expect the more restricted types 1–3 of grammars to correspond to more special machines. This is in fact that case and in this section we shall define the types of machines corresponding to these grammars and use them to derive some of their properties.

A *sequential machine* M is given by three sets and two functions describing its action. There is a set S of *states* as well as two finite alphabets: an *input* X and an *output* Y. The action is described by a *transition function* $\delta : S \times X \to S$ and an *output function* $\lambda : S \times X \to Y$. To operate the machine we start from a given state s and input x; then the machine passes to the state $\delta(s, x)$ and produces the output $\lambda(s, x)$. In general the input will not just be a letter but a word w on X. The machine reads w letter by letter and gives out $y \in Y$ according to the output function λ, while passing through the different states in accordance with the transition function δ. The output is thus a word on Y, of the same length as w, obtained as follows. Define mappings $\delta' : S \times X^* \to S$, $\lambda' : S \times X^* \to Y^*$ by the equations

$$\delta'(s, 1) = s, \quad \delta'(s, ux) = \delta(\delta'(s, u), x) \quad s \in S, x \in X, u \in X^*, \tag{11.3.1}$$

$$\lambda'(s, 1) = 1, \quad \lambda'(s, ux) = \lambda'(s, u)\lambda(\delta'(s, u), x). \tag{11.3.2}$$

These equations define δ', λ' by induction on the length of words. It is clear that δ', λ' extend δ, λ respectively and so we may without risk of confusion omit the primes from δ', λ'. From (11.3.1) it is clear that

$$\delta(s, 1) = s, \quad \delta(s, uv) = \delta(\delta(s, u), v) \quad s \in S, u, v \in X^*, \tag{11.3.3}$$

so the mapping δ just defines an action of the free monoid X^* on S. We note that this holds even though no conditions were imposed on δ.

From this definition it is clear that a machine is completely specified by the set of all quadruples of the form $(s, x, \lambda(s, x), \delta(s, x))$. Sometimes it is preferable to start from a more general notion. Let S, X, Y be as before and define an *automaton* A as a set of quadruples

$$P = P(A) \subseteq S \times X \times Y \times S. \qquad (11.3.4)$$

The members of P are called its *edges*; each edge (s, x, y, s') has an initial state s, input x, output y and final state s'. For a sequential machine each pair $(s, x) \in S \times X$ determines a unique edge (s, x, y, s') and whenever our set P of edges is such that

C. *for each pair* $(s, x) \in S \times X$ *there exists a unique* $y \in Y$ *and* $s' \in S$ *such that* $(s, x, y, s') \in P$,

then we can define λ, δ by writing $y = \lambda(s, x)$, $s' = \delta(s, x)$ and we have a sequential machine. Two edges are *consecutive* if the final state of the first edge is also the initial state of the second. By a *path* for A we understand a sequence $u = (u_1, \ldots, u_n)$ of consecutive edges

$$u_i = (s_{i-1}, x_i, y_i, s_i).$$

Its *length* is n, s_0 is the initial and s_n the final state, $x_1 \ldots x_n$ its *input label* and $y_1 \ldots y_n$ its *output label.* It is clear how an automaton can be represented by a graph with the set S of states as vertex set, each edge being labelled by its input and output. Sometimes one singles out two subsets I, F of S; a path is called *successful* if its initial state is in I and its final state in F. The set $L(A)$ of all input labels of successful paths is a subset of X^*, called the *behaviour* of A, or also the set *accepted* by A. We note that the output does not enter into the behaviour; when the output is absent (so that P now consists of triples (s, x, s')), A is called an *acceptor*. As an example consider an acceptor with states s_0, s_1, s_2, input x,y and transition function

δ	x	y
s_0	s_1	s_2
s_1	s_2	s_1
s_2	s_2	s_2

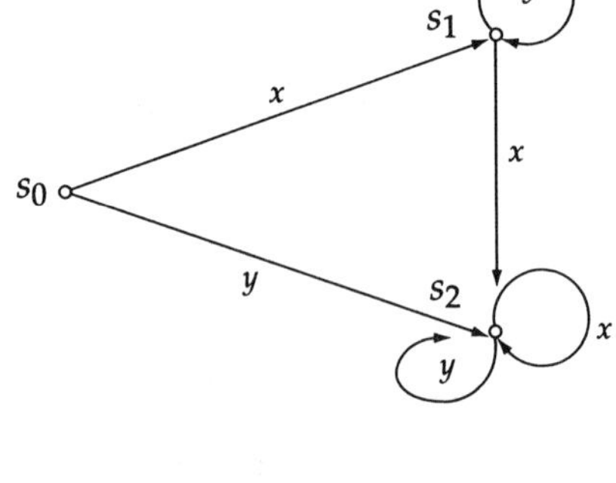

The graph is as shown. If $I = \{s_0\}$, $F = \{s_1\}$, then the behaviour is xy^*; for $I = \{s_0\}$, $F = \{s_2\}$ the behaviour is $xy^*xX^* \cup yX^*$.

An automaton is said to be *complete* if it satisfies condition **C** above (so that we have a sequential machine), and the set I of initial states consists of a single state. To operate a complete acceptor we take any word in X^* and use it as input with the machine in state I; this may or may not lead to a successful path, i.e. a path ending in F. We shall be interested in its behaviour, i.e. the set of input labels corresponding to successful paths. Thus the above example is a complete acceptor; we note how the graph makes it very easy to compute its behaviour. In constructing an acceptor it is usually convenient not to demand completeness, although complete acceptors are easier to handle. Fortunately there is a reduction allowing us to pass from one to the other:

Proposition 11.3.1. *For each acceptor A there is a complete acceptor C with the same behaviour. If A is finite (i.e. with a finite set of states), then so is C.*

Proof. Let the set of states for A be S, with initial state I and final state F. We take C to be on the same alphabet as A, with state set the set of all subsets of S, initial state $\{I\}$ and final set of states all sets meeting F. The transition function for C is given by $\delta(U, x) = V$, where V consists of all states v such that (u, x, v) is an edge in A for some $u \in U$. It is clear that C has the same behaviour as A, and it is easily seen to be complete. ∎

We remark that every subset Y of X^* is the behaviour of some acceptor; we take X^* as state set, 1 as initial state and Y as final set of states, with right multiplication as transition function. Our aim is to describe the behaviour of finite acceptors; we shall find that this consists precisely of all regular languages. To prove this result we shall need to construct a 'minimal' acceptor for a given language.

We shall use the notation (S, i, F) for a complete acceptor, where S is the set of states, i is the initial state and F is the set of final states; the alphabet is usually denoted by X and so will not be mentioned explicitly, and the transition function is indicated by juxtaposition; thus instead of $\delta(s, x) = s'$ we write $sx = s'$. A state s in an acceptor $A = (S, i, F)$ is said to be *accessible* if there is a path from i to s, *coaccessible* if there is a path from s to a state in F. As far as the behaviour of A is concerned, we can clearly neglect any states that are not both accessible and coaccessible. If every state of A is both accessible and coaccessible, then A is called *trim*.

Given two acceptors A, A' with state sets S, S', we define a *state homomorphism* from A to A' as a map $f : S \to S'$ such that $(s, x, t) \in A \Rightarrow (sf, x, tf) \in A'$. If f has an inverse which is also a state homomorphism, f is called an *isomorphism*. To give an example, let us put

$$L_s = \{v \in X^* | s.v \in F\};$$

then L_s consists of the set of words which give a successful path with s as initial state. Two states s, t are called *separable* if $L_s \neq L_t$, *inseparable* otherwise. If any two distinct states are separable, the acceptor is said to be *reduced*. Every acceptor (finite or infinite) has a homomorphic image which is reduced and has the same behaviour; to obtain it we simply identify all pairs of inseparable states.

For every subset Y of X^* we can define a reduced acceptor $A(Y)$ whose behaviour is Y. The states of $A(Y)$ are the non-empty sets

$$u^{-1}Y = \{v \in X^* | uv \in Y\},$$

where u ranges over X^*. The initial state is $1^{-1}Y = Y$ and the final states are the states $u^{-1}Y$ containing 1. The transition function is defined by

$$Z.u = u^{-1}Z, \quad \text{where } u \in X. \tag{11.3.5}$$

This is a partial function (i.e. not everywhere defined) since $u^{-1}Z$ may be empty, but it is single-valued. If for u we take a word in X, we have, by induction on the length of u,

$$w \in Z.ux = (Z.u)x \Leftrightarrow wx \in u^{-1}Z \Leftrightarrow uxw \in Z.$$

This shows that (11.3.5) holds for any $u \in X^*$. As a consequence we have

$$w \in L(A(Y)) \Leftrightarrow 1 \in Y.w \Leftrightarrow w \in Y,$$

which shows that the behaviour of $A(Y)$ is indeed Y. We shall call $A(Y)$ the *minimal acceptor* for Y; its properties follow from

Theorem 11.3.2. *Let $A = (S, i, F)$ be a trim acceptor and put $Y = L(A)$, the behaviour of A. Then there is a state homomorphism $\varphi : A \to A(Y)$ to the minimal acceptor for Y which is surjective on states, given by*

$$\varphi : s \mapsto L_s = \{v \in X^* | s, v \in F\}. \tag{11.3.6}$$

Proof. Since A is trim, any state s in S is accessible, so $iu = s$ for some $u \in X^*$; further, s is coaccessible, so $sv \in F$ for some $v \in X^*$ and it follows that L_s defined by (11.3.6) is non-empty. Thus φ is well-defined. To show that it is a homomorphism we have to verify that when $s.x = t$, then $L_s.x = L_t$. But we have

$$w \in L_s.x \Leftrightarrow xw \in L_s \Leftrightarrow s.xw \in F \Leftrightarrow w \in L_t;$$

so φ is indeed a homomorphism. It is surjective, because if $u^{-1}Y \neq \emptyset$, then there is a successful path in A with label uv, where $v \in Y$. Now

$$v \in u^{-1}Y \Leftrightarrow uv \in Y \Leftrightarrow sv = iuv \in F \Leftrightarrow v \in L_s;$$

thus $u^{-1}Y = L$ and this shows (11.3.6) to be surjective. ∎

As we have seen, for any subset Y of X^* there is a reduced acceptor with behaviour Y; taking this to be A in Theorem 11.3.2, we find φ in this case to be an isomorphism, by the definition of 'reduced'. It follows that A must be reduced.

Corollary 11.3.3. *The minimal acceptor for any subset of X^* is reduced.* ∎

Of course this is also not hard to verify directly.
We can now establish

Theorem 11.3.4. *A language is regular if and only if it is the precise set accepted by a finite acceptor.*

Proof. Let $A = (S, i, F)$ be a finite acceptor with behaviour Y and write the transition function as δ for clarity. For our grammar $G = \{X, V, \rightarrow\}$ we take X to be the input of A and $V = S$, the set of states, with $\sigma = i$, the initial state. For each state α in S and $x \in X$ we include in G the rule

$$\alpha \rightarrow x\beta \quad \text{if } \delta(\alpha, x) = \beta, \tag{11.3.7}$$

and for each final state ω we include the rule $\omega \rightarrow 1$. Given any word $w = x_1 \ldots x_r$, let us put $\delta(i, x_1) = s_1, \ldots, \delta(s_{r-1}, x_r) = s_r$. Then the rules (11.3.7) include $i \rightarrow x_1 s_1, \ldots, s_{r-1} \rightarrow x_r s_r$, hence $\sigma \rightarrow x_1 s_1 \rightarrow x_1 x_2 s_2 \rightarrow \ldots \rightarrow x_1 \ldots x_r s_r$. If $s_r \in F$, then $s_r \rightarrow 1$ and $x_1 \ldots x_r$ is included in $L(G)$. On the other hand, if $x_1 \ldots x_r \in L(G)$, consider the rules in G: they are all of the form $\alpha \rightarrow x\beta$ or $\alpha \rightarrow 1$ and the number of variables is constant in the application of the former rule and decreases by 1 when the latter is applied. Thus any derivation of $x_1 \ldots x_r$ must be of the form $i \rightarrow x_1 s_1$, $s_1 \rightarrow x_2 s_2, \ldots, s_{r-1} \rightarrow x_r s_r$, $s_r \rightarrow 1$. This means that $\delta(s_{i-1}, x_i) = s_i$ $(i = 1, \ldots, r)$ and $s_r \in F$, so $x_1 \ldots x_r$ is accepted by A.

Conversely, let G be a regular grammar, with derived language $L(G)$. For our acceptor A we take the alphabet X of G as input and the set V of variables as state set, with σ as initial state and a triple (α, x, β) for each rule $\alpha \rightarrow x\beta$, while the final state set consists of all α such that $\alpha \rightarrow 1$. Then it is clear that the derivations of G correspond precisely to the successful paths in A; the details may be left to the reader. Hence $L(G)$ is the set accepted by A. ∎

The acceptor constructed in this proof may not be complete, but it is trim provided that any superfluous variables have been removed from G. It follows by Theorem 11.3.2 that for a regular language the minimal acceptor is finite. This provides a practical way of determining whether a language is regular: Y is a regular language iff its minimal acceptor $A(Y)$ is finite. We illustrate this result by some examples.

Examples

1. $\{xy^n \mid n \geq 0\}$. The minimal acceptor has states $s_0 = xy^*$ and $s_1 = y^*$, and its operation is given by the table:

	s_0	s_1
x	s_1	
y		s_1

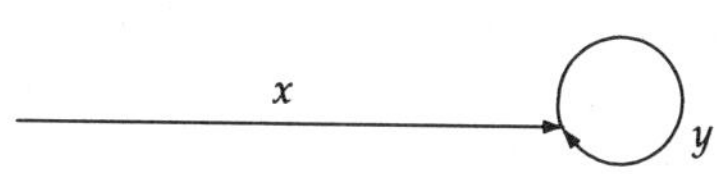

	s_0	s_1
x	s_1	
y		s_1

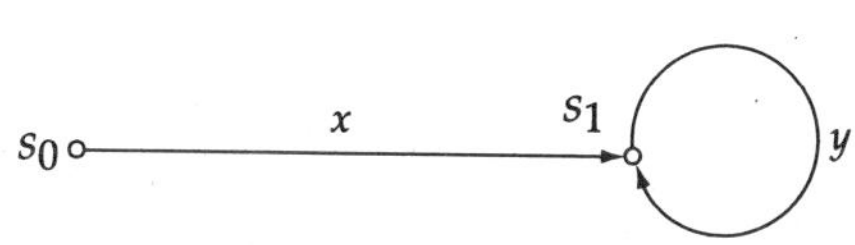

The initial state is s_0 and the final state is s_1. The behaviour can be read off from the graph.

2. $\{x^n y | n \geq 0\}$. Here the states are $s_0 = x^* y$ and $s_1 = \{1\}$, with initial state s_0 and final state s_1:

	s_0	s_1
x	s_0	
y	s_1	

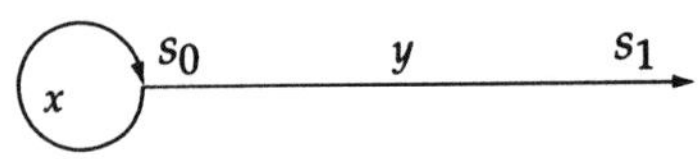

3. $\{x^n y^n | n \geq 0\}$. We have the states $s_0 = \{x^n y^n | n \geq 0\}$, $s_1 = \{x^n y^{n+1} | n \geq 0\}$, $s_2 = \{x^n y^{n+2} | n \geq 0\}, \ldots, t_0 = \{1\}$, $t_1 = \{y\}$, $t_2 = \{y^2\}, \ldots$. The initial state is s_0 and the final states are s_0, t_0, while the operations are given by the table below:

	s_0	s_1	s_2	$\cdots$	t_0	t_1	t_0	$\cdots$
x	s_1	s_2	s_3					
y		t_0	t_1			t_0	t_1	

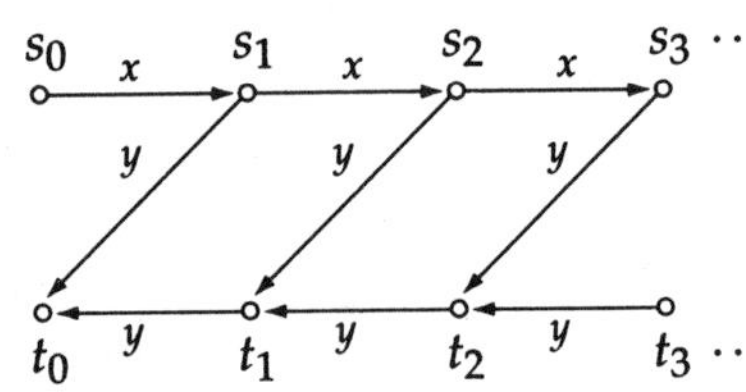

It should be clear from these examples how the behaviour of an acceptor may be read off from its graph. We note that in none of the cases is the acceptor complete; the transition function, though single-valued, is not everywhere defined. But this does not impair its usefulness; in any case we could replace it by a complete acceptor, using Proposition 11.3.1. Sometimes it is easier to test for regularity by means of the following necessary condition:

Proposition 11.3.5. *Let L be an infinite regular language on X. Then there exist $w', y, w'' \in X^*, y \neq 1$, such that $w'y^n w'' \in L$ for $n = 1, 2, \ldots$.*

Proof. Since L is regular, there is an acceptor A for L, with a finite number, m say, of states. The language L is infinite on a finite alphabet, so it contains a word w of length $> m$. The acceptor reads w letter by letter, moving from state to state as it reads. Since there are more than m steps, two of these states must be the same; say s_1 occurs twice. If y is the portion read between the first time and the second time in s_1, then $w = w'yw''$ and $y \neq 1$. Clearly our machine will also accept $w'y^2w''$, and generally $w'y^n w''$; thus $w'y^n w'' \in L$.

For context-free languages there is a condition similar to that in Proposition 11.3.5; this is sometimes known as the *pumping lemma*:

Proposition 11.3.6. *Let G be a CF-grammar. Then there exist integers p, q such that every word w of length greater than p in $L(G)$ can be written as $w = w'uzvw''$, where $uv \neq 1$, $|uzv| \leq q$ and $w'u^n zv^n w'' \in L(G)$ for all $n \geq 1$.*

Proof. The rules of G are all of the form $\alpha \to t$, $t \in A^*$. Suppose the number of variables is k and choose p so large that every word $w \in L(G)$ of length $\geq p$ has more than k steps in its derivation. Each of these steps has the form $\alpha \to t$; hence some variable occurs twice, so the part of the derivation from the first to the second occurrence of α reads $\alpha \to \ldots \to u\alpha v$, where $uv \neq 1$. It follows that $w = w'uzvw''$, where α occurs only once in the derivation $\alpha \to \ldots \to z$, which therefore has at most k steps. Therefore uzv has bounded length, and by repeating the steps between α and $u\alpha v$ we obtain $w'u^n zv^n w''$ for all $n \geq 1$. ∎

Proposition 11.3.5 shows that the language $\{x^n zy^n\}$ which we saw to be context-free, is not regular, and Proposition 11.3.6 shows that $\{x^n y^n z^n\}$ is not context-free.

Let us return to the example $\{x^n zy^n\}$; we have just seen that it is not regular, and so cannot be obtained from a finite acceptor. Intuitively we can see that a finite acceptor does not have the means of comparing the exponents of x and y. To make such a comparison requires a memory of some kind, and we shall now describe a machine with a memory capable of accepting CF-languages. The memory to be described is of a rather simple sort, a 'first-in, last-out' store, where we only have access to the last item in the store.

A *pushdown acceptor* (PDA for short) is an acceptor which in addition to its set of states S and input alphabet X has a set Σ of *store symbols*, with initial symbol λ_0 and a transition function $\delta : S \times X \times \Sigma \to S \times \Sigma^*$; but for a given triple of arguments there may be several or no values. At any stage the machine is described by a triple (s_i, w, α), where $s_i \in S$, $w \in X^*$, $\alpha \in \Sigma^*$. We apply δ to the triple consisting of s_i, the first letter of w and the last letter of α. If $w = xw'$, $\alpha = \alpha'\lambda$ say, and (s_j, β) is a value of $\delta(s_i, x, \lambda)$, then

$$(s_i, xw', \alpha'\lambda) \to (s_j, w', \alpha'\beta)$$

is a possible move. Thus the effect of δ is to move into a state s_j, remove the initial factor x from w and replace the final letter λ of α by β. We say that a word w on X is *accepted* by the machine if, starting from (s_0, w, λ_0) there is a series of moves to take us to $(s_r, 1, \gamma)$, where s_0 is the initial and s_r a final state. With this definition we have

Theorem 11.3.7. *The context-free languages constitute the precise class of sets accepted by pushdown acceptors.* ∎

We shall not give the proof here (see e.g. Arbib (1969)), but as an example we describe a PDA for $\{x^n y^n | n \geq 1\}$. Its states are s_0, s_1, s_2, where s_0 is initial and s_2 final. The store symbols are λ (initial symbol), μ, ν. We give the values for $\delta(s_i,.,.)$ in the form of a table for each s_i:

s_0	λ	μ	ν
x	$s_0\mu$	$s_0\mu\nu$	$s_0\nu^2$
y		$s_2\mu$	s_1

s_1	λ	μ	ν
x			
y		$s_2\mu$	s_1

Blanks and the remaining values (for s_2) remain undefined. To see how $x^n y^n$ is accepted, but no other strings, we note how the store acts as a memory, remembering how many factors x have been taken off. If we think of the store as arranged vertically, at each stage we remove the topmost symbol and add a number of symbols at the top, rather like a stack of plates in a cafeteria; this explains the name.

By a somewhat more elaborate process, with a tape on which the input is written (a 'linear-bounded' automaton) one can devise a class of machines which accept precisely all the CS-languages (see Landweber [1963]). These machines are more special than Turing machines in that their tape length is bounded by a linear function of the length of the input word.

Finally we note the following connexion with monoids:

Theorem 11.3.8. *A language L on X is regular if and only if there is a homomorphism $f : X^* \to M$ to a finite monoid M such that L is the complete inverse image of a subset N of $M : L = f^{-1}(N)$.*

Proof. Given a regular language L on X, we have a finite acceptor A which accepts precisely L. Now A defines an action of X^* on the set S of states of A. Thus we have a homomorphism $f : X^* \to \text{Map}(S)$. If P is the subset of Map(S) of all mappings taking s_0 into the set F of final states, then $w \in L$ iff $wf \in P$; thus $L = f^{-1}(P)$, and by definition Map(S) is a finite monoid. Thus the condition is satisfied.

Conversely, given $f : X^* \to M$ with $L = f^{-1}(N)$ for some $N \subseteq M$, we consider the acceptor A consisting of the alphabet X, state set M, action $\delta(a, x) = a.(xf)$, $(a \in M, x \in X)$, with neutral element 1 as initial state and N as set of final states. The language accepted by A is just L. ∎

Exercises

1. Find the languages generated by the following grammars: (i) $\sigma \to \sigma^2; x; y$, (ii) $\sigma \to \sigma^3; x; y$, (iii) $\sigma \to \sigma^2; x\sigma y; y\sigma x; 1$, (iv) $\sigma \to x\sigma x; xyx; x$.
2. Find a CF-grammar on x, y generating the set of all words in which x is immediately followed by y.
3. Find a grammar on x, y generating the set of all words in which each left factor has at least as many x's as y's. Is this a CF-language?
4. A grammar with rules of the form $\alpha \to x$, $\alpha \to \beta x$ is sometimes called *left regular*, while a grammar with rules of the form $\alpha \to x$, $\alpha \to x\beta$ is called *right regular*. Show that every language generated by a left regular grammar can also be generated by a right regular grammar. (Hint. Interpret the words of the language as circuits in the acceptor graph; a left and a right regular grammar correspond to the two senses of traversing these loops.)
5. Show that every CF-language in one letter is regular.
6. Show that a language in a one-letter alphabet $\{x^n \mid n \in I\}$ is regular iff the set I of exponents is ultimately periodic.

7. Construct a PDA for the set of all palindromes with 'centre marker', i.e. $L(G)$, where $G : \sigma \to x\sigma x; y\sigma y; z$ (Hint. Put one half of the word in store and then match the other half.)
8. Find a PDA for the set of all even palindromes $G : \sigma \to x\sigma x;\ y\sigma y;\ 1$. (Hint. Construct a PDA to 'guess' the centre.)
9. An automaton is called *deterministic* (resp. *total*) if for each $s \in S$, $x \in X$ there exists at most (resp. least) one pair $y \in Y$, $s' \in S$ such that $(s, x, y, s') \in P(A)$. For any A define its *reverse* A^0 as the automaton with state set S, input Y, output X and $P(A^0)$ as the set of all $(s', y, x, s) \in A$. Show that A^0 is deterministic whenever A is reduced.
10. A complete automaton A with N states $s_1, \ldots, s_N$ can be described by a set of $N \times N$ matrices $P(x|y)$ $(x \in X, y \in Y)$ where the (i, j)-entry of $P(x|y)$ is 1 if $\delta(s_i, x) = y$ and $\lambda(s_i, x) = s_j$, and 0 otherwise. Define $P(u|v)$ recursively for $u \in X^*$, $v \in Y^*$ by $P(ux|vy) = P(u|v)P(x|y)$, $P(u|v) = 0$ if $|u| \neq |v|$. Show that $P(uu'|vv') = P(u|v)P(u'|v')$. Further put $P(x) = \sum_y P(x|y)$, write π for the row vector whose i-th component is 1 if s_i is the initial state and 0 otherwise and write f for the column vector with component 1 for a final and 0 for a non-final state. Show that for any $u \in X^*$, $\pi P(u) f$ is 1 if u is accepted and 0 otherwise.
11. With the notation of Exercise 10, put $T = \sum_x P(x)$; verify that the (i, j)-entry of T^n is the number of words of length n in X which give a path from s_i to s_j. Show that if $\lambda(n)$ denotes the number of words of length n accepted, then the length generating function $L(t) = \sum \lambda(n)t^n$ satisfies $L(t) = \pi(I - tT)^{-1}f$. Use the characteristic equation of T to find a recursion formula for $\lambda(n)$.

11.4 Variable-length codes

The codes studied in Chapter 10 were block codes, where each code word has the same length. However, in practice different letters occur with different frequencies and an efficient code will represent the more frequently occurring letters by the shorter code word, e.g. in Morse code, the letters e, t which are among the most commonly occurring are represented by a dot and a dash respectively. For this reason it is of interest to have codes in which the code words have varying lengths. In this section we shall describe such codes; the main problem is to design the code so as to ensure that messages can be uniquely decoded. Of course we can only take the first steps in the subject, but these will include results which are of interest and importance in general coding theory, besides leading to a better understanding of free monoids and free algebras.

Let $X = \{x_1, \ldots, x_r\}$ be our alphabet and X^* the free monoid on X. Any subset A of X^* generates a submonoid, which we write as $\langle A \rangle$. By a *code* on X we understand a subset A of X^* such that every element of $\langle A \rangle$ can be uniquely factorized into elements of A; in other words, $\langle A\rangle$ is the free monoid on A as free generating set. Thus if $w \in \langle A\rangle$ and

$$w = a_1 \dots a_m = b_1 \dots b_n, \quad a_i, b_j \in A,$$

then $m = n$ and $a_i = b_i$, $i = 1, \dots, n$.

For example, X itself is a code; more generally, X^n, the set of all products of length n (for any given $n \geq 1$) is a code. Further, any subset of a code is a code. The set $\{x, xy, yx\}$ is *not* a code, because the word $xyx = xy.x = x.yx$ has two distinct factorizations.

Our first problem is how to recognize codes. If A is not a code, then we have an equality between two distinct words in A, and by cancellation we may take this to be of the form

$$au = bv, \quad \text{where } a, b \in A,\ u, v \in \langle A\rangle.$$

If we express everything in terms of X we find by the rigidity of X^*,

$$\text{either } a = bz \text{ or } b = az, \quad \text{for some } z \in X^*.$$

If $a = bz$, we say that b is a *prefix* of a. Let us define a *prefix set* as a non-empty subset of X^* in which no element is a prefix of another. For example, $\{1\}$ is a prefix set; any prefix set $A \neq \{1\}$ cannot contain 1, because 1 is a prefix of any other element of X^*.

What we have just found is that if A is not a code and $A \neq \{1\}$, then A is not a prefix set; thus we have

Proposition 11.4.1. *Every prefix set $\neq \{1\}$ is a code.* ■

By a *prefix code* we shall understand a prefix set $\neq \{1\}$. To give an example, $\{y, xy, x^2y, x^3y\}$ is a prefix set and hence a prefix code. We can think of this example as the alphabet $1, x, x^2, x^3$ with y as place marker.

By symmetry we define a *suffix set* as a non-empty subset of X^* in which no element is a suffix, i.e. right-hand factor, of another. Now the left-right symmetry of the notion of code shows that every suffix set $\neq \{1\}$ is a code; such a code will be called a suffix code. Since there exist suffix codes which are not prefix, e.g. $\{x, xy\}$, we see that the converse of Proposition 11.4.1 is false. In fact there exist procedures for determining whether a given subset of X^* is a code, but they are quite lengthy and will not be given here (the Sardinas–Patterson algorithm, see Lallement (1979), Berstel and Perrin (1985)). In any case, prefix codes are of particular interest in coding theory, since any message in a prefix code an be deciphered reading letter-by-letter from left to right (it is a 'zero-delay' code). This property actually characterizes prefix codes, for if a code is not prefix, say $a = bz$, where a, b are both code words, then at any occurrence of b in a message we have to read past this point to find out whether a or b is intended.

On any monoid M we can define a preordering by left divisibility:

$$u \leq v \Leftrightarrow v = uz \text{ for some } z \in M. \tag{11.4.1}$$

Clearly this relation is reflexive and transitive; we claim that when M is conical, with cancellation, then '$\leq$' is antisymmetric, so that we have a partial ordering. For if $u \leq v$, $v \leq u$, then $v = uz$, $u = vz'$, hence $v = uz = vz'z$, so $z'z = 1$ by cancellation, and since M is conical, we conclude that $z = z' = 1$.

We shall be particularly interested in the ordering (11.4.1) on free monoids. In that case the set of left factors of any element u is totally ordered, by rigidity, and since the length of chains of factors is bounded by $|u|$, the ordering satisfies the minimum condition. In terms of the ordering (11.4.1) on a free monoid, a prefix set is just an anti-chain, and by BA, Proposition 3.2.8 there is a natural bijection between anti-chains and lower segments. Here a 'lower segment' is a subset containing with any element all its left factors; such a set, if non-empty, is called a *Schreier set*; it is clear that every Schreier set contains 1.

Let us describe this correspondence between prefix sets and Schreier sets more explicitly: if C is a prefix set in X^*, then the corresponding Schreier set is the complement of CX^* in X^*; for a Schreier set P the corresponding prefix set is the set of all minimal elements in the complement of P.

A prefix set C is said to be *right large* if CX^* meets wX^* for every $w \in X^*$. By the rigidity of X^* this just amounts to saying that every element of X^* is comparable with some element of C. Hence the Schreier set corresponding to a right large prefix set C consists precisely of the proper prefixes of elements of C. To sum up these relations we need one more definition. In any monoid a product AB of subsets A, B is said to be *unambiguous* if each element of AB can be written in just one way as $c = ab$, where $a \in A$, $b \in B$.

Proposition 11.4.2. *Let X^* be the free monoid on a finite alphabet X. Then there is a natural bijection between prefix sets and Schreier sets: to each prefix set C corresponds $P = X^* \backslash CX^*$ and to each Schreier set P corresponds $C = PX \backslash P$, i.e. the set of minimal elements in $X^* \backslash P$, and we have the unambiguous product*

$$X^* = C^*P. \tag{11.4.2}$$

Moreover, P is finite if and only if C is finite and right large.

Proof. The description of each of C, P in terms of the other follows by BA, Proposition 3.2.8. Thus C is the set of minimal elements in $X^* \backslash P$; since $1 \in P$, any such minimal element must have the form px ($p \in P, x \in X$), and moreover, $px \notin P$. Thus

$$C = PX \backslash P = \{px | p \in P, x \in X, px \notin P\}. \tag{11.4.3}$$

Now to establish (11.4.2), take $w \in X^*$; either $w \in P$ or w has a maximal proper prefix p in P. In the latter case $w = pxu$, where $x \in X$ and $px \notin P$; therefore $px \in C$ by (11.4.3). Further $|u| < |w|$ and $1 \in P$, so by induction on the length, $u \in C^*P$, hence $w \in C^*P$ and (11.4.2) follows. Now if

$$w = u_1 \dots u_r p = v_1 \dots v_s q, \quad \text{where } p, q \in P, u_i, v_j \in C,$$

then since C is prefix, $u_1 = v_1$, so we can cancel u_1 and conclude by induction on r that $r = s$, $u_i = v_i$, $i = 1, \dots, r$, $p = q$. This shows (11.4.2) to be unambiguous.

Suppose that P is finite; then so is C, by (11.4.3). Given $w \in X^*$, either $w \in P$; then some right multiple of w is not in P, because the length of the elements in P is bounded. A least such element c is a right multiple of w in C, so $w < c$. Or $w \notin P$; then since $1 \in P$, there is a minimal prefix c of w not in P and this is again in C, so $c \leq w$. This shows C to be right large.

Conversely, suppose that C is finite right large and let P be the corresponding Schreier set. Given $w \in X^*$, either $w \geq c$ or $w < c$ for some $c \in C$, and the first alternative is excluded for members of P. Thus P consists of all proper prefixes of elements of C and this is again a finite set. ∎

For a closer study of codes it is useful to have a numerical measure for the elements of X^*. By a *measure* on X^* we understand a homomorphism μ of X^* into the multiplicative monoid of positive real numbers, such that

$$\sum_{x \in X} \mu(x) = 1. \tag{11.4.4}$$

Clearly $\mu(1) = 1$ and the value of μ on X can be assigned arbitrarily as positive real numbers, subject only to (11.4.4); once this is done, μ is completely determined on X^* by the homomorphism property. For example, writing $m(x) = r^{-1}$, we obtain the *uniform measure* on X^*:

$$m(w) = r^{-|w|}.$$

Any measure μ on X^* can be extended to subsets by putting

$$\mu(A) = \sum_{a \in A} \mu(a).$$

We note that $\mu(A)$ is a positive real number or ∞, and $\mu(X) = 1$ by (11.4.4). For a product of subsets we have

$$\mu(AB) \leq \mu(A)\mu(B), \tag{11.4.5}$$

with equality if the product is unambiguous. To prove (11.4.5), let us first take A, B finite, say $A = \{a_1, \ldots, a_m\}$, $B = \{b_1, \ldots, b_n\}$. We have

$$\sum \mu(a_i b_j) = \sum \mu(a_i)\mu(b_j) = \left(\sum \mu(a_i)\right)\left(\sum \mu(b_j)\right),$$

and here each member of AB occurs just once if AB is unambiguous, and otherwise more than once, so we obtain (11.4.5) in this case, with equality in the unambiguous case. In general (11.4.5) holds for any finite subsets A', B' of A, B by what has been shown. Therefore $\mu(A'B') \leq \mu(A)\mu(B)$, and now (11.4.5) follows by taking the limit.

In particular, for any code C, the product CC is unambiguous by definition, hence on writing $C^2 = CC$ etc., we have

$$\mu(C^n) = \mu(C)^n, \quad n = 1, 2, \ldots. \tag{11.4.6}$$

Let us apply (11.4.6) to X; we have $\mu(X) = 1$ by (11.4.4) and X is clearly a code, hence we find

$$\mu(X^n) = 1 \text{ for all } n \geq 1. \tag{11.4.7}$$

We shall need an estimate of $\mu(A)$ for finite sets A. We recall that $X^+ = X^* \backslash \{1\}$.

Lemma 11.4.3. *Let μ be any measure on X^*. Then for any finite subset A of X^+,*

$$\mu(A) \leq \max\{|a| \mid a \in A\}. \tag{11.4.8}$$

Proof. Since A is finite, the right-hand side of (11.4.8) is finite, say it equals d. Then $A \subseteq X \cup X^2 \cup \ldots \cup X^d$, and so by (11.4.7),

$$\mu(A) \leq \mu(X) + \mu(X^2) + \ldots + \mu(X^d) = d.$$ ■

From this lemma we can obtain a remarkable inequality satisfied by codes, which shows that to be a code, a set must not be too large.

McMillan Inequality. *Let C be any code on X^*. Then for any measure μ on X we have*

$$\mu(C) \leq 1. \tag{11.4.9}$$

Proof. Consider first the case where C is finite and let $\max\{|c| \mid c \in C\} = d$. Then by (11.4.6), (11.4.8),

$$\mu(C)^n = \mu(C^n) \leq nd,$$

since the elements of C^n have length at most nd. Taking n-th roots, we find that $\mu(C) \leq (nd)^{1/n}$. Here d is fixed; letting $n \to \infty$, we have $(nd)^{1/n} \to 1$, therefore $\mu(C) \leq 1$. In the general case every finite subset of C is a code and so satisfies (11.4.9), hence this also holds for C itself. ■

Of course the condition (11.4.9) is by no means sufficient for a code, since any set C will satisfy (11.4.9) for a suitable measure, if we choose X large enough.

A code on X is said to be *maximal* if it is not a proper subset of a code on X. Maximal codes always exist by Zorn's lemma, since the property of being a code is of finite character. The above inequality provides a convenient test for maximality:

Proposition 11.4.4. *Let C be a code on X. If $\mu(C) = 1$ for some measure μ, then C is a maximal code.*

Proof. Suppose that C is not a maximal code. Then we can find a code B containing C and another element b, say. We have $\mu(B) \geq \mu(C) + \mu(b) > 1$, and this contradicts (11.4.9). ■

For example, X and more generally X^n for any n is a maximal code.

Although the inequality (11.4.9) is not sufficient to guarantee that C is a code, there is a sense in which this inequality leads to a code. This is expressed in the next result, which gives a construction for codes with a prescribed uniform measure.

Theorem 11.4.5. *Let $n_1, n_2, \ldots$ be any sequence of positive integers. Then there exists a code $A = \{a_1, a_2, \ldots\}$ with $|a_i| = n_i$ in an alphabet of r letters if and only if*

$$r^{-n_1} + r^{-n_2} + \ldots \leq 1 \quad \text{(Kraft–McMillan inequality).} \tag{11.4.10}$$

We remark that the left of (11.4.10) is just the uniform measure of A.

Proof. The necessity of (11.4.10) is clear by (11.4.9) and the above remark. Conversely, assume that (11.4.10) holds, take the n_i to be ordered by size: $n_1 \leq n_2 \leq \ldots$, and let $X = \{0, 1, \ldots, r-1\}$ be the alphabet. Define the partial sums of (11.4.10):

$$s_k = r^{-n_1} + \ldots + r^{-n_{k-1}},$$

and for each s_k define an integer

$$p_k = r^{n_k} s_k = r^{n_k - n_1} + \ldots + r^{n_k - n_{k-1}}.$$

Each p_k is an integer and since $s_k < 1$ by (11.4.10), we have

$$0 \leq p_k < r^{n_k}.$$

Now take a_k to be the element of X^* formed by expressing p_k in the scale of r, with enough 0's prefixed to bring the length up to n:

$$a_k = \alpha_1\alpha_2 \ldots \alpha_{n_k} = \alpha_1 r^{n_k - 1} + \alpha_2 r^{n_k - 2} + \ldots + \alpha_{n_k - 1} r + \alpha_{n_k}, \quad \alpha_i \in X.$$

We claim that $A = \{a_1, a_2, \ldots\}$ is a code of the required type. The lengths are right by construction, and we shall complete the proof by showing that A is a prefix code. If a_j is a prefix of a_i, $j < i$, then a_j is obtained from a_i by cutting off the last $n_i - n_j$ digits. Thus p_j is the greatest integer in the fraction

$$\frac{p_i}{r^{n_i - n_j}} = r^{n_j} s_i \geq r^{n_j}(s_j + r^{-n_j}) = p_j + 1.$$

But this is a contradiction. Thus A is indeed a prefix code. ■

For example, take $r = 3$ and consider the sequence 1, 1, 2, 2, 3, 3, 3. We have $\mu(A) = 1/3 + 1/3 + 1/9 + 1/9 + 1/27 + 1/27 + 1/27 = 1$, so we have a maximal code. It is given by the table:

k	1	2	3	4	5	6	7
n_k	1	1	2	2	3	3	3
p_k	0	1	6	7	24	25	26
a_k	0	1	20	21	220	221	222

The construction of a_k given in Theorem 11.4.5 can be described by the rule: choose the least number in the ternary scale which is not a prefix of 222 and which has no a_i $(i < k)$ as a prefix.

We have seen that codes are certain subsets of X^* that are not too large; we now introduce a class of subsets that are not 'too small', in order to study the interplay between these classes. A subset A of X^* is said to be *complete* if the submonoid generated by it meets every ideal of X^*, i.e. every word in X^* occurs as a factor in some word in $\langle A \rangle$:

$$X^*wX^* \cap \langle A \rangle \neq \varnothing \quad \text{for all } w \in X^*.$$

Proposition 11.4.6. *Let A be a finite complete subset of X^* and let m be the uniform measure on X. Then $m(A) \geq 1$.*

Proof. Let L be the set of prefixes, R the set of suffixes and F the set of factors of members of A. Since A is finite, L, R and F are all finite. We claim that

$$R\langle A \rangle L \cup F = X^*. \tag{11.4.11}$$

For, given $w \in X^*$, we have, by the completeness of A,

$$pwq = a_1 a_2 \ldots a_n,$$

where $a_i \in A$. Now either w occurs as a factor in some a_i and so $w \in F$, or two or more of the a_i form part of w. In that case w consists of a word in A with a suffix of some a_i on the left and a prefix of some a_j on the right, and this is just (11.4.11). Now (11.4.11) shows that $m(X^*) = m(R)m(\langle A \rangle)m(L) + m(F)$, and since $m(X^*)$ is infinite, while $m(F)$, $m(R)$, $m(L)$ are finite, it follows that $m(\langle A \rangle)$ is infinite. Thus

$$\infty = m(\langle A \rangle) \leq \sum m(A^n) \leq \sum m(A)^n.$$

If $m(A) < 1$, this is a contradiction, therefore $m(A) \geq 1$, as claimed. ■

We next establish a connexion with codes:

Theorem 11.4.7 (Schützenberger). *Any maximal code is complete.*

Proof. Let A be a code which is not complete; we shall show how to enlarge it. If $|X| = 1$, any non-empty set is complete, so we have $A = \varnothing$ and then X is a larger code. When $|X| > 1$, we have to find $b \notin A$ such that $A \cup \{b\}$ is a code. Since A is not complete, there is a word $c \in X^*$ such that $X^*cX^* \cap \langle A \rangle = \varnothing$. One might be tempted at this point to adjoin c to A, but this leads to problems because c might intersect itself; we shall construct b to avoid this. Let $|c| = \gamma$ and put $c = xc'$, where $x \in X$. Choose $y \neq x$ in X and put

$$b = cxy^\gamma = xc'xy^\gamma. \tag{11.4.12}$$

From the definition of c it is clear that

$$X^*bX^* \cap \langle A \rangle = \varnothing. \tag{11.4.13}$$

We claim that $A \cup \{b\}$ is a code. For if not, then we have an equation

$$a_1 \dots a_r = a'_1 \dots a'_s, \qquad \textit{where } a_i, a'_j \in A \cup \{b\}, \tag{11.4.14}$$

and (11.4.14) is non-trivial. Since A is a code, b must occur in (11.4.14), and by (11.4.13) it must occur on both sides of (11.4.14), say $a_i = a'_j = b$ for some i, j. We take i, j minimal; if the two occurrences of b do not overlap, we have a contradiction to (11.4.13) (see diagram)

b

$\langle A \rangle \quad b$

If there is an overlap, it must be at least γ letters, by (11.4.12), but that is impossible, because in b letters 1 and $\gamma + 1$ are x, whereas the last γ letters of b are y. ■

We note that the converse result does not hold (see Exercise 5). Our final result clarifies the relation between complete sets and codes.

Theorem 11.4.8 (Boë, de Luca and Restivo [1980]). *Let A be a finite subset of X. Then any two of the following imply the third:*

(a) *A is a code,*
(b) *A is complete,*
(c) $m(A) = 1$.

Proof. (a, b) $\Rightarrow$ (c). If A is a complete code, then $m(A) = 1$ by Proposition 11.4.6 and McMillan's inequality. (b, c) $\Rightarrow$ (a). If $m(A) = 1$, but A is not a code, then for some n, $m(A^n) < m(A)^n = 1$, hence A^n is not complete, and so neither is A. (c, a) $\Rightarrow$ (b). If $m(A) = 1$ and A is a code, then A is a maximal code and so it is complete, by Theorem 11.4.7. ■

The situation is reminiscent of what happens for bases in a vector space. A basis is a linearly independent spanning set, and of the following conditions on a set A of vectors in a vector space V any two imply the third:

(a) A is linearly independent,
(b) A is a spanning set,
(c) $|A| = \dim V$.

In the proof the exchange axiom plays a vital role: it has the consequence that any spanning set contains a linearly independent set which still spans. The analogue here is false; there are minimal complete sets that are not codes. For example, take $X = \{x, y\}$, $A = \{x^3, x^2yx, x^2y, yx, y\}$. It is easily verified that A is minimal complete. The subset $A_1 = \{x^3, x^2yx, yx, y\}$ is not a code and $m(A_1) = 1$, while all other proper subsets of A are codes. What can be shown is the following (see Boë, de Luca and Restivo [1980]):

A minimal complete set is a code iff all its proper subsets are codes.

For finite codes the converse of Theorem 11.4.7 holds: every finite complete code is maximal (this follows from Theorem 11.4.8 and Proposition 11.4.4), but there are infinite complete codes which are not maximal (see Exercise 5); this also shows that Theorem 11.4.8 does not extend to infinite sets.

Exercises

1. Determine which of the following are codes: (i) $\{xy, xy^2, y^2\}$, (ii) $\{x^2, xy, x^2y, xy^2, y^2\}$, (iii) $\{x^2, xy^2, x^2y, xy^3, y^2, yx\}$.
2. Construct a code for $r = 2$ and the sequence $1, 2, 3, \ldots$. Likewise for $r = 4$ and the sequence $1, 1, 1, 2, 2, 2, 3, 3, 3, \ldots$.
3. Let X be an alphabet, G a group and $f : X^* \to G$ a homomorphism. Show that for any subgroup H of G, Hf^{-1} is a free submonoid of X^*, and its generating set is a maximal code which is *bifix* (i.e. prefix and suffix). Such a code is called a *group code*.
4. Let X be an alphabet and $w_x(u)$ the x-length of $u \in X^*$. Show that for any integer m the mapping $f : u \mapsto w_x(u) \pmod{m}$ is a homomorphism from X^* to $\mathbf{Z}/m$ and describe the group code $0f^{-1}$. Show that for $m = 2$, $X = \{x, y\}$, $0f^{-1} = \{y\} \cup \{xy^*x\}$; find $0g^{-1}$ where $g : u \mapsto w_x(u) \pmod{3}$.
5. Let $X = \{x, y\}$. Show that $\delta(u) = w_x(u) - w_y(u)$ is a homomorphism to $\mathbf{Z}$. Describe the corresponding group code D (this is known as the *Dyck code*). Show that D is complete and remains complete if one element is omitted.
6. Let $f : X^* \to Y^*$ be an injective homomorphism of free monoids. Show that if A is a code in X^*, then Af is a code in Y^*; if B is a code in Y^*, then Bf^{-1} is a code in X^*.
7. Let A, B be any codes in X^*. Show that A^n is a code for any $n \geq 1$, but AB need not be a code.
8. Let $X = \{x, y\}$, $\mu(x) = p$, $\mu(y) = q$, where $p, q \geq 0$, $p + q = 1$. Show that $C = \{xy, yx, xy^2x\}$ is not a code, even though it satisfies (11.4.9).

11.5 Free algebras and formal power series rings

There is a further way of describing languages, namely as formal power series. This is in some respects the simplest and most natural method. Let X be a finite alphabet and k a commutative field. By a *formal power series* in X over k we understand a function f on X^* with values in k. The value of f at $u \in X^*$ is denoted by (f, u) and f itself may be written as a series

$$f = \sum (f, u)u. \tag{11.5.1}$$

Here (f, u) is called the *coefficient* of u, in particular, $(f, 1)$ is called the *constant term* of f.

Series are added and multiplied by the rules

$$(f+g,u)=(f,u)+(g,u), \tag{11.5.2}$$

$$(fg,u)=\sum_{yz=u}(f,y)(g,z). \tag{11.5.3}$$

Since each element of X^* has only a finite number of factors, the sum in (11.5.3) is finite, so fg is well-defined. The set of all these power series is denoted by $k\langle\langle X\rangle\rangle$; it is easily seen to form a k-algebra with respect to these operations. For each power series f its *support* is defined as

$$D(f)=\{u\in X^*|(f,u)\neq 0\}.$$

Thus u lies in the support of f precisely if it occurs in the expression (11.5.1) for f. The elements of finite support are called *polynomials* in X; they are in fact just polynomials, i.e. k-linear combinations of products of elements of X, but care must be taken to preserve the order of the factors, since the elements of X do not commute. These polynomials form a subalgebra $k\langle X\rangle$, called the free k-algebra on X (see Section 8.7). We remark that $k\langle X\rangle$ can also be defined as the monoid algebra of the free monoid X^*, in analogy to the group algebra.

For each power series f we define its order $o(f)$ as the minimum of the lengths of terms in its support. The order is a positive integer or zero, according as $(f,1)$ is or is not zero. For a polynomial f we can also define its degree $d(f)$; it is the maximum of the lengths of terms in its support. If all terms of f have the same length r, so that $o(f)=d(f)=r$, then f is said to be *homogeneous* of degree r.

We remark that if u is a series of positive order, we can form the series

$$u^*=1+u+u^2+\dots.$$

The infinite series on the right 'converges' because if $o(u)=r$, then for any given d, u^n contributes only if $rn\leq d$. Thus in calculating the terms of degree d in u^* we need only consider u for $n=0,1,\dots,[d/r]$. We also note that u^* satisfies the equations

$$u^*u=uu^*=u^*-1.$$

Hence $(1-u)u^*=u^*(1-u)=1$, so u^* is the inverse of $1-u$:

$$(1-u)^{-1}=u^*=\sum_0^\infty u^n. \tag{11.5.4}$$

It is easily verified that $k\langle X\rangle$ has the familiar universal property: every mapping $\varphi:X\to A$ into a k-algebra A can be extended in just one way to a homomorphism $\bar\varphi:k\langle X\rangle\to A$. As a consequence every k-algebra can be written as a homomorphic image of a free k-algebra, possibly on an infinite alphabet. Of course free algebras on an infinite alphabet are defined in exactly the same way; for power series rings there are several possible definitions, depending on the degrees assigned to the variables, but this need not concern us here, as we shall only consider the case of a finite alphabet.

The free algebra $k\langle X\rangle$ may be regarded as a generalization of the polynomial ring $k[x]$, to which it reduces when X consists of a single element x. The polynomial ring is of course well known and has been thoroughly studied. The main tool is the Euclidean algorithm; this allows one to prove that $k[x]$ is a principal ideal domain (PID) and a unique factorization domain (UFD) (see BA, Section 10.2). The UF-property extends to polynomials in several (commuting) variables (BA, Section 10.3), but there is no analogue to the principal ideal property in this case. For the non-commutative polynomial ring $k\langle X\rangle$ the UF-property persists, albeit in a more complicated form, and we shall have no more to say about it here (see Exercise 3 below and Cohn (1985), Chapter 3). The principal ideal property generalizes as follows. We recall from Section 8.7 that a free right ideal ring, right fir for short, is a ring R with invariant basis number (IBN) in which every right ideal is free as right R-module; left firs are defined similarly and a left and right fir is called a fir. In the commutative case a fir is just a PID and the fact that $k[x]$ is a PID generalizes to the assertion that $k\langle X\rangle$ is a fir. This is usually proved by the weak algorithm, a generalization of the Euclidean algorithm, to which it reduces in the commutative case. We shall not enter into the details (see Cohn (1985), Chapter 2), but confine ourselves below to giving a direct proof that $k\langle X\rangle$ is a fir. This method, similar to the technique used to prove that subgroups of free groups are free, is due to Jacques Lewin; our exposition follows essentially Berstel and Reutenauer (1988) (see also Cohn (1985) Chapter 6).

Theorem 11.5.1. *Let $F = k\langle X\rangle$ be the free algebra on a finite set X over a field k and let $\mathfrak{a}$ be any right ideal of F. Then there exists a Schreier set P in X^* which is maximal linearly independent* (mod $\mathfrak{a}$). *If C is the corresponding prefix set, determined as in Proposition 11.4.2, then for each $c \in C$ there is an element of $\mathfrak{a}$:*

$$f_c = c - \sum \alpha_{c,p}p \quad (p \in P, \alpha_{c,p} \in k)$$

where the sum ranges over all $p \in P$, but has only finitely many non-zero terms for each $c \in C$, such that $\mathfrak{a}$ is free as right F-module on the $f_c (c \in C)$ as basis. Similarly for left ideals, and F has IBN, so it is a fir.

Proof. The monoid X^* is a k-basis of F, hence its image in $F/\mathfrak{a}$ is a spanning set and it therefore includes a basis of $F/\mathfrak{a}$ as k-space. Moreover, we can choose this basis to be a Schreier set, by building it up according to length. Thus if P_n is a Schreier set which forms a basis for all elements of F of degree at most n (mod $\mathfrak{a}$), then the set P_nX spans the space of elements of degree at most $n+1$ and by choosing a basis from it we obtain a Schreier set P_{n+1} containing P_n and forming a basis (mod $\mathfrak{a}$) for the elements of degree at most $n+1$. In this way we obtain a Schreier set $P = \cup P_n$ which is maximal k-linearly independent (mod $\mathfrak{a}$) and hence a k-basis. Let C be the corresponding prefix set. For each $c \in C$ the set $P \cup \{c\}$ is still a Schreier set, but by the maximality of P it is linearly dependent mod $\mathfrak{a}$, say

$$f_c = c - \sum \alpha_{c,p}p \in \mathfrak{a}, \tag{11.5.5}$$

where the sum ranges over P and almost all the $\alpha_{c,p}$ vanish. We claim that every $b \in F$ can be written as

$$b = \sum f_c g_c + \sum \beta_p p, \quad \text{where } g_c \in F,\ \beta_p \in k, \tag{11.5.6}$$

and the sums range over C and P respectively. By linearity it is enough to prove this when b is a monomial. When $b \in P$, this is clear; we need only take $\beta_p = 1$ for $p = b$ and the other coefficients zero. When $b \notin P$, it has a prefix in C by Proposition 11.4.2, say $b = cu$, where $c \in C$ and $u \in F$. By (11.5.5) we have

$$b = cu = f_c u + \sum \alpha_{c,p} pu. \tag{11.5.7}$$

For any $p \in P$, either $pu \in P$ or $pu = c_1 u_1$ where $c_1 \in C$ and hence $|p| < |c_1|$, $|u_1| < |u|$. In the first case we have achieved the form (11.5.6); in the second case we use induction on $|u|$ to express $c_1 u_1$ in the same form. Thus we can reduce all the terms on the right of (11.5.7) to the form (11.5.6) and the conclusion follows.

We claim that the elements (11.5.5) form the desired basis of $\mathfrak{a}$. To show that they generate $\mathfrak{a}$, let us take $b \in \mathfrak{a}$ and apply the natural homomorphism $F \to F/\mathfrak{a}$. Writing the image of r as $\bar{r}$, we have

$$0 = \bar{b} = \sum \beta_p \bar{p}.$$

Since the $\bar{p}$ are linearly independent by construction, we have $\beta_p = 0$, so $b = \sum f_c g_c$ and it follows that the f_c generate a. To prove their independence over F, assume that $\sum f_c g_c = 0$, where not all the g_c vanish. Then by (11.5.5)

$$\sum c g_c = \sum \alpha_{c,p} p g_c. \tag{11.5.8}$$

Take a word w of maximal length occurring in some g_c, say in $g_{c'}$. Since C is a prefix code, $c'w$ occurs with a non-zero coefficient λ on the left of (11.5.8). Hence

$$\lambda = \sum \alpha_{c,p} \mu_{c,p},$$

where $\mu_{c,p}$ is the coefficient of $c'w$ in pg_c. Now the relation $c'w = pu$ can hold only when p is a proper prefix of c', hence $|p| < |c'|$, $|u| > |w|$ and this contradicts the definition of w. This contradiction shows that the f_c are linearly independent over F, so they form a basis of $\mathfrak{a}$, which is therefore a free right ideal. By symmetry every left ideal is free, and F clearly has IBN, since we have a homomorphism $F \to k$, obtained by setting $X = 0$. This shows F to be a fir. ∎

We recall the equation $X^* = C^*P$ obtained in Proposition 11.4.2. In the power series ring this may be written as $(1 - X)^{-1} = (1 - C)^{-1}P$, where X, C, P are now the sums of the corresponding sets. On multiplying up, we obtain $1 - C = P(1 - X)$, or

$$C = PX - (P - 1). \tag{11.5.9}$$

This tells us again that the prefix set C consists of all products px $(p \in P, x \in X)$, which are not in P. By Proposition 11.4.2, P is finite iff C is finite and right large, but in our case this just means that $\mathfrak{a}$ is finitely generated and large as right ideal, while the finiteness of P means that $\mathfrak{a}$ has finite codimension in F. By replacing the elements of X by 1 in (11.5.9), we thus obtain

Corollary 11.5.2. *Let $F = k\langle X\rangle$ be the free algebra as in Theorem 11.5.1 and $\mathfrak{a}$ a right ideal of F. Then $\mathfrak{a}$ has finite codimension in F if and only if $\mathfrak{a}$ is finitely generated and large as right ideal. If $|X| = d$, $\mathfrak{a}$ has codimension r and has a basis of n elements, then*

$$n - 1 = r(d - 1). \qquad \blacksquare \qquad (11.5.10)$$

We remark that (11.5.10) is analogous to Schreier's formula for the rank of a subgroup of a free group (see Section 3.4); it is known as the *Schreier–Lewin formula.*

We now turn to consider the power series ring. To describe the structure of $k\langle\langle X\rangle\rangle$ we recall that a *local ring* is a ring R in which the set of all non-units forms an ideal $\mathfrak{m}$. Clearly $\mathfrak{m}$ is then the unique maximal ideal of R, and $R/\mathfrak{m}$ is a skew field, called the *residue class field* of R.

Proposition 11.5.3. *The power series ring $k\langle\langle X\rangle\rangle$ on any finite set X over a field k is a local ring with residue class field k. Its maximal ideal consists of all elements with zero constant term.*

Proof. The mapping $X \to 0$ defines a homomorphism of $k\langle\langle X\rangle\rangle$ onto k, hence the kernel $\mathfrak{m}$, consisting of all elements with zero constant term, is an ideal in $k\langle\langle X\rangle\rangle$. It follows that $k\langle\langle X\rangle\rangle/\mathfrak{m} \cong k$, and $\mathfrak{m}$ contains no invertible element. Any element f not in $\mathfrak{m}$ has non-zero constant term λ and so $\lambda^{-1}f = 1 - u$, where $o(u) > 0$. By (11.5.4) we have $(1-u)^{-1} = u^*$, hence $f^{-1} = \lambda^{-1}u^*$. $\blacksquare$

A power series f is called *rational* if it can be obtained from the elements of $k\langle X\rangle$ by a finite number of operations of addition, multiplication and inversion of series with non-zero constant terms. The rational series form a subring of $k\langle\langle X\rangle\rangle$, denoted by $k\langle X\rangle_{\text{rat}}$, as we shall see in Proposition 11.5.4 below, and the method of Proposition 11.5.3 shows that $k\langle X\rangle_{\text{rat}}$ is again a local ring.

Let us note that any square matrix A over $k\langle X\rangle_{\text{rat}}$ is invertible provided that its constant term is invertible over k. To prove this fact, we write $A = A_0 - B$, where A_0 is over k and B has zero constant term. By hypothesis A_0 is invertible over k and on writing $A_0^{-1}A = I - A_0^{-1}B$ we reduce the problem to the case where $A_0 = I$. As in the scalar case we can now write

$$A^{-1} = (I - A_0^{-1}B)^{-1}A_0^{-1} = (A_0^{-1}B)^*A_0^{-1},$$

which makes sense because all the terms of $A_0^{-1}B$ have positive order. Strictly speaking this does not make it clear that all the entries lie in $k\langle X\rangle$; to see this we need to invert the entries of $I - U$, where the terms of U have positive orders, term by term. Since the diagonal entries are units, while the non-diagonal entries are non-units, this is always possible. This method also yields a criterion for the rationality of power series.

Proposition 11.5.4 (Schützenberger). *The set $k\langle X\rangle_{\text{rat}}$ of all rational series is a subalgebra of $k\langle\langle X\rangle\rangle$. Moreover, for any $f \in k\langle\langle X\rangle\rangle$ the following conditions are equivalent:*

(a) *f is rational,*
(b) *$f = u_1$ is the first component of the solution of a system*

$$u = Bu + b, \tag{11.5.11}$$

where B is a matrix and b is a column over $k\langle X\rangle$, B is homogeneous of degree 1 and b has degree at most 1,
(c) *$f = u_1$ is a component of a system*

$$Fu = b, \tag{11.5.12}$$

where F is a matrix with invertible constant term and b is a column over $k\langle X\rangle$.

We remark that (11.5.11) can also be written as

$$(I - B)u = b; \tag{11.5.13}$$

thus it has the form (11.5.12), where F now has constant term I and has no terms of degree higher than 1.

Proof. (a) $\Rightarrow$ (b). We shall show that the set of all elements satisfying (b) forms a subalgebra of $k\langle\langle X\rangle\rangle$ containing $k\langle X\rangle_{\text{rat}}$, in which every series of order 0 is invertible. It then follows that this subalgebra contains $k\langle X\rangle_{\text{rat}}$.

It is clear that $a \in k \cup X$ is the solution of $u_1 = a$. Given f, g, suppose that $f = u_1$, where u is the solution of (11.5.11) and $g = v_1$, where v is the solution of $v = Cv + c$, where C satisfies the same conditions as B in (11.5.11). We shall rewrite these equations as $(I - B)u = b$, $(I - C)v = c$. Then $f - g$ is the first component of the solution of the system

$$\begin{pmatrix} I - B & e_1 - B_1 & 0 \\ 0 & & I - C \end{pmatrix} w = \begin{pmatrix} b \\ c \end{pmatrix}, \tag{11.5.14}$$

where $e_1 = (1, 0, \dots, 0)^T$ and B_1 is the first column of B; for (11.5.14) is satisfied by $w = (u_1 - v_1, u_2, \dots, u_m, v_1, \dots, v_n)^T$. In (11.5.14) the matrix on the left is not of the required form, but it can be brought to the form of a matrix with constant term I (and no terms of degree higher than 1) by subtracting row $m + 1$ from row 1 to get rid of the coefficient 1 in the $(1, m + 1)$-position.

Similarly fg is the first component of the solution of

$$\begin{pmatrix} I - B & b & 0 \\ 0 & & I - C \end{pmatrix} w = \begin{pmatrix} 0 \\ c \end{pmatrix}, \tag{11.5.15}$$

for (11.5.15) is satisfied by $w = (u_1v_1, u_2v_1, \dots, u_mv_1, v_1, \dots, v_n)^T$. If b has a non-zero constant term, we can bring (11.5.15) to the required form by subtracting appropriate multiples of row $m + 1$ from row $1, \dots,$ row m.

It remains to invert a series of order 0 or, what comes to the same thing (after what has been shown), a series with constant term 1. Let f have zero constant term and suppose that $f = u_1$, where u satisfies (11.5.11). We shall invert $1 + f$ by finding an equation for g, where $(1-g)(1+f)=1$. We may assume that $b=(0,\ldots,0,1)^T$ by writing (11.5.11) in the form

$$\begin{pmatrix} I-B & -b \\ 0 & 1 \end{pmatrix}\begin{pmatrix} u \\ 1 \end{pmatrix}=\begin{pmatrix} 0 \\ 1 \end{pmatrix},$$

subtracting multiples of column $2,\ldots,n-1$ from column n to reduce the constant term of b to 0, and adding corresponding multiples of 1 to the components of u. This leaves u_1 unchanged and it is sufficient: b_1 already has zero constant term because this is true of u_1. Thus our system now has the form (after a change of notation)

$$(I-B)u=e_n,$$

and here $n>1$, because u_1 has zero constant term. Let E_{n1} be the matrix with $(n,1)$-entry 1 and 0 elsewhere. Then we have

$$(I-B+E_{n1})u=e_n(1+u_1). \tag{11.5.16}$$

But we can also solve the system

$$(I-B+E_{n1})v=e_n, \tag{11.5.17}$$

and we can again bring the matrix on the left to the required form by subtracting the first row from the last, without affecting the solution. Comparing (11.5.16) and (11.5.17), we find that

$$u=v(1+u_1).$$

In particular, $u_1=v_1(1+u_1)$, hence $(1-v_1)(1+u_1)=1+u_1-u_1=1$, so we have found a left inverse for $1+u_1$. By uniqueness it is a two-sided inverse, which is what we had to find. This then shows that the set of components of solutions of systems (11.5.11) is a ring containing $k\langle X\rangle$ and admitting inversion when possible.

(b) $\Rightarrow$ (c) is clear, and to prove (c) $\Rightarrow$ (a) we must show that the solution of (11.5.12) has rational components. We have seen that the matrix $I-B$ has an inverse whose entries are rational, hence the same is true of $u=(I-B)^{-1}b$.

We have now shown (a)–(c) to be equivalent, and the elements satisfying (11.5.11) form a subring containing $k\langle X\rangle$, hence a subalgebra. It follows that $k\langle X\rangle_{\text{rat}}$ is a subalgebra in which all series of order 0 have inverses. ∎

With every language L on the alphabet X we associate a power series f_L, its *characteristic series*, defined by

$$(f_L,u)=\begin{cases}1 & \text{if } u\in L,\\ 0 & \text{if } u\notin L.\end{cases}$$

In this way the language is described by a single element of $K\langle\langle X\rangle\rangle$. Our main object will be to characterize the regular and context-free languages.

Theorem 11.5.5 (Schützenberger). *A language on X is regular if and only if its characteristic series is rational.*

Proof. Let L be a regular language. Then L is generated by a grammar with rules of the form

$$\alpha \to x\beta, \quad \alpha \to y, \tag{11.5.18}$$

moreover, each word of L has a single derivation. Let us number the variables of the grammar as $u_1, \ldots, u_n$, where $u_1 = \sigma$, and number the terminal letters as $x_1, \ldots, x_r$. The rules (11.5.18) may be written as

$$\alpha = \sum x\beta + \sum y, \tag{11.5.19}$$

where the summations are over all the right-hand sides of rules with α on the left. If we express (11.5.19) in terms of the u's and x's, we find

$$u_i = \sum b_{ij}u_j + b_i; \tag{11.5.20}$$

to generate the language we replace $\sigma = u_1$ by the right-hand side of (11.5.20) and continue replacing each u_i by the corresponding right-hand side. We thus obtain power series $f_1, \ldots, f_n$ such that $u_1 = f_1$ satisfies (11.5.20), and f_1 is the characteristic series of L; clearly f_1 is rational.

Conversely, if the characteristic series for L is rational, then it is given as a component of the solution of a system of the form (11.5.20), where the b_{ij} are linear homogeneous in the x_i and the b_i are of degree at most 1. We take the grammar of L in the form $x_i \to x_k u_j$ if x_k occurs in b_{ij} and $u_i \to 1$ if b_i has a non-zero constant term. This ensures that the derivations give precisely the words of L, thus L is regular. ∎

For example, consider the language $\{xy^n\}$, with grammar $\sigma \to \sigma y; x$. Its characteristic series is obtained by solving the equation $u = uy + x$, i.e. $u = x(1-y)^{-1} = \sum xy^n$.

Next we consider the problem of characterizing context-free languages. For this purpose we define another subalgebra of the power series ring. An element f of $k\langle\langle X\rangle\rangle$ is said to be *algebraic* if it is of the form $f = \alpha + u_1$, where $\alpha \in k$ and u_1 is the first component of the solution of a system of equations

$$u_i = \varphi_i(u, x), \quad i = 1, \ldots n, \tag{11.5.21}$$

where φ_i is a (non-commutative) polynomial in the u's and x's without constant term or linear term in the u's. The set of all algebraic elements is denoted by $k\langle X\rangle_{\text{alg}}$; we shall show that it is a subalgebra of $k\langle\langle X\rangle\rangle$.

Proposition 11.5.6. *Any system (11.5.21), where φ is a polynomial without constant term or linear term in the u's, has a unique solution in $k\langle\langle X\rangle\rangle$ with components of positive order, and the set $k\langle X\rangle_{\text{alg}}$ of all such elements is a subalgebra and a local ring, each element of order zero being invertible.*

Proof. Writing $u_i^{(\nu)}$ for the component of degree ν of u_i, we find by equating homogeneous components in (11.5.21),

$$u_i^{(\nu)} = \varphi_i^{(\nu)}(u, x).$$

Here $\varphi_i^{(\nu)}$ is the sum of all terms of degree ν in φ_i. By hypothesis, for any term $u_j^{(\mu)}$ occurring in $\varphi_i^{(\nu)}$ we have $\mu < \nu$, so the components of $u_i^{(\nu)}$ are uniquely determined in terms of the $u_j^{(\mu)}$ with $\mu < \nu$, while $u_i^{(o)} = 0$, again by hypothesis. Thus (11.5.21) has a unique solution u_i of positive order in the x's.

If $u_i = \varphi_i(u, x)$ $(i = 1, \ldots, m)$, $v = \psi_j(u, x)$ $(j = 1, \ldots, n)$ are two such systems, then to show that $u_1 - v_1$, $u_1 v_1$, $\sum u_1^\nu = (1 - u_1)^{-1}$ are algebraic we combine the above systems of equations for u_i, v_j with the equations $w = \varphi_1 - \psi_1$, $w = \varphi_1 \psi_1$, $w = \varphi_1 + u_1 w$ respectively. This shows that we have a subalgebra. Moreover. the elements of order 0 are invertible, so we have a local ring. ■

It is clear that we have the inclusions

$$k\langle X\rangle \subset k\langle X\rangle_{\text{rat}} \subset k\langle X\rangle_{\text{alg}} \subset k\langle\langle X\rangle\rangle; \tag{11.5.22}$$

that the inclusions are strict is easily seen, by considering the case where X consists of a single letter.

We now come to the promised characterization of context-free languages:

Theorem 11.5.7 (Schützenberger). *A language L in an alphabet X is context-free if and only if its characteristic series is algebraic.*

Proof. Let L be context-free, generated by a grammar with the rules $u_i \to w_{ik}$, where w_{ik} is a word in the u's and x's, and $u_1 = \sigma$ is the sentence symbol. If there is a rule of the form $u_i \to u_j$, we replace it by the rules $u_i \to f$, where f runs over the right-hand sides of all the rules $u_j \to w_{jk}$. We now write

$$u_i = \varphi_i(u, x), \tag{11.5.23}$$

where $\varphi_i(u, x)$ is the sum of the right-hand sides of all the rules with u_i on the left. On solving the system (11.5.23) we obtain for u_1 the series f_L, hence f_L is algebraic.

Conversely, assume that f_L is algebraic, given by u_1, where u is the solution of (11.5.23). Then the language L is obtained by applying all the rules $u_i \to w$, where w runs over all the words in the support of $\varphi_i(u, x)$; hence L is context-free, as we had to show. ■

To give an example, the language $\{x^n y^n\}$ has the characteristic series $\sum x^n y^n$, which is obtained by solving the equation

$$u = 1 + xuy.$$

Exercises

1. Adapt the proof of Theorem 11.5.1 to the case of an infinite alphabet.
2. Verify that the prefix set associated with a right ideal of finite codimension in a free algebra is a maximal code. To what extent does the finiteness condition of Corollary 11.5.2 apply to general (non-free) k-algebras?
3. Factorize the element $xyzyx + xyz + zyx + xyx + x + z$ of $F = k\langle x, y, z\rangle$ in all possible ways. (It can be shown that any two complete factorizations of an

element of F have the same number of terms, and these terms can be paired off in such a way that corresponding terms are 'similar', where a, b are similar iff $F/aF \cong F/bF$.)

4. Show that in $\mathbf{R}\langle x, y\rangle$ the element $xy^2x + xy + yx + x^2 + 1$ is an atom, but it does not remain one under extension of $\mathbf{R}$ to $\mathbf{C}$.
5. Show that every CF-language on one letter is regular.
6. Show that the inclusions in (11.5.22) are strict.
7. Define the *Hankel matrix* of a power series f as the infinite matrix $H(f)$ indexed by X^*, whose (u, v)-entry is (f, uv). Show that f is rational iff $H(f)$ has finite rank.

Further exercises on Chapter 11

1. Let A be an infinite set and $M(A)$ the set of all injective mappings of A into itself such that the complement of the image is infinite. Show that $M(A)$ is a semigroup admitting right cancellation and right division (writing mappings on the right), i.e. given $\alpha, \beta \in M(A)$, the equation $\alpha x = \beta$ has a unique solution.
2. Show that two elements of a free monoid commute iff they can be written as powers of the same element.
3. Let F be a free monoid. Show that if $uv = vw$, then there exist $a, b \in F$ such that $u = ab$, $w = ba$, $v = (ab)^r a = a(ba)^r$ for some $r \geq 0$.
4. Let C be the monoid on a, b as generating set with defining relation $ba = 1$. Show that each element of C can be written uniquely as $a^r b^s$, $r, s \geq 0$. (C is called the *bicyclic monoid.* Hint. Define C as set of mappings of $\mathbf{N}^2$ into itself by the rules $(m, n)a = (m, n-1)$ if $n \geq 1$, $(m, 0)a = (m+1, 0)$, $(m, n)b = (m, n+1)$.)
5. Show that any CF-language can be generated by a grammar whose rules are all of the form $\alpha \to \beta\gamma$ or $\alpha \to x$ ($\alpha, \beta, \gamma \in V$, $x \in X$). (Hint. Use induction on the lengths of the right-hand sides of all rules not of this form; this is called the *Chomsky normal form.*)
6. A grammar is called *self-embedding* if it includes a derivation $\alpha \to u\alpha v$, where $uv \neq 1$ and $\alpha \in V$. Show that a language L is regular iff there is a CF-grammar generating L which is not self-embedding.
7. Show that every language of type 0, 2 or 3 is closed under substitution: if L has type $\nu\,(=0, 2, 3)$ with alphabet $X = \{x_1, \ldots, x_r\}$ and x_1 is replaced by a language of type ν with an alphabet Y disjoint from X, the resulting language is of type ν on $X \cup Y$. A CS-language is closed under substitution provided that the language substituted is proper.
8. Show that a CF-language satisfies the following strengthening of the pumping lemma (sometimes called the *iteration lemma*): If L is context-free, there exists an integer p such that for any word w of length $|w| \geq p$ and any partition $(\nu_1, \ldots, \nu_5)$ of $|w|$ such that $\nu_3 > 0$ and either $\nu_2 > 0$ or $\nu_4 > 0$, there is a factorization $w = v_1 \ldots v_5$ with $|v_i| = \nu_i$ and $v_1 v_2^n v_3 v_4^n v_5 \in L$ for all n. Use the result to verify that $L = \{a^* bc\} \cup \{a^p ba^n ca^n | p \text{ prime}, n \geq 0\}$ is not context-free but satisfies the conditions of the pumping lemma (see Berstel (1979)).

9. Show that the generating set of a maximal free submonoid of a free monoid is a maximal code.
10. Show that a subset C of X^* is a code iff the submonoid generated by C has the characteristic series $(1 - C)^{-1}$.
11. Verify that the power series ring on X may be interpreted as the incidence algebra of X^* when the latter is ordered by right divisibility (see BA, Section 5.6). (Hint. Define $f_{u,v} = (f, z)$ if $u = zv$ and 0 otherwise.) Hence deduce Proposition 11.5.3 from the fact that the matrix in the incidence algebra is invertible iff all its diagonal elements are invertible (see BA, Proposition 5.6.1).

Bibliography

This is primarily a list of books where the topics are pursued further (and which were often used as sources). A second list contains articles having a bearing on the text as well as those referred to in the text. References in the text are by name and date, the latter enclosed in round brackets for books and square brackets for papers.

I. Books

Albert, A.A. (1939), Structure of Algebras, AMS Colloquium Publications 24, Providence, RI.

Arbib, M.A. (1969), Theories of Abstract Automata, Prentice-Hall, Englewood Cliffs, NJ.

Artin, E. (1957) Geometric Algebra, Interscience, New York.

Artin, E. (1965), Collected Papers, Addison-Wesley, Reading, MA.

Barbilian, D. (1956), Teoria Aritmetica a Idealilor (in inele necomutative), Ed. Acad. Rep. Pop. Romine, Bucharest.

Barwise, J. (Ed.) (1977), Handbook of Mathematical Logic, North-Holland, Amsterdam.

Bass, H. (1962), The Morita Theorems, Oregon Lectures.

Bass, H. (1968), Algebraic K-theory, Benjamin, New York.

Bell, J.L. and Slomson, A.B. (1971), Models and Ultraproducts: An Introduction, North-Holland, Amsterdam.

Berstel, J. (1979), Transductions and Context-free Languages, Teubner, Stuttgart.

Berstel, J. and Perrin, D. (1985), Theory of Codes, Academic Press, New York.

Berstel, J. and Reutenauer, C. (1988), Rational Series and their Languages, Springer, Heidelberg.

Bourbaki, N. (1974), Algebra I, Chapters 1–3, Addison-Wesley, Reading, MA.

Bourbaki, N. (1990), Algebra II, Chapters 4–7, Springer, Heidelberg.

Burnside, W. (1911, 1955), Theory of Groups of Finite Order, Dover, New York.

Chevalley, C. (1951), Introduction to the Theory of Algebraic Functions of One Variable, AMS Math. Surveys 6, New York.

Cohn, P.M. (1966), Morita Equivalence and Duality, Queen Mary College Math. Notes.

Cohn, P.M. (1977), Skew Field Constructions, LMS Lecture Note Series 27, Cambridge University Press.

Cohn, P.M. (1981) Universal Algebra, 2nd edn. Reidel, Dordrecht.

Cohn, P.M. (1985), Free Rings and their Relations, 2nd edn., LMS Monographs 19, Academic Press, London.

Cohn, P.M. (1991), Algebraic Numbers and Algebraic Functions, Chapman & Hall.

Cohn, P.M. (1994), Elements of Linear Algebra, Chapman & Hall, London.

Cohn, P.M. (1995), Skew Fields, Theory of General Division Rings, Vol. 57, Encyclopedia of Mathematics and its Applications, Cambridge University Press.

Cohn, P.M. (2000), Introduction to Ring Theory, SUMS, Springer, London.

Conway, J.H. and Sloane, N.J.A. (1988), Sphere Packings, Lattices and Groups, Grundlehren d. math. Wiss. 290, Springer, Berlin.

Curtis, C.W. and Reiner, I. (1981) Methods of Representation Theory I, John Wiley & Sons, New York.

Davis, M. (1958), Computability and Unsolvability, McGraw-Hill, New York.

Dickson, L.E. (1901, 1958), Linear Groups, with an Exposition of the Galois Field Theory, Dover, New York.

Draxl, P. (1983), Skew Fields, LMS Lecture Note Series 83, Cambridge University Press.

Eilenberg, S. (1974–78), Automata, Languages and Machines, A–C, Academic Press, New York.

Eisenbud, D. (1995), Commutative Algebra, with a View to Algebraic Geometry, Springer-Verlag, New York.

Faith, C. (1981) Algebra I: Rings, Modules and Categories, Grundlehren d. math. Wiss. 190, Springer, Heidelberg.

Feit, W. (1982), The Representation Theory of Finite Groups, North-Holland, Amsterdam.

Goodearl, K.R. and Warfield Jr., R.B. (1989), An Introduction to Non-commutative Noetherian Rings, LMS Student Texts 16, Cambridge University Press.

Gorenstein, D. (1982), Finite Simple Groups, Plenum, New York.

Hall Jr., M. (1959), The Theory of Groups, Macmillan, New York.

Hartshorne, R. (1977), Algebraic Geometry, Graduate Texts in Math. 52, Springer, Heidelberg.

Herman, G.T. and Rozenberg, G. (1975), Developmental Systems and Languages, North-Holland, Amsterdam.

Herstein, I.N. (1968), Noncommutative Ring Theory, Carus Math. Monographs 15, Math. Association of America.

Herstein, I.N. (1976), Rings with Involution, Chicago Lectures in Math., Chicago University Press.

Hill, R. (1985), Introduction to Coding Theory, Oxford University Press.

Hilton, P.J. and Stammbach, U. (1971), A Course in Homological Algebra, Graduate Texts in Math. 4, Springer, Heidelberg.

Huppert, B. (1967), Endliche Gruppen I, Grundlehren d. math. Wiss. 134, Springer, Berlin.

Jacobson, N. (1953), Lectures in Abstract Algebra II, Linear Algebra, Van Nostrand, New York.

Jacobson, N. (1956, 1964), Structure of Rings, AMS Colloquium Publs. 37, Providence, RI.

Jacobson, N. (1985, 1989), Basic Algebra I, II (2nd edn.) Freeman, San Francisco.

Jacobson, N. (1996), Finite-dimensional Division Algebras over Fields, Springer, Berlin.

James, G. and Kerber, A. (1981), The Representation Theory of the Symmetric Group, Vol. 16, Encyclopedia of Mathematics and its Applications, Addison-Wesley, Reading, MA.

Jategaonkar, A.V. (1986), Localization in Noetherian Rings, LMS Lecture Note Series 98, Cambridge University Press.

Lallement, G. (1979), Semigroups and Combinatorial Applications, John Wiley & Sons, New York.

Lam, T.Y. (1978), Serre's Conjecture, Lecture Notes in Math..635, Springer, Berlin.

Lint, J.H. van (1982), Introduction to Coding Theory, Graduate Texts in Math. 86, Springer, Berlin.

Lothaire, M. (1983, 1997), Combinatorics on Words, Cambridge Math. Library, Cambridge University Press.

Mac Lane, S. (1963), Homology, Grundlehren d. math. Wiss. 114, Springer, Berlin.

McConnell, J.C. and Robson, J.C. (1987), Non-commutative Noetherian Rings, John Wiley & Sons, Chichester.

McEliece, R.J. (1977), The Theory of Information and Coding, Encyclopedia of Mathematics and its Applications, Addison-Wesley, Reading, MA.

Mitchell, B. (1965), Theory of Categories, Academic Press, New York.

Peano, G. (1889), Arithmeticas Principia, Novo Methodo Exposito, Torino.

Pierce, R.S. (1982), Associative Algebras, Graduate Texts in Math. 88, Springer, Heidelberg.

Procesi, C. (1973), Rings with a Polynomial Identity, Dekker, New York.

Reiner, I. (1976), Maximal Orders, LMS Monographs 5, Academic Press, London.

Robinson, A. (1963), Introduction to Model Theory and the Metamathematics of Algebra, North-Holland, Amsterdam.

Rowen, L.H. (1980), Polynomial Identities in Ring Theory, Academic Press, New York.

Rowen, L.H. (1988), Ring Theory I, II, Academic Press, New York.

Schofield, A.H. (1985), Representations of Rings over Skew Fields, LMS Lecture Notes 92, Cambridge University Press.

Serre, J.-P. (1967, 1971), Représentations Linéaires des Groupes Finis, Hermann, Paris.

Stroyan, K.D. and Luxemburg, W.A.J. (1976), Introduction to the Theory of Infinitesimals, Academic Press, New York.

Weber, H. (1894, 1896, 1906), Lehrbuch der Algebra I–III (1960 reprint), Chelsea, New York.

Weil, A. (1967), Basic Number Theory, Grundlehren d. math. Wiss. 144, Springer, Berlin.

Weyl, H. (1939), The Classical Groups, Princeton University Press (2nd edn. 1946).

II. Papers

Amitsur, S.A. [1965], Generalized polynomial identities and pivotal polynomials, Trans. Amer. Math. Soc. 114, 210–226.

Amitsur, S.A. [1966], Rational identities and applications to algebra and geometry, J. Algebra 3, 304–359.

Amitsur, S.A. [1972], On central division algebras, Isr. J. Math. 12, 408–420.

Auslander, M. and Goldman, O. [1960], The Brauer group of a commutative ring, Trans. Amer. Math. Soc. 97, 367–409.

Azumaya, G. [1950], Corrections and supplementaries to my paper concerning Krull–Schmidt's theorem, Nagoya Math. J. 1, 117–124.

Bass, H. [1960], Finitistic dimension and a generalization of semiprimary rings, Trans. Amer. Math. Soc. 95, 466–488.

Bergman, G.M. [1974a], Modules over coproducts of rings, Trans. Amer. Math. Soc. 200, 1–32.

Bergman, G.M. [1974b], Coproducts and some universal ring constructions, Trans. Amer. Math. Soc. 200, 33–88.

Bergman, G.M. [1978], The diamond lemma in ring theory, Adv. in Math. 29, 178–218.

Boë, J.M., de Luca, A,. and Restivo, A. [1980] Minimal completable sets of words, Theor. Comput. Sci. 12, 325–332.

Chase, S.U. [1960], Direct products of modules, Trans. Amer. Math. Soc. 97, 457–473.

Chevalley, C. [1955], Sur certains groupes simples, Tohoku Math. J. 7, 14–66.

Cohn, P.M. [1956], Embeddings in semigroups with one-sided division, J. London Math. Soc. 31, 169–181.

Cohn, P.M. [1961], Quadratic extensions of skew fields, Proc. London Math. Soc. (3) 11, 531–556.

Cohn, P.M. [1966], On the structure of the $\mathbf{GL}_2$ of a ring, Publ. Math. IHES, 30, 5–53.

Cohn, P.M. [1987], Valuations in free fields, in 'Algebra, some current trends', Proc. Varna 1986, eds. L.L. Avramov and K.B.Tchakerian, Springer Lecture Notes in Math. 1352, 75–87.

Cohn, P.M. [1989], The construction of valuations on skew fields, J. Indian Math. Soc. 54, 1–45.

Dieudonné, J. [1943], Les déterminants sur un corps non-commutatif, Bull. Soc. Math. France 71, 27–45.

Gelfand, I.M. and Retakh, V. [1997], Quasideterminants I, Sel. math. New ser. 3, 517–546.

Goldie, A.W. [1958], The structure of prime rings under ascending chain conditions, Proc. London Math. Soc. (3) 8, 589–608.

Hall, P. [1928], A note on soluble groups, J. London Math. Soc. 3, 89–105.

Hall, P. [1937], A characteristic property of soluble groups, J. London Math. Soc. 12, 198–200.

Henkin, L. [1960], On mathematical induction, Amer. Math. Monthly 67, 323–338.

Landweber, P.S. [1963], Three theorems on phrase-structure grammars of type 1, Inform. Control 6, 131–136.

Merkuryev, A.S. and Suslin, A.A. [1986], On the structure of Brauer groups of fields, Math. USSR Izvestiya 27, 141–155.

Nagata, M. [1957], A remark on the unique factorization theorem, J. Math. Soc. Japan, 9, 143145.

Newman, M.H.A. [1942], On theories with a combinatorial definition of 'equivalence', Ann. of Math. 43, 223–243.

Ore, O. [1931], Linear equations in non-commutative fields, Ann. Math. 32, 463–477.

Roganov, Yu.V. [1975], The dimension of a tensor product on a projective bimodule (Russian), Mat.Zametki 18, 895–902.

Saşiada, E. and Cohn, P.M. [1967], An example of a simple radical ring, J. Algebra 5, 373–377.

Schofield, A.H. [1985], Artin's problem for skew field extensions, Math. Proc. Camb. Phil. Soc. 97, 16.

Shannon, C.E. [1948], A mathematical theory of communication, Bell Syst. Tech. J. 27, 379–423, 623–656. Reprinted in Slepian (ed.) Key Papers in the Development of Information Theory, 1974, IEEE Press, New York.

Sierpiński, W. [1945], Sur les fonctions de plusieurs variables, Fund. Math. 7, 33, 169–173.

Smoktunowicz, A. [2002], A simple nil ring exists, Comm. Alg. 30(1), 27–59.

Webber, D.B. [1970], Ideals and modules in simple Noetherian rings, J. Algebra 16, 239–242.

List of Notations

The number indicates the page where the term is first used or defined. Terms used only or mainly in one location are not included. When no page number is given, the term is defined in BA.

Number systems

$\mathbf{N}$	the natural numbers, 24
$\mathbf{N}_0$	the natural numbers with 0
$\mathbf{Z}/m$	the numbers mod m
$\mathbf{Z}$	the integers 2
$\mathbf{Q}$	the rational numbers
$\mathbf{R}$	the real numbers
$\mathbf{C}$	the complex numbers

Set theory

$\varnothing$	the empty set
$\lvert X\rvert$	the cardinal of the set X
$\mathscr{P}(X)$	the power set (set of all subsets) of X 6
$X\backslash Y$	the complement of Y in X xi
Y^X or $\mathrm{Map}(X, Y)$	the set of all mappings from X to Y, 2
$\mathrm{Map}(X)$	set of all mappings of X into itself, 4
$\ker f$	kernel of a correspondence f, 5
$\aleph_0$	aleph-null, the cardinal of $\mathbf{N}$

Number theory

$\max(a, b)$	the larger of a and b
$\min(a, b)$	the smaller of a and b
$a\mid b$	a divides b

(a, b)	highest common factor (HCF) of a and b
$[a, b]$	least common multiple (LCM) of a and b
$[x]$	greatest integer not exceeding x
δ_{ij}	Kronecker delta
$\mu(n)$	Möbius function
$\varphi(m)$	Euler function

Group theory

Sym_n or S_n	symmetric group of degree n
Alt_n or A_n	alternating group of degree n
$\mathrm{sgn}\,\sigma$	sign of the permutation σ
St_p	stabilizer of the point p, 119
$\mathbf{C}_n$	cyclic group of order n
$\mathbf{D}_m$	dihedral group of order $2m$, 93
$(G : H)$	index of H in G
$(x, y) = x^{-1}y^{-1}xy$	commutator of x and y
G'	commutator subgroup (derived group) of G
$\mathrm{Aut}(G)$	automorphism group of G
$\mathrm{Inn}(G)$	group of inner automorphisms of G
$N \lhd G$	N is a normal subgroup of G
$A \rtimes G$	semidirect product, 93
$N_G(H)$	normalizer of H in G
$\mathbf{GL}_n(R)$	general linear group over a ring R 116
$\mathbf{E}_n(R)$	subgroup generated by elementary matrices
$\mathbf{U}_n(\mathbf{C})$	unitary group over $\mathbf{C}$, 242
$\mathbf{Sp}_{2m}(K)$	symplectic group over K, 121
$\mathbf{O}_n(K)$, $\mathbf{SO}_n(K)$	orthogonal group over K, 126
$\mathbf{K}_1(A)$, $\mathbf{SK}_1(A)$	Whitehead group of A, 196

Rings and modules

${}^mV^n$	space of all $m \times n$ matrices over V
${}^mV(= {}^mV^1)$	space of m-component column vectors over V
$V^n(= {}^1V^n)$	space of n-component row vectors over V
$\mathfrak{M}_n(R)$ or R_n	$n \times n$ matrix ring over R
E_{ij}	matrix unit, 116
$B_{ij}(a)$	elementary matrix, 116
$\mathrm{diag}(d_1, \ldots, d_r)$	diagonal matrix, 271
$\mathrm{Hom}(U, V)$	set of all homomorphisms from U to V
$\mathrm{End}(U)$	ring of all endomorphisms of U
$U \otimes V$	tensor product of U and V
tM	torsion submodule of M
$\rho(a)$	rank of an endomorphism a, 310

$\mathbf{P}(M)$	projective cover of M, 141
$\mathbf{T}(M)$	top of M, 141
$Z(M)$	singular submodule of M, 286
λ_a, ρ_a	left, right multiplication by a, 179
R^0	opposite ring of R
$\mathbf{J}(R)$	Jacobson radical of R
A_E	algebra obtained from A by extending the ground field to E
$\mathbf{B}_k$	Brauer group of the field k, 188
$\mathbf{B}(F/k)$	relative Brauer group, 206
Deg A	degree of A, 189
$(a, b; k)$, $(a, b; k\|$	quaternion algebra, 201
$(F/k, \sigma, \alpha)$	cyclic algebra, 215
$R^{\times}$	set of all non-zero elements in an integral domain
R^{ab}	abelianization of R, 80
A^1	augmented algebra, 323
$\mathbf{U}(A)$	group of units of A, 118
Ann(X)	annihilator of X
$R[x]$	polynomial ring in x over R
$R[x; \sigma, \alpha]$	skew polynomial ring, 276
$R[[x; \alpha]]$	formal power series ring in x over R, 279
$R((x; \alpha))$	ring of formal Laurent series, 280
$\mathrm{A}_1[K]$	Weyl algebra, 278
$k\langle X\rangle$	free k-algebra on a set X, 78
$k\langle\langle X\rangle\rangle$	free power series k-algebra on a set X
$\mathbf{T}_A(U)$	tensor A-ring over U, 78
X^*	free monoid on a set X, 396
X^+	$X^*\backslash\{1\}$

Field theory

$[V:k]$	dimension of V over k
$N_{E/F}(x)$	norm of x from E to F
$T_{E/F}(x)$	trace of x from E to F
$d(f)$	degree of polynomial f, 275
$\mathscr{F}(R)$	field of fractions of an integral domain R
$\mathbf{F}_q$	field of q elements
$\mathbf{Q}_p$	field of p-adic numbers
$\mathbf{H}$	Hamilton quaternions, 202

Category theory

Ob$(\mathscr{A})$	class of all A-objects
$\mathscr{A}(X, Y)$	set of all maps from X to Y
Ens	category of sets

Author index

Subject index